INTERNATIONAL PROGRAMME ON CHEMICAL SAFETY
INTERNATIONAL AGENCY FOR RESEARCH ON CANCER
COMMISSION OF THE EUROPEAN COMMUNITIES
INSTITUTE OF OCCUPATIONAL HEALTH, FINLAND

# MONITORING HUMAN EXPOSURE TO CARCINOGENIC AND MUTAGENIC AGENTS

Proceedings of a joint symposium
held in Espoo, Finland
12-15 December 1983

EDITORS

A. BERLIN M. DRAPER K. HEMMINKI H. VAINIO



IARC Scientific Publications No. 59

INTERNATIONAL AGENCY FOR RESEARCH ON CANCER
LYON

1984

EC-EUR 9591 EN

Distributed by
Oxford University Press, Walton Street, Oxford OX2 6DP

London New York Toronto Delhi Bombay Calcutta Madras Karachi Kuala Lumpur Singapore
Hong Kong Tokyo Nairobi Dar es Salaam Cape Town Melbourne Auckland

and associated companies in Beirut Berlin Ibadan Mexico City Nicosia

Oxford is a trade mark of Oxford University Press

Distributed in the United States by Oxford University Press, New York

ISBN 0 19 723056 3
ISBN 92 832 1159 6 (Publisher)



The authors alone are responsible for the views expressed in the signed articles in this publication. None of the organizers of the symposium nor any person acting on their behalf is responsible for the use which might be made of the following information.



PRINTED IN FRANCE

# CONTENTS

**SESSION II**

Chairman: A.E. Bennett
Rapporteur: P. Lohman

## SESSION III

Chairman: P. Westerholm
Rapporteur: M. Sorsa



## OVERVIEW



## REPORTS OF PANEL DISCUSSIONS

## LIST OF PARTICIPANTS

| | |
|---|---|
| J. ADAMKIEWICZ | Institute for Cell Biology, University of Essen, Hufelandstrasse 55, D-4300 Essen 1, FRG |
| G. AHLBORG | Regional Hospital Örebro, Department of Occupational Medicine, S-Örebro, Sweden |
| K. AHLMAN | Outokumpu Oy, PL 280, SF-00101 Helsinki 10, Finland |
| A. AITIO | Institute of Occupational Health, Haartmaninkatu 1, SF-00290 Helsinki 29, Finland |
| L. ALESSIO | University Institute of Occupational Medicine, Via San Barnaba 8, I-20122 Milan, Italy |
| J. ALEXANDER | National Institute of Public Health, Geitmyrsveien 75, N-Oslo 1, Norway |
| D. ANDERSON | The British Industrial Biological Research Association, Woodmansterne Road, Carshalton, Surrey SM5 4EF, UK |
| J. ANTONIADES | Lankenau Hospital, City Line & Lancaster Avenues, Philadelphia, PA 19151, USA |
| L. ARINGER | National Board of Occupational Safety and Health, Arbetarskyddsstyrelsen FMM, S-17184 Solna, Sweden |
| E.G. ASTRUP | Elkem A/S, Department of Occupational Health, Postboks 4224 - Torshov, N-Oslo 4, Norway |
| J.-C. AUBRUN | Rhone-Poulenc SA, 25 Quai Paul Doumer, F-92408 Courbevoie, France |
| T. AUNE | National Institute of Public Health, Geitmyrsveien 75, N-Oslo 1, Norway |
| H. AUTRUP | Laboratory of Environmental Carcinogenesis, NDR Frihavnsgade 70, DK-2100 Copenhagen Ø, Denmark |
| H. BARTSCH | International Agency for Research on Cancer, 150 cours Albert Thomas, F-69372 Lyon Cedex 08, France |
| A.E. BENNETT | Health and Safety Directorate, Commission of the European Communities, Luxembourg, Grand Duchy of Luxembourg |
| A. BERLIN | Health and Safety Directorate, Commission of the European Communities, Luxembourg, Grand Duchy of Luxembourg |
| P.A. BERTAZZI | Institute of Occupational Health, 8 Via San Barnaba, I-20122 Milan, Italy |
| F.W. BEST | ICI Europa, Everslaan 45, B-3078 Everberg, Belgium |

| | |
|---|---|
| H.M. BOLT | University of Dortmund, Institute for Occupational Physiology, Ardeystrasse 67, D-4600 Dortmund 1, FRG |
| R. BOS | Institute of Pharmacology/Toxicology, Geert Grooteplein Noord 21, Nijmegen, The Netherlands |
| R. BROUNS | RBGD - Rotterdam, Langs de Baan 10, Hoogvliert-RT, The Netherlands |
| J. BUSTAMANTE | European Trade Union Confederation, 37 Warmoesberg, B-1000 Brussels, Belgium |
| A. CARERE | Institute of Health, Viale Regina Elena 299, I-00161 Rome, Italy |
| A.N. CHEBOTAREV | Institute of Medical Genetics, Academy of Medical Sciences of the USSR, Kashirskoje shosse, 6 A, Moscow 115 478, USSR |
| J. CLENCH-AAS | Norwegian Institute Air Research, PO Box 130, N-2001 Lilleström, Norway |
| E. CLONFERO | Institute of Occupational Health, Via Facciolati 71, I-35100 Padua, Italy |
| E. COSTA | National Asbestos Centre, Piazza della Vittoria 11/6, I-16121 Genoa, Italy |
| R. CREBELLI | Institute of Health, Viale Regina Elena 299, I-00161 Rome, Italy |
| T. DEILHAUG | National Institute of Public Health, Geitmyrsveien 75, N-Oslo 1, Norway |
| V. DIGERNES | Dyno Industries A/S, N-2000 Lilleström, Norway |
| P. DOLARA | Institute of Pharmacology, Viale Morgagni 65, I-50134 Florence, Italy |
| R. VAN DOORN | Bedrijfsgezondh. Dienst, Midden-Ijssel, van Calcarstraat 10, 7415 CK Deventer, The Netherlands |
| F.R. DUURT | S.I.P.M. The Hague, PO Box 162, 2501 AN The Hague, The Netherlands |
| E. DYBING | National Institute of Public Health, Geitmyrsveien 75, N-Oslo 1, Norway |
| L. EHRENBERG | Wallenberg Laboratory, University of Stockholm, S-106 91 Stockholm, Sweden |
| P. EINISTÖ | Institute of Occupational Health, Haartmaninkatu 1, SF-00290 Helsinki 29, Finland |

| | |
|---|---|
| E. ELOVAARA | Institute of Occupational Health, Haartmaninkatu 1, SF-00290 Helsinki 29, Finland |
| G. ERICSSON | Arbetarskyddsstyrelsen, S-17184 Solna, Sweden |
| I. FARKAS | WHO Regional Office for Europe, 8 Scherfigsvej, DK-2100 Copenhagen Ø, Denmark |
| P.B. FARMER | Medical Research Council Toxicology Unit, Woodmansterne Road, Carshalton, Surrey SM5 4EF, UK |
| M. FAVERO | University Estadual de Campinas, Rua Dr Quirino 1856, Campinas, Brazil |
| A. FORNI | Institute of Occupational Health, 8 Via San Barnaba , I-20122 Milan, Italy |
| R.C. GARNER | University of York, Cancer Research Unit, Heslington, York YO1 5DD, UK |
| D. GOMPERTZ | Health & Safety Executive, 403 Edgware Road, London NW2 6LN, UK |
| P. GORDON | National Institute of Public Health, Geitmyrsveien 75, N-Oslo 1, Norway |
| R. GRAFSTRÖM | Karolinska Institute, Department of Forensic Medicine - Toxicology, S-10401 Stockholm, Sweden |
| P.L. GROVER | Chester Beatty Research Institute, Fulham Road, London SW3, UK |
| J. HANKE | Institute of Occupational Medicine, Tevesy 8, Lódž, Poland |
| M.L. HÄNNINEN | College of Veterinary Medicine, Hämeentie 57, SF-00550 Helsinki 55, Finland |
| C. HARRIS | National Cancer Institute, Laboratory of Human Carcinogenesis, Bethesda, MD 20205, USA |
| Y. HASEGAWA | World Health Organization, Avenue Appia, CH-1211 Geneva 27, Switzerland |
| Å. HAUGEN | National Institute of Public Health, Geitmyrsveien 75, N-Oslo 1, Norway |
| T. HEINONEN | EFLAB, Pulttitie 9-11, SF-00810 Helsinki 81, Finland |
| S.S. HELDAAS | Norsk Hydro, Porsgrunn Fabrikker, N-3900 Porsgrunn, Norway |
| H. HEMMING | National Institute of Environmental Health, Box 60208, S-10401 Stockholm, Sweden |

| | |
|---|---|
| K. HEMMINKI | Institute of Occupational Health, Haartmaninkatu 1, SF-00290 Helsinki 29, Finland |
| D. HENSCHLER | Institute of Toxicology, Versbacher Strasse 9, D-8700 Würzburg, FRG |
| S. HERNBERG | Institute of Occupational Health, Haartmaninkatu 1, SF-00290 Helsinki 29, Finland |
| E. HESELTINE | International Agency for Research on Cancer, 150 cours Albert Thomas, F-69372 Lyon Cedex 08, France |
| E. HIDLE | A/S Jotungruppen, PO Box 400, N-3200 Sandefjord, Norway |
| C. HOGSTEDT | National Board of Occupational Safety and Health, Department of Occupational Medicine, S-17184 Solna, Sweden |
| L.W. HOLM | National Board of Occupational Safety and Health, S-17184 Solna, Sweden |
| B. HOLMBERG | National Board of Occupational Safety and Health, S-17184 Solna, Sweden |
| J.A. HOLME | National Institute of Public Health, Geitmyrsveien 75, N-Oslo 1, Norway |
| J. HONGSLO | National Institute of Public Health, Geitmyrsveien 75, Oslo 1, Norway |
| S. HONKASALO | National Institute of Environmental Health, Box 60208, S-10401 Stockholm, Sweden |
| W. HOWE | Imperial Chemical Industries, PLC C.T.L., Alderley Park, Macclesfield, Cheshire, UK |
| C. HUGOD | National Board of Health - Denmark, St Kongensgade 1, DK-1264 Copenhagen, Denmark |
| M.S. HUUSKONEN | Uusimaa Regional Institute of Occupational Health, Arinatie 3, SF-00370 Helsinki 37, Finland |
| M. IKEDA | Tohoku University School of Medicine, Seiryo-Cho 2, Sendai, Japan |
| B. ISOMAA | Åbo Academy - Institute of Biology, Porthansg. 3, SF-20500 Åbo 50, Finland |
| M. JAROSZEWSKI | Directorate of Labour, Inspectorate Service, Ryesgade 113, DK-2100 Copenhagen Ø, Denmark |
| J. JÄRVISALO | Institute of Occupational Health, Arinatie 3, SF-00370 Helsinki 37, Finland |

| | |
|---|---|
| G. DE JONG | Shell Nederland, Raffinaderij BV, PO Box 7000 (via dept. PRS/22), 3000 HA Rotterdam, The Netherlands |
| F. JONGENEELEN | Institute of Pharmacology/Toxicology, Geert Grooteplein Noord 21, Nijmegen, The Netherlands |
| S.K. KASHYAP | National Institute of Occupational Health, Meghani Nagar, Ahmedabad-380016, India |
| T. KAUPPINEN | Institute of Occupational Health, Haartmaninkatu 1, SF-00290 Helsinki 29, Finland |
| H. KIVISTÖ | Institute of Occupational Health, Arinatie 3, SF-00370 Helsinki 37, Finland |
| L. KNUDSEN | Labour Inspection Office, Rosenvängets Alle 16-18, DK-2100 Copenhagen Ø, Denmark |
| V. KODAT | Ministry of Health of the CSR, W. Piecka 98, Prague 10, Czechoslovakia |
| J.J. KOLK | Akzo N.V., Post Bus 186, 6800 LS Arnhem, The Netherlands |
| M. KOUROS | University of Düsseldorf, Medical Institute for Environmental Hygiene, Gurlittstrasse 52, D-4000 Düsseldorf 1, FRG |
| E. KRIEK | The Netherlands Cancer Institute, 121 Plesmanlaan, Amsterdam, The Netherlands |
| H. KRUSE | Unt. Such.st. für Umwelt und Toxikologie, Fleckenstrasse, D-Kiel, FRG |
| J. LÄHDETIE | University of Turku, Kiinanmyllynkatu 10, SF-20520 Turku 52, Finland |
| B. LAMBERT | Karolinska Hospital, Department of Clinical Genetics, S-10401 Stockholm, Sweden |
| P. LANDRIGAN | National Institute for Occupational Safety & Health, 4676 Columbia Parkway, R.A. Taft Laboratory, Cincinnati, OH 45226, USA |
| M. LANG | EFLAB, Pulttitie 9-11, SF-00810 Helsinki 81, Finland |
| R. LAUWERYS | University of Louvain, 30.54 Clos Chapelle aux Champs, B-1200 Brussels, Belgium |
| F. LAVAL | Gustave Roussy Institute, Rue Camille Desmoulins, F-94800 Villejuif, France |
| B. LAVENIUS | Götaverken Företagshälsovärd AB, Box 8713 - Halsocentr. Gothia, S-40275 Göteborg, Sweden |

| | |
|---|---|
| M. LAX | Institute of Occupational Health, Haartmaninkatu 1, SF-00290 Helsinki 29, Finland |
| M.S. LEGATOR | University of Texas, Medical Branch at Galveston, 24 Keiller Building F-19, Galveston, TX, USA |
| C. LEGRAVEREND | EFLAB, Pulttitie 9-11, SF-00810 Helsinki 81, Finland |
| J. LEWALTER | Bayer AG, Bayerwerk - Ambulanz Nord, D-Leverkusen, FRG |
| P.H.M. LOHMAN | Medical Biological Laboratory TNO, PO Box 45, 2280 AA Rijswijk, The Netherlands |
| E. LONGSTAFF | Imperial Chemical Industries, PLC C.T.L. Alderley Park, Macclesfield, Cheshire, UK |
| I. LUNDBERG | National Board of Occupational Safety and Health, S:t Eriksg. 130 A III, S-11343 Stockholm, Sweden |
| P. LUNDBERG | National Board of Occupational Safety and Health, S-17184 Solna, Sweden |
| M. LUOTAMO | Institute of Occupational Health, Haartmaninkatu 1, SF-00290 Helsinki 29, Finland |
| E. LYNGE | The Danish Cancer Registry, 66 Landskronagade, DK-Copenhagen, Denmark |
| D.G. MACPHEE | La Trobe University, Bundoora, Victoria 3083, Australia |
| K. MÄKELÄ | Kemira Oy, Box 330, SF-00101 Helsinki 10, Finland |
| K. MÄKIPAJA | Starckjohann-Teko Oy, PL 54, 15101 Lahti 10, Finland |
| H. MALKER | Arbetarskyddstyrelsen, S-17184 Solna, Sweden |
| C.N. MARTIN | University of York, Cancer Research Unit, Heslington, York YO1 5DD, UK |
| L. MELDGAARD | National Agency of Environment, Protection, Strandgade 29, DK-1401 Copenhagen K, Denmark |
| M. MERCIER | International Programme on Chemical Safety, World Health Organization, CH-1211 Geneva 27, Switzerland |
| R. MONTESANO | International Agency for Research on Cancer, 150 cours Albert Thomas, F-69372 Lyon Cedex 08, France |
| G. MOWE | Institute of Occupational Health, Gydas Vei 8, N-Oslo 3, Norway |
| H.-G. NEUMANN | Institute for Toxicology, Versbacher Strasse 9, D-8700 Warzburg, FRG |

| | |
|---|---|
| L. NIEMINEN | Farmos Group Ltd/Research Center, PO Box 425, SF-20101 Turku 10, Finland |
| Å. NORDSTRÖM | National Board of Occupational Safety and Health, Box 6104, S-Umeå, Sweden |
| H. NORPPA | Institute of Occupational Health, Haartmaninkatu 1, SF-00290 Helsinki 29, Finland |
| T. NORSETH | Institute of Occupational Health, Gydas Vei 8, N-Oslo 3, Norway |
| R. NYSTEN | Neste Oy Porvoo Works, Terveystalo, SF-06850 Kulloo, Finland |
| F. OESCH | Institute of Pharmacology, University of Mainz, D-6500 Mainz 1, FRG |
| H. OHSHIMA | International Agency for Research on Cancer, 150 cours Albert Thomas, F-69372 Lyon Cedex 08, France |
| S. ÖVREBÖ | Institute of Occupational Health, Gydas Vei 8, PO Box 8149 Dep., N-Oslo 1, Norway |
| H. PALVA | Kemira Oy, PL 330, SF-00101 Helsinki 10, Finland |
| K. PEKARI | Institute of Occupational Health, Haartmaninkatu 1, SF-00290 Helsinki 29, Finland |
| O. PELKONEN | University of Oulu, Department of Pharmacology, SF-90220 Oulu 22, Finland |
| T. VAN PETEGHEM | N.V. Sidmar, Kennedylaan 51, B-9020 Ghent, Belgium |
| J. PIETERS | Chief Medical Office, Dr. Reijersstraat 8, Leidschendam, The Netherlands |
| B.L. POOL | German Cancer Research Institute, Toxicology/Chemistry, Im Neuenheimer Feld 280, D-6900 Heidelberg, FRG |
| W. PRETSCH | Society for Radiation and Environmental Research, Ingolstädter Landstrasse 1, D-8042 Neuherberg, FRG |
| W.K. DE RAAT | MT-TNO Department of Biology, Schoemakerstraat 97, 2628 VK Delft, The Netherlands |
| E. RAFFN | Labour Inspection Office, Rosenvängets Alle 16-18, DK-2100 Copenhagen Ø, Denmark |
| K. RANDERATH | Baylor College of Medicine, Department of Pharmacology, Houston, TX 77030, USA |

| | |
|---|---|
| J. RANTANEN | Institute of Occupational Health, Haartmaninkatu 1, SF-00290 Helsinki 29, Finland |
| C. REUTERWALL | National Board of Occupational Safety and Health, S-17184 Solna, Sweden |
| J. RHEINGANS | Bayer Ag, Drug Division, Rheinuferstrasse 7, D-4150 Krefeld 11, FRG |
| A. RINGSTRÖM | Association of Swedish Chemical Industries, Box 5501, S-11485 Stockholm, Sweden |
| H. ROELFZEMA | Directorate General of Labour/Toxicology, Balen van Andelplein 2, Voorburg, The Netherlands |
| P. ROTO | Tampere Regional Institute of Occupational Health, PL 486, SF-33101 Tampere 10, Finland |
| I. ROZOV | World Health Organization, Avenue Appia, CH-1211 Geneva 27, Switzerland |
| E.D. RUBERY | Health & Social Security, R 904 Hannibal House, Elephant & Castle, London SE1 6TE, UK |
| N. RUME | Ministry of Health, Occupational Medicine, 22 rue Goethe, 1637 Luxembourg, Grand Duchy of Luxembourg |
| F. RUSPOLINI | Inail Technical Consultancy, Via Nomentana 74, I-Rome, Italy |
| S. SALOMAA | Institute of Occupational Health, Haartmaninkatu 1, SF-00290 Helsinki 29, Finland |
| C.G. SANDBERG | Swedish Employers' Confederation, S-10330 Stockholm, Sweden |
| R. SARACCI | International Agency for Research on Cancer, 150 cours Albert Thomas, F-69372 Lyon Cedex 08, France |
| A. SARRIF | Du Pont Company, Elkton Road, Newark, DE 19711, USA |
| F. SARTO | Institute of Occupational Health, Via Facciolati 71, I-35100 Padua, Italy |
| K. SAVELA | Institute of Occupational Health, Haartmaninkatu 1, SF-00290 Helsinki 29, Finland |
| S. SELEVAN | NIOSH/Institute of Occupational Health, Haartmaninkatu 1, SF-00290 Helsinki 29, Finland |
| M. SHELBY | National Institute of Environmental Health Sciences, PO Box 12233, Research Triangle Park, NC 27709, USA |

| | |
|---|---|
| L. SIMONATO | International Agency for Research on Cancer, 150 cours Albert Thomas, F-69372 Lyon Cedex 08, France |
| N.J. VAN SITTERT | Shell International Petroleum, Maatschappij BV, Van Hogenhoucklaan 60, The Hague, The Netherlands |
| E. SKYTTÄ | Technical Research Centre of Finland, Food Laboratory, Biologinkuja 1, SF-02150 Espoo 15, Finland |
| E. SÖDERLUND | National Institute of Public Health, Geitmyrsveien 75, N-Oslo 1, Norway |
| K.A. SOLBERG | Norwegian Employers' Confederation, Box 6710 St Olavsplass, N-Oslo 1, Norway |
| M. SORSA | Institute of Occupational Health Academy of Finland, Haartmaninkatu 1, SF-00290 Helsinki 29, Finland |
| L. SPAROS | University of Athens, 121 Vassilisis Sofias, Athens, Greece |
| F.G. STENBÄCK | University of Oulu, Department of Pathology, Kajaanintie 52 D, SF-90220 Oulu 22, Finland |
| M. STREICHER | Ste Sanofi, 195 Route d'Espagne, F-31036 Toulouse, France |
| E. SUNDQUIST | National Board of Labour Protection, PO Box 536, SF-33101 Tampere 10, Finland |
| H. SUUTARINEN | Finnish Employers' Confederation, Eteläranta 10, SF-00130 Helsinki 13, Finland |
| S.R. TANNENBAUM | Massachusetts Institute of Technology, 77 Massachusetts Avenue 56-311, Cambridge, MA 02139, USA |
| G. THOMAS | Unit of Clinical Pharmacology, 200 rue du Faubourg St Denis, F-75010 Paris, France |
| L. TIKKANEN | Technical Research Centre of Finland, Food Laboratory, Biologinkuja 1, SF-02150 Espoo 15, Finland |
| J.J.B. TINKLER | Health & Safety Executive, Baynards House, Chepstow Place, London W2 4TF, UK |
| H.H. TJÖNN | Directorate of Labour Inspection, Fridtjof Nansens Vei 14, N-Oslo, Norway |
| L. TOMATIS | International Agency for Research on Cancer, 150 cours Albert Thomas, F-69372 Lyon Cedex 08, France |
| W.F. TORDOIR | Shell International Petroleum Mij. BV, PO Box 162, 2501 AN The Hague, The Netherlands |

| | |
|---|---|
| A. ULANDER | Regional Hospital Örebro, Department of Occupational Medicine, S-Örebro, Sweden |
| M. VAAHTORANTA | Huhtamäki Oy, PO Box 406, SF-20101 Turku 10, Finland |
| H. VAINIO | International Agency for Research on Cancer, 150 cours Albert Thomas, F-69372 Lyon Cedex 08, France |
| P. VENIER | Institute of Animal Biology, Via Loredan 10, I-35100 Padua, Italy |
| S. VENITT | Institute of Cancer Research Pollards Wood, Nightingales Lane, Chalfont St Giles HP8 4SP, UK |
| T. VERGIEVA | Medical Academy, Sofia, Bulgaria |
| N.P.E. VERMEULEN | University of Leiden, PO Box 9502, 2300 RA Leiden, The Netherlands |
| K. WAHLBERG | National Board of Occupational Safety and Health, S-171 84 Solna, Sweden |
| P. WARDENBACH | State Institute for Labour Protection and Research on Accidents, Vogelpothsweg 50, D-4600 Dortmund 17, FRG |
| P. WESTERHOLM | Swedish Trade Union Confederation (LO), Barnhusgatan 18, S-10553 Stockholm, Sweden |
| M.I. WILLEMS | Institute CVO-Toxicology and Nutrition TNO, PO Box 360, 3700 AJ Zeist, The Netherlands |
| G.N. WOGAN | Massachusetts Institute of Technology, 77 Massachusetts Avenue 16-333, Cambridge, MA 02139, USA |
| M.F. WOODER | Shell International Petroleum Co. Ltd, York Road, London SE1 7NA, UK |
| B.H. WOOLLEN | ICI Central Toxicology Laboratory, Alderley Park, Macclesfield, Cheshire, UK |
| M.J. WRAITH | Shell Research Ltd, Sittingbourne Research Centre, Sittingbourne, Kent, UK |
| A.J. WYROBEK | University of California, L. Livermore National Laboratory, PO Box 5507, Livermore, CA 94550, USA |
| A. YARDLEY-JONES | Shell UK Oil, PO Box 3 Stanlow Refinery, Ellesmere Port/S. Wirral L65 4HB, UK |

## EDITORIAL BOARD

# FOREWORD

**J. Rantanen**

*Institute of Occupational Health, Helsinki, Finland*

In many ways chemicals are a basic factor in the well-being and effective functioning of present-day societies. The quality of life at both the social and individual levels would be severely restricted if we had to live without petroleum products, plastics, synthetic drugs, synthetic textiles and man-made mineral fibres or without the synthetic materials needed to produce modern electronic devices.

As in the case of drugs, the use of other chemicals implies not only the desired benefits but also adverse effects. We are compelled to compromise between costs and benefits. When the desired and the adverse effects of the use of chemicals are weighed, it is pertinent to ask whether we really need all the 2000 new chemicals synthesized annually and whether all the 60 000-70 000 chemicals used daily in industrialized societies are truly essential. Some people have wondered how long our ecosystem can survive and how long the biosystem, man included, can tolerate the doubling of the consumption of chemicals that occurs every seven years.

To make the actual situation at the national level more concrete, I would like to use Finland as an example. The host country of this seminar produces about four million tonnes of chemicals annually. Their total consumption amounts to 3.6 million tonnes a year or 0.74 tonnes *per caput*. The total annual volume of toxic chemical wastes comes to 0.5 million tonnes, or about 100 kg *per caput*. The annual consumption of plastics is approximately 100 kg *per caput*. The national register of toxic chemicals used in industries presently contains some 24 000 entities, and 45% of our labour force is exposed occupationally to chemicals during stages of production, packing and transport, in the specific use of chemicals or in the handling of chemical wastes. On average, some 3.5% of the labour force is exposed to carcinogens, but in the small industries of this province, Uusimaa, 22% of the workers face the possibility of exposure to carcinogens and some 46% to allergens. Fortunately, our data show that exposure levels in the work environment have decreased during the 1970s: only about 10% of some 25 000 hygienic measurements done by the Institute of Occupational Health detected levels above the present hygienic standards.

In addition to the work environment, other environments also are affected by increasing numbers and amounts of chemicals. The Finnish population's total load of chemicals from various environments has been calculated recently, and the amounts to which the population is exposed were unexpectedly high. This is in spite of the fact that Finland certainly is neither the most 'chemicalized' nor the most polluted country on this continent.

In fact, all nations within the United Nations family are faced today with growing problems of 'chemicalization', and questions of weighing risks and benefits have become more and more relevant. This is because all societies have become increasingly dependent on chemicals. Severe and system-wide damage to the ecosystem has been recognized, such as damage due to acid rain, the pollution of soil and water and the increasing rates of certain types of cancer and allergies. Severe chemical accidents or epidemics, such as the polychlorinated biphenyls (PCBs) epidemics in Japan, Taiwan and the USA, the dioxin accident in Italy, the mercury epidemics in Iraq and Japan and the wide food-oil epidemic in Spain, have also emphasized the importance of chemical safety, even in highly developed societies.

Scientific communities have long warned us about the global pollution of the ecosystem, including man, for instance, by PCBs and their derivatives. The concentrations of PCB in human fat and milk are now surprisingly high, and we are still surrounded by facilities and apparatus containing PCBs. In Finland, for example, there are more than three million pieces of electrical apparatus containing more than 2000 tonnes of PCBs.

'Chemicalization' entails problems that were not perceived by decision-makers when early warnings were issued; now these problems are appearing in concrete form before administrators, and effective solutions are urgently needed. To a great extent the problems are international in character and, consequently, the solutions require international action and collaboration.

Several governments and political and economic associations, including the European Economic Community, the Council of Mutual Economic Assistance and the Organization for Economic Cooperation and Development (OECD), and international organizations within the United Nations, have undertaken specific programmes to meet the problems caused by chemicals. The most global approach has been adopted by the International Programme on Chemical Safety (IPCS), one of the agencies responsible for the organization of this seminar.

All the international and national actions undertaken thus far demonstrate the need for governments to regulate the growing problems caused by 'chemicalization' of both developed and developing countries. Furthermore, an increasing number of international conventions and agreements have been compiled to prevent transboundary effects of chemicals. One of these agreements comprises joint action for the protection of the Baltic Sea, which surrounds the island on which our seminar is being held. Monitoring of the exposure of both the ecosystem and the population always plays an important role in these programmes and agreements.

Several means can be applied to prevent chemical hazards. The most effective one would be to refrain from using hazardous substances and to replace them with non-hazardous ones. In practice, this principle is at least partly followed, for example, in the chemical programme of the OECD. Not all hazardous exposure, however, can be avoided, because we simply do not know all the possible risks, despite extensive testing programmes. So far, appropriate toxicity testing has been done only for about 10% of the chemicals used daily and in the work environment; for example, only 1% of the industrial chemicals have exposure limits.

Concepts of biological and environmental monitoring may be described as follows (after Zielhuis, 1980):

*Biological monitoring*: Systematic, continuous or repeated activity by which biological samples are collected for analysis of concentrations of pollutants, metabolites or specific non-adverse biochemical-effect parameters for immediate application, with the objective of assessing exposure and health risk to exposed subjects by comparing the findings with a reference level, leading - if necessary - to corrective action.

*Environmental monitoring*: Systematic, continuous or repeated activity by which environmental samples are collected for analysis of pollutant concentrations, with the objective of assessing exposure and the health risk to exposed subjects by comparing the findings with a reference level, leading - if necessary - to corrective action.

Although it is a tool of secondary prevention, follow-up of chemical exposure by environmental or biological monitoring will be needed far into the future. Firstly, several hazardous chemicals are persistent; and, even in cases where their use has been stopped, exposure may occur for years. Secondly, the effectiveness of primarily preventive actions must be followed by monitoring. Thirdly, new, unrecognized risks may be revealed as studies are done using monitoring data.

For the most long-acting chemicals, that is, chemical carcinogens, primary prevention is even more important than in the case of so-called conventional chemicals. Human populations, however, will continue for a long time in the future - if not forever - to be exposed to carcinogens. Exposure to carcinogens has attracted much interest at both national and international levels, and collaborative programmes have been commenced to develop methodologies for risk identification and risk estimation. An international Convention on Occupational Cancer (No. 139) was adopted by the International Labour Conference in 1974. The Convention was followed by the Occupational Cancer Recommendation (No. 147) by the ILO. The implementation of these recommendations requires monitoring of carcinogenic exposure.

Monitoring carcinogenic exposure is a difficult and demanding task for several reasons. Firstly, carcinogen policies in most countries, and according to the ILO recommendation, imply the reduction of exposure as soon as it is detected. After reductive measures, monitoring takes place at concentration levels close to the lowest limits of detection of the monitoring methods. Secondly, the control of exposure may be expensive and may cause considerable action in the environment; thus, the monitoring data on which decisions for action are based must be highly reliable. Thirdly, the detection of carcinogenic exposure may have great juridical value even decades after the monitoring data have been collected. Hence, the development of monitoring methods, the ascertainment of their validity and reliability and their application in practice are questions of utmost importance.

The IPCS, the IARC and the CEC are to be gratefully acknowledged for organizing this seminar jointly with the Finnish Institute of Occupational Health. I have every reason to believe that this seminar will be an important step towards chemical safety in both the work environment and the general environment, not only in the countries represented here but also, *via* the IPCS, in all the member countries of the United Nations.

# FOREWORD

## L. Tomatis

*International Agency for Research on Cancer, Lyon, France*

The Institute of Occupational Health of Helsinki is to be congratulated for having chosen an extremely important subject for this meeting and for convening it at a very timely period. In fact, while nobody questions the great potential for improving public health measures in monitoring the exposure and the initial effects of chemicals, there is considerable confusion about what this actually means. There is confusion about the extent to which a methodology that can be used routinely has been developed and, what is most important, about the extent to which monitoring of exposure to toxic agents and of early effects can actually be related to the risk of long-term adverse effects. Among these long-term effects, cancer is the most widely discussed, in part because it is one of the less difficult end-points to be observed, followed by spontaneous abortions and congenital malformations; but we should also not forget that exposure to chemicals that are labelled, on the basis of experimental data, as carcinogens or mutagens, may have other medium- or long-term effects.

There is little doubt that one of the main limitations in the evaluation of cancer risks has been the difficulty, and most often the impossibility, of obtaining information on the level of exposure to a carcinogen. Traditionally, most human carcinogens have been recognized *a posteriori* following retrospective studies. In these studies, with only very rare and in any case partial exceptions, data on exposure were not available. Where some data were available, they were of an indirect nature; that is, they were based on environmental measurements, mainly of ambient air, providing some idea of possible exposure levels of a particular population group, but providing no information on individual exposure.

During this meeting, we shall learn what is known about individual monitoring of exposure and of early effects, and how it may be applied to epidemiological surveys. Of all the methods available for measuring an effect, the best known, and possibly that which has been used most widely, is the search for chromosomal aberrations in peripheral lymphocytes.

To this, other methods have been added recently, such as the search for sister chromatid exchanges and micronuclei. But, even for chromosomal aberrations, it is not entirely clear how specific they are in predicting a long-term adverse health effect. It is very important, therefore, and by now very urgent also, that this meeting engage in a thorough discussion of the capacity of these methods to predict, qualitatively and quantitatively, long-term effects. The same applies to the monitoring of exposure.

Measurement of early effects includes the possible measuring of both the actual individual level of exposure and the individual level of response. Thus, it would also provide a way of assessing individual variability of response or, if we would like to go a little further, of measuring individual susceptibility, with all the implications, some of which are rather dangerous, that this could have. Needless to say, the laboratory methods that aim at providing such important information have to be discussed and verified with extreme care before they can be recommended for routine use.

As you all certainly realize, it is only if we succeed in validating these methods, first for their qualitative and then for their quantitative predictive capacity, that new tools will be provided by which to build an epidemiology which fulfills its main objective, that is, to prevent long-term effects before they occur. That is a truly public-health-oriented epidemiology.

Epidemiologists have been blamed in the past, and still are, for using methods that are so coarse and imprecise as to prevent them from identifying any risk that is not a very high one. If experimentalists want to contribute efficiently to bulding a better epidemiology, they must work to assess exposure levels precisely, with routinely applicable methods, and they must develop methods for assessing early biological effects that can be rigorously validated before being accepted as predictors of a late adverse health effect. It is clear that, in order to achieve this, experimentalists and epidemiologists must work closely together in a mutually enriching, integrated approach. This meeting, by bringing together scientists of different disciplines - but all public-health oriented - seems to provide an excellent forum for progress in this direction. We must not forget, however, that monitoring of exposure and of early effects of chemicals, important as it is, still represents a second level of protection, the first, and wherever possible obligatory, being the avoidance of exposures already identified as carcinogenic and/or mutagenic.

## FOREWORD

**A.E. Bennett**

*Health and Safety Directorate, Commission of the European Communities, Luxembourg, Grand Duchy of Luxembourg*

On behalf of the Commission of the European Communities, may I say it is a great pleasure to join with the Institute of Occupational Health, the International Programme on Chemical Safety and the International Agency for Research on Cancer (IARC) in the organization of this symposium.

The topic is an important one for the Commission. In the field of occupational health, the Commission has a varied role. Simultaneously, it is engaged in sponsoring and conducting research, evaluating the outcome of research in terms of significance to and impact on health, evaluating performance in the control of occupational risks and, lastly, proposing action in the form of regulations, hence the importance of the discussions and conclusions of scientific seminars such as this one.

This year represents the tenth anniversary of our cooperation with the World Health Organization and IARC in the field of chemical safety. Soon after the 1972 Stockholm 'Conference on the Environment', a symposium was held in Paris on 'The Health Effects of Environmental Chemicals'. This was followed three years later by a meeting in Luxembourg on 'The Use of Biological Specimens for the Assessment of Human Exposure to Environmental Chemicals'. Three years after this, in 1980, a seminar on 'Assessment of Toxic Agents at the Workplace' was held, again in Luxembourg. Most recently, a workshop was held on 'Manpower Needs and Training in Toxicology and Chemical Safety'. One of the recommendations from this workshop is of special interest to us today, namely, that an ethical code be established for toxicologists to ensure that findings indicating adverse health effects of chemicals are fully reported and made known.

Thus the route to Helsinki today has led through Stockholm, Paris and Luxembourg, and I am sure that you will agree with me that it is difficult to think of a better place, or better hosts than the Institute of Occupational Health, to consider the need to develop an appropriate approach to the problem of carcinogenic or mutagenic agents in the workplace or environment.

In conclusion, I should like to add, that to ensure the success of this symposium we must weigh very carefully the short- and long-term implications of any conclusions. The panel discussions could prove of particular importance. These should seek to establish the rationale and application of indicators of genotoxic effects in occupational health. The concept of monitoring in relation to human biological media is not readily acceptable to workers who are, after all, the main interest group. The differences of understanding concealed in the words we use should lead us to ensure that we clarify issues rather than confuse our thinking.

# FOREWORD


## M. Mercier

*International Programme on Chemical Safety, Geneva, Switzerland*


The International Programme on Chemical Safety (IPCS) is a tripartite cooperative venture of the United Nations Environment Programme, the International Labour Organisation and the World Health Organization. So, let me, on behalf of Dr Mahler, Director-General of WHO, Dr Tolba, Executive Head of UNEP and Mr Blanchard, Director-General of the International Labour Organization, convey the pleasure and gratitude of IPCS to the hosts of this International Seminar on 'Methods of Monitoring Human Exposure to Carcinogenic and Mutagenic Agents'.

It is through the dedicated and hard work of all members of the National Organizing Committee of the seminar - its Chairman, Professor Rantanen, the Secretary-General, Dr Hemminki, and other colleagues - that we are able, during this week, to discuss and elaborate on one of the most pressing problems in toxicology today.

I should like to point out also that it is no coincidence that Finland has been chosen as the site for the seminar. Though a small country, Finland has already hosted, with great generosity and scientific excellence, a number of important World Health Organization meetings during recent years. It has unfailingly encouraged and actively participated in IPCS from the days of the programme's inception. Special reference should be made to the support provided to the programme by the Helsinki Institute of Occupational Health as headed by its Director-General, Professor Rantanen.

It is a special pleasure to note that in organizing the seminar, IPCS was able to join forces with the International Agency for Research on Cancer (IARC) and the Commission for the European Communities (CEC). The number of co-sponsors and the geographical spread of the participants of the seminar make it a truly international venture.

I should like to join Professor Rantanen in welcoming all the participants and to express our appreciation of the fact that you have found time in your undoubtedly busy schedules to come to Espoo and share your knowledge of the problem under review with us.

We are all aware of the need to develop biological monitoring techniques in order to assess the potential for chemically induced cancer and genetic damage in humans. The seminar will have the very important task of critically reviewing currently available techniques for the assessment of human exposure to genotoxic agents, and evaluating their significance and usefulness to health.

It is hoped that the participants of this seminar will make recommendations upon the validation procedures which may be undertaken and also identify areas where further research is required.

I wish you a very successful meeting and look forward to the outcome of your discussions.

## SUMMARY

Participants in the symposium reviewed the basic principles of monitoring for carcinogenic and mutagenic agents. They emphasized that the following aspects were of great importance and required further study and development:

- assessment of human exposure to carcinogenic and mutagenic agents by direct chemical measurements and by other approaches;

- the toxicokinetics and metabolic pathways of these agents;

- dose-response relationships and individual variation in metabolism and susceptibility to these agents

The methods currently available or under development were considered from two main aspects:

- as indicators of exposure,

- as indicators of biological end-points.

The following factors were evaluated in judging the usefulness of individual methods:

- technical considerations, such as the chemical specificity, detection limits, precision, quality control, interlaboratory reproducibility, possibility of sample storage, and ease of execution;

- assessment of whether exposure data obtained by the method are quantitatively accurate; give information on duration and other time variables of exposure, on the dose that reached the target site and on total internal dose; and whether there exist background levels of the parameter being measured;

- assessment of whether effects measured by the method are adverse or of no known significance;

- whether the results obtained by the method can be interpreted on an individual or only on a group basis;

- whether the method can be used currently for epidemiological research only or for biological monitoring in occupational health practice.

The following methods were reviewed:

(1) *Determination of chemicals and their metabolites in biological samples.* Such methods have a long history in biological monitoring and usually have been or can be developed to a satisfactory technical level. However, adequate analysis of low exposure levels may necessitate the use of sophisticated methods, and the biological matrix may complicate the analyses. These methods are usually specific for individual chemicals and allow interpretation of the data on an individual basis provided that the kinetic behaviour of the chemical in the body is known.

(2) *Determination of urinary thioethers.* These methods have been used to estimate the overall level of exposure to compounds that conjugate with glutathione. The available methods, which measure total thiols, are non-specific, diet-dependent and insensitive due to fluctuating and high 'normal' background levels. However, in certain situations they can be of practical use. The development of methods that determine specific thioethers was considered an approach of considerable interest.

(3) *Detection of mutagenic activity in excreta.* These methods have been used as non-specific indicators of exposure to mutagenic compounds. However, it was recognized that a number of technical difficulties exist, such as:

- decreases or increases in mutagenic activity during sample preparation;

- the possibility of obtaining spurious positive results with some techniques, due, e.g., to the presence of varying amounts of amino acids in the sample.

The method should be developed further, but technical improvements and validation are required before it can be used for extensive routine monitoring of human exposures.

(4) *Detection of chromosomal aberrations.* Analysis of chromosomal aberrations has been used to detect early changes at an individual level that may or may not correlate with adverse health effects; chromosomal aberrations may also be indicators of adverse effects on a group basis. Rigorous study design is necessary, since many factors may affect cytogenetic parameters. More control populations should be studied in order that the effects of the many confounding factors can be quantified. Intra- and interlaboratory comparisons and 'blind' reading of slides are mandatory in order to improve and maintain high quality in the results. As cytogenetic methods are complicated and exacting, and thus expensive, they should be used only after other approaches have indicated the possible existence of a suspect exposure.

(5) *Detection of sister chromatid exchanges.* This method has been used successfully in experimental research programmes to measure an indicator of exposure. Until now, its effective use in monitoring occupational exposure has not been fully exploited; however, in a number of situations, e.g., exposure to ethylene oxide, the method seems to be promising.

(6) *Detection of micronuclei.* This test is at too early a stage of development to be evaluated for its potential use in monitoring occupational exposure.

(7) *Detection of changes in sperm morphology.* This method has been used to indicate possible adverse health effects in a limited number of situations. The relevance of various parameters of morphology is under experimental study and needs further validation. Questions of appropriate sampling methods and control populations are highly pertinent, and further studies are needed.

(8) *Determination of covalent adducts in proteins and nucleic acids.* These methods may be sensitive and specific for the determination of exposure, but they are still in an experimental stage. Measurement of protein adducts may be made more readily because of their relative abundance, but measurement of DNA adducts may be more relevant as a measure of health hazard.

(9) *Determination of protein variants in blood and measurement of point mutations in lymphocytes.* These are both promising methods, but at present, they are in an early experimental stage of development.

(10) *Detection of DNA repair in somatic cells.* This method may indicate damage inflicted by an exposure, but it is still in an experimental stage.

(11) *Detection of tumour markers.* This method is used in the diagnosis and treatment of cancer, and many marker proteins are induced in experimental animals exposed to carcinogens. At the present stage, assays of tumour markers do not seem to be appropriate for biological monitoring.

Several examples were presented in which application of some of the methods described above has led to specific action by governmental regulatory agencies, industry and hospitals to eliminate or minimize exposure to toxic agents.

The participants at the symposium recognized fully the important potential usefulness of these methods, but pointed out a number of limitations that must be overcome.

The variability of the results obtained by these methods was discussed both in terms of methodology and individual biological differences. It was agreed that considerable improvements could be made in methodology. Individuals may differ in genotype, such as in the degree of inherent susceptibility to an agent; in their susceptibility to enzyme induction; in immunological competency; and in haematological profile. A better understanding of the quantitative effects of confounding factors can lead to better experimental design and to the development of statistical procedures that may overcome, in many instances, difficulties in interpreting the results.

It was recognized that a certain confusion has existed in distinguishing 'biological monitoring' for exposure from 'health surveillance', and that the two represent separate components of a continuum which ranges from the measurement of agents in the body to detection of exposure to early signs of disease.

A better understanding of the health significance of the biological endpoints measured by these methods is needed before they can be recommended for use in the health surveillance of workers occupationally exposed to carcinogenic and mutagenic agents.

In order to assess the usefulness for public health of these methods, it is essential:

- to study whether their endpoints predict disease in experimental animals;
- to determine the persistence over time of these endpoints and whether they reflect the pattern of exposure;
- to assess, qualitatively and quantitatively, the possible relationship between the measured parameter and long-term health effects in exposed workers, using epidemiological studies based on quantitatively defined risks and the informed consent of the workers;
- to study the magnitude of non-occupational confounding factors.

Practical implementation of these methods must still be undertaken with caution. Some of the methods described above may be useful when there is reason to suspect exposure to unknown carcinogenic and mutagenic agents; in all such cases, only group results can be evaluated, although appropriate attention must be paid to the individual. In this respect, ethical considerations are of paramount importance, and the informed consent of workers must be obtained. Some of these methods may help to assess the risk associated with particular working conditions, but they should never be considered substitutes for implementing measures to lower exposure.

## CONCLUSIONS

There is an urgent need to develop methods to assess exposure to carcinogens and mutagens and to establish a set of criteria by which the methods currently or potentially available can be validated. The factors to be considered in setting such criteria should include the following:

1. Appropriate for:
   exposure assessment
   health effect assessment

2. Results valid for:
   individual
   group

3. Reproducibility within and between laboratories

4. Accuracy (specificity, recovery)

5. Detection limit

6. Inter- and intra-individual variations in non-exposed reference populations (due to race, sex, age, etc.)

7. Effect of possible interfering factors (diet, smoking, alcohol, etc.)

8. Absence of background levels

9. Simplicity

10. Possibility of sample storage

These criteria should be refined and expanded and then applied to evaluate the methods considered to be most promising at present. This would make possible a common understanding of the significance, current stage of development and potential use of these methods.

The following methods appear to be those most suited for development and limited use, with appropriate precautions;

- determination of chemicals and their metabolites in biological fluids and tissues,
- determination of thioethers in urine,
- detection of mutagenic activity in urine,
- detection of chromosomal aberrations,
- detection of sister chromatid exchange,
- testing for micronuclei in lymphocytes and/or epithelial cells,
- determination of sperm morphology (in selected situations).

A number of methods are promising but require extensive development and validation before they can be used:

- detection of protein and DNA adducts,
- detection of protein variants in blood,
- detection of point mutations in blood cells,
- investigation of DNA repair in somatic cells,
- detection of tumour markers.

Comparative programmes and collaborative studies at the international level are encouraged in order to improve the techniques and thus to reduce variability due to methodology.

Once, a method has been properly validated from a technical point of view, baseline reference values should be established, keeping in mind, however, the special problems of selecting control groups.

Research should be encouraged, in particular, to obtain data on the quantitative importance of non-occupational confounding factors and to determine the extent to which biological endpoints measured in the above methods are predictors of carcinogenicity and of reproductive effects.

Finally, it would be desirable to establish, at the international level, the scientific relevance of these methods in order to provide guidance for researchers and occupational physicians who may be required to provide such information in the workplace. It is strongly emphasized that such information must be given to individuals participating in any epidemiological investigation or monitoring programme.

The Editors

# SESSION I

Chairman: D. Henschler
Rapporteur: R. Lauwerys

# BASIC CONCEPTS OF MONITORING HUMAN EXPOSURE

**R. Lauwerys**

*Unité de Toxicologie Industrielle et Médicale, Université Catholique de Louvain, Brussels, Belgium*

## SUMMARY

Environmental and biological monitoring are two complementary approaches for evaluating human exposure to genotoxic agents. Environmental monitoring involves, usually, measurement of chemicals either by physico-chemical techniques or *via* biological activities (e.g., mutagenic activity of airborne pollutants). Biological monitoring relies on two groups of tests: those based on determination of the substances or their biotransformation products in various biological media and those based on detection and, possibly, quantification of those biological changes that result from the reaction of the organism to exposure.

The application of tests for exposure and the interpretation of the results require a careful evaluation of several issues, mainly the sensitivity, specificity and health significance of the selected monitoring parameters, the precision and accuracy of the analytical technique and the ethical aspects.

## INTRODUCTION

This paper comments upon some basic principles of monitoring human exposure to foreign chemicals and considers, very briefly, how these principles may apply to the main methods currently available for monitoring exposure to genotoxic agents. The advantages and limitations of each test are not discussed in detail, but a general framework is outlined.

Traditionally, human exposure to an exogenous chemical has been approached by direct measurement of its concentration or of the amount present in one or several environmental compartments. For example, in industry, measurement of the pollutant concentrations in ambient air has long been, and is still, a classical method for assessing exposure. Likewise, in the field of environmental medicine, techniques such as market-basket surveys and duplicate meal studies are still used frequently to monitor exposure. It should be realized, that for any estimate of the external dose of a xenobiotic, i.e., the amount offered to the body per unit of time over a specified time-interval, more information is needed than the chemical concentration in environmental compartments (air, food, water, etc.). Other parameters, such as duration of exposure, pulmonary ventilation, food consumption, etc., must be known in order to estimate the external dose.

Studies on the fate of xenobiotics in the human organism and on their biological effects have led to other approaches to exposure monitoring, grouped under the name: biological monitoring of exposure.

Environmental and biological monitoring of exposure should not be regarded as opposing but, on the contrary, as truly complementary methods to prevent excessive exposure to exogenous chemicals. Both approaches have their own advantages and disadvantages, which have been reviewed elsewhere (Zielhuis, 1979; Lauwerys, 1983). The greatest advantage of biological monitoring is that, usually, the biological parameter of exposure is related more directly than any environmental measurement to the adverse health effects that are the subject of prevention. Therefore, it may offer a better estimate of the risk than ambient monitoring. Biological monitoring considers absorption by all the routes and from various sources (environmental and occupational exposure), also personal hygiene habits and individual variations in the absorption rate and metabolism. However, for genotoxic chemicals that act locally, e.g., some inhaled lung carcinogens, the determination of the chemical in inspired air or, possibly, the mutagenic activity of the material collected from ambient air, may be more relevant for risk assessment than any biological monitoring method based on blood or urine analysis (Monarca *et al.*, 1982).

## DISCUSSION

For the chemicals that exert their toxic effects after penetration of the circulation, knowledge of their fate in the organism (Fig. 1) may be the basis of several biological monitoring methods. The amount of chemical that enters the organism is distributed between several compartments within blood and tissues; it may give rise to inactive or active metabolites. The metabolites, also, may be distributed between various compartments and, like the parent compound, may be excreted by different routes. Acting directly, or indirectly through its active metabolites, the chemical binds to critical or noncritical target molecules. Repair mechanisms may release degradation products that may be detected in various tissues or excreta, but if the repair mechanisms are inadequate or insufficient, cell lesions may occur.

**Fig. 1. Fate of xenobiotics in the organism**

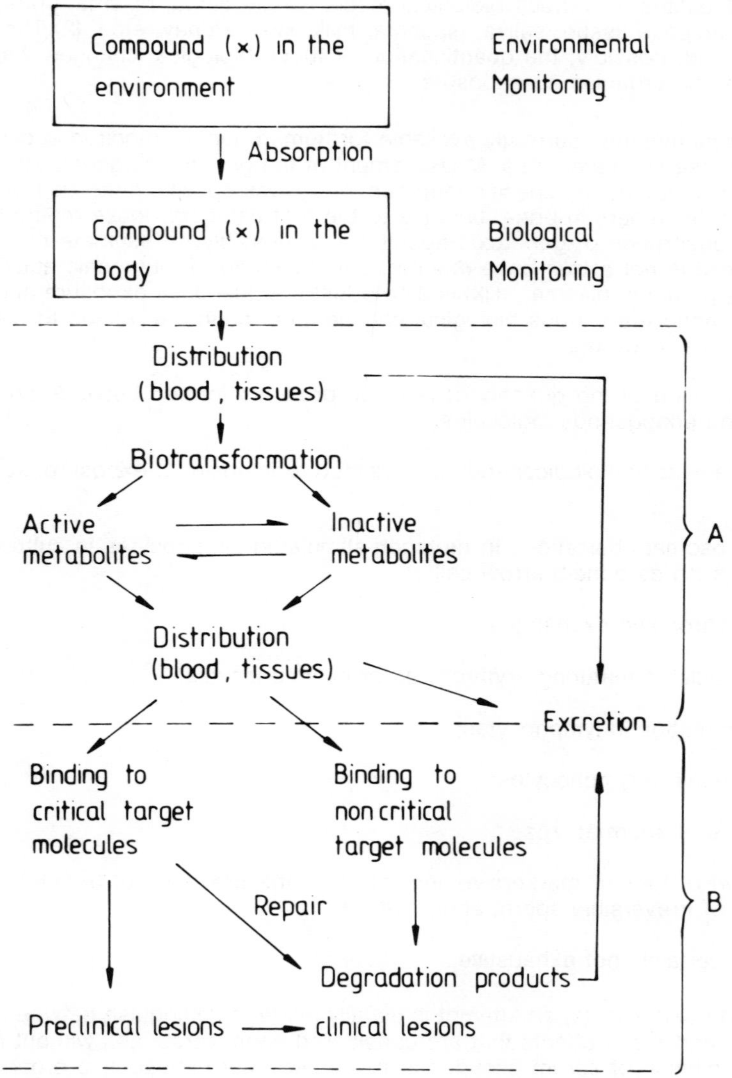


Depending on the biological parameter selected, and the relationship of time of sampling to that of exposure, different information may be obtained; data may reflect the amount of the chemical recently absorbed, the amount already stored in one or several compartments of the organism (body burden) or the amount of the active chemical species bound to the sites of action.

Biological monitoring methods can be classified in two broad categories (Fig. 1). (1) Those based on determination of the concentration of the substance itself or its biotransformation products (metabolites) in various biological media (whole blood, plasma, urine, expired air, sputum, hair, adipose tissue, saliva, placenta, milk, liver, kidney, etc.). (2) Those based on the detection and, possibly, the quantification of those biological changes that result from the reaction of the organism to exposure.

The biological methods currently available for human genetic monitoring can be distributed between these two categories. Measurement of the genotoxic agent and/or its metabolites in biological material by specific methods, or by less specific ones such as the measurement of total thioethers in urine, belongs to the first category; these methods all attempt to detect the absorption of genotoxic agent(s) and, possibly, to estimate the internal dose also. Although it is not a true dose monitor, determination of mutagenic activity in biological material (e.g., urine, plasma, milk) is also a test that identifies exposure and not a biological effect. Among the various biological changes that might be related to the intensity of exposure, are the following:

(1) Determination of the quantity of reaction products formed between an active genotoxic agent and endogenous molecules.

(2) Measurement of biological endpoints related to intensity of exposure, such as (Vainio *et al.*, 1983):

- chromosomal aberrations in mutagen-stimulated lymphocytes in culture or in other cells such as bone-marrow cells;
- sister chromatid exchanges;
- micronuclei in maturing erythrocytes or in lymphocytes;
- point mutation in lymphocytes;
- DNA repair in lymphocytes;
- analysis of spermatozoa

(3) Detection of tumour markers as indicators of the presence of transformed cells and the detection of irreversible sperm abnormalities.

This list is certainly not exhaustive.

For nongenotoxic agents, an attempt is usually made to distinguish between adverse biological effects and those effects that are considered nonadverse, i.e., without health significance (e.g., inhibition of serum pseudocholinesterase after exposure to organophosphorus pesticides). It is mainly the latter that are useful for monitoring exposure. Indeed, the main goal of biological monitoring of exposure is either to ensure that current or past human exposure to xenobiotics is 'safe' (i.e., does not entail an unacceptable health risk) or to detect potential excessive exposure before the occurrence of detectable adverse health effects. It is, essentially, a preventive medical activity. In this regard, it must be distinguished from the detection of adverse biological effects in exposed subjects (e.g., increased levels of cytolytic enzymes in plasma after exposure to hepatotoxic chemicals, increased proteinuria after exposure to nephrotoxic agents, etc.) that, when present, indicate that exposure is, or has been, excessive. The latter biological analyses should be part of a programme for early detection of health impairment due to chemicals, and not of a programme for biological monitoring of exposure. Of course, in practice, both programmes are often applied concomitantly, since an internal exposure that is considered safe by present knowledge may yet cause some harmful effects in susceptible individuals. Furthermore, the distinction between adverse and nonadverse biological effects is not always clear-cut. At an Environmental Protection Agency - World Health Organization - Commission of the European Communities symposium on biological monitoring (Berlin *et al.*, 1979), it was proposed that a biological effect should be considered adverse if there is an impairment of functional capacity, if there is a decreased ability to compensate for additional stress, if there is a decreased ability to maintain homeostasis, if there is an enhanced susceptibility to other environmental influences or if any such impairment is likely to become manifest in the near future.

Application of these criteria to the early biological effects induced by genotoxic agents is not always straightforward. Monitoring methods based on the determination of adducts between active genotoxic species and proteins or nucleic acids attempt to estimate the internal dose in various body compartments. For these parameters, it might be possible to propose a threshold value below which the risk of adverse effects is negligible. It is accepted that detection of tumour markers in biological fluids or of irreversible sperm abnormalities indicates that an adverse effect has already occurred. If the effects are detectable, the harm has already been done, although it may still be at an early stage. Between these two extremes, there is a group of effects (e.g., chromosomal aberrations, sister chromatid exchanges, micronuclei), for which the health significance may not be immediately apparent. Should they be considered as adverse effects, predictive of impending health impairments, or simply as exposure parameters, or both? It is one of the objectives of this symposium to summarize present knowledge on the health significance of these various biological endpoints.

Before a biological monitoring programme is implemented one must also consider whether the specificity of the selected method(s) allows evaluation of the results on an individual basis or on a group basis. In other words, are the monitoring methods recommended mainly for epidemiological studies or are they useful for individual risk assessment, also? One should know, also, whether the methods permit a quantitative or only a qualitative estimate of exposure. Some human genetic monitoring methods (e.g., chromosomal analysis) are probably not specific enough to evaluate the risk of excessive exposure for individuals. Furthermore, the absence of well-defined dose-effect relationships precludes any quantitative estimate of exposure. However, when applied under well-defined conditions (e.g., inclusion of a properly selected control group, matched particularly for confounding variables such as ethanol and drug consumption, smoking, etc.), these methods may be valuable for identifying groups at risk of exposure to genetically active substances and, hence, in justifying the implementation of better preventive measures.

Some attention must also be given to the ethical aspect of monitoring, particularly to the type of information that should be provided to a group of persons and to individuals when biological changes are detected that suggest exposure to genotoxic agents.

Finally, before any monitoring method is routinely applied, there are important technical problems to be considered, such as the precision and the accuracy of the methods, as well as more practical considerations such as the stability of the sample, the degree of sophistication of the analytical technique (need of specially trained personnel, elaborate equipment) and, not least, the cost.

## REFERENCES

Berlin, A., Wolff, A.H. & Hasegawa, Y., eds (1979) *The Use of Biological Specimens for the Assessment of Human Exposure to Environmental Pollutants*, The Hague, Martinus Nijhoff

Lauwerys, R.R. (1983) *Industrial Chemical Exposure: Guidelines for Biological Monitoring*, Davis, CA, Biomedical Publications

Monarca, S., Pasquini, R., Sforzolini, G.S., Viola, V. & Fagioli, F. (1982) Application of the *Salmonella* mutagenicity assay and determination of polycyclic aromatic hydrocarbons in workplaces exposed to petroleum pitch and petroleum coke. *Int. Arch occup. environ. Health*, *49*, 223-239

Vainio, H., Sorsa, M. & Hemminki, K. (1983) Biological monitoring in surveillance of exposure to genotoxicants. *Am. J. ind. Med.*, *4*, 87-103

Zielhuis, R.L. (1979) *General aspects of biological monitoring*. In: Berlin, A., Wolff, A.H. & Hasegawa, Y., eds, *The Use of Biological Specimens for the Assessment of Human Exposure to Environmental Pollutants*, The Hague, Martinus Nijhoff, pp. 341-359

# HUMAN EXPOSURE TO POTENTIALLY CARCINOGENIC COMPOUNDS

**K. Hemminki & H. Vainio***

*Institute of Occupational Health, Helsinki, Finland and*
**International Agency for Research on Cancer, Lyon, France*

**SUMMARY**

Average exposures to suspected carcinogens were surveyed for the population in Finland. Alcohol, food and tobacco constitute the largest sources of exposure. Occupational populations may be more exposed than the population at large, by many orders of magnitude. For example, occupational exposure to airborne asbestos and trichloroethylene may be 10 000 times higher than exposure outside workplaces. Application of exposure data to health surveillance and epidemiological research is discussed.

## INTRODUCTION

Carcinogenic and mutagenic compounds are commonly found in the human environment, in food, drinking-water, beverages, tobacco and ambient air (Nagao *et al.*, 1978; Hemminki *et al.*, 1983). Food has attracted particular attention as being a major source of carcinogens and mutagens (Hirono, 1981; Sugimura, 1982; Ames, 1983). All populations are thus exposed to carcinogenic and mutagenic agents at some level. The present article reviews estimates of the levels of exposure of the population in Finland to compounds known or suspected to be carcinogenic to experimental animals or man. Average population exposures and heavy occupational exposures are discussed separately. These data are limited by lack of knowledge about actual exposures and about the carcinogenic potency of some of the chemicals in humans. It is striking, however, that the exposures tend to be focussed in certain occupational subpopulations.

## AVERAGE EXPOSURE OF THE FINNISH POPULATION

The average exposure of the population of Finland to suspected chemical carcinogens and mutagens (Hemminki *et al.*, 1983) was estimated recently by extrapolating from relevant measurements. These levels are shown in Table 1 according to origin and type, and concern compounds found to be possibly carcinogenic to man or to experimental animals. References to studies of carcinogenicity and mutagenicity are given in the comprehensive review by Hemminki *et al.* (1983). Table 1 is supplemented here by evaluations of the International Agency for Research on Cancer (IARC) working groups for particular exposures (IARC, 1982a,b), and the compounds or exposures listed in the table are categorized according to the IARC criteria as being: (1) causally associated with cancer in humans; (2) probably carcinogenic to humans; and (3) carcinogenic ('sufficient evidence') in experimental animals, if not in groups 1 or 2. For most of the other compounds or exposures that were considered by IARC working groups, the evidence was judged to be limited or insufficient for a number of reasons.

The largest sources of exposure in the Finnish population are alcohol (in the form of acetaldehyde, 10 g/person per day), food ingredients (400 mg/person per day), tobacco tar (50 mg/person per day), indoor air (1 mg/person per day), urban air (0.2 μg/person per day) and drinking water (0.07 mg/person per day), as shown in Figure 1. When the suspected carcinogens in food are analysed separately, natural plant ingredients, particularly flavonoids and tannins, stand out. Although many other mutagens have been found in plants, their carcinogenicity is poorly established (Ames, 1983). The strength of the evidence for carcinogenicity of flavonoids (IARC, 1983) and tannins (Hemminki *et al.*, 1983) is weak also;

**Fig. 1. Daily average exposure to suspected carcinogens of the population in Finland**

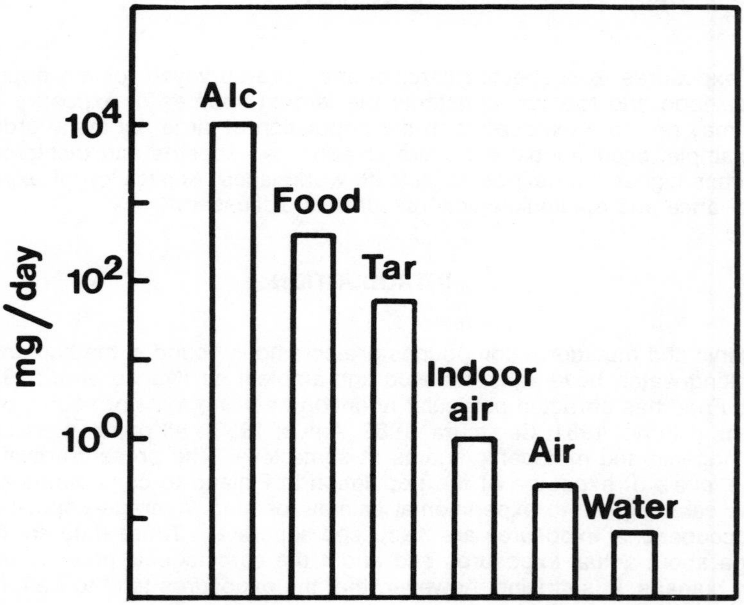


Alc = alcohol

**Table 1. Estimated exposure of the Finnish population to potential genotoxins**

| Category | Compound | Exposure (μg/day) |
|---|---|---|
| FOOD | | |
| Plant ingredients | coumarin | ? |
| | flavonoids, mutagenic | 50 000 |
| | hydrazine[b] derivatives | 1 |
| | parasorbic acid | ? |
| | safrole[c] | ? |
| | derivatives | 150 000 |
| | tannin | |
| Metals | arsenic[a] | 60 |
| | cadmium[b] | 13 |
| | chromium[a] | 30 |
| | lead[c] | 70 |
| | nickel[b] | 130 |
| Mycotoxins | aflatoxins[b] | 0.01 |
| | *Fusarium* toxins | ? |
| | patulin | 10 |
| Organic contaminants | chlordane | 4 |
| | chlorinated phenols[b] | 5 |
| | DDT derivatives[b] | 2 |
| | hexachlorobenzene[c] | 0.5 |
| | polychlorinated biphenyls[b] | 7 |
| | polycyclic aromatic hydrocarbons[b] | 1.5 |
| | toxaphene[c] | 4 |
| Food additives | agar | 15 000 |
| | carrageenan[b] | 50 000 |
| | cyclamate | 21 000 |
| | fast green FCF | 5 |
| | hexamethylenetetramine | 65 |
| | nitrate | 16 000 |
| | nitrite | 3 500 |
| | saccharin[c] | 15 000 |
| | sorbic acid | 37 000 |
| Pesticides | benomyl | 38 |
| | DDT derivatives[b] | 0.02 |
| | other chlorinated hydrocarbons | 2 |
| | dithiocarbamates | 29 |
| | *o*-phenylphenol | 20 |
| | organophosphates | 11 |
| Preparation and storage | amino acid pyrolysates | <3 |
| | - from glutamic acid | <3 |
| | - from tryptophan | 100 |
| | malonaldehyde | 1 |
| | nitrosamines | 2 |
| | polycyclic aromatic hydrocarbons[b] | |
| WATER | nitrite | 20 |
| | polycyclic aromatic hydrocarbons[b] | 0.2 |
| | trihalomethanes | 50 |
| | - chloroform | 25 |
| COMMUNITY AIR | | |
| Metals | cadmium[b] | 0.02 |
| | lead[b] | 6 |
| Other | asbestos[a] | 0.01-0.1 |
| | benzene[a] | 100 |
| | formaldehyde[b] | 100 |
| | halogenated hydrocarbons | 20 |
| | polycyclic aromatic hydrocarbons[b] | 0.4 |
| HOME ENVIRONMENT | formaldehyde[b] | 1000 |
| | other variable contaminants | ? |
| ALCOHOL | acetaldehyde | 10 000 000 |
| TOBACCO | tar | 50 000 |
| | acetaldehyde | 2000 |
| | benzene[a] | 200 |
| | cadmium[b] | 1-4 |
| | formaldehyde[b] | 50 |
| | *N*-nitroso compounds | 2 |
| | polycyclic aromatic hydrocarbons[b] | 1 |

[a] Causally associated with cancer in humans (IARC, 1982a)

[b] Probably carcinogenic to humans

[c] Carcinogenic to experimental animals, if not in [a] or [b]

studies on these groups of compounds are further hampered by the heterogeneity of the naturally occurring species. Other major sources of suspected carcinogens are food additives such as carrageenan, sorbic acid, nitrate/nitrite, agar and the artificial sweetening agents. Even in this group, the strength of the evidence for carcinogenicity is usually weak (IARC, 1982a).

This type of estimation presents several problems. The first is the limitation posed by incomplete information, both for the level of exposure and the carcinogenic potency of the chemicals. Little is known about levels of exposure to natural products, as emphasized by Ames (1983); data on exposure to flavonoids and tannins and on their carcinogenicity are especially meagre. Since many plant ingredients are consumed in large quantities, further research is essential. Analytical knowledge is also lacking with regard to mycotoxins, combustion products, amino acid pyrolysates, asbestos in drinking-water and technochemical products. It is interesting that the carcinogenicity of many food additives, coffee and tea - to which many people are highly exposed - has not been tested adequately. Coffee is known to contain several mutagens, such as chlorogenic acid, glyoxal, methylglyoxal and diacetyl, in considerable quantities (Sugimura, 1982; Ames, 1983).

A second major problem is the limited understanding of the mechanisms of human cancer. Hence, methods for testing for carcinogenicity may not be relevant to all aspects of carcinogenesis; they are well suited to detecting tumour initiators, but less suited for the detection of promoters and cocarcinogens (Weisburger & Williams, 1981). Yet man is exposed continually to mixtures of tumour initiators, promoters, cocarcinogens and inhibitors. A major limitation is knowledge about carcinogenic potency. Because carcinogenesis is a multi-faceted process, no single parameter is likely to predict, quantitatively, the carcinogenic potency of a given compound. The carcinogenic potency of individual chemicals ranges over many orders of magnitude, aflatoxin $B_1$, for example, being about 10 million times more potent than trichloroethylene in causing tumours in mouse liver; thus, the levels of exposure listed in Table 1 cannot be compared among themselves, given the different potencies of the exposures. Although it is commonly estimated that tobacco smoke causes somewhere between 20 and 30% of cases of cancer in man (Doll & Peto, 1981), the quantity of known carcinogens in tobacco smoke does not appear to be overwhelming (Table 1). Thus, either the information on relative potency and the effects of combined exposure is limited, or there are groups of carcinogens still to be discovered.

## EXPOSED POPULATIONS - OCCUPATIONAL EXPOSURES

The estimated exposures given in Table 1 are averages calculated for the total population of Finland. Such calculations are always theoretical, and the major exposures occur mainly in smaller subgroups. Typically, acetaldehyde derived from ethanol affects only those subgroups that consume alcohol. Most of the adult Finnish population consumes very little alcohol, but 10% - the heavy drinkers - consume 5 to 10 times more than the average daily dose of 12 g pure ethanol. Other types of exposure, also, tend to be focussed on smaller subgroups, e.g., fish eaters ingest organic pollutants and mercury, ice-cream eaters ingest carrageenan, fruit eaters ingest pesticide residues and urban residents breathe many types of pollutants. Some exposure, e.g. to medicinal drugs and occupational chemicals, typically affects small populations only. Some quantified occupational exposures are discussed in detail below.

The Institute of Occupational Health in Finland has carried out extensive measurements on the concentrations of selected carcinogenic and mutagenic chemicals in the ambient air of workplaces, and this information has been used to estimate the daily exposure of workers. For most exposures, data are available both for all workers considered to be exposed and for those considered to be heavily exposed. Table 2 (from Kauppinen *et al.*, 1982) lists these data, showing the number of workers in each category; for heavy exposures, the type of work is listed also. For comparison, the estimated average exposure of the total population is given.

The group heavily exposed to metals is more exposed than the total population by less than 10 times for arsenic, cadmium, and nickel, but for chromium the factor is 100. The size of the heavily exposed populations varies from 200 (cadmium and lead) to 2000 (chromium and nickel). Since the population of Finland is 4.8 million, of whom 2.1 million are economically active and 0.5 million are employed in industrial occupations, these numbers represent a small proportion and do not contribute essentially to the exposure of the total population.

It is estimated that asbestos workers in general breathe 200 μg of fibre per working day and the heavily exposed subpopulation up to 1000 μg per day. These levels massively exceed the exposure of the general population, which is 0.01-0.1 μg, contributed by pollution of community air. The calculation does not include the possible presence of asbestos in drinking water.

Exposure to organic carcinogens and mutagens varies extensively in absolute quantity and in reference to the total population. For some organic chemicals, e.g., styrene, tetrachloroethylene, 1,1,1-trichloroethane, and trichloroethylene, the exposure of the total population, as far as is known, is very low (less than 5 μg per day), compared with the massive occupational exposures of 0.1 to 3 g per day. For other organic chemicals, such as formaldehyde and polycyclic aromatic hydrocarbons, the difference between the exposure of the total population and that of heavily exposed workers is less dramatic - about 10-fold. Food provides relatively high background levels of polycyclic aromatic hydrocarbons and indoor air has similar levels of formaldehyde.

Examples of heavy occupational exposures compared with total population exposures are shown in Figure 2. The smallest differences shown are about 10-fold and the largest about 10 000-fold.

## EXPOSURE DATA - RELEVANCE TO CANCER EPIDEMIOLOGY

Recently, an IARC working group re-evaluated the strength of the evidence for carcinogenicity for all chemicals, for groups of chemicals and for occupational exposures, considered in *Monographs* 1 to 29, for which some carcinogenicity data on humans were available (IARC, 1982a). The working group concluded that seven industrial processes or occupational exposures were associated with cancer in humans and that one industrial process was probably carcinogenic to humans. Twenty-three chemicals or groups of chemicals were considered carcinogenic to humans, and a further thirteen highly probable. However, even where there was a causal association between an exposure and an excess of cancer, quantitative data on the level of human exposure were available for only a few substances, i.e., for arsenic, asbestos, benzene, chromium, nickel and vinyl chloride. It is striking that, in all

**Table 2. Focussing of industrial exposures among worker populations in Finland**

| Exposure | Exposed population (in 1000 persons) | | Estimated daily exposure (μg) | Type of work or of exposure |
|---|---|---|---|---|
| Arsenic | total population[a] | | 60 | food |
| | exposed workers | (5) | 100 | - |
| | heavily exposed | (1) | 500 | ore refining, foundry work |
| Cadmium | total population | | 13 | food |
| | exposed workers | (2) | 30 | - |
| | heavily exposed | (0.2) | 200 | hard soldering, smelting |
| Chromium | total population | | 30 | food |
| | exposed workers | (30) | 1000 | - |
| | heavily exposed | (2) | 3000 | ferrochromium production, stainless steel welding, steel foundry work |
| Lead | total population | | 75 | food, community air |
| | exposed workers | (20) | 200 | - |
| | heavily exposed | (0.2) | 1500 | smelting, battery production |
| Nickel | total population | | 130 | food |
| | exposed workers | (10) | 300 | - |
| | heavily exposed | (2) | 1000 | nickel refining, steel foundry work |
| Asbestos | total production | | 0.01-0.1 | community air |
| | exposed workers | (3) | 200 | - |
| | heavily exposed | (0.3) | 1000 | production of asbestos goods |
| Benzene | total population | | 100 | community air |
| | exposed workers | (10) | 5000 | - |
| | heavily exposed | (0.1) | 20 000 | benzene production, transportation |
| Chlorophenols | total population | | 5 | food |
| | exposed workers | (5) | 500 | - |
| | heavily exposed | (1) | 2000 | antimould treatment in sawmills |
| Formaldehyde | total population | | 1100 | room and community air |
| | exposed workers | (20) | 3000 | - |
| | heavily exposed | (2.5) | 10 000 | particle-board and glue production, foundry work |
| Polycyclic aromatic hydrocarbons | total population | | 4 | food, community air |
| | exposed workers | (100) | 4 | - |
| | heavily exposed | (6) | 50 | iron foundry work, chimney sweeping |
| Styrene | total population | | ? | |
| | exposed workers | (3) | 2 000 000 | |
| | heavily exposed | (2) | 3 000 000 | reinforced plastics production |
| Tetrachloroethylene | total population | | 1-5 | community air |
| | exposed workers | (2) | 70 000 | - |
| | heavily exposed | (0.1) | 500 000 | dry-cleaning |
| 1,1,1-Trichlorethane | total population | | ? | grease removal |
| | exposed workers | (4) | 100 000 | |
| Trichlorethylene | total population | | 1-5 | community air |
| | exposed workers | (1) | 300 000 | - |
| | heavily exposed | (0.5) | 500 000 | grease removal |

[a] Total population of Finland is 4.8 million

**Fig. 2. Comparison of occupational and average exposures**

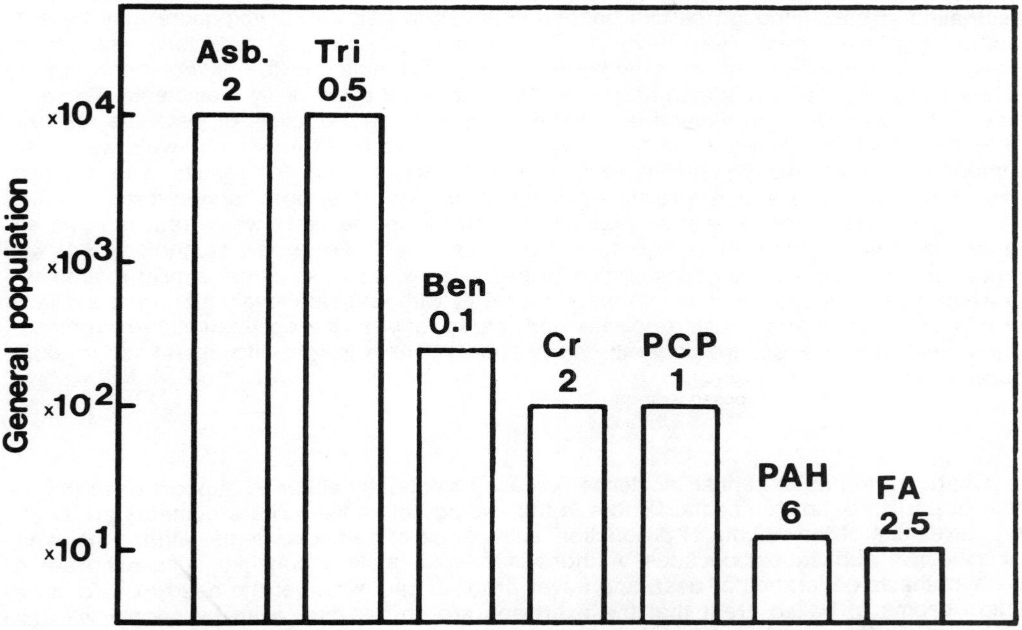


Asb, asbestos; Tri, trichloroethylene; Ben, benzene; Cr, chromium; PCP, polychlorinated phenols; PAH, polycyclic aromatic hydrocarbons; FA, formaldehyde. The figures above the bars indicate the size, in thousands, of the occupational population exposed.

cases, the estimated exposure levels were much higher even than the heavy industrial exposure presently encountered in Finland (cf. 'heavily exposed' in Table 2). On the basis of the present levels of exposure in Finland and the risk ratios given in relevant epidemiological studies, it is calculated that known human occupational carcinogens may account for 60-70 new cases of cancer annually in that country (Vainio *et al.*, 1982); this represents about 0.5% of all new cancer cases in Finland presently.

Most of these epidemiological studies give risk ratios between 2 and 10 for levels of exposure much higher than those prevalent in Finland. Much higher risk ratios have been detected, but only for rare types of cancer. Improvements in industrial hygiene and technology, for health reasons in general, have decreased the levels of exposure, decreasing both the cancer risk and the chance of observing it. Thus, recent epidemiological studies of occupational cancer frequently report risk ratios near one. An encouraging aspect of the development of occupational health standards and services, however, is that data on exposure are becoming more extensive.

Analyses of exposure data can also interpret trends in cancer incidence within a population or between populations. If available, exposure data over extended periods can be compared with cancer incidence data in order to formulate hypotheses about increases or decreases in risk. Although certain general changes in patterns of exposure may be anticipated, the levels themselves may still be surprising. For example, it has long been recognized that polyhalogenated aromatic hydrocarbons accumulate in the environment, but the levels in daily food and in human tissues and milk have not been easily predictable. Two analyses were carried out in Finland, at five-year intervals, on the levels of pesticide residues in Finnish food (cf. Hemminki *et al.*, 1982). About a doubling of exposure levels was seen. Although there are many plausible explanations for this finding, the results remain provocative and call into question a relatively new use of pesticides, post-harvest treatment (i.e., use of pesticides after harvest as opposed to use before harvest), which results in an apparent increase in the level of exposure. This is an example of a new technology that was introduced without serious consideration of health consequences. Other similar examples - in which exposure patterns have changed markedly without safety evaluation - are available in food, cosmetics and housing policies and, particularly, in the occupational environment. Thus, analyses of exposure patterns can provide valuable insight into alterations in exposure.

Changing trends in cancer incidence are used frequently either to support or refute the role of particular environmental factors in the etiology of cancer. The arguments are based on correlation of the volume of production, sale or use of particular items with the incidence of cancer in subsequent decades. Authorities in epidemiology consider such exercises to be hypothesis-generating at best, and never proof of causality; yet the need for this correlation seems to be so great that the exercises are carried out, even by people who are aware of the inherent fallacies.

When exposure is related to the incidence of a cancer, it is of primary importance to establish the true level of exposure. An increase in the number of new chemicals is no proof of increasing exposure; measurements of exposure are necessary. The very likelihood of the fact that over the years, some exposures have increased and others decreased within a population, is a major contra-indication for correlation analyses. For example, in many industrialized populations, the following exposures have increased over the last two or three decades (although it must be emphasized that the agents themselves have changed markedly during that period): pesticide residues and halogenated environmental contaminants, food additives (with the possible exception of nitrate/nitrite), cosmetic products, automobile exhausts and emissions from synthetic household wares. Analogously, the following exposures are likely to have decreased in most industrialized populations: toxins in spoilt and mouldy food, pyrolysis products in food and fumes of burning wood and coal both in homes and in the environment. When it is considered, in addition, that the variety of food has changed considerably - possibly resulting in modifications of responses to environmental carcinogens - it is clear that any correlation study focussing on an individual parameter will be a gross oversimplification of the complexity of the exposure panorama. Any attempt to resolve these problems requires systematic and long-term collection of results from a variety of exposure measurements.

## CONCLUSIONS

Some estimates of the exposure of the Finnish population to carcinogenic and mutagenic agents have been reviewed. Despite the crudeness of the estimates, it is clear that extensive differences in exposure exist between these agents and among the subpopulations studied. Alcohol, some food constituents and tobacco represent important exposures, even when calculated per total population. Occupational exposure tends to be confined to small subgroups, although the extent of exposure may be high.

It is clear that not all exposures are equally harmful; carcinogenic and mutagenic agents vary in potency in comparable test systems by factors as high as $10^7$. Before there can be any meaningful interpretation of exposure in relation to human health, many more data must be obtained on the carcinogenicity of the relevant compounds and on their fate in man.

## REFERENCES

Ames, B. (1983) Dietary carcinogens and anti-carcinogens. Oxygen radicals and degenerative diseases. *Science*, *221*, 1256-1264

Doll, R. & Peto, R. (1981) The causes of cancer: quantitative estimates of avoidable risks of cancer in the United States today. *J. natl Cancer Inst.*, *66*, 1191-1265

Hemminki, K., Vainio, H., Sorsa, M. & Salminen, S. (1983) An estimation of the exposure of the population in Finland to suspected chemical carcinogens. *J. environ. Sci. Health*, *C1*, 55-95

Hirono, I. (1981) Natural carcinogenic products of plant origin. *CRC Crit. Rev. Toxicol.*, *8*, 235-277

IARC (1982a) *IARC Monographs on the Evaluation of the Carcinogenic Risk of Chemicals to Humans*, Suppl. 4, *Chemicals, Industrial Processes and Industries Associated with Cancer in Humans*, Lyon

IARC (1982b) *IARC Monographs on the Evaluation of the Carcinogenic Risk of Chemicals to Humans*, Vol. 30, *Miscellaneous pesticides*, Lyon

IARC (1983) *IARC Monographs on the Evaluation of the Carcinogenic Risk of Chemicals to Humans*, Vol. 31, *Some Food Additives, Feed Additives and Naturally Occurring Substances*, Lyon

Kauppinen, T., Tossavainen, A., Vainio, H. & Kalliokoski, P. (1982) Suomalaisten työperäinen altistuminen. *Ympäristö ja Terveys*, *2-3*, 207-212

Nagao, M. & Sugimura, T. (1978) Environmental mutagens and carcinogens. *Ann. Rev. Genet.*, *12*, 117-159

Sugimura, T. (1982) Mutagens, carcinogens, and tumor promoters in our daily food. *Cancer*, *49*, 1970-1984

Vainio, H., Kauppinen, T., Hemminki, K., Tossavainen, A. & Rantanen, J. (1982) Työperäinen syöpä Suomessa. *Suom. Lääkäril 37*, 2685-2691

Weisburger, J.H. & Williams, G.M. (1981) Carcinogen testing: current problems and new approaches. *Science*, *214*, 401-407

# BIOLOGICALLY-ACTIVE AND CHEMICALLY-REACTIVE POLYCYCLIC HYDROCARBON METABOLITES

**D.H. Phillips & P.L. Grover**

*Chester Beatty Laboratories, Institute of Cancer Research: Royal Cancer Hospital, London, UK*

## SUMMARY

The mechanism by which polycyclic hydrocarbons produce tumours in mammalian tissues exposed to them involves biotransformation of the compounds to chemically-reactive species that covalently modify cellular informational macromolecules. In all cases known, epoxides of some form are the reactive species involved. The most common pathway is the formation of vicinal diol-epoxides, the reactive centre of the molecule commonly being adjacent to the 'bay-region'. With some hydrocarbons, the involvement in DNA binding of non-'bay-region' diol-epoxides, of a phenol epoxide and of a 'bay-region' diol-epoxide containing a phenolic function (a triol-epoxide) has also been demonstrated. The relative importance to the carcinogenic process of the different pathways leading to DNA-binding products may be reflected by the biological activities of the intermediates involved.

## INTRODUCTION

Polycyclic aromatic hydrocarbons have been studied for many years by researchers in disciplines related to toxicology and carcinogenesis. Their current prominence as research tools derives both from their historical importance - compounds of this class were the first pure chemical carcinogens identified (Phillips, 1983) - and from their widespread distribution in the environment as byproducts of the combustion of fossil fuels and other organic materials (Grimmer, 1979). In addition, since the class includes both potently carcinogenic and inactive compounds, the polycyclic hydrocarbons have been a useful series for the study of structure-activity relationships. Since it is now established that they must undergo metabolism in order to exert their various biological effects in living systems, studies on the biotransformation of polycyclic hydrocarbons, by mammalian cells and cell fractions, have become central to an understanding of the action of these compounds, and are of prime importance in assessing the risk that the compounds pose to human health.

This paper gives a brief review of polycyclic hydrocarbon metabolism, with particular emphasis on pathways of activation, and summarizes current knowledge on the interaction of activated metabolites with cellular constituents, particularly DNA, the covalent modification of which is widely held to be a critical early event in carcinogen-induced tumour initiation (Grover, 1979). Wherever possible, reference is made to recent comprehensive review articles, since the original research articles are too numerous to cite individually here.

The structures of some representative hydrocarbons are shown in Figure 1.

**Fig. 1. Parent structures of the polycyclic hydrocarbons and derivatives mentioned in the text**

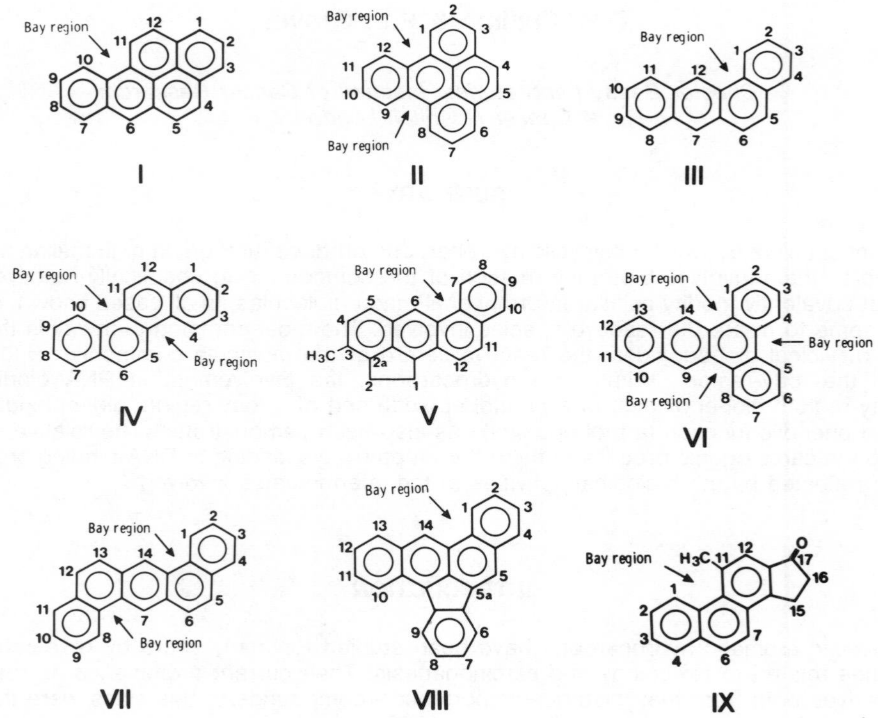


I, benzo[*a*]pyrene; II, benzo[*e*]pyrene; III, benz[*a*]anthracene; IV, chrysene; V, 3-methylcholanthrene; VI, dibenz[*a,c*]anthracene; VII, dibenz[*a,h*]anthracene; VIII, dibenzo[*a,e*]fluoranthrene; IX, 15,16-dihydro-11-methylcyclopenta[*a*]phenanthren-17-one

## GENERAL ASPECTS OF POLYCYCLIC HYDROCARBON METABOLISM

In common with many other classes of xenobiotics, polycyclic hydrocarbons undergo oxidative metabolism to more polar, and hence more readily excretable, products. Examples of the major types of phase I metabolites are shown in Figure 2 for benzo[*a*]pyrene, the most extensively studied compound in this class. The enzymes responsible for the initial introduction of oxygen into a hydrocarbon molecule are the NADPH-dependent monooxygenases that are present in cells of most mammalian tissues, although the highest activity is obtained with liver preparations. Cell fractionation studies have shown that the activity resides mainly in the endoplasmic reticulum (microsomal fraction), although some activity is also present in nuclei. All of the metabolites shown in Figure 2 are thought to arise *via* the formation of one common type of intermediate, viz., an epoxide. The addition of an oxygen atom across an aromatic double bond, to form a simple epoxide, is a two-electron process, and the enzymes specifically involved in the oxygen transfer are the cytochrome P-450 family of enzymes, and NADPH-cytochrome P-450 reductase. In practice, epoxides are not normally isolated from metabolizing systems, as they readily undergo further enzymic or nonenzymic processes, and their formation is generally inferred from the nature of the more stable end-products isolated.

**Fig. 2. Major types of polycyclic hydrocarbon phase-I metabolites, illustrated here as products of benzo[*a*]pyrene metabolism**

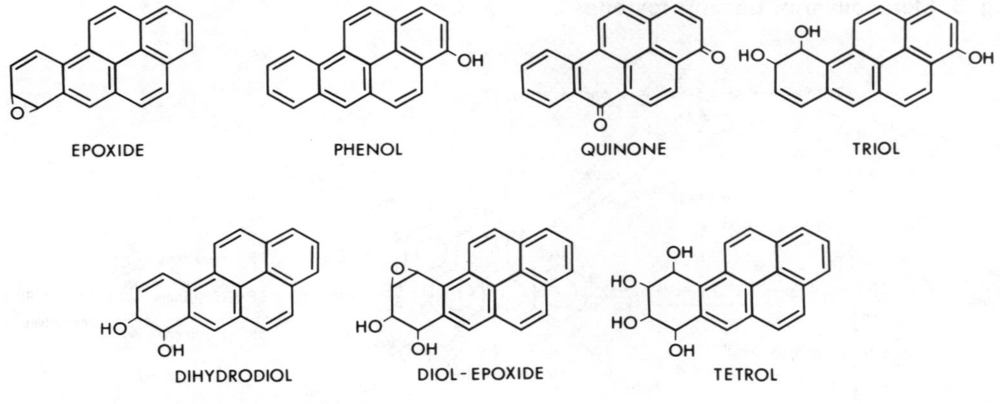


Firstly, there are the phenols, which arise from the nonenzymic rearrangement of epoxides. The phenols, in turn, may be oxidized nonenzymically to quinones, they may be conjugated with glucuronic and sulphuric acids, or they may be further metabolized through epoxide formation on other double bonds in the molecule (see below). Secondly, the action of glutathione S-transferases on hydrocarbon epoxides converts them to glutathione conjugates that are able to undergo subsequent conversion to mercapturic acids. Thirdly, epoxides are also substrates for the microsomal enzyme epoxide hydrolase, which catalyses the addition of water and the formation of dihydrodiols. In all but a few cases, the dihydrodiols so formed have the *trans* configuration.

Dihydrodiols, themselves, may also undergo further metabolism by monooxygenases, to form diol-epoxides. Where the epoxide function is introduced in the same benzo ring as the dihydrodiol group (a vicinal diol-epoxide), two types of diol-epoxide may be formed, depending on whether the epoxide is on the same face of the ring as the distal hydroxyl group (a *syn*-isomer) or is on the opposite face (an *anti*-isomer). The conformation of these substituents, relative to each other, dramatically influences the chemical and biological properties of the metabolites. Diol-epoxides, like simple epoxides, are too reactive to be isolated from metabolizing systems, and their formation is inferred from the detection of their hydrolysis products, tetrols, or by their covalent binding to nucleic acids (see below).

Reduction of dihydrodiols to catechols is also known to occur; dihydrodiol dehydrogenase may, therefore, compete with the monooxygenase pathway for further metabolism of dihydrodiols.

The metabolism of polycyclic hydrocarbons, and, in particular, of benzo[*a*]pyrene, has been extensively reviewed (Sims & Grover, 1974; Gelboin, 1980; Cooper *et al.*, 1983); Figure 3 shows the known metabolic pathways for benzo[*a*]pyrene. It can be seen that all regions of the molecule undergo metabolism, a feature common to all hydrocarbons. In the case of compounds possessing methyl groups, metabolism to hydroxymethyl derivatives also occurs.

**Fig. 3. Metabolism of benzo[*a*]pyrene**

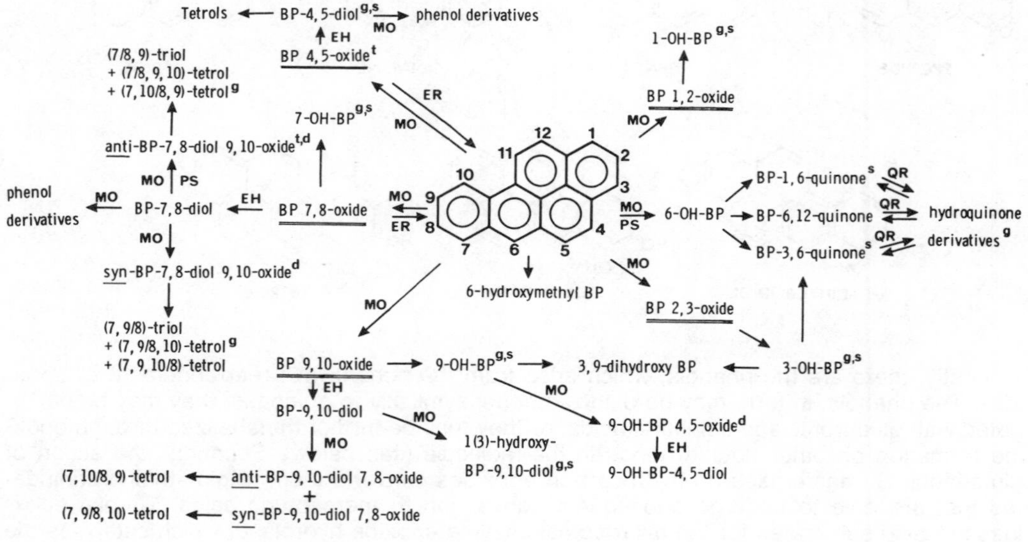


MO, monooxygenase; EH, epoxide hydrolase; ER, epoxide reductase; QR, quinone reductase; PS, prostaglandin synthetase. *t*, can be converted into a glutathione conjugate; *s*, can be converted into a sulphuric acid conjugate; *g*, can be converted into a glucuronic acid conjugate; *d*, contributes to the covalent binding of hydrocarbon to nucleic acids in cells and tissues that have been treated with benzo[*a*]pyrene. Reproduced from Cooper *et al.* (1983), with permission

## PATHWAYS OF ACTIVATION

Until about 10 years ago, available information on the biological properties of hydrocarbon metabolites indicated that dihydrodiols and conjugates with glutathione, glucuronic acid and sulfate were purely detoxication products, and that phenols, although cytotoxic, had little carcinogenic or mutagenic activity. This is still valid for the conjugates, except that certain bacteria in the gastrointestinal tract can hydrolyse benzo[*a*]pyrene conjugates and may, thus, reactivate the compound (Renwick & Drasar, 1976). At that time, simple epoxides were considered to be likely candidates for the biological activity of the parent compounds, and to be the intermediates by which covalent binding to nucleic acids was mediated *in vivo* and in metabolizing systems *in vitro*. However, Baird (1979) demonstrated convincingly that the chromatographic properties of the hydrocarbon-nucleoside adducts, present in hydrolysates of DNA isolated from tissues or cells that had been treated with a hydrocarbon, were not the same as those formed when DNA was treated *in vitro* with an epoxide. Only then was attention turned to the further metabolism of dihydrodiols. Subsequently, it was shown that further metabolism of a dihydrodiol of benzo[*a*]pyrene, the 7,8-dihydrodiol, in a microsomal system, yielded a species that reacted extensively with DNA, a reaction that was shown to be due to metabolism of the dihydrodiol to a vicinal diol-epoxide, benzo[*a*]pyrene 7,8-diol 9,10-epoxide (reviewed by Gelboin, 1980; Conney, 1982; Cooper *et al.*, 1983). The nucleoside adducts derived from the reaction of this metabolite with DNA were then shown to have chromatographic properties identical to those present in DNA hydrolysates isolated from hamster embryo cells and mouse skin that had been treated with benzo[*a*]pyrene.

In order to determine the pathway of metabolic activation of a hydrocarbon, several approaches are currently adopted in parallel. Firstly, studies must be undertaken to identify the metabolites formed in the mammalian tissue that may be a target for tumour initiation by the compound. Secondly, the biological activities of known metabolites, in a variety of in-vivo and in-vitro systems, are determined; these include mammalian and bacterial mutagenicity assays, cell transformation assays and in-vivo studies on the activity of the metabolites as tumour initiators or complete carcinogens. The third approach involves the elucidation of the structures of hydrocarbon-nucleoside adducts, isolated from the DNA or RNA of cells or tissues that have been treated with the hydrocarbon.

When these approaches were applied to the activation of benzo[*a*]pyrene, a wealth of evidence was obtained that suggested that benzo[*a*]pyrene is metabolically activated predominantly *via* the 7,8-dihydrodiol to the *anti*-7,8-diol 9,10-epoxide (Figure 4) (Gelboin & Ts'o, 1978; Conney, 1982; Cooper *et al.*, 1983; Phillips, 1983). The diol-epoxide has the epoxy function adjacent to the bay-region of the molecule, and this led to the proposal that other polycyclic hydrocarbons are also activated *via* pathways that result in the formation of bay-region epoxides (Jerina & Daly, 1977). The idea was based on theoretical calculations that carbonium ions formed at the bay-region of hydrocarbon molecules will be more reactive than those formed at other parts of the molecules.

**Fig. 4. Major pathway of metabolic activation of benzo[*a*]pyrene (BP) showing the absolute stereochemistry of each metabolite**

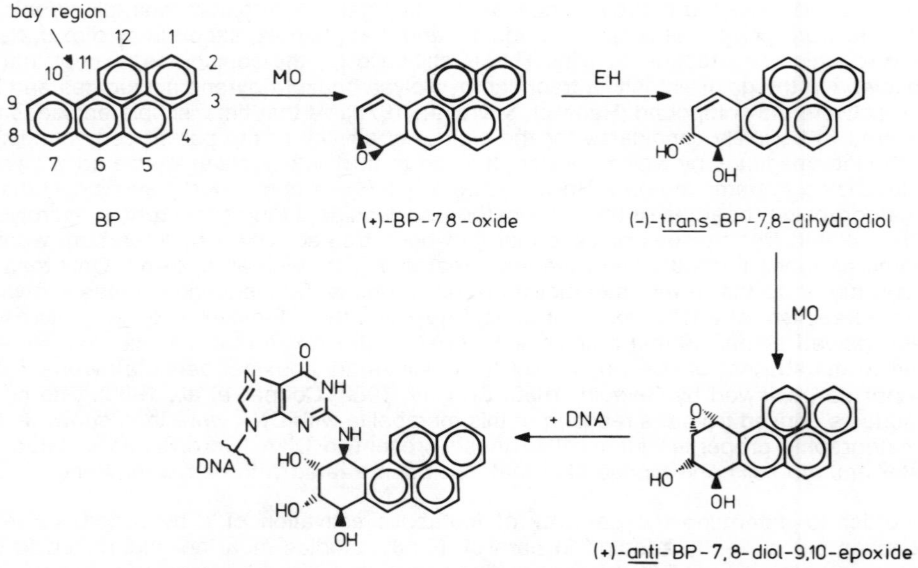


The structure of the major BP-DNA guanine adduct in DNA is also shown. MO, monooxygenase; EH, epoxide hydrolase; ER, epoxide reductase; QR, quinone reductase; PS, prostaglandin synthetase. *t*, can be converted into a glutathione conjugate; *s*, can be converted into a sulphuric acid conjugate; *g*, can be converted into a glucuronic acid conjugate; *d*, contributes to the covalent binding of hydrocarbon to nucleic acids in cells and tissues that have been treated with benzo[*a*]pyrene. Reproduced from Phillips (1983), with permission.

The following sections review the evidence for the formation of hydrocarbon-DNA adducts *via* bay-region diol-epoxides and *via* other types of reactive intermediate. Attention is drawn, also, to those instances where intermediates in the pathways to reactive metabolites have been shown to possess biological activity in test systems.

## BAY-REGION DIOL-EPOXIDES

As already mentioned, benzo[*a*]pyrene is activated by metabolism, principally *via* the bay-region 7,8-diol 9,10-epoxides. The pathway is stereoselective because benzo[*a*]pyrene is metabolised preferentially to the (-)-*trans*-7,8-dihydrodiol, *via* the (+)-7,8-oxide, each of which is more active than its enantiomer (Phillips, 1983) (Fig. 4). The DNA adducts that are formed in tissues or cells treated with the hydrocarbon are derived predominantly from the (+)-*anti*-diol-epoxide, and, in some cases, to a lesser extent from the (+)-*syn*-diol-epoxide. The (+)-*anti*-isomer and the (-)-*syn*-isomer are, biologically, more active than their respective enantiomers.

Benz[*a*]anthracene (III, Fig. 1) has only weak activity as a tumour initiator, but its 3,4-dihydrodiol, which could give a bay-region diol-epoxide by further metabolism, has been found to be 10 times more tumorigenic than benz[*a*]anthracene, on mouse skin and in newborn mice (Conney, 1982). Furthermore, only the 3,4-dihydrodiol, of the five possible dihydrodiols, showed greater mutagenicity in the microsome-mediated *S. typhimurium* assay than the parent compound. This strongly suggests that the ultimately carcinogenic form of benz[*a*]anthracene is one or more 3,4-diol 1,2-epoxides. Both the *syn*-isomer and the *anti*-isomer are highly mutagenic, although the former is the more active in bacterial assays and the latter more active in the Chinese hamster V-79 cell assay. In addition, both enantiomers have greater tumour-initiating activity than benz[*a*]anthracene on mouse skin and in newborn mice, the *anti*-isomer being the more active. The involvement of the bay-region in the activation of this hydrocarbon is further supported by the identification of nucleoside adducts in hydrolysates of DNA, isolated from either hamster embryo cells or mouse skin treated with benz[*a*]anthracene, that were identical chromatographically to those derived from DNA that had been reacted with the *anti*-3,4-diol 1,2-epoxide (Sims & Grover, 1981). However, it has been shown that non-bay-region diol-epoxides, also, are involved in the binding of benz[*a*]anthracene to DNA in mouse skin and hamster embryo cells and this is discussed in the next section.

Methylation of benz[*a*]anthracene, at certain positions, results in compounds with considerably greater carcinogenic activity. The most active monomethyl derivative is 7-methylbenz[*a*]anthracene, while 7,12-dimethylbenz[*a*]anthracene is the most active disubstituted derivative and one of the most potently carcinogenic polycyclic hydrocarbons known (Dipple, 1976). All five possible dihydrodiols of 7-methylbenz[*a*]anthracene are formed when the hydrocarbon is incubated with mouse skin in organ culture, or metabolized by rat-liver microsomal fractions (Sims & Grover, 1981). When they were investigated for their biological activity, only the 3,4-dihydrodiol, which would be expected to yield a bay-region diol-epoxide by further metabolism, was more active than the parent compound. The suggestion that this dihydrodiol is an intermediate in the activation of 7-methylbenz[*a*]anthracene is supported, also, by demonstration that the major nucleoside adduct that is obtained from DNA isolated from hydrocarbon-treated mouse skin is similar to that obtained from DNA reacted with a bay-region 3,4-diol 1,2-epoxide, and different from products obtained with either the 1,2-diol 3,4-epoxide or the 8,9-diol 10,11-epoxide.

Similar studies with 7,12-dimethylbenz[*a*]anthracene have shown that 3,4-dihydrodiol, the potential precursor of bay-region diol-epoxides, is the only dihydrodiol to possess biological activity significantly greater than the parent hydrocarbon (Sims & Grover, 1981). Analysis of the nucleoside adducts that are present in hydrolysates of DNA isolated from hydrocarbon-treated mouse skin or mouse embryo cells, indicates that 7,12-dimethylbenz[*a*]anthracene is activated to both the *syn*- and the *anti*-isomers of the bay-region 3,4-diol 1,2-epoxides (Dipple *et al.,* 1983a). 7-Hydroxymethyl-12-methylbenz[*a*]anthracene is a major metabolite of this hydrocarbon and has moderately high biological activity; some of the minor adducts formed in mouse skin are reported to be derived from this metabolite, presumably *via* one or both of its 3,4-diol 1,2-epoxides (DiGiovanni *et al.*, 1983).

3-Methylcholanthrene (V, Fig. 1), which may be considered to be a trisubstituted derivative of benz[*a*]anthracene, is a potent carcinogen that also appears to be activated *via* a bay-region diol-epoxide. Thus the 9,10-dihydrodiol is considerably more active than the hydrocarbon itself, or than the 4,5-, 7,8- and 10,11-dihydrodiols and the 2a,3-diol, in a number of biological assays (Sims & Grover, 1981). In addition, metabolites in which the methylene bridge is oxygenated or hydroxylated have significant biological activity (Conney, 1982); the evidence available suggests that some of the DNA adducts that are formed in mouse skin are derived from the covalent binding of one or both 9,10-diol 7,8-epoxides, while others may be formed from 1- or 2-hydroxy derivatives of 9,10-diol 7,8-epoxides.

Benzo[*e*]pyrene (II, Fig. 1) is a five-ring symmetrical hydrocarbon that possesses little or no carcinogenic activity in the mouse. Theoretical calculations predict, however, that the bay-region diol-epoxides, the 9,10-diol 10,11-epoxides, should be reactive. Indeed, synthetic samples of these compounds are weakly active tumour initiators and the *anti*-isomer is reactive towards DNA (Sims & Grover, 1981; Conney 1982). However, metabolism studies in hamster embryo cells and with rat-liver microsomal fractions show that only trace amounts are formed of the 9,10-dihydrodiol, the precursor of the bay-region diol-epoxides, from the parent hydrocarbon, and no conversion of the dihydrodiol to the bay-region diol-epoxides has been detected. Thus, the failure of benzo[*e*]pyrene to become metabolised to any significant extent *via* the expected activating pathway can be considered the reason for its lack of carcinogenic activity.

All three possible dihydrodiols of dibenz[*a,h*]anthracene (VII, Fig. 1), a symmetrical hydrocarbon that has two identical bay-regions and is moderately carcinogenic, have been detected as metabolites in mouse skin in organ culture and in rat-liver microsomal incubations (Sims & Grover, 1981; Conney, 1982). Only the dihydrodiol that can give rise to bay-region diol-epoxides on further metabolism, i.e., the 3,4-dihydrodiol, has significant tumorigenic activity on mouse skin and is strongly activated to mutagenic products in the Ames test. Although the *syn*- and *anti*-3,4-diol 1,2-epoxides showed no significant tumour-initiating activity when applied to mouse skin, the overall results suggest, nevertheless, that dibenz[*a,h*]anthracene is metabolically activated *via* one or more bay-region diol-epoxides. As yet, no direct study has been reported on the nature of the hydrocarbon-nucleoside adducts obtained from DNA hydrolysates from mouse skin treated with the hydrocarbon.

The weak tumour initiator, chrysene (IV, Fig. 1), has two identical bay-regions. Substitution with a methyl group, at a position adjacent to one of them, produces a substantially more potent compound, 5-methylchrysene (Dipple, 1976). The bay-region diol-epoxides of chrysene, and their precursor, the 1,2-dihydrodiol, have been shown to possess higher mutagenic activity than chrysene, and the dihydrodiol and *anti*-diol-epoxide are also more tumorigenic than the hydrocarbon itself (Conney, 1982). Furthermore, it has been determined that only the (-)-enantiomer of the 1,2-dihydrodiol has tumour-initiating activity and that, of the four possible isomers of chrysene 1,2-diol 3,4-epoxide, only the (+)-*anti*-enantiomer is active (Chang *et al.*, 1983). In mutagenicity studies, both optical isomers of *anti*-1,2-diol 3,4-epoxide were highly active (Wood *et al.*, 1982). The major nucleoside adducts that are present in hydrolysates of DNA isolated from hamster embryo cells treated with chrysene have been shown to be identical to products of the reaction of DNA with *anti*-chrysene-1,2-diol 3,4-epoxide (Hodgson *et al.*, 1982). When the adducts formed in mouse skin were analysed, an adduct derived from the *anti*-bay-region diol-epoxide was again found, but a second, more polar, adduct was also present that appears to be derived from a triol-epoxide (Hodgson *et al.*, 1983a) (see below).

Studies on 5-methylchrysene have demonstrated the high biological activity of the 1,2-dihydrodiol and the involvement of the bay-region 1,2-diol 3,4-epoxide in covalent binding of the hydrocarbon to DNA in mouse skin (Melikian *et al.*, 1982).

The compound 15,16-dihydro-11-methylcyclopent[*a*]phenanthren-17-one (IX, Fig. 1) is a potent carcinogen which, like 7,12-dimethylbenz[*a*]anthracene and 5-methylchrysene, has a methyl group adjacent to the bay-region. Coombs *et al.* (1979) and Coombs and Bhatt (1982) have demonstrated that the 3,4-dihydrodiol has high mutagenic and tumour-initiating activity, and that the *trans*-3,4-diol 1,2-epoxide, a bay-region epoxide, is the intermediate in the binding of the parent compound to DNA in mouse skin.

The weakly carcinogenic benzo[*c*]phenanthrene is predicted, from theoretical calculations, to be activated to a bay-region 3,4-diol 1,2-epoxide *via* the 3,4-dihydrodiol. Supporting evidence has been obtained, showing that the 3,4-dihydrodiol is more tumorigenic on mouse skin than the parent compound and than either the 1,2- or 5,6-dihydrodiols, and is metabolized to products that are more mutagenic towards bacterial and mammalian cells; in addition, the *syn*- and *anti*-3,4-diol 1,2-epoxides are highly mutagenic in the same systems (Conney, 1982).

As well as these results, studies on the hexacyclic hydrocarbons, dibenzo[*a*,*h*]pyrene and dibenzo[*a*,*i*]pyrene, have demonstrated mutagenic and carcinogenic activity for their bay-region diol-epoxides and precursor dihydrodiols (Conney, 1982). None of the dihydrodiols of phenanthrene, the simplest hydrocarbon with a bay-region and a compound with only very weak activity as a tumour initiator, was metabolized to products mutagenic to *S. typhimurium*; the *syn*- and *anti*-diol-epoxides were moderately mutagenic to *S. typhimurium* but inactive in tumorigenicity studies (Conney, 1982). For all three, there are, as yet, no reports on synthesis nor on tests for biological activity of diol-epoxides formed at sites other than bay-regions.

## NON-BAY-REGION DIOL-EPOXIDES

While the 3,4-dihydrodiol and 3,4-diol 1,2-epoxides of benz[*a*]anthracene are highly active as mutagens and as tumour initiators, and the *anti*-isomer of the latter is involved in the binding of benz[*a*]anthracene to DNA in mouse skin and hamster embryo cells (see above), other adducts, formed in roughly equal amounts, are derived from reaction of the *anti*-8,9-diol 10,11-epoxide with DNA (Sims & Grover, 1981). Thus, both a bay-region and a non-bay-region diol-epoxide contribute to the DNA binding of benz[*a*]anthracene *in vivo*, but only in the former case do the intermediates in the activation pathway show substantially greater biological activity than the parent hydrocarbon. Nevertheless, after metabolic activation, the 8,9-dihydrodiol was more active than benz[*a*]anthracene in inducing mutations in *S. typhimurium* TA100, although less active in tumorigenicity studies on mouse skin and in newborn mice (Sims & Grover, 1981; Conney, 1982). Similarly, the *anti*-8,9-diol 10,11-epoxide was mutagenic towards *S. typhimurium* but 35-fold less active than the bay-region *anti*-3,4-diol 1,2-epoxide (Conney, 1982).

Dibenz[*a*,*c*]anthracene (VI, Fig. 1) is a very weak carcinogen in mouse skin, and possesses three bay-regions. Only small amounts of the potential precursors of bay-region diol-epoxides, the 1,2- and 3,4-dihydrodiols, were detected when the hydrocarbon was incubated with hepatic microsomal fractions or with mouse skin in organ culture, the major product being the 10,11-dihydrodiol in both cases (Sims & Grover, 1981). In mutagenicity tests *in vitro*, none of the dihydrodiols exhibited greater activity than dibenz[*a*,*c*]anthracene, although both the 1,2- and 10,11-dihydrodiols were more active than the hydrocarbon as tumour initiators on mouse skin (Chouroulinkov *et al.*, 1983). The 10,11-dihydrodiol is further metabolized by rat-liver microsomal fractions to a non-bay-region diol-epoxide, *anti*-10,11-diol 12,13-epoxide. This compound reacts with DNA *in vitro* to give products which, on hydrolysis, were indistinguishable, chromatographically from some of those obtained from hydrolysates of DNA from hamster embryo cells that were treated with dibenz[*a*,*c*]anthracene (Hewer *et al.*, 1981); other, unidentified adducts were also present, however. When DNA isolated from mouse skin that had been treated with dibenz[*a*,*c*]anthracene was similarly hydrolysed and analysed, no products with the chromatographic properties of hydrocarbon-nucleoside adducts were observed. Thus, although a non-bay-region diol-epoxide is known to be formed by metabolism, and reacts with DNA in cultured cells and in microsomal metabolizing systems, there is, so far, no evidence that this is the pathway of activation in mouse skin.

## PHENOL-EPOXIDES AND TRIOL-EPOXIDES

In addition to the well-documented bay-region diol-epoxide pathway of benzo[*a*]pyrene activation, additional DNA adducts have been observed in some biological systems, including mouse skin, cultured hepatocytes and microsomal incubations performed in the presence of DNA, that are thought to be formed from the 9-hydroxy-4,5-oxide of benzo[*a*]pyrene (Cooper *et al.*, 1983). 9-Hydroxybenzo[*a*]pyrene, however, does not exhibit marked biological activity in carcinogenicity tests and its activity in mutagenicity assays is considerably less than that of the 7,8-dihydrodiol (Gelboin, 1980; Conney, 1982; Cooper *et al.*, 1983). Thus the significance of the formation of adducts from the 9-hydroxy-4,5-oxide and their possible contribution to the biological activity of benzo[*a*]pyrene remain to be established.

As discussed above, chrysene is metabolically activated in hamster embryo cells *via* the bay-region *anti*-1,2-diol 3,4-epoxide (Hodgson *et al.*, 1982). However, in mouse skin a more polar adduct was observed, in addition to those identical to adducts formed in hamster embryo cells (Hodgson *et al.*, 1983a). When either chrysene 1,2-dihydrodiol or 3-hydroxychrysene was incubated with DNA, in a rat-liver microsomal metabolizing system, an adduct was obtained from hydrolysates, in each case, that was identical to the more polar mouse skin DNA adduct. These results suggest that a triol-epoxide, 9-hydroxychrysene-1,2-diol 3,4-oxide, is the reactive intermediate involved. The biological activities of chrysene phenols have not yet been reported.

Hulbert and Grover (1983) propose that phenolic OH-groups present in epoxides will activate the epoxide moieties and permit their rearrangement to quinone-methides. Only epoxides that also possess phenolic OH-groups in certain positions will form these resonance-stabilized carbonium ions; these include the 9-position of a benzo[*a*]pyrene 4,5-oxide and the 9-position of the chrysene bay-region 1,2-diol 3,4-epoxide.

Metabolic studies of the carcinogen, dibenzo[*a,e*]fluoranthene (VIII, Fig. 1), have revealed a profile with products of greater complexity than those obtained with other hydrocarbons (Saguem *et al.*, 1983a,b). A number of the major metabolites involve the formation of a dihydrodiol in one region of the molecule and a phenol at a distant site. The metabolites involved in the DNA binding of this hydrocarbon have not been identified, and it is possible that the further metabolism of one of these triols could yield a reactive triol-epoxide.

Although it has not been detected as a metabolite of benzo[*a*]pyrene, 2-hydroxybenzo[*a*]pyrene has significant biological activity (Gelboin, 1980; Cooper *et al.*, 1983). The fluorescence spectrum of the deoxyribonucleoside adduct that was formed when 2-hydroxybenzo[*a*]pyrene was metabolized by rat-liver microsomal fractions in the presence of DNA indicates activation in the 7,8,9,10-ring (Dock *et al.*, 1978). If a diol-epoxide formed in this ring was able to form a quinone-methide through the 2-phenol, theory predicts that it would be the non-bay-region 9,10-diol 7,8-epoxide and not a bay-region diol-epoxide.

## SITES OF REACTION WITH DNA

The principal sites of DNA modification by activated metabolites of polycyclic hydrocarbons are the guanine residues (Phillips & Sims, 1979); adenine residues are modified to a lesser extent, with very minor involvement of cytosine moieties in some instances.

When the bay-region 7,8-diol 9,10-epoxides of benzo[*a*]pyrene react with DNA *in vitro*, the major adduct is formed through covalent reaction between the 2-amino group of guanine and the 10-position of benzo[*a*]pyrene (Phillips & Sims, 1979; Cooper *et al.*, 1983); the diol-epoxide principally involved in this interaction *in vivo* is the (+)-*anti*-isomer (Fig. 4). Reaction occurs, also, with the exocyclic amino group of adenine residues in DNA, and this accounts for about 3% of DNA modification *in vivo*. Studies of the in-vitro reaction of benzo[*a*]pyrene diol-epoxides with DNA have demonstrated other minor products, including reaction at the $O^6$ and *N*-7 positions of guanine (Cooper *et al.*, 1983).

The major products of the reaction of the ($\pm$)*anti*-8,9-diol 10,11-epoxide of benz[*a*]anthracene with poly G were, similarly, found to involve linkage of the 2-amino group of guanine residues to the 11-position of the hydrocarbon moiety (Sims & Grover, 1981). The same position in guanine was modified, also, when the ($\pm$)*anti*-1,2-diol 3,4-epoxide of chrysene was reacted with poly G (Hodgson *et al.*, 1983b).

In contrast to in-vivo studies with benzo[*a*]pyrene, the DNA binding of 7,12-dimethylbenz[*a*]anthracene, mediated *via* the bay-region 3,4-diol 1,2-epoxides (see above), involves a much larger contribution both of the *syn*-isomer and of adenine residues (Dipple *et al.*, 1983a). In fact, in mouse skin, an adduct formed from the reaction of the *syn*-isomer with adenine can account for up to 40% of the total DNA binding (Dipple *et al.*, 1983b). Since 7,12-dimethylbenz[*a*]anthracene is a more potent tumour initiator on mouse skin, compared to benzo[*a*]pyrene, than their relative DNA-binding levels would predict, it has been argued that modifications of adenine residues in DNA may be, potentially, more carcinogenic than guanine modifications.

## CONCLUSIONS AND PERSPECTIVES

The overwhelming weight of evidence now points to epoxides as the activated intermediates in the carcinogenicity, mutagenicity and DNA-binding of polycyclic hydrocarbons. Although other mechanisms have been proposed, including the formation of benzylic esters (Flesher & Sydnor, 1973; Watabe *et al.*, 1982), so far, DNA adducts derived from such a pathway have not been isolated from any treated cells or animal tissue. Extension of the diol-epoxide theory to other hydrocarbons has been considerably successful and the predictions of the bay-region theory, concerning the sites of epoxide formation resulting in greatest reactivity, have been generally supported by studies of the biological activity of their intermediates. However, there are cases, most notably among the weakly carcinogenic hydrocarbons, of non-bay-region diol-epoxides contributing to the overall DNA binding in target tissues, as well as triol-epoxides and, in the case of the potent carcinogen benzo[*a*]pyrene, a phenol-epoxide. As the metabolic precursors of these activated species, where investigated, have not shown appreciable biological activity, it remains an open question whether DNA adducts formed *via* bay-region diol-epoxides are potentially carcinogenic and mutagenic while those formed from other epoxides are not.

However, no satisfactory theory has yet emerged to predict the biological activity of a polycyclic hydrocarbon. The theory would have to encompass not only predictions of the reactivity of putative ultimately-reactive metabolites, but also the extent to which the parent compound is metabolized *via* pathways leading to those species. The ease with which mammalian cells are able to detoxify reactive metabolites, once formed, will also be of prime importance in determining the biological activity of a compound. Further, there is, as yet, no satisfactory explanation of why methyl substitution at certain positions of weak carcinogens, such as benz[*a*]anthracene and chrysene, results in considerable enhancement of their potency.

The major benzo[*a*]pyrene-DNA adducts in most systems, including a number of human tissues, have been shown to be derived from isomers of the bay-region 7,8-diol 9,10-epoxides that are covalently bound to the 2-amino group of guanine residues. However, anomalous activation appears to occur in a number of tissues in the rat. In mammary tissue, which is susceptible to benzo[*a*]pyrene carcinogenesis, a complex pattern of hydrocarbon-deoxyribonucleoside adducts was observed, none of which showed the chromatographic properties of adducts derived from the bay-region diol-epoxides (Grover *et al.*, 1983). Similar studies in lung and liver (Boroujerdi *et al.*, 1981) reveal little or no contribution by this pathway to the DNA-bound products observed. On the other hand, the *syn*- and *anti*- bay-region diol-epoxides account for most of the DNA binding in cultured rat colon incubated with benzo[*a*]pyrene (Autrup *et al.*, 1980), and the same effect is seen in rat skin, which, paradoxically, is refractory to benzo[*a*]pyrene carcinogenesis (Weston *et al.*, 1982). The existence of these interspecies differences emphasizes the importance of a thorough understanding of the mechanisms involved in the metabolism and metabolic activation of polycyclic hydrocarbons in animal tissues known to be susceptible or resistant to carcinogenesis by these agents. In attempting to extrapolate these data to human tissues that are suspected of susceptibility to hydrocarbon carcinogenesis, it must be clearly established that the animal models used are appropriate.

## REFERENCES

Autrup, H., Schwartz, R.D., Essigmann, J.M., Smith, L., Trump, B.F. & Harris, C.C. (1980) Metabolism of aflatoxin $B_1$, benzo[*a*]pyrene and 1,2-dimethylhydrazine by cultured rat and human colon. *Teratog. Carcinog. Mutag.*, *1*, 3-13

Baird, W.M. (1979) *The use of radioactive carcinogens to detect DNA modifications*. In: Grover, P.L., ed., *Chemical Carcinogens and DNA*, Vol. 1, Boca Raton, FL, CRC Press, pp. 59-83

Boroujerdi, M., Kung, H.-C., Wilson, A.G.E. & Anderson, M.W. (1981) Metabolism and DNA binding of benzo[*a*]pyrene *in vivo* in the rat. *Cancer Res.*, *41*, 951-957

Chang, R.L., Levin, W., Wood, A.W., Yagi, H., Tada, M., Vyas, K.P., Jerina, D.M. & Conney, A.H. (1983) Tumorigenicity of enantiomers of chrysene 1,2-dihydrodiol and of the diastereomeric bay-region chrysene 1,2-diol-3,4-epoxides on mouse skin and in newborn mice. *Cancer Res.*, *43*, 192-196

Chouroulinkov, I., Coulomb, H., MacNicoll, A.D., Grover, P.L. & Sims, P. (1983) Tumour-initiating activities of dihydrodiols of dibenz[*a,c*]anthracene. *Cancer Lett.*, *19*, 21-26

Conney, A.H. (1982) Induction of microsomal enzymes by foreign chemicals and carcinogenesis by polycyclic aromatic hydrocarbons: G.H.A. Clowes memorial lecture. *Cancer Res.*, *42*, 4875-4917

Coombs, M.M. & Bhatt, T.S. (1982) High skin tumour initiating activity of the metabolically derived *trans*-3,4-dihydro-3,4-diol of the carcinogen 15,16-dihydro-11-methylcyclopenta-[*a*]phenanthren-17-one. *Carcinogenesis, 3,* 449-451

Coombs, M.M., Kissonerghis, A.-M., Allen, J.A. & Vose, C.W. (1979) Identification of the proximate and ultimate forms of the carcinogen 15,16-dihydro-11-methylcyclopenta[*a*]phenanthren-17-one. *Cancer Res.*, *39*, 4160-4165

Cooper, C.S., Grover, P.L. & Sims, P. (1983) The metabolism and activation of benzo[*a*]pyrene. *Prog. Drug Metab.*, *7*, 295-396

DiGiovanni, J., Nebzydoski, A.P. & Decina, P.C. (1983) Formation of 7-hydroxymethyl-12-methylbenz[*a*]anthracene-DNA adducts from 7,12-dimethylbenz[*a*]anthracene in mouse epidermis. *Cancer Res.*, *43*, 4221-4226

Dipple, A. (1976) *Polynuclear aromatic carcinogens*. In: Searle, C.E., ed., *Chemical Carcinogens*, Washington, D.C., ACS Monograph 173, pp. 245-314

Dipple, A., Pigott, M., Moschel, R.C. & Constantino, N. (1983a) Evidence that binding of 7,12-dimethylbenz[*a*]anthracene to DNA in mouse embryo cell cultures results in extensive substitution of both adenine and guanine residues. *Cancer Res.*, *43*, 4132-4135

Dipple, A., Sawicki, J.T., Moschel, R.C. & Bigger, C.A.H. (1983b) *7,12-Dimethylbenz[a]anthracene - DNA interactions in mouse embryo cell cultures and mouse skin*. In: Rydström, J., Montelius, J. & Bengtsson, M., eds, *Extrahepatic Drug Metabolism and Chemical Carcinogenesis*, Amsterdam, Elsevier, pp. 439-448

Dock, L., Undeman, O., Gräslund, A. & Jernström, B. (1978) Fluorescence study of DNA-complexes formed after metabolic activation of benzo[*a*]pyrene derivatives. *Biochem. Biophys. Res. Commun.*, *85*, 1275-1282

Flesher, J.W. & Sydnor, K.L. (1973) Possible role of 6-hydroxymethylbenzo[*a*]pyrene as a proximate carcinogen of benzo[*a*]pyrene and 6-methylbenzo[*a*]pyrene. *Int. J. Cancer*, *11*, 433-437

Gelboin, H.V. (1980) Benzo[*a*]pyrene metabolism, activation and carcinogenesis: role and regulation of mixed-function oxidases and related enzymes. *Physiol. Rev.*, *60*, 1107-1166

Gelboin, H.V. & Ts'o, P.O.P., eds (1978) *Polycyclic Hydrocarbons and Cancer*, Vols 1 & 2, New York, Academic Press

Grimmer, G. (1979) *Sources and occurrence of polycyclic aromatic hydrocarbons*. In: Castegnaro, M., Bogovski, P., Kunte, H. & Walker, E.A., eds, *Environmental Carcinogens Selected Methods of Analysis*, Vol. 3, *Analysis of Polycyclic Aromatic Hydrocarbons in Environmental Samples* (*IARC Publications No. 29*), Lyon, International Agency for Research on Cancer, pp. 31-54

Grover, P.L., ed. (1979) *Chemical Carcinogens and DNA*, Vols. 1 & 2, Boca Raton, FL, CRC Press

Grover, P.L., Phillips, D.H., Cooper, C.S., Swallow, W.H., Weston, A., Vigny, P., O'Hare, M., Neville, A.M. & Sims, P. (1983) *Metabolism and activation of polycyclic hydrocarbons in mammary and other tissues*. In: Rydström, J., Montelius, J. & Bengtsson, M., eds, *Extrahepatic Drug Metabolism and Chemical Carcinogenesis*, Amsterdam, Elsevier, pp. 429-438

Hewer, A., Cooper, C.S., Ribeiro, O., Pal, K., Grover, P.L. & Sims, P. (1981) The metabolic activation of dibenz[*a,c*]anthracene. *Carcinogenesis*, *2*, 1345-1352

Hodgson, R.M., Pal, K., Grover, P.L. & Sims, P. (1982) The metabolic activation of chrysene by hamster embryo cells. *Carcinogenesis*, *3*, 1051-1056

Hodgson, R.M., Weston, A. & Grover, P.L. (1983a) Metabolic activation of chrysene in mouse skin: Evidence for the involvement of a triol-epoxide. *Carcinogenesis*, *4*, 1639-1643

Hodgson, R.M., Cary, P.D., Grover, P.L. & Sims, P. (1983b) Metabolic activation of chrysene by hamster embryo cells: evidence for the formation of a 'bay-region' diol-epoxide-$N^2$-guanine adduct in RNA. *Carcinogenesis*, *4*, 1153-1158

Hulbert, P.B. & Grover, P.L. (1983) Chemical rearrangement of phenol-epoxide metabolites of polycyclic aromatic hydrocarbons to quinone-methides. *Biochem. biophys. Res. Commun.*, *117*, 129-134

Jerina, D.M. & Daly, J.W. (1977) *Oxidation at Carbon*. In: Parke, D.V. & Smith, R.L., eds, *Drug Metabolism: From Microbe to Man*, London, Taylor and Francis, pp. 13-32

Melikian, A.A., LaVoie, E.J., Hecht, S.S. & Hoffmann, D. (1982) Influence of a bay-region methyl group on formation of 5-methylchrysene dihydrodiol epoxide: DNA adducts in mouse skin. *Cancer Res.*, *42*, 1239-1242

Phillips, D.H. (1983) Fifty years of benzo[*a*]pyrene. *Nature*, *303*, 468-472

Phillips, D.H. & Sims, P. (1979) *Polycyclic aromatic hydrocarbon metabolites: their reactions with nucleic acids*. In: Grover, P.L., ed., *Chemical Carcinogens and DNA*, Vol. 2, Boca Raton, FL, CRC Press, pp. 29-57

Renwick, A.G. & Drasar, B.S. (1976) Environmental carcinogens and large bowel cancer. *Nature*, *263*, 234-235

Saguem, S., Mispelter, J., Perin-Roussel, O., Lhoste, J.M. & Zajdela, F. (1983a) Multi-step metabolism of the carcinogen dibenzo[*a,e*]fluoranthene. I. Identification of the metabolites from rat microsomes. *Carcinogenesis*, *4*, 827-835

Saguem, S., Perin-Roussel, O., Mispelter, J., Lhoste, J.M. & Zajdela, F. (1983b) Multi-step metabolism of the carcinogen dibenzo[*a,e*]fluoranthene. II. Metabolic pathways. *Carcinogenesis*, *4*, 837-842

Sims, P. & Grover, P.L. (1974) Epoxides in polycyclic aromatic hydrocarbon metabolism and carcinogenesis. *Adv. Cancer Res.*, *20*, 165-274

Sims, P. & Grover, P.L. (1981) *Involvement of dihydrodiols and diol epoxides in the metabolic activation of polycyclic hydrocarbons other than benzo[a]pyrene*. In: Gelboin, H.V. & Ts'o, P.O.P., eds, *Polycyclic Hydrocarbons and Cancer*, Vol. 3, New York, Academic Press, pp. 117-181

Watabe, T., Ishizuka, T., Isobe, M. & Ozawa, N. (1982) A 7-hydroxymethyl sulphate ester as an active metabolite of 7,12-dimethylbenz[*a*]anthracene. *Science*, *215*, 403-405

Weston, A., Grover, P.L. & Sims, P. (1982) Metabolism and activation of benzo[*a*]pyrene by mouse and rat skin in short-term organ culture and *in vivo*. *Chem.-biol. Interact.*, *42*, 233-250

Wood, A.W., Chang, R.L., Levin, W., Yagi, H., Tada, M., Vyas, K.P., Jerina, D.M. & Conney, A.H. (1982) Mutagenicity of the optical isomers of the diastereomeric bay-region chrysene 1,2-diol-3,4-epoxides in bacterial and mammalian cells. *Cancer Res.*, *42*, 2972-2976

# METABOLISM OF GENOTOXIC AGENTS: HALOGENATED COMPOUNDS

**H.M. Bolt**

*Abteilung Toxikologie und Arbeitsmedizin, Institut für Arbeitsphysiologie an der Universität Dortmund, Dortmund, Federal Republic of Germany*

## SUMMARY

Most halogenated compounds showing genotoxicity do not themselves react with macromolecules but are transformed to reactive metabolites. In order to explain the widely different genotoxicities of halogenated ethylenes, the 'optimum stability' theory of epoxides has been developed recently. An epoxide must be reactive to alkylate DNA, but stable enough to reach the target from the place of its formation. Also, pharmacokinetic aspects are important in accounting for differences in genotoxicities of closely related compounds; examples are provided. Reactive metabolites may bind to proteins, to lipids and coenzymes, and to DNA, but binding to targets other than DNA is far more common than DNA alkylation. An example is provided by investigations on 2,2'-dichlorodiethyl ether from which chloroacetaldehyde is formed. In contrast to vinyl chloride, which generates chloroethylene oxide (chlorooxirane) as well as chloroacetaldehyde, this compound leads to extensive covalent protein binding of metabolites, but not to DNA binding.

## INTRODUCTION

Halogenated, especially chlorinated, compounds are of great industrial importance, both because of their technical properties as solvents, plastics monomers, biocides, etc., and because technical processes need those amounts of chlorine that are generated during NaCl electrolysis for NaOH production.

There are problems in the metabolism of some halogenated compounds to genotoxic metabolites. Based on a recent review of the topic (Laib, 1982), this paper discusses the basic principles by which simple organic molecules gain genotoxicity on halogen substitution.

## MECHANISMS OF GENOTOXICITY

Halogen substitution may lead to the effects outlined in Figure 1.

**Fig. 1. DNA binding by (a) alkyl halides, (b) ethylene dihalides and (c) halo-olefins**

(a) <u>Direct reactivity towards DNA</u>

$CH_3I > CH_3Br > CH_3Cl$

$Cl-CH_2-O-CH_2-Cl$: bifunctional alkylating agent (α-haloether)

(b) <u>1,2-Dihaloethanes and glutathione</u>

$$X-CH_2-CH_2-X \xrightarrow{GSH} GS-CH_2-CH_2-X \longrightarrow G-S^{\oplus}\begin{smallmatrix}CH_2\\|\\CH_2\end{smallmatrix} + X^{\ominus}$$

DNA alkylation

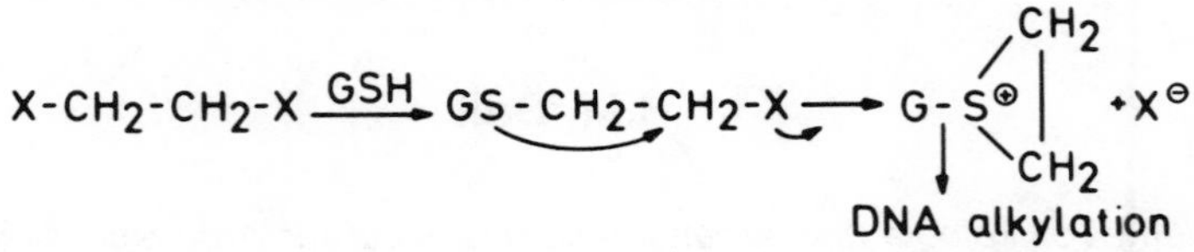

(c) <u>Oxidative metabolism of haloolefins</u>

$$\text{\textbackslash}C=C\text{/} \longrightarrow \text{\textbackslash}C\overset{O}{-}C\text{/} \xrightarrow[\text{detoxification}]{\text{DNA alkylation}}$$

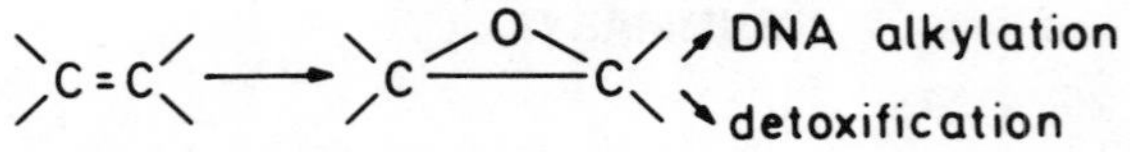

Halogen substitution affects
- epoxidation rate
- epoxide stability and reactivity

(1) The direct reactivity of haloalkanes generally decreases from iodo- to bromo- and chloro-substituted compounds. The reactivity of chlorinated alkanes is lowest; a carcinogenicity of methyl chloride is presently under discussion. Alpha-haloethers are especially reactive, the most prominent example being *bis*-chloromethyl ether which is a bifunctional alkylating agent (van Duuren *et al.*, 1969). When it 'alkylates' water, it decomposes rapidly to formaldehyde and hydrochloric acid.

(2) The mechanism of genotoxicity of 1,2-dihaloethanes (ethylene dibromide/dichloride) has been the subject of some outstanding recent contributions. On conjugation with glutathione, a reactive thiiranium ion is formed (van Bladeren *et al.*, 1981). As the role of conjugation in genotoxicity will be discussed extensively in the following paper (F. Oesch), I will not further exemplify this interesting topic.

(3) Haloolefins are generally biotransformed by microsomal monooxygenase(s) under formation of epoxide intermediates (Bolt *et al.*, 1982a). Structural theories have been put forward to explain differences in the genotoxicity of compounds in this series. In general, halogen substitution is expected to affect both the rate of epoxidation of a given olefin and the reactivity of the epoxide, based on the assumption that only the epoxide, and not other concurrent or successive metabolites, bind to DNA. That this is true, is experimentally demonstrated for the most important carcinogen of this series, vinyl chloride (see below).

**STRUCTURE-ACTIVITY RELATIONSHIPS FOR HALOETHENES**

There is agreement that halogenated ethylenes are uniformly metabolized *via* primary epoxidation (Bonse & Henschler, 1976; Bartsch *et al.*, 1979; Bolt *et al.*, 1982a); the primary epoxides (oxiranes) rearrange to halogenated aldehydes or acyl halides (Fig. 2). These, in part, are converted to halogenated acetic acids. At different metabolic levels, conjugation with glutathione may take place (not shown in Fig. 2), finally resulting in excretion of sulfur-containing metabolites; such metabolites have been isolated after exposure of humans and of experimental animals to vinyl chloride and vinylidene chloride (Müller *et al.*, 1976; Watanabe *et al.*, 1976; Jones & Hathway, 1978).

**Fig. 2. Initial metabolic steps for haloethylenes**

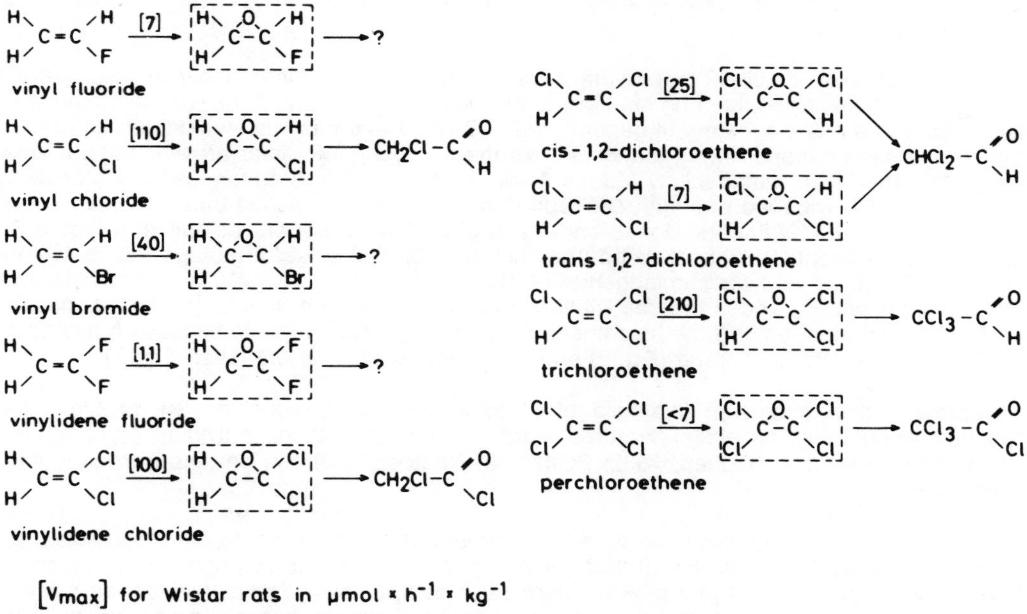


The maximal metabolic rates, $v_{max}$, are given in brackets

Using in-vitro experiments, Guengerich *et al.* (1981) suggested that only the epoxides of haloethylenes, not their rearrangement products, alkylate DNA. This has been confirmed *in vivo* for vinyl chloride n below). Pharmacokinetic studies (Filser & Bolt, 1979, 1981) have revealed that the rates of metabolism of individual haloethenes differ by more than two orders of magnitude (see numbers in brackets in Fig. 2). Therefore, the carcinogenicity of haloethylenes is determined by (1) the rate of epoxidation and (2) factors of epoxide stability and reactivity.

The first approach to a structure-activity relationship has been published by Henschler and associates (Bonse *et al.*, 1975; Greim *et al.*, 1975; Bonse & Henschler, 1976; Henschler

& Bonse, 1979). They suggested that reactivities, and hence toxicities, of individual chlorooxirane intermediates depend on the type of chlorine substitution, in that symmetric substitution renders the epoxide more stable and not mutagenic, while asymmetric substitution causes unstable and mutagenic epoxides. Trichloroethylene was exempted from this rule because its epoxide (2,2,3-trichlorooxirane) was transformed immediately to chloral at the cytochrome P-450 site. In general, the theory implied that minimum stability of halogenated epoxides, coinciding with maximum reactivity, would lead to the highest genotoxicity.

In 1982, this 'minimum stability' theory was modified (Bolt *et al.*, 1982b). After comparison of the oncogenic effects of haloethylenes in relation to their quantitative metabolism, it appeared that an 'optimum stability' of the epoxides should determine the genotoxic response. Among the haloethylenes, the epoxide of vinyl chloride (chlorooxirane), in quantitative terms, appeared to represent an optimum between the extremes of stability and reactivity in both reaching the DNA target and reacting with it, after being formed at the monooxygenase site. A further decrease in stability was thought to render the oxirane too short-lived to reach the target.

These 'optimum stability' arguments were corroborated by Jones and Mackrodt (1982, 1983). They defined epoxide ring stability as the two-centre energy of the weaker of the two C-O bonds in a given epoxide. In general, the C-O bond that has less substituted carbon in the halogenated oxirane ring is the weaker of the two bonds (Politzer *et al.*, 1981). It was suggested that both mutagenicity (Jones & Mackrodt, 1982) and carcinogenicity (Jones & Mackrodt, 1983) required optimum *C-O* bond energies. The combined data of Greim *et al.* (1975), Bolt *et al.* (1982a) and Jones and Mackrodt (1982, 1983) are compiled in Figure 3. Jones and Mackrodt (1983) postulated that 'threshold bands' exist for mutagenicity between -14.5 and -12.8 eV and for carcinogenicity between -14.1 and -12.9 eV, and that substituted ethylenes with epoxides that fall within these limits should be deemed potentially hazardous. Vinyl chloride, among the haloethenes, is especially carcinogenic because it has both rapid metabolism to its oxirane (Fig. 2) and 'optimum stability' of the latter (Fig. 3).

Further achievements in this field of structure-activity relationships will probably be reached when other molecular features, such as the strength of individual carbon-halogen bonds (Politzer & Hedges, 1982; Politzer & Proctor, 1983) can be taken properly into account, also.

With regard to oncogenicity, the data for vinylidene chloride did not fit the theoretical prediction; the reason for this was probably the high acute hepatotoxic potential of this compound that allows only long-term experiments with comparatively low doses (Jones & Mackrodt, 1983). Predictions for the epoxides derived from ethene, butadiene and acrylonitrile (Jones & Mackrodt, 1983) were in favour of mutagenicity and carcinogenicity for these compounds. Mutagenicity of the epoxide of acrylonitrile (glycidonitrile) has been shown experimentally (Cerná *et al.*, 1981; Peter *et al.*, 1983); butadiene (Bolt *et al.*, 1983) and ethene (Filser & Bolt, 1983) are biotransformed *in vivo* to their (mutagenic) epoxides. However, any risk estimate for ethene must take primary account of the fact that the metabolism to its (mutagenic and carcinogenic) epoxide is very slow; hence, carcinogenicity of ethene is likely to be found in a range where classical animal bioassays will fail (Bolt & Filser, 1984).

## DNA AND PROTEIN ALKYLATION

A prerequisite for the connection of molecular orbital features of epoxides with the carcinogenicity of parent ethenes is that only this epoxide, and no other metabolite of the ole-

**Fig. 3. The 'optimum stability' theory for halogenated oxiranes, according to the data of Greim et *al.* (1975), Bolt et *al.* (1982a), and Jones and Mackrodt (1982, 1983)**

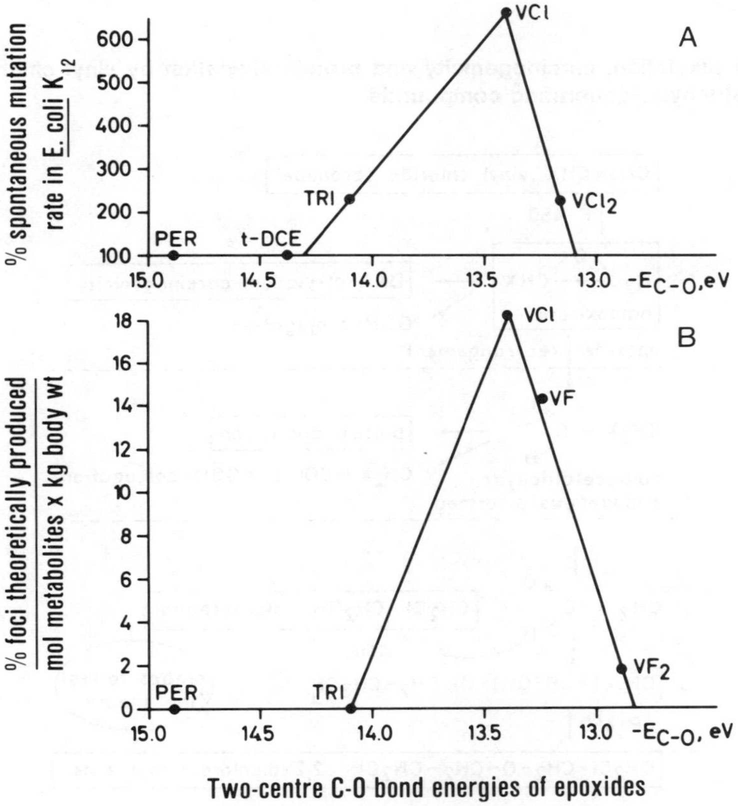


A, mutagenicity; PER, perchloroethene; t-DCE, *trans*-1,2-dichloroethene; TRI, trichloroethene; VCl, vinyl chloride; $VCl_2$, vinylidene chloride

B, carcinogenicity; VF, vinyl fluoride; $VF_2$, vinylidene fluoride

fin, binds irreversibly to DNA. That this is true, has been proven *in vitro* by Guengerich *et al.* (1981) for vinyl bromide and vinyl chloride. Now, proof has been obtained *in vivo* for vinyl chloride (Gwinner *et al.*, 1983).

The role of chlorooxirane and chloroacetaldehyde in the carcinogenicity of vinyl chloride was studied by comparing biological effects of vinyl chloride exposure with those of 2,2'-dichlorodiethyl ether (Fig. 4), as a metabolic precursor of chloroacetaldehyde. In both cases, the haloaldehyde is formed within the liver cell. Consistent with the theory of Guengerich *et al.* (1981), DNA alkylation (at N-7 of guanine), and formation of preneoplastic enzyme-altered hepatic foci, was detected only after vinyl chloride exposure, not after exposure to 2,2'-dichlorodiethyl ether. Also, chloroethanol did not induce enzyme-altered foci. Alkylation of proteins (in liver and other organs), however, was observed after vinyl chloride as well as after 2,2'-dichlorodiethyl ether exposure. This shows that metabolically derived chloroacetaldehyde is capable of alkylating protein structures, but not DNA; DNA alkylation after vinyl chloride exposure must be due to the epoxide.

**Fig. 4. DNA alkylation, carcinogenicity and protein alkylation by vinyl chloride and other chloroacetaldehyde-generating compounds**

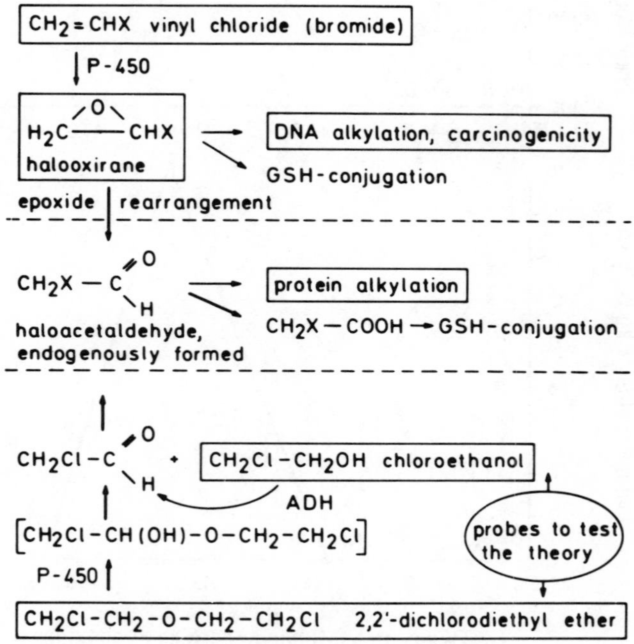


## SPECIFIC HEPATOTOXIC METABOLITES OF HALOGENATED COMPOUNDS ?

The importance of specific epoxide binding to DNA in carcinogenicity of halogenated compounds should not lead us to ignore other, possibly toxic, metabolites. Laib (1982) has shown that a common feature of the hepatotoxins chloroform, carbon tetrachloride, 1,1,2-trichloroethane and vinylidene chloride, which they share only with perchloroethylene, is generation of halogenated, acylating intermediates like phosgene and chloroacetyl chloride (Fig. 5). An investigation into the metabolic behaviour of vinylidene chloride in rats (Reichert *et al.*, 1979) has revealed urinary excretion of methylthioacetyl-aminoethanol; this suggests covalent interaction of a metabolite (probably chloroacetyl chloride) with ethanolamine moieties of lipids. Laib (1982) has discussed formation of 'activated' haloacyl-coenzyme A derivatives by direct chemical coenzyme A acylation. The possibility of a covalent binding to coenzyme A has been outlined by Bolt *et al.* (1982).

**Fig. 5. Formation of acylating metabolites from haloethanes, haloethenes and halomethanes**

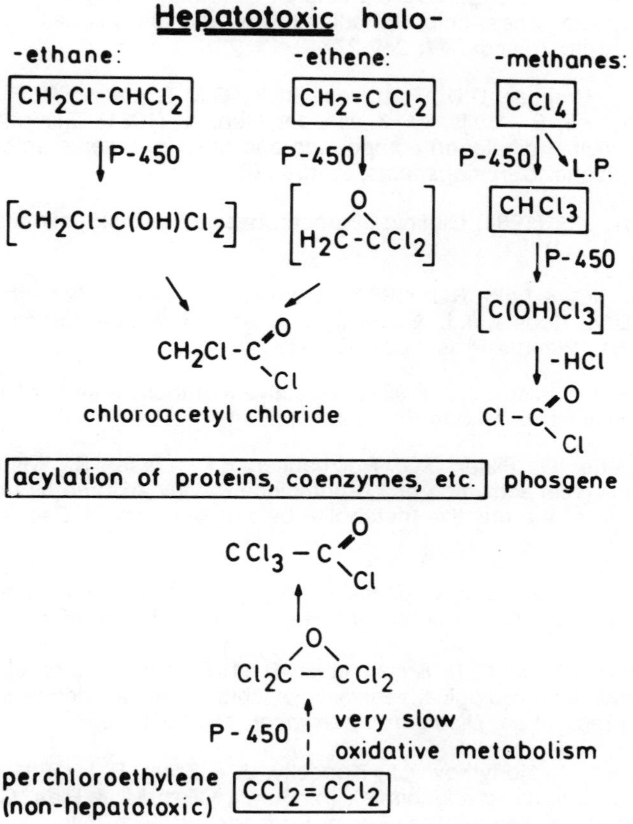


Figure 5 shows the pathways of formation of acylating metabolites from halogenated compounds. Although forming a haloacylating metabolic intermediate, perchloroethylene does not share the acute toxic effects of the other compounds (Fig. 5) due to its low metabolic rate (Fig. 2). Pharmacokinetic aspects, especially quantitative determinations of the amounts of xenobiotics metabolized *via* individual pathways, are therefore of prime importance for comparative studies of toxicity, including genotoxicity, of halogenated compounds.

## ACKNOWLEDGEMENT

The author thanks the 'Deutsche Forschungsgemeinschaft' for continuous support of the work of his group which is part of this review.

## REFERENCES

Bartsch, H., Malaveille, C., Barbin, A. & Planche, G. (1979) Mutagenic and alkylating metabolites of haloethylenes, chlorobutadienes and dichlorobutenes by rodent or human liver tissue. *Arch. Toxicol.*, *41*, 249-277

van Bladeren, P.J., Breimer, D.D., Rotteveel-Smijs, G.M.T., de Knijff, P., Mohn, G.R., van Meeteren-Wälchli, B., Buijs, W. & van der Gen, A. (1981) The relation between the structure of vicinal dihalogen compounds and their mutagenic activation *via* conjugation to glutathione. *Carcinogenesis*, *2*, 499-505

Bolt, H.M. & Filser, J.G. (1984) Olefinic hydrocarbons: a first risk estimate for ethene. *Toxicol. Pathol.*, *12*, (1)

Bolt, H.M., Filser, J.G. & Laib, R.J. (1982a) *Covalent binding of haloethylenes*. In: Snyder, R., Parke, D.V., Kocsis, J.J. & Jollow, D.J., eds, *Biological Reactive Intermediates II*, New York, NY, Plenum Press, pp. 667-683

Bolt, H.M., Laib, R.J. & Filser, J.G. (1982b) Reactive metabolites and carcinogenicity of halogenated ethylenes. *Biochem. Pharmacol.*, *31*, 1-4

Bolt, H.M., Schmiedel, G., Filser, J.G., Rolzhäuser, H.P., Lieser, K., Wistuba, D. & Schurig, V. (1983) Biological activation of 1,3-butadiene to vinyl oxirane by rat liver microsomes and expiration of the reactive metabolite by exposed rats. *J. Cancer Res. clin. Oncol.*, *106*, 112-116

Bonse, G. & Henschler, D. (1976) Chemical reactivity, biotransformation, and toxicity of polychlorinated aliphatic compounds. *Crit. Rev. Toxicol.*, *5*, 395-409

Bonse, G., Urban, T., Reichert, D. & Henschler, D. (1975) Chemical reactivity, metabolic oxirane formation and biological reactivity of chlorinated ethylenes in the isolated perfused liver preparation. *Biochem. Pharmacol.*, *24*, 1829-1834

Cerná, M., Kocisova, J., Kodytkova, J., Kopecký, J. & Sram, R.J. (1981) *Mutagenic activity of oxiranecarbonitrile (glycidonitrile)*. In: Gut,J., Cikrt, M. & Plaa, G.L., eds, *Industrial and Environmental Xenobiotics*, Berlin, Springer, pp. 251-254

van Duuren, B.L., Sivak, A., Goldschmidt, B.M., Katz, C. & Melchionne, S. (1969) Carcinogenicity of halo-ethers. *J. natl Cancer Inst.*, *43*, 481-486

Filser, J.G. & Bolt, H.M. (1979) Pharmacokinetics of halogenated ethylenes in rats. *Arch. Toxicol.*, *42*, 123-136

Filser, J.G. & Bolt, H.M. (1983) Exhalation of ethylene oxide by rats on exposure to ethylene. *Mutat. Res.*, *120*, 57-60

Greim, H., Bonse, G., Radwan, Z., Reichert, D. & Henschler, D. (1975) Mutagenicity *in vitro* and potential carcinogenicity of chlorinated ethylenes as a function of metabolite oxirane formation. *Biochem. Pharmacol.*, *24*, 2013-2017

Guengerich, F.P., Mason, P.S., Stott, W.T., Fox, T.R. & Watanabe, P.G. (1981) Roles of 2-haloethylene oxides and 2-haloacetaldehydes derived from vinyl bromide and vinyl chloride in irreversible binding to protein and DNA. *Cancer Res.*, *41*, 4391-4398

Gwinner, L.M., Laib, R.J., Filser, J.G. & Bolt, H.M. (1983) Identification of chloroethylene oxide as the reactive metabolite of vinyl chloride towards DNA: comparative studies with 2,2'-dichlorodiethyl ether. *Carcinogenesis*, *4*, 1483-1486

Henschler, D. & Bonse, G. (1979) *Metabolic activation of chlorinated ethylene derivatives.* In: Cohen, Y., ed., *Advances in Pharmacology and Therapeutics, Proc. 7th Int. Congr. Pharmacol., Paris, 1978.* Vol. 9, *Toxicology*, Oxford, Pergamon Press, pp. 123-130

Jones, B.K. & Hathway, D.E. (1978) The biological fate of vinylidene chloride in rats. *Chem.-biol. Interact.*, *20*, 27-41

Jones, R.B. & Mackrodt, W.C. (1982) Structure-mutagenicity relationship for chlorinated ethylenes: a model based on the stability of the metabolically derived epoxides. *Biochem. Pharmacol.*, *31*, 3710-3713

Jones, R.B. & Mackrodt, W.C. (1983) Structure-genotoxicity relationship for aliphatic epoxides. *Biochem. Pharmacol.*, *32*, 2359-2362

Laib, R.J. (1982) Specific covalent binding and toxicity of aliphatic halogenated xenobiotics. *Rev. Drug Metab. Drug Interact.*, *4*, 1-48

Müller, G., Norpoth, K. & Eckard, R. (1976) Identification of two urine metabolites of vinyl chloride by GC-MS investigations. *Int. Arch. occup. environ. Health*, *38*, 69-75

Peter, H., Schwarz, M., Mathiasch, B., Appel, K.E. & Bolt, H.M. (1983) A note on synthesis and reactivity towards DNA of glycidonitrile, the epoxide of acrylonitrile. *Carcinogenesis*, *4*, 235-237

Politzer, P. & Hedges, W.L. (1982) A study of the reactive properties of the chlorinated ethylenes. *Int. J. quantum Chem.*, *9*, 307-319

Politzer, P. & Proctor, T.R. (1983) Calculated properties of some possible vinyl chloride metabolites. *Int. J. quantum Chem.*, *10*

Politzer, P., Trefonas, P., Politzer, I.R. & Elfman, B. (1981) Molecular properties of the chlorinated ethylenes and their epoxide metabolites. *Ann. NY Acad. Sci.*, *367*, 478-492

Reichert, D., Werner, H.W., Metzler, M. & Henschler, D. (1979) Molecular mechanism of 1,1-dichloroethylene toxicity: excreted metabolites reveal different pathways of reactive intermediates. *Arch. Toxicol.*, *42*, 159-169

Watanabe, P.G., McGowan, G.R., Madrid, E.O. & Gehring, P.J. (1976) Fate of [$^{14}$C]vinyl chloride following inhalation exposure in rats. *Toxicol. appl. Pharmacol.*, *37*, 49-59

# METABOLISM OF GENOTOXIC AGENTS: CONTROL OF REACTIVE EPOXIDES BY HYDROLASE AND TRANSFERASE REACTIONS

**F. Oesch**

*Institute of Toxicology University of Mainz, Mainz, Federal Republic of Germany*

## SUMMARY

Hydrolase and transferase reactions play dual roles in the control of carcinogenic and mutagenic species. In some instances, they play an activating or coactivating role. However, as far is known, in most cases they are wholly or predominantly inactivating mechanisms. The important hydrolase and transferase enzymes that are involved in the control of reactive epoxides are particularly well studied. These include epoxide hydrolases and glutathione transferases that react directly with electrophilic epoxides, as well as conjugating enzymes, such as glucuronosyl transferases and sulfotransferases, that sequester nucleophilic precursors of complex epoxides such as phenols and dihydrodiols. These, and other enzymes that are involved in biosynthesis and the further metabolism of reactive metabolites, are an important contributing factor to differences in susceptibility, since they differ in quantity and, sometimes, also in substrate specificity between organs, developmental stages, sexes and animal species. Knowledge of these variables is, therefore, required for a rational extrapolation to humans of the toxicity data obtained in available test systems; the rational interpretation of data obtained by biomonitoring requires similar knowledge.

## INTRODUCTION

A wide array of functionally diverse enzymes is involved in the biotransformation of carcinogenic and mutagenic compounds. The important group of enzymes that is responsible for the control of reactive epoxides has been particularly well studied (Ullrich *et al.*, 1975; Wood *et al.*, 1976; Oesch, 1979a,b). Many natural as well as man-made foreign compounds possess olefinic or aromatic double bonds, and can be transformed to epoxides by microsomal monooxygenases. By virtue of their electrophilic reactivity, these epoxides may react spontaneously with nucleophilic centres in the cell and thus covalently bind to deoxyribonucleic acid, ribonucleic acid and protein. Such alterations of critical cellular macromolecules may disturb the normal biochemistry of the cell and lead to cytotoxic, allergic, mutagenic and/or carcinogenic effects. Whether the effects will be manifested depends on the chemical reactivity as well as on other properties (geometry, lipophilicity) of the epoxide in question on the one hand, and enzymes controlling the concentration of such epoxides on the other. There are several microsomal monooxygenases that differ in activity and substrate specificity. With large substrates, some monooxygenases attack preferentially at one specific site, different from that attacked by others. Some of these pathways lead to reactive products; others are detoxification pathways. Moreover, the hydrolase and transferase enzymes that metabolize the epoxides provide a further determining factor. Finally, enzymes which sequester precursors of complex epoxides play an important role (Glatt *et al.*, 1981), which is frequently overlooked. It is these hydrolase and transferase reactions that are discussed in this paper.

## DETERMINATION OF THE ROLE OF SECOND-STEP ENZYMES BY THE PATTERN OF FIRST-STEP ENZYMES

The role of hydrolase and transferase enzymes in the control of metabolically-formed epoxides is highly dependent on the pattern of microsomal monooxygenases present in a particular situation. The reason is that several of the individual monooxygenases differ substantially from each other with respect to preferential site of oxidative attack in substrate molecules. This leads to various sets of reactive metabolites, some of which may be detoxified by second-step enzymes, while others are converted to even more reactive secondary or tertiary metabolites by the same second or third-step enzyme(s); thus the enzymes have dual roles (Bentley *et al.*, 1977). Benzo[*a*]pyrene is metabolized by monooxygenase to the reactive 7,8-oxide, which is inactivated by microsomal epoxide hydrolase to the corresponding 7,8-dihydrodiol. This is, however, the precursor molecule for a second monooxygenation step, which reintroduces an epoxide moiety leading to a dihydrodiol bay-region epoxide. According to chemical quantum calculations by Lehr and Jerina (1977), the latter epoxide is especially reactive chemically and has been shown to be an ultimate carcinogen (Slaga *et al.*, 1977; Kapitulnik *et al.*, 1978a,b; Levin *et al.*, 1978; Slaga *et al.*, 1979).

Thus, a single enzyme can play a multiple role in inactivating some metabolites while producing precursors for other reactive species. To generate information for risk estimation, therefore, we need to know not only the differences in these enzyme activities between test systems and the system for which we need the information, but also the exact role of these enzymes. To study this question, we used in-vitro systems that can monitor a toxic effect quantitatively, e.g., bacterial mutagenicity (Oesch & Glatt, 1976). Enzymes with a characteristic cofactor can be characterized by either removal or addition of that cofactor. Enzymes that share their cofactor with others can be monitored for their relative importance only if they are isolated to apparent homogeneity. This applies, also, to enzymes that do not have a characteristic cofactor, such as epoxide hydrolase which merely adds the elements of water (Oesch, 1973, 1979a,b). Frequently, several reactive metabolites are involved in a given toxic response. They may be toxicologically different from one another and, since they are reactive and short-lived, it is often difficult to quantitate or characterize them chemically. Therefore, using various strains of bacteria, we monitored certain synthetically-prepared reactive metabolites for their potency in causing mutations. When the K-region epoxide, benzo[*a*]pyrene 4,5-oxide, was monitored with *Salmonella typhimurium* TA98 and

TA1537, strain 98 was mutated somewhat more efficiently, by a factor of 1.6. Following in-situ bioactivation of the chemically synthesized 7,8-dihydrodiol to the corresponding dihydrodiol bay-region epoxide, strain TA98 was much more efficiently reverted than TA1537, by a factor of 15. This ratio was different again from that observed after in-situ bioactivation of the 9,10-dihydrodiol, etc. So, for each of the various reactive metabolites that are derived from the same parent compound, there is a characteristic ratio controlling the reversion of various strains of bacteria. This ratio can be used to determine which reactive metabolites have been predominantly responsible for a given mutagenic effect.

When microsomal epoxide hydrolase, which had been isolated to apparent homogeneity (Bentley & Oesch, 1975), was added to benzo[*a*]pyrene activated by liver microsomes from untreated mice, the mutagenicity was decreased to 1-2% of the original rate and could not be reduced further by adding more enzyme (Fig. 1). This remaining mutagenicity is caused by the epoxide hydrolase-resistant portion of reactive metabolites. By the ratios of the mutagenic potency towards various strains, it was shown that the K-region 4,5-oxide was predominantly responsible for the major (i.e., epoxide hydrolase-sensitive) portion of the mutagenic effect under these conditions. The situation was fundamentally different when liver microsomes from mice pretreated with 3-methylcholanthrene were used, and a different pattern of monooxygenase isoenzymes was produced. These isoenzymes created a different pattern of primary reactive metabolites, on which the epoxide hydrolase had, at first, a weak but activating effect (Bentley *et al.*, 1977). This occurred because dihydrodiol epoxides had become predominantly responsible for the mutagenic effect. While the microsomes that were used to create the metabolites do, themselves, possess epoxide hydrolase and can generate dihydrodiol epoxides, the addition of more epoxide hydrolase led to a small but significant increase in mutagenicity, since more of the dihydrodiol epoxides were produced. When the amount of epoxide hydrolase was increased further, a small decrease and then a small increase in mutagenicity was observed (Bentley *et al.*, 1977). The effect was multi-phasic, since several metabolites were contributing measurably to the observed mutagenicity.

## THE ROLE OF MICROSOMAL AND CYTOSOLIC EPOXIDE HYDROLASE AND DIHYDRODIOL DEHYDROGENASE

The metabolites of polycyclic hydrocarbons that seem to be responsible for most of the carcinogenic and mutagenic effects induced by these compounds are vicinal dihydrodiol epoxides (Sims *et al.*, 1974; Huberman *et al.*, 1976; Newbold & Brookes, 1976; Slaga *et al.*, 1976; Wislocki *et al.*, 1976; Hecht *et al.*, 1978; Levin *et al.*, 1978; Vigny *et al.*, 1980; MacNicoll *et al.*, 1981). Mutagenicity and DNA-binding experiments with *trans*-7,8-dihydro-7,8-dihydroxybenzo[*a*]pyrene indicate that some inactivation is caused by the presence of glutathione (Glatt & Oesch, 1977; Guenthner *et al.*, 1980; Glatt *et al.*, 1981) but not by microsomal epoxide hydrolase (Glatt, 1976; Wood *et al.*, 1976; Bentley *et al.*, 1977; Glatt *et al.*, 1981). The inability of microsomal epoxide hydrolase to inactivate the *anti*-isomer of the corresponding bay-region diol-epoxide, together with the very weak effect on the activity of the *syn*-isomer (Wood *et al.*, 1976), confirm the latter observation. These negative findings may be a result of the short half-life of the diol-epoxide in an aqueous environment, although it may not necessarily be the same in a biological membrane.

Not all vicinal diol-epoxides are of low stability. The non-bay-region diol-epoxide, *r*-8,*t*-9-dihydroxy-*t*-10,11-oxy-8,9,10,11-tetrahydrobenz[*a*]anthracene (BA-8,9-diol 10,11-oxide), has a half-life of many hours and is, therefore, useful for metabolic studies. It is mutagenic (Malaveille *et al.*, 1977; Wood *et al.*, 1977), it is often the major DNA-binding

**Fig. 1. Effect of apparently homogeneous microsomal epoxide hydrolase on the number of revertant colonies from *Salmonella typhimurium* TA1537 and TA98 by metabolically activated benzo[*a*]pyrene**

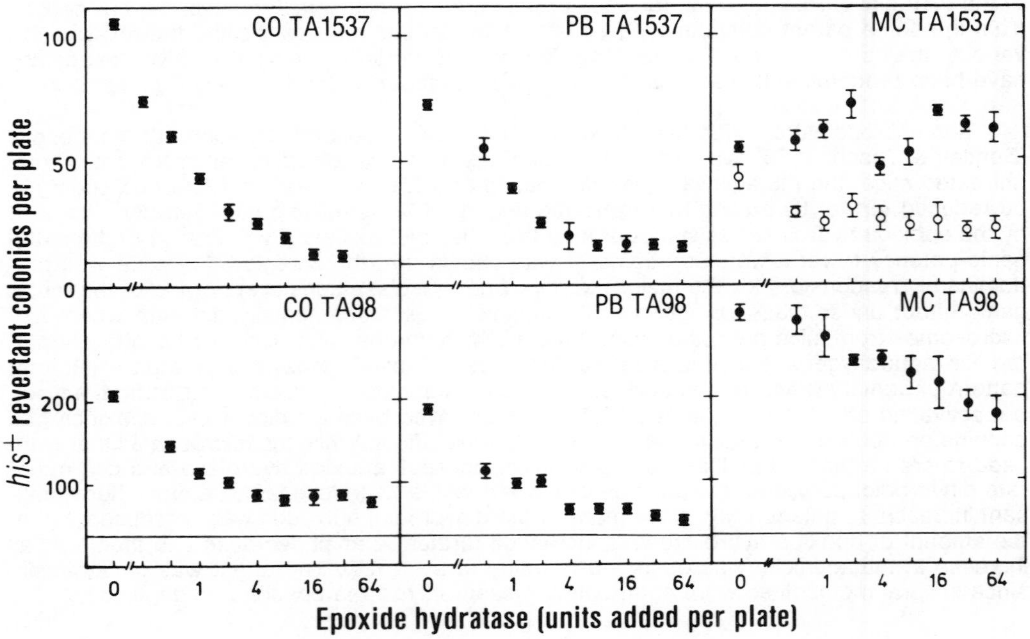


CO, control liver microsomes from untreated C3H mice; PB, pretreatment with phenobarbital; MC, pretreatment with 3-methylcholanthrene. Horizontal lines represent the mean of spontaneous mutations. From Oesch (1979a), reprinted by permission of Springer Verlag, Heidelberg.

species formed from benz[*a*]anthracene *in vivo* and *in vitro* (Cooper *et al.*, 1980a,b; Vigny *et al.*, 1980; MacNicoll *et al.*, 1981) and it serves as a model for less stable diol-epoxides. The results reported below show that a diol-epoxide can be metabolically inactivated by dihydrodiol dehydrogenase, but not by microsomal or cytosolic epoxide hydrolase. If it is assumed that the activities of the enzymes investigated *in vivo* are comparable to those in the experiments *in vitro*, the role of inactivation of the diol-epoxide by dihydrodiol dehydrogenase would be slower than the rate of inactivation of the K-region oxide by microsomal epoxide hydrolase, but still sufficiently rapid to affect the diol-epoxide concentrations substantially, in mammalian systems.

To study their role in metabolic inactivation, the three enzymes, microsomal and cytosolic epoxide hydrolase and cytosolic dihydrodiol dehydrogenase, were purified, and bacterial mutagenicity was used as an indicator of their effects on the mutagenicity of both BA-8,9-diol 10,11-oxide and benz[*a*]anthracene 5,6-oxide (BA 5,6-oxide), the K-region epoxide. As expected from its substrate specificity (Bentley *et al.*, 1976; Jerina *et al.*, 1977), microsomal epoxide hydrolase readily inactivated BA 5,6-oxide (Fig. 2). No significant effect on the mutagenicity of BA-8,9-diol 10,11-oxide was obtained, even at 100-fold of that required for complete inactivation of the K-region oxide. Relatively large amounts of cytosolic epoxide hydrolase were required to inactivate BA 5,6-oxide. Cytosolic epoxide hydrolase, like microsomal epoxide hydrolase, did not inactivate the diol-epoxide (Fig. 2). However, the diol-epoxide was inactivated by dihydrodiol dehydrogenase.

**Fig. 2. Influence of purified microsomal and cytosolic epoxide hydrolase and dihydrodiol dehydrogenase on the mutagenicity of benz[*a*]anthracene (BA) 5,6-oxide (•) and BA-8,9-diol 10,11-oxide (o) for *Salmonella typhimurium* TA 100**

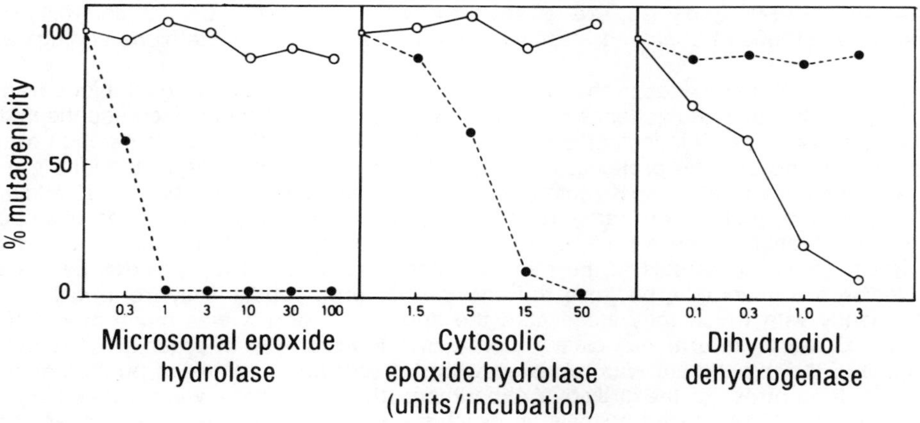


The number of mutants above solvent control, induced by 1 μg of BA 5,6-oxide or 3 μg of BA-8,9-diol 10,11-oxide in the presence of various amounts of purified enzymes, is given as the percentage of the corresponding value without enzyme.

Comparatively high amounts of dihydrodiol dehydrogenase were needed for inactivation of the diol-epoxide, whereas a low amount of microsomal epoxide hydrolase was sufficient to inactivate BA 5,6-oxide. A 50% inactivation of 1 μg of BA 5,6-oxide was achieved, either with 0.4 units of purified rat microsomal epoxide hydrolase, equivalent to 1.3 mg of liver, or with 7 units of purified rabbit cytosolic epoxide hydrolase, equivalent to 28 mg of liver. These are relatively small enzyme quantities. Microsomal epoxide hydrolase equivalent to 330 mg of rat liver and cytosolic epoxide hydrolase equivalent to 200 mg of rabbit liver failed to inactivate the diol-expoxide, whereas with dihydrodiol dehydrogenase, an amount equivalent to 200 mg of liver was required to obtain a 50% inactivation. The effective but moderate rate of inactivation of vicinal diol-epoxides suggests that differences in dihydrodiol dehydrogenase activity, among species, organs, and physiological states, are likely causes for differences in susceptibility to the effects of polycyclic aromatic hydrocarbons.

## THE ROLE OF GLUTATHIONE TRANSFERASES

Several glutathione transferases are known to occur in rat-liver cytosol. The forms AA, A, B, C, D, and E are named in the reverse order of their elution from a carboxymethylcellulose column (Habig *et al.*, 1974). A seventh form has been named glutathione transferase M, because of its ability to react with menaphthylsulfate (Gilham, 1971). We have recently discovered and purified an additional enzyme with distinct properties, that we have termed glutathione transferase X (Friedberg *et al.*, 1983). This form, and the glutathione transferases A, B, and C, which are the most abundant forms present in rat liver in terms of the amount of protein (Jakoby *et al.*, 1976), have been purified to apparent homogeneity and investigated for their abilities to inactivate the two prototype epoxides discussed above, BA 5,6-oxide and BA-8,9-diol 10,11-oxide. Vicinal diol-epoxides appear to be poor substrates for epoxide hydrolases (Wood *et al.*, 1976, 1977; Glatt *et al.*, 1982), but mutagenicity and DNA-binding experiments that have been carried out, using *trans*-7,8-dihydro-7,8-dihydroxybenzo[*a*]pyrene and an activating system, have shown that some inactivation occurs when glutathione is added (Glatt & Oesch, 1977; Hesse *et al.*, 1980; Glatt *et al.*, 1981). With *r*-7,*t*-8-dihydroxy-*t*-9,10-oxy-7,8,9,10-tetrahydrobenzo[*a*]pyrene and BA-8,9-diol 10,11-oxide, Cooper *et al.* (1980a,b) observed enzyme-catalysed formation of glutathione conjugates.

Incubation of the epoxides in the presence of increasing amounts of glutathione transferases A, C and X, but in the absence of glutathione, did not significantly decrease the mutagenicity of the epoxides. This indicates that there is no appreciable covalent reaction of these epoxides with the enzyme proteins. When glutathione was added, the glutathione transferases A, C and X inactivated the K-region epoxide and the diol-epoxide. About 1000-fold higher concentrations of glutathione transferase were required for inactivation of the diol-epoxide than for inactivation of the K-region epoxide. This finding was independent of the enzyme form used. These similarities in the ratio of inactivation of the two epoxides by the three enzymes were remarkable, because the enzymes differed substantially from each other in the efficiency with which they inactivated the epoxides. Form X was much more efficient than form C which, in turn, was clearly more active than form A. In order to verify this striking finding, the experiment was repeated with different preparations of purified enzymes. Moreover, in contrast to the previous experiment, these enzymes were purified from the livers of Aroclor 1254-treated animals, in which the enzymes are present in higher concentrations. Glutathione transferase B was included in these mutagenicity experiments, also. The experiment confirmed both the difference in inactivation of the two epoxides by the glutathione transferases A, C and X and their relative efficiencies. Glutathione transferase B, also, inactivated the K-region epoxide more efficently than the diol-epoxide; its activity towards the diol-epoxide was similar to that of enzyme A, which was the least active of the other glutathione transferases (Fig. 3).

## CONCLUSIONS

The amounts of the various purified enzymes that are required for 50% inactivation of mutagenicity of the two prototype epoxides BA 5,6-oxide (a K-region epoxide) and BA-8,9-diol 10,11-oxide (a vicinal dihydrodiol non-bay-region epoxide), as related to the amounts of enzyme present in the liver, are summarized in Table 1. When intrinsic enzyme activities and the relative amounts of enzymes present in the liver are considered, and subcellular compartmentalization is disregarded, the glutathione transferases can play a more important role in the inactivation of BA 5,6-oxide and BA-8,9-diol 10,11-oxide than dihydrodiol dehydrogenase or the epoxide hydrolases, although the latter are specific enzymes for the hydrolysis of epoxides. Among the glutathione transferases, there are large differences in efficiency of detoxification, and this holds true even for forms A, C and X, which are closely related immunologically (Jakoby *et al.*, 1976; Friedberg *et al.*, 1983), and which differ in efficiency by more than an order of magnitude. With regard to the enzymes present in rat

**Fig. 3. Influence of glutathione transferases on the mutagenicity of benz[*a*]anthracene (BA) 5,6-oxide (•) and BA-8,9-diol 10,11-oxide (o)**

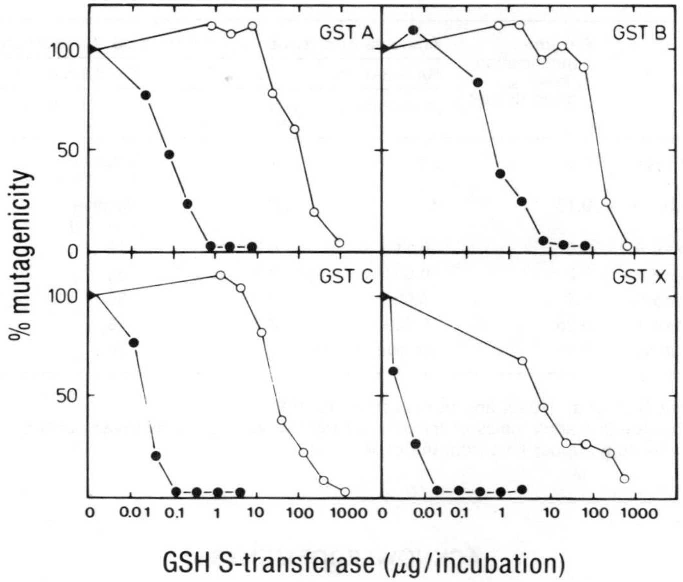


The number of mutants above solvent control, induced by 1 μg of BA 5,6-oxide or by 3 μg of BA-8,9-diol 10,11-oxide in the presence of various amounts of the purified glutathione transferases, is given as the percentage of the corresponding value without enzyme.

liver, it appears that forms C and X are able to contribute most to the inactivation of the two epoxides examined here, form C because of its quantitative abundance in rat liver and form X because of its high efficiency in inactivating these epoxides. Such an estimate appears rather crude when different types of enzymes are compared, because of differences in cofactor concentration, in pH optima and in other environmental factors that may lead to substantial differences between enzyme activity *in vivo* and under experimental conditions. Microsomal epoxide hydrolase was tested in the mutagenicity experiments as the free purified enzyme, whereas *in vivo*, it is situated in the endoplasmic reticulum and in other membranes (Stasiecki *et al.*, 1980). This may be of great advantage in comparison with cytosolic enzymes, since there are increased opportunities for reaction with the epoxides that are generated in these membranes and tend to stay there because of their relative lipophilicity (Glatt & Oesch, 1977). Dihydrodiol dehydrogenase may not only inactivate diol-epoxides, but may also sequester precursor dihydrodiols, an effect that has not been taken into account in the experimental model used. In spite of these limitations, the data presented here, showing a more efficient inactivation of epoxides *in vitro* by glutathione transferases than by the other enzymes, indicate that glutathione transferases may play an important role in the inactivation of epoxides *in vivo*, also. Further, the data indicate that glutathione transferase X is especially important, at least for the two prototype epoxides investigated, but, in all likelihood, also for the two groups of epoxides for which they represent prototypes.

**Table 1. Relative contribution of various purified enzymes to the inactivation of two prototype epoxides, compared with the relative amounts of these enzymes in the liver[a]**

| Enzyme | Enzyme concentration in liver[b] (µg/mg tissue) | Amounts of enzyme required for a 50 % reduction in mutagenicity | | | |
|---|---|---|---|---|---|
| | | BA-5,6-oxide | | BA-8,9-diol 10,11-oxide | |
| | | µg/incubation | mg liver equivalents | µg/incubation | mg liver equivalents |
| Microsomal epoxide hydrolase | 0.5 | 0.7 | 1.3 | inactive (>> 170) | >> 300[c] |
| Cytosolic epoxide hydrolase | 0.16 | 4 | 30 | inactive (>> 30) | >>200[c] |
| Glutathione transferase A | 0.5 | 0.11 | 0.2 | 110 | 200 |
| Glutathione transferase B | 2.2 | 0.5 | 0.2 | 130 | 60 |
| Glutathione transferase C | 1.1 | 0.02 | 0.017 | 30 | 20 |
| Glutathione transferase X | 0.25 | 0.003 | 0.011 | 6 | 20 |
| Dihydrodiol dehydrogenase | 0.45 | inactive (>> 3) | >> 1000[c] | 70 | 170 |

[a] Data taken from Glatt *et al.* (1982) and from Glatt *et al.* (1983)
[b] Values refer to untreated adult males of the species from which the enzyme was purified
[c] Determinable only as an upper limit from the experiment

## ACKNOWLEDGEMENTS

I thank my collaborators for their work presented in this summary and the Fonds der Chemischen Industrie for financial support.

## REFERENCES

Bentley, P. & Oesch, F. (1975) Purification of rat liver epoxide hydratase to apparent homogeneity. *FEBS Lett.*, *59*, 291-295

Bentley, P., Schmassmann, H.U., Sims, P. & Oesch, F. (1976) Epoxides derived from various polycyclic hydrocarbons as substrates of homogeneous and microsome-bound epoxide hydratase: a general assay and kinetic properties. *Eur. J. Biochem.*, *69*, 97-103

Bentley, P., Oesch, F. & Glatt, H.R. (1977) Dual role of epoxide hydratase in both activation and inactivation. *Arch. Toxicol.*, *39*, 65-75

Cooper, C.S., MacNicoll, A.D., Ribeiro, O., Gervasi, G.P., Hewer, A., Walsh, C., Pal, K., Grover, P.L. & Sims, P. (1980a) The involvement of a non-'bay-region' diol-epoxide in the metabolic activation of benz[*a*]anthracene in hamster embryo cells. *Cancer Lett.*, *9*, 53-59

Cooper, C.S., Hewer, A., Ribeiro, O., Grover, P.L. & Sims, P. (1980b) The enzyme-catalysed conversion of *anti*-benzo[*a*]pyrene-7,8-diol 9,10-oxide into a glutathione conjugate. *Carcinogenesis*, *1*, 1975-1980

Friedberg, T., Milbert, U., Bentley, P., Guenthner, T.M. & Oesch, F. (1983) Purification and characterization of a new cytosolic glutathione S-transferase (glutathione S-transferase X) from rat liver. *Biochem. J.*, *215*, 617-625

Gilham, B. (1971) The reaction of aralkyl sulphate esters with glutathione catalysed by rat liver preparations. *Biochem. J.*, *121*, 667-672

Glatt, H.R. (1976) Die Bedeutung verschiedener aktivierender und inaktivierender Stoffwechselschritte für die Mutagenität des Karzinogens Benzo[*a*]pyren. Thesis of University of Basel

Glatt, H.R. & Oesch, F. (1977) Inactivation of electrophilic metabolites by glutathione transferases and limitation of the system due to subcellular localization. *Arch. Toxicol.*, *39*, 87-96

Glatt, H.R., Billings, R., Platt, K.L. & Oesch, F. (1981) Improvement of the correlation of bacterial mutagenicity with carcinogenicity of benzo[*a*]pyrene and four of its major metabolites by activation with intact liver cells instead of cell homogenate. *Cancer Res.*, *41*, 270-277

Glatt, H.R., Cooper, C.S., Grover, P.L., Sims, P., Bentley, P., Mertes, M., Waechter, F., Vogel, K., Guenthner, T.M. & Oesch, F. (1982) Inactivation of a diol-epoxide by dihydrodiol dehydrogenase, but not by two epoxide hydrolases. *Science*, *215*, 1507-1509

Glatt, H.R., Friedberg, T., Grover, P.L., Sims, P. & Oesch, F. (1983) Inactivation of a diol-epoxide and a K-region epoxide by glutathione S-transferases: high efficiency of the new form X. *Cancer Res.*, *43*, 5713-5717

Guenthner, T.M., Jernström, B. & Orrenius, S. (1980) On the effect of cellular nucleophiles on the binding of metabolites of 7,8-dihydroxy-7,8-dihydrobenzo[*a*]pyrene and 9-hydroxy-benzo[*a*]pyrene to nuclear DNA. *Carcinogenesis*, *1*, 407-418

Habig, W.H., Pabst, M.J. & Jakoby, W.B. (1974) Glutathione *S*-transferases. The first enzymatic step in mercapturic acid formation. *J. biol. Chem.*, *249*, 7130-7139

Hecht, S.S., LaVoie, E., Mazzorese, R., Amin, S., Bedenko, V. & Hoffmann, D. (1978) 1,2-Dihydro-1,2-dihydroxy-5-methylchrysene, a major activated metabolite of the environmental carcinogen 5-methylchrysene. *Cancer Res.*, *38*, 2191-2198

Hesse, S., Jernström, B., Martinez, M., Guenthner, T., Orrenius, S., Christodoulides, L. & Ketterer, B. (1980) Inhibition of binding of benzo[*a*]pyrene metabolites to nuclear DNA by glutathione and glutathione S-transferase B. *Biochem. Biophys. Res. Commun.*, *94*, 612-617

Huberman, E., Sachs, L., Yang, S.K. & Gelboin, H.V. (1976) Identification of mutagenic metabolites of benzo[*a*]pyrene in mammalian cells. *Proc. natl Acad. Sci. USA*, *73*, 607-611

Jakoby, W.B., Ketley, J.N. & Habig, W.H. (1976) Rat glutathione S-transferases: binding and physical properties. In: Arias, I.M. & Jakoby, W.B., eds, *Glutathione: Metabolism and Function*, Vol. 6, New York, Raven Press, pp. 213-220

Jerina, D.M., Dansette, P.M., Lu, A.Y.H. & Levin, W. (1977) Hepatic microsomal epoxide hydrolase: a sensitive radiometric assay for hydration of arene oxides of carcinogenic aromatic hydrocarbons. *Mol. Pharmacol.*, *13*, 342-351

Kapitulnik, J., Wislocki, P.G., Levin, W., Yagi, H., Jerina, D.M. & Conney, A.H. (1978a) Tumorigenicity studies with diol-epoxides of benzo[*a*]pyrene which indicate that (±)-*trans*-7,8-dihydroxy-9,10-epoxy-7,8,9,10-tetrahydrobenzo[*a*]pyrene is an ultimate carcinogen in newborn mice. *Cancer Res.*, *38*, 354-358

Kapitulnik, J., Wislocki, P.G., Levin, W., Yagi, H., Thakker, D.R., Akagi, H., Koreeda, M., Jerina, D.M. & Conney, A.H. (1978b) Marked differences in the carcinogenic activity of optically pure (+)- and (-)-*trans*-7,8-dihydrobenzo[*a*]pyrene in newborn mice. *Cancer Res.*, *38*, 2661-2665

Lehr, R.E. & Jerina, D.M. (1977) Metabolic activations of polycyclic hydrocarbons. Structure-activity relationships. *Arch. Toxicol.*, *39*, 1-6

Levin, W., Wood, A.W., Wislocki, P.G., Chang, R.L., Kapitulnik, J., Mah, H.D., Yagi, H., Jerina, D.M. & Conney, A.H. (1978) Mutagenicity and carcinogenicity of benzo[*a*]pyrene and benzo[*a*]pyrene derivatives. In: Gelboin, H.V. & Ts'o, P.O.P., eds, *Polycylic Hydrocarbons and Cancer*, Vol. I, New York, Academic Press, pp. 189-194

MacNicoll, A.D., Cooper, C.S., Ribeiro, O., Pal, K., Hewer, A., Grover, P.L. & Sims, P. (1981) The metabolic activation of benz[*a*]anthracene in three biological systems. *Cancer Lett.*, *11*, 243-249

Malaveille, C., Kuroki, T., Sims, P., Grover, P.L. & Bartsch, H. (1977) Mutagenicity of isomeric diol-epoxides of benzo[*a*]pyrene and benz[*a*]anthracene in *S. typhimurium* TA98 and TA100 and in V79 Chinese hamster cells. *Mutat. Res.*, *44*, 313-326

Newbold, R.F. & Brookes, P. (1976) Exceptional mutagenicity of benzo[*a*]pyrene diol epoxide in cultured mammalian cells. *Nature*, *261*, 52-54

Oesch, F. (1973) Mammalian epoxide hydrases: inducible enzymes catalyzing the inactivation of carcinogenic and cytotoxic metabolites derived from aromatic and olefinic compounds. *Xenobiotica*, *3*, 305-340

Oesch, F. (1979a) Enzymes as regulators of toxic reactions by electrophilic metabolites. In: Chambers, P.L. & Günzel, P., eds, *Mechanism of Toxic Action on Some Target Organs*, Heidelberg, Springer-Verlag, pp. 215-227

Oesch, F. (1979b) Enzymes as regulators of toxic reactions by electrophilic metabolites. *Arch. Toxicol.*, *Suppl. 2*, 215-227

Oesch, F. & Glatt, H.R. (1976) Evaluation of the relative importance of various enzymes involved in the control of mutagenic and cytotoxic metabolites. In: Montesano, R., Bartsch, H. & Tomatis, L., eds, *Screening Tests in Chemical Carcinogenesis* (*IARC Scientific Publications No. 12*), Lyon, International Agency for Research on Cancer, pp. 255-274

Sims, P., Grover, P.L., Swaisland, A., Pal, K. & Hewer, A. (1974) Metabolic activation of benzo[*a*]pyrene proceeds by a diol-epoxide. *Nature*, *252*, 226-228

Slaga, T.J., Viaje, A., Berry, D.L., Bracken, W.M., Buty, S.G., Scribner, J.D. (1976) Skin-tumor initiating ability of benzo[*a*]pyrene 4,5-, 7,8- and 9,10-oxide, 7,8-diol-9,10-epoxides and 7,8-diol. *Cancer Lett.*, *2*, 115-122

Slaga, T.J., Viaje, A., Bracken, W.M., Berry, D.L., Fischer, S.M., Miller, D.R. & Leclerc, S.M. (1977) Skin-tumor-initiating ability of benzo[*a*]pyrene-7,8-diol-9,10-epoxide (anti) when applied topically in tetrahydrofuran. *Cancer Lett.*, *3*, 23-30

Slaga, T.J., Bracken, W.J., Gleason, G., Levin, W., Yagi, H., Jerina, D.M. & Conney, A.H. (1979) Marked differences in the skin tumor-initiating activities of the optical enantiomers of the distereomeric benzo[*a*]pyrene 7,8-diol-9,10-oxides. *Cancer Res.*, *39*, 67-71

Stasiecki, P., Oesch, F., Bruder, G., Jarasch, E.D. & Franke, W.W. (1980) Distribution of enzymes involved in metabolism of polycyclic aromatic hydrocarbons among rat liver endomembranes and plasma membranes. *Eur. J. Cell Biol.*, *21*, 79-92

Ullrich, V., Weber, P. & Wollenberg, P. (1975) Tetrahydrofurane - an inhibitor for ethanol-induced liver microsomal cytochrome P-450. *Biochim. Biophys. Res. Commun.*, *64,* 808-813

Vigny, P., Kindts, M., Duquesne, M., Cooper, C.S., Grover, P.L. & Sims, P. (1980) Metabolic activation of benz[*a*]anthracene: fluorescence spectral evidence indicates the involvement of a non-'bay-region' diol epoxide. *Carcinogenesis*, *1*, 33-41

Wislocki, P.G., Wood, A.W., Chang, R.L., Levin, W., Yagi, H., Hernandez, O., Jerina, D.M. & Conney, A.H. (1976) High mutagenicity and toxicity of a diol epoxide derived from benzo[*a*]pyrene. *Biochem. Biophys. Res. Commun.*, *68*, 1006-1012

Wood, A.W., Levin, W., Lu, A.Y.H., Yagi, H., Hernandez, O., Jerina, D.M. & Conney, A.H. (1976) Metabolism of benzo[*a*]pyrene derivatives to mutagenic products by highly purified hepatic microsomal enzymes. *J. biol. Chem.*, *251*, 4882-4890

Wood, A.W., Chang, R.L., Levin, W., Lehr, R.E., Schaefer-Ridder, M., Karle, J.M., Jerina, D.M. & Conney, A.H. (1977) Mutagenicity and cytotoxicity of benz[*a*]anthracene diol epoxides and tetrahydroepoxides: exceptional activity of the bay region 1,2-epoxides. *Proc. natl Acad. Sci. USA*, *74*, 2746-2750

# INTERINDIVIDUAL VARIATION IN CARCINOGEN METABOLISM, DNA DAMAGE AND DNA REPAIR

**K. Vahakangas, H. Autrup & C.C. Harris[1]**

*Laboratory of Human Carcinogenesis, National Cancer Institute, National Institutes of Health, Bethesda, MD 20205, USA*

## SUMMARY

A number of laboratory approaches are currently available to measure interindividual variation in carcinogen metabolism, amounts of carcinogen-DNA adducts and DNA repair capacity. Wide interindividual variations have been found among strains of experimental inbred animals and among individuals in outbred populations, including humans. The quantitative relationships, if any, between these parameters and the human cancer risk will be determined by biochemical and molecular epidemiological studies, of high and low cancer risk populations and of cancer-prone families.

## INTRODUCTION

Both susceptibility to cancer and the latency from the carcinogen exposure to a clinically evident cancer differ greatly between individuals. Causes of this variation in cancer risk include the amount and frequency of carcinogen exposure, as well as inherited and acquired host factors (reviewed by Harris *et al.*, 1980b; Perera & Weinstein, 1982; Harris, 1983). Some of the facts that relate these questions to interindividual variation in carcinogen metabolism, in formation of carcinogen-DNA adducts and in DNA repair, are:

(1) Many chemical carcinogens require metabolic activation in order to exert their oncogenic effects (reviewed by Miller & Miller, 1981; Pelkonen & Vahakangas, 1980).

[1]To whom correspondence should be addressed

(2) Genetic polymorphism is well known in xenobiotic metabolism (reviewed by Idle & Ritchie, 1983).

(3) Metabolism, formation of DNA-adducts and carcinogenicity are positively correlated in animal studies (reviewed by Pelkonen *et al.*, 1980; Pelkonen & Nebert, 1982).

(4) A deficiency in DNA repair, e.g., xeroderma pigmentosum, may be a predisposition to cancer (reviewed by Cleaver *et al.*, 1982).

## EXAMPLES OF ENZYMES INVOLVED IN CARCINOGEN METABOLISM

The most important enzyme system involved in the metabolism of chemical carcinogens is the ubiquitous, cytochrome-P-450-containing monooxygenase system. On a biochemical basis, the relative nonspecificity and the organ and individual variation in xenobiotic metabolism are caused, at least partly, by the multiple forms of cytochrome P-450 present in target tissues (Lu, 1979). For instance, in rabbit liver, at least five isolated forms (P-450 LM 2-6) have been identified, on the basis of different molecular weights, different but overlapping substrate specificities and different sensitivity to inducing agents (references in Ekstrom *et al.*, 1982). Recent studies have implicated multiple forms of cytochrome P-450 also in human liver (Ekstrom *et al.*, 1982) and placenta (Pasanen & Pelkonen, 1981). Attempts to isolate cytochrome P-450 in human lung have failed so far, although monooxygenase-associated activities are present (Prough *et al.*, 1977; McManus *et al.*, 1980).

Other important enzyme systems have been found in human tissues, as well, like multiple forms of both epoxide hydrolase, which catalyses the metabolism of epoxides and arene oxides to dihydrodiols (Glatt *et al.*, 1984), and conjugating enzymes, which convert lipid-soluble molecules into more water-soluble products. Glucuronyl transferase, sulfotransferase and glutathione transferase are the most important conjugating enzymes. In liver, the activity of glucuronyl transferase is higher than sulfotransferase and glutathione transferase activities. In several extrahepatic tissues, including colon, bronchus and oesophagus, the major conjugation products of benzo[*a*]pyrene are the sulfate esters and glutathione conjugates (Harris *et al.*, 1982a).

## POLYMORPHISM IN XENOBIOTIC METABOLISM

The best-known genetic condition, that affects the kinetics of several drugs, e.g., isoniazid, hydralazine and procainamide, is *N*-acetylation polymorphism. Acetylation status is controlled by two autosomal alleles at a single locus; rapid acetylation is a dominant trait and slow acetylation is recessive. The heterozygote genotype is probably the genetic basis for an intermediate phenotype that has been found recently (Chapron *et al.*, 1980). Approximately half the Western population are slow acetylators. Slow acetylation leads to increased plasma levels of the drug and is associated, also, with certain side-effects of these compounds, e.g., systemic lupus erythematosus (Reidenberg & Drayer, 1978).

Some of the aromatic amines are carcinogens, and the acetylation phenotype might determine the susceptibility to these carcinogens, both in rabbits (McQueen *et al.*, 1983) and in humans (Lower *et al.*, 1979; Cartwright *et al.*, 1982). Cartwright and coworkers found an excess of slow acetylators among bladder cancer patients who had been exposed to *N*-substituted aryl compounds. There were more slow acetylators, also, among patients with more advanced disease.

Polymorphic oxidation of debrisoquine, also, has been well characterized. Humans can be divided into extensive metabolizers and poor metabolizers, on the basis of the ratio of unchanged drug to the main metabolite, 4-hydroxydebrisoquine, excreted in urine (Idle & Ritchie, 1983). The trait for poor metabolism is autosomal recessive. Patients with bronchial carcinoma have a lower debrisoquine metabolic ratio than controls, even if the lowest-cigarette-consuming cancer patients are compared to the highest-consuming control group (Hetzel *et al.*, 1982). This is true, also, for liver cancer patients, who are considerably faster hydroxylators of debrisoquine than controls (Idle & Ritchie, 1983).

The difference in inducibility of one of the monooxygenase activities, aryl hydrocarbon hydroxylase (AHH), between mouse type C57BL/6 (responsive to induction by polynuclear aromatic hydrocarbons) and DBA/2 (nonresponsive), can be explained almost totally by a difference at a single gene locus, called *Ah* (Nebert & Atlas, 1978). Homozygous $Ah^bAh^b$ and heterozygous $Ah^bAh^d$ are responsive; homozygous $Ah^dAh^d$ are nonresponsive. On the other hand, at least six alleles in more than one loci were necessary to explain the AHH-induction in crosses between responsive and nonresponsive animals from a dozen inbred mouse strains. Nebert and Atlas (1978) concluded, from studies in twins, that there is enough evidence, also, for a heritable variation of AHH-inducibility in man. A recent study on AHH-inducibility in cultured lymphocytes (Borresen *et al.*, 1981) confirms the heritability, and suggests even trimodal distribution of AHH-inducibility. Most pharmacogenetic traits show a unimodal Gaussian distribution, but this does not necessarily indicate multifactorial inheritance (Motulsky, 1978). In many instances, family and twin studies have not been performed, as yet.

## VARIATION IN CARCINOGEN METABOLISM

In studies on human xenobiotic drug metabolism, a wide interindividual variation in the rate of metabolism is observed generally. For example, AHH activity varies 20-300-fold among individuals (Table 1). The variation is mainly quantitative and generally of the same order of magnitude as in pharmacogenetic studies. Qualitative differences are either fewer or non-detectable. Compared to AHH, less variation has been found in epoxide hydrolase activity in the tissues studied (less than 10-fold in most of the tissues, Table 1). This observation, together with results from studies on inbred experimental animals (Nebert & Atlas, 1978), indicate that AHH and epoxide hydrolase are under different genetic and/or environmental control.

**Table 1. The highest interindividual variation reported for aryl hydrocarbon hydroxylase (AHH) and epoxide hydrolase (EH) activities in human tissues**

| Organ | Fold variation | | | |
|---|---|---|---|---|
| | AHH (No. of cases) | Reference | EH (No. of cases) | Reference |
| Liver | 76 (32) | Kapitulnik *et al.* (1977) | 63 (166) | Glatt *et al.* (1983) |
| Lung | 20 (76) | Sabadie *et al.* (1981) | 4 (57) | Glatt *et al.* (1980) |
| Bronchus | 20 (10) | Kahng *et al.* (1981) | 3 (4) | Harris *et al.* (1977b) |
| Placenta | 350 (24) | Kapitulnik *et al.* (1976) | 8 (21) | Vaught *et al.* (1979) |
| Skin | 7 (13) | Shuster *et al.* (1980) | 3 (6) | Oesch *et al.* (1978) |
| Kidney | 12 (23) | Yamasaki *et al.* (1977) | NT[a] | |

[a] NT, not tested

At different levels of biological organization, e.g., tissue *vs* subcellular fractions, there are both quantitative and qualitative differences in benzo[*a*]pyrene (BP) metabolism (Pelkonen & Nebert, 1982). The reasons for this are quite obvious. Microsomal fractions may not contain soluble enzyme activities, like those of glutathione transferase and cytosolic epoxide hydrolase. The in-vitro optimized conditions may very well differ profoundly from those in a living cell (Pelkonen & Nebert, 1982). For instance, in microsomes from human peripheral lung, the pattern of BP metabolites formed differs strikingly from that formed in peripheral lung cultures (Cohen *et al.*, 1979; Sabadie *et al.*, 1981; Autrup *et al.*, 1982).

BP metabolism has been analysed extensively in cultured human bronchus, colon, trachea and oesophagus in our laboratory (Harris *et al.*, 1982b; Autrup & Harris, 1983). The variation between individuals, in the formation of organic-soluble metabolites, as well as in total metabolism, is highest in colon and lowest in trachea. Of the organic-soluble metabolites, BP-7,8-diol (the proximal carcinogen) and the tetrols (as an indication of formation of the ultimate form, BP diol-epoxide) are of most interest. In bronchus, colon and oesophagus, interindividual variation in the amount of BP-7,8-diol formed is very high but still much lower than in trachea (Harris & Autrup, 1980). BP tetrols are found in all cultured organs studied. Both the amount and the interindividual variation in tetrol formation are highest in gastrointestinal tissues (Autrup *et al.*, 1982). Endogenous factors like the hormonal state can affect BP metabolism, also, as demonstrated by Mass *et al.* (1981) in human endometrial cultures.

Differences occur, also, in cancerous tissue compared to normal tissue from the same patient. AHH activity is lower in cancerous than in nontumorous lung tissue (Sabadie *et al.*, 1981). In normal human peripheral lung cultures, sulfate esters are the major conjugates of BP, but in lung tumours, glucuronides are the major conjugates (Cohen *et al.*, 1979). A similar shift towards glucuronides has been found, also, in colon tumours (Cohen *et al.*, 1983).

For several cancers, smokers are at higher risk than nonsmokers. The clearest effect of smoking on carcinogen metabolism is the induction of AHH in placenta. In addition to the increased mean value of placental AHH in smokers, interindividual variation is greater also. At the same time, there is no increase in the activity or variability of epoxide hydrolase (Vaught *et al.*, 1979; Pelkonen *et al.*, 1981). Gurtoo *et al.* (1983) described a positive correlation between the number of cigarettes smoked per day and the amount of interindividual variation in placental AHH-activity; variation was 85-fold among those who smoked 10 cigarettes per day, 1000-fold for 20 and over 1500-fold among people smoking more than 30 cigarettes per day. Induction in liver and lung is much lower (Harris *et al.*, 1982b). Regardless of small, or absent, induction in the in-vitro parameters of liver drug metabolism (AHH, cytochrome P-450; Boobis *et al.*, 1980; Vahakangas *et al.*, 1983), the lower mean of antipyrine half-lives and the higher mean of antipyrine clearances indicate that in-vivo antipyrine-metabolism is increased by smoking (Hart *et al.*, 1976; Vahakangas *et al.*, 1983).

## VARIATION IN CARCINOGEN BINDING TO DNA

Interindividual variation in the binding of BP to DNA, in human tissue explant cultures, is about 50-200-fold (Harris *et al.*, 1982b). Other carcinogens have been studied less, but differences of the same range have been found (Table 2).

**Table 2. The highest variation reported for carcinogen-DNA-binding in human tissues of BP, benzo[*a*]pyrene; AFB, aflatoxin $B_1$; DMNA, dimethylnitrosamine; 1,2-DMH, 1,2-dimethylhydrazine; 2-AAF, 2-acetylaminofluorene; DMBA, dimethylbenzanthracene**

| Organ | Fold variation | | | | | |
|---|---|---|---|---|---|---|
| | BP | AFB | DMNA | 1,2-DMH | 2-AAF | DMBA |
| Oesophagus | 99[a] | 70[a] | 90[a] | | | |
| Bronchus | 75[a] | 120[a] | 60[a] | 10[b] | 18[c] | 50[d] |
| Peripheral lung | 3[e] | | | | | |
| Liver | | 12[f] | | | | |
| Colon | 130[a] | 150[a] | 145[a] | 80[c] | | |
| Bladder | 68[a] | 127[c] | | | 114[c] | |
| Endometrium | 70[a] | | | | | |

[a] Harris *et al.* (1982d)
[b] Harris *et al.* (1977a)
[c] Daniel *et al.* (1984)
[d] Harris (1976)
[e] Stoner *et al.* (1978)
[f] Booth *et al.* (1981)

There is also a wide variation in the protein-binding of carcinogens in various tissues: bladder (Stoner *et al.*, 1982), bronchus (Harris *et al.*, 1977a; Stoner *et al.*, 1982), colon (Autrup *et al.*, 1977), oesophagus (Harris *et al.*, 1979a). The ratio of BP protein to BP DNA, within a sample, varies between explants from different individuals and between different doses of carcinogen (Harris *et al.*, 1977b; Harris *et al.*, 1979a; Autrup *et al.*, 1977).

BP is used widely as a model compound for the carcinogenic polynuclear aromatic hydrocarbons, and it is important to study whether BP-binding levels have any predictive value for DNA binding of other chemical classes of carcinogens. Daniel *et al.* (1984) showed a positive correlation between binding of BP and 2-acetylaminofluorene, in both bladder and bronchus explants from the same individual. Autrup *et al.* (1980) found a weak but statistically significant positive correlation between BP and 1,2-dimethylhydrazine binding to DNA, in cultured human colon. No correlation was shown between aflatoxin $B_1$-DNA and BP-DNA binding in bladder, bronchus (Daniel *et al.*, 1984) and colon (Autrup *et al.*, 1980), nor between BP and *N*-nitrosodimethylamine in bronchus, colon, duodenum and oesophagus (Autrup, 1982).

In addition to in-vitro studies, several research groups are developing methods to detect the minute amounts of carcinogen-DNA adducts formed *in vivo* in individuals exposed to carcinogens. The molecular mechanism behind tumour initiation, as well as the proof and dosimetry of carcinogen exposure, should be pursued by such methods (Müller & Rajewsky, 1981). Three principal ways to detect carcinogen-DNA adducts in humans are being developed currently: (1) Immunological methods: immunoassays (reviewed by Poirier, 1981; Harris *et al.*, 1982c) and immunohistochemical techniques (Heyting *et al.*, 1983; Eggset *et al.*, 1983); (2) fluorescence measurements (Daudel *et al.*, 1975; Rahn *et al.*, 1980); and (3) postlabelling of the DNA-adducts with a radioactive marker (Randerath *et al.*, 1981; Gupta *et al.*, 1982). Antibodies towards BP-DNA, aflatoxin $B_1$-DNA, acetylaminofluorene-DNA and $O^6$-methyl, -ethyl and -butyl guanine adducts already exist (reviewed by Müller & Rajewsky, 1981). Using a polyclonal antibody, raised towards BP diol-epoxide-modified DNA (Poirier *et al.*, 1980), Perera *et al.* (1982) found BP-DNA adducts in lung tissue and/or blood cells of five lung cancer patients in a group of fifteen. Using the same antibody in the ultrasensitive enzymatic radioimmunoassay (Harris *et al.*, 1979b), Shamsuddin *et al.* (1983) found some

BP-DNA-positive samples among roofers and foundry workers. There was a 60-fold variation in the amount of modified DNA. However, many of the samples did not contain any detectable BP-DNA adducts.

Cancer patients, treated with anticancer drugs, form an interesting group for in-vivo studies. Antibody towards *cis*-platinum-DNA has been found (Poirier *et al.*, 1982). The amount of *cis*-platinum given to patients is large compared to any environmental exposure, so that, by studying these patients, valuable information will be obtained on the applicability of this approach and the relationship between exposure and DNA binding.

## VARIATION IN DNA REPAIR

DNA binding is an endpoint that is related to the metabolism of a carcinogen to an ultimate carcinogenic metabolite. The amount of DNA binding is not totally determined, however, by the activity of the metabolizing enzymes. The availability of metabolite in the target tissue (metabolism within the target tissue *vs* transportation of metabolites, e.g., from liver to lung) and repair of DNA are other important variables. Besides the extensive defects in DNA repair that are found in some genetic diseases, e.g., xeroderma pigmentosum, variation in the repair capacity can be found also in normal populations (Setlow, 1983).

There are many ways to detect DNA repair. Ultra-violet damage-induced repair has been studied primarily by measuring unscheduled DNA synthesis. A wide interindividual variation (4.5-fold) was found in human leucocytes (Lambert *et al.*, 1979). Wide variation is found, also, in the bromodeoxyuridine photolysis test in normal fibroblast cultures (Setlow, 1983). The measure of repair, here, is the number of single-strand breaks caused by photosensitive bromodeoxyuridine incorporated into DNA during repair. The variation in DNA repair is different among cell strains. For instance, leucocyte cultures show more variation than fibroblasts.

$O^6$-Alkylguanine is one of the putative precarcinogenic DNA lesions that are caused by a number of *N*-nitroso-compounds. This alkylation product is removed by $O^6$-alkylguanine-DNA transalkylase. Because one alkyl group is bound per enzyme molecule and the binding is irreversible, the number of acceptor proteins can be measured in cell extracts, using DNA containing radioactive $O^6$-alkylguanine (Setlow, 1983). Myrnes *et al.* (1983) compared this enzyme activity in different human tissues and found high interindividual differences. Liver had highest activity and an eight-fold variation. Colon and small intestine showed even higher variation (10- and 40-fold, respectively). No correlation in the tissues was found between the activities of this and another DNA repair enzyme, uracil DNA glycosylase. The interindividual variation for uracil DNA glycosylase was three-fold in liver and normal stomach, 5.5-fold in colon and 60-fold in small intestine. $O^6$-Alkylguanine-DNA transalkylase activity also varies in normal *vs* tumour tissue (Setlow, 1983). In two colon tumours, Myrnes *et al.* (1983) found higher $O^6$-alkylguanine-DNA transalkylase activity than in normal colon from the same individuals, while another repair enzyme, uracil DNA glycosylase, was equal in tumorous and normal tissues.

## IMPLICATIONS OF INTERINDIVIDUAL VARIATION

Covalent binding of electrophilic carcinogens to DNA is considered an important event in the initiation of chemically induced cancer in experimental animals (Miller & Miller, 1981; Pelkonen & Nebert, 1982). The studies that have been made so far indicate a wide interindividual variation among outbred animals, including humans, in the amount of carcinogen bound to DNA *in vitro* under identical carcinogen exposure (Harris *et al.*, 1982a). This is not surprising, because the amount of DNA binding depends on metabolism, affinity to other macromolecules and DNA repair capacity, which all show variation as well. If highly sensitive assays are available to measure the amounts of carcinogen-DNA adducts in target tissues from people exposed to environmental carcinogens, e.g., BP and aflatoxin $B_1$, the relationships between this type of carcinogen damage and cancer risk will be more easily determined.

As previously discussed, drugs may be useful as probes, to predict interindividual differences in carcinogen metabolism and, possibly, cancer risk. Debrisoquine is a promising example of a drug probe that may be useful in phenotyping individuals for these investigations.

AHH activity has been studied as a possible marker, and some investigators have reported differences between cancer patients and healthy individuals/noncancer patients. In lung cancer patients, Kellermann *et al.* (1978) showed increased saliva antipyrine elimination, which they consider a better measure of metabolic differences between these groups than AHH-inducibility in cultured lymphocytes. The latter has yielded controversial results (Kellermann *et al.*, 1973; Paigen *et al.*, 1977, 1978). Interestingly enough, in a very carefully controlled study, Kouri *et al.* (1982) showed a strong positive correlation between pulmonary carcinoma and a high benzanthracene-induced AHH level in cryopreserved peripheral blood lymphocytes from 57 patients. Rudiger *et al.* (1980) described enhanced BP-metabolism and DNA binding in cultured blood monocytes of lung cancer patients. Higher binding of BP to DNA has been found, also, in macroscopically normal cultured bronchi, from patients with nonglandular lung cancer compared to noncancerous patients and those with glandular mucous differentiated cancers (Harris *et al.*, 1980a). In addition to the increased risk of skin cancer in xeroderma pigmentosum patients (Cleaver *et al.*, 1982), there is evidence to suggest that patients with a genetic predisposition to colorectal cancer have a reduced DNA-repair capacity (Pero *et al.*, 1983).

Aflatoxin-guanine adducts, putative DNA repair products, can be found in urine of aflatoxin-$B_1$-exposed rats (Bennet *et al.*, 1981) and humans (Autrup *et al.*, 1983). Only a minority (6/81) of the people living in an endemic area of liver cancer, Kenya, had detectable aflatoxin $B_1$-DNA adducts in their urine. On the other hand, all the six positives were from the two areas in Murang'a district that are economically more depressed, where improper handling and storage of food is most likely and where aflatoxin $B_1$ contamination has been shown previously (Autrup *et al.*, 1983). An increased amount of adducts in the urine might mean either increased binding to DNA or more efficient repair or both. Whether it has any predictive value for cancer risk remains to be determined.

## REFERENCES

Autrup, H. (1982) Carcinogen metabolism in human tissues and cells. *Drug. Metab. Rev.*, *13*, 603-646

Autrup, H. & Harris, C.C. (1983) *Metabolism of chemical carcinogens by human tissues*. In: Harris, C.C. & Autrup, H., eds, *Human Carcinogenesis*, New York, Academic Press, pp. 169-194

Autrup, H., Harris, C.C., Stoner, G.D., Jesudason, M.L. & Trump, B.F. (1977) Binding of chemical carcinogens to macromolecules in cultured human colon. *J. natl Cancer Inst.*, *59*, 351-354

Autrup, H., Schwartz, R.D., Essigman, J.M., Smith, L., Trump, B.F. & Harris, C.C. (1980) Metabolism of aflatoxin $B_1$, benzo[*a*]pyrene and 1,2-dimethylhydrazine by cultured rat and human colon. *Teratog. Carcinog. Mutag.*, *1*, 3-13

Autrup, H., Grafstrom, R.C., Brugh, M., Lechner, J.F., Haugen, A., Trump, B.F. & Harris, C.C. (1982) Comparison of benzo[*a*]pyrene metabolism in bronchus, esophagus, colon and duodenum from the same individual. *Cancer Res.*, *42*, 934-938

Autrup, H., Bradley, K., Shamsuddin, A.K.M., Wakhisi, J. & Wasunna, A. (1983) Detection of putative adduct with fluorescence characteristics identical to 2,3-dihydro-2-(7'-guanyl)-3-hydroxyaflatoxin $B_1$ in human urine collected in Murang'a district, Kenya. *Carcinogenesis*, *4*, 1193-1195

Bennett, R.A., Essigman, J.M. & Wogan, G.N. (1981) Excretion of an aflatoxin-guanine adduct in the urine of aflatoxin $B_1$-treated rats. *Cancer Res.*, *41*, 650-654

Boobis, A.R., Brodie, M.J., Kahn, G.C., Fletcher, D.R., Saunders, J.H. & Davies, D.S. (1980) Monooxygenase activity of human liver in microsomal fractions of needle biopsy specimens. *Br. J. clin. Pharmacol.*, *9*, 11-19

Booth, S.C., Bosenberg, H., Garner, R.C., Hertzog, P.J. & Norpoth, K. (1981) The activation of aflatoxin $B_1$ in liver slices and in bacterial mutagenicity assays using livers from different species including man. *Carcinogenesis*, *2*, 1063-1068

Borresen, A.-L., Berg, K. & Magnus, P. (1981) A twin study of aryl hydrocarbon hydroxylase (AHH) inducibility in cultured lymphocytes. *Clin. Genet.*, *19*, 281-289

Cartwright, R.A., Glashan, R.W., Rogers, H.J., Ahmad, R.A., Barham-Hall, D., Higgins, E. & Kahn, M.A. (1982) Role of *N*-acetyltransferase phenotypes in bladder carcinogenesis: a pharmacogenetic epidemiological approach to bladder cancer. *Lancet*, *ii*, 842-846

Chapron, D.J., Kramer, P.A. & Mercik, S.A. (1980) Kinetic discrimination of three sulfamethazine acetylation phenotypes. *Clin. Pharmacol. Ther.*, *27*, 104-113

Cleaver, J.E., Bodell, W.J., Gruenert, D.C., Kapp, L.N., Kaufmann, W.K., Park, S.D. & Zelle, B. (1982) *Repair and replication abnormalities in various human hypersensitive diseases*. In: Harris, C.C. & Cerutti, P.A., eds, *Mechanisms of Chemical Carcinogenesis*, New York, Alan R. Liss, pp. 409-418

Cohen, G.M., Mehta, R. & Meredith-Brown, M. (1979) Large interindividual variations in metabolism of benzo[*a*]pyrene by peripheral lung tissue from lung cancer patients. *Int. J. Cancer*, *24*, 129-133

Cohen, G.M., Grafstrom, R.C., Gibby, E.M., Smith, L., Autrup, H. & Harris, C.C. (1983) Metabolism of benzo[*a*]pyrene and 1-naphthol in cultured human tumorous and nontumorous colon. *Cancer Res.*, *43*, 1312-1315

Daniel, F.B., Stoner, G.D. & Schut, H.A.J. (1984) *Interindividual variation in the DNA binding of chemical genotoxins following metabolism by human bladder and bronchus explants*. In: deSerres, F.J. & Pero, R., eds, *Individual Susceptibility to Genotoxic Agents in Human Population*, New York, Plenum Press (in press)

Daudel, P., Duquesne, M., Vigny, P., Grover, P.L. & Sims, P. (1975) Fluorescence spectral evidence that benzo[*a*]pyrene-DNA products in mouse skin arise from diol-epoxides. *FEBS Lett.*, *57*, 250-253

Eggset, G., Volden, G. & Krokan, H. (1983) U.V.-induced DNA damage and its repair in human skin *in vivo* studied by sensitive immunohistochemical methods. *Carcinogenesis*, *4*, 745-750

Ekstrom, G., von Bahr, C., Glaumann, H. & Ingelman-Sundberg, M. (1982) Interindividual variation in benzo[*a*]pyrene metabolism and composition of isoenzymes of cytochrome P-450 as revealed by SDS-gel electrophoresis of human liver microsomal fractions. *Acta pharmacol. toxicol.*, *50*, 251-260

Glatt, H.R., Lorenz, J., Fleischmann, R., Remmer, H., Ohnhans, E.E., Kaltenbach, E., Tegtmeyer, F., Rudiger, H. & Oesch, F. (1980) *Interindividual variations of epoxide hydratase activity in human liver and lung biopsies, lymphocytes and fibroblast cultures*. In: Coon, M.J., Conney, A.J., Estabrook, R.W., Gelboin, H.V., Gillette, J.R. & O'Brien, P.J., eds, *Microsomes, Drug Oxidations and Chemical Carcinogenesis*, Vol. 2, New York, Academic Press, pp. 651-654

Glatt, H.R., Mertes, I., Wolfel, T. & Oesch, F. (1984) *Epoxide hydrolases in laboratory animals and man*. In: *Biochemical Basis of Chemical Carcinogenesis*, New York, Raven Press, pp. 107-121

Gupta, R.C., Reddy, M.V. & Randerath, K. (1982) $^{32}$P-postlabelling analysis of non-radioactive aromatic carcinogen-DNA adducts. *Carcinogenesis*, *3*, 1081-1092

Gurtoo, H.L., Williams, C.J., Gottlieb, K., Mulhern, A.I., Caballes, L., Vaught, J.B., Marinello, A.J. & Bansal, S.K. (1983) Population distribution of placental benzo[*a*]pyrene metabolism in smokers. *Int. J. Cancer*, *31*, 29-37

Harris, C.C. (1976) Chemical carcinogenesis and experimental models using human tissues. *Beitr. Path. Bd.*, *158*, 389-404

Harris, C.C. (1983) *Concluding remarks: Role of carcinogens, cocarcinogens, and host factors in cancer risk*. In: Harris, C.C. & Autrup, H., eds, *Human Carcinogenesis*, New York, Academic Press, pp. 941-970

Harris, C.C. & Autrup, H. (1980) Interspecies, interindividual and intertissue variations in benzo[*a*]pyrene metabolism. *VDI-Ber.*, *358*, 293-300

Harris, C.C., Autrup, H., Stoner, G.D., McDowell, E.M., Trump, B.F. & Schafer, P. (1977a) Metabolism of dimethylnitrosamine and 1,2-dimethylhydrazine in cultured human bronchi. *Cancer Res.*, *37*, 2309-2311

Harris, C.C., Autrup, H., Stoner, G., Yang, S.K., Leutz, J.C., Gelboin, H.V., Selkirk, J.K., Connor, R.J., Barrett, L.A., Jones, R.T., McDowell, E. & Trump, B.F. (1977b) Metabolism of benzo[*a*]pyrene and 7,12-dimethylbenz[*a*]anthracene in cultured human bronchus and pancreatic duct. *Cancer Res.*, *37*, 3349-3355

Harris, C.C., Autrup, H., Stoner, G.D., Trump, B.F., Hillman, E., Schafer, P.W. & Jeffrey, A.M. (1979a) Metabolism of benzo[*a*]pyrene, *N*-nitrosodimethylamine, and *N*-nitrosopyrrolidine and identification of the major carcinogen-DNA adducts formed in cultured human esophagus. *Cancer Res.*, *39*, 4401-4406

Harris, C.C., Yolken, R.H., Krokan, H. & Hsu, I.C. (1979b) Ultrasensitive enzymatic radioimmunoassay: application to detection of cholera toxin and rotavirus. *Proc. natl Acad. Sci. USA*, *76*, 5336-5339

Harris, C.C., Autrup, H., Trump, B.F., McDowell, E.M, Apostolides, A. & Schafer, P. (1980a) Benzo[*a*]pyrene metabolism in cultured bronchi from patients with or without lung cancer. (Abstract). *Proc. Am. Assoc. Cancer Res.*, *21*, 117

Harris, C.C., Mulvihill, J.J., Thorgeirsson, S.S. & Minna, J.D. (1980b) Individual differences in cancer susceptibility. *Ann. int. Med.*, *92*, 809-825

Harris, C.C., Trump, B.F., Grafstrom, R. & Autrup, H. (1982a) Differences in metabolism of chemical carcinogens in cultured human epithelial tissues and cells. *J. Cell. Biochem.*, *18*, 285-294

Harris, C.C., Trump, B.F., Autrup, H., Hsu, I.C., Haugen, A. & Lechner, J. (1982b) *Studies of host factors in carcinogenesis using cultured human tissues and cells*. In: Armstrong, B. & Bartsch, H., eds, *Host Factors in Human Carcinogenesis* (*IARC Scientific Publications No. 39*), Lyon, International Agency for Research on Cancer, pp. 497-514

Harris, C.C., Yolken, R.H. & Hsu, I.C. (1982c) *Enzyme immunoassays: applications in cancer research*. In: Busch, H. & Yeoman, L.C., eds, *Methods in Cancer Research*, New York, Academic Press, pp. 213-243

Harris, C.C., Grafstrom, R.C., Lechner, J.F. & Autrup, H. (1982d) *Metabolism of N-nitrosamines and repair of DNA damage in cultured human tissues and cells*. In: Magee, P.N., ed., *Nitrosamines and Human Cancer* (*Banbury Report 12*), Cold Spring Harbor, NY, Cold Spring Harbor Laboratory, pp. 121-139

Hart, P., Farrell, G.C., Cooksley, W.G.E. & Powell, L.W. (1976) Enhanced drug metabolism in cigarette smokers. *Br. med. J.*, *ii*, 147-149

Hetzel, M.R., Law, M., Keal, E.E., Sloan, T.P., Idle, J.R. & Smith, R.L. (1982) *Inborn susceptibility/resistance to lung cancer*. In: Cumming, G. & Bonsignore, G., eds, *Cellular Biology of the Lung*, New York, Plenum Press, pp. 448-457

Heyting, C., van der Laken, C.J., van Raamsdonk, W. & Pool, C.W. (1983) Immunohistochemical detection of $O^6$-ethyldeoxyguanosine in the rat brain after in-vivo applications of *N*-ethyl-*N*-nitrosourea. *Cancer Res.*, *43*, 2935-2941

Idle, J.R. & Ritchie, J.C. (1983) *Probing genetically variable carcinogen metabolism using drugs*. In: Harris, C.C. & Autrup, H., eds, *Human Carcinogenesis*, New York, Academic Press, pp. 857-881

Kahng, M.W., Smith, M.W. & Trump, B.F. (1981) Aryl hydrocarbon hydroxylase in human bronchial epithelium and blood monocyte. *J. natl Cancer Inst.*, *66*, 227-232

Kapitulnik, J., Lewin, W., Poppers, P.J., Tomaszewski, J.E., Jerina, D.M. & Conney, A.H. (1976) Comparison of the hydroxylation of zoxazolamine and benzo[*a*]pyrene in human placenta. Effect of cigarette smoking. *Clin. Pharmacol. Ther.*, *20*, 557-564

Kapitulnik, J., Poppers, P.J. & Conney, A.H. (1977) Comparative metabolism of benzo[*a*]pyrene and drugs in human liver. *Clin. Pharmacol. Ther.*, *21*, 166-176

Kellermann, G., Shaw, C.R. & Luyten-Kellermann, M. (1973) Aryl hydrocarbon hydroxylase inducibility and bronchogenic carcinoma. *New Engl. J. Med.*, *289*, 934-937

Kellermann, G., Luyten-Kellermann, M., Jett, J.R., Moses, H.L. & Fontana, R.S. (1978) Aryl hydrocarbon hydroxylase in man and lung cancer. *Human Genet.*, *Suppl. 1*, 161-168

Kouri, R.E., McKinney, C.E., Slomiany, D.J., Snodgrass, D.R., Wray, N.P. & McLemore, T.L. (1982) Positive correlation between high aryl hydrocarbon hydroxylase activity and primary lung cancer as analyzed in cryopreserved lymphocytes. *Cancer Res.*, *42*, 5030-5037

Lambert, B., Ringborn, U. & Skoog, L. (1979) Age-related decrease of ultraviolet light-induced DNA repair synthesis in human peripheral leukocytes. *Cancer Res.*, *39*, 2792-2795

Lower, G.M., Jr, Nilsson, T., Nelson, C.E., Wolf, H., Gamsky, T.E. & Bryan, G.T. (1979) *N*-Acetyltransferase phenotype and risk in urinary bladder cancer: approaches in molecular epidemiology. Preliminary results in Sweden and Denmark. *Environ. Health Perspect.*, *29*, 71-79

Lu, A.Y.H. (1979) Multiplicity of liver drug metabolizing enzymes. *Drug Metab. Rev.*, *10*, 187-208

Mass, M.J., Rodgers, N.T. & Kaufman, D.G. (1981) Benzo[*a*]pyrene metabolism in organ cultures of human endometrium. *Chem.-biol. Interact.*, *33*, 195-205

McManus, M.E., Boobis, A.R., Pacifici, G.M., Frempong, R.Y., Brodie, M.J., Kahn, G.C., Whyte, C. & Davies, D.S. (1980) Xenobiotic metabolism in the human lung. *Life Sci.*, *26*, 481-487

McQueen, C.A., Maslansky, C.J. & Williams, G.M. (1983) Role of the acetylation polymorphism in determining susceptibility of cultured rabbit hepatocytes to DNA damage by aromatic amines. *Cancer Res.*, *43*, 3120-3123

Miller, E.C. & Miller, J.A. (1981) Mechanisms of chemical carcinogenesis. *Cancer*, *47*, 1055-1064

Motulsky, A.G. (1978) Multifactorial inheritance and heritability in pharmacogenetics. *Human Genet.*, *Suppl. 1*, 7-11

Müller, R. & Rajewsky, M.F. (1981) Antibodies specific for DNA components structurally modified by chemical carcinogens. *J. Cancer Res. clin. Oncol.*, *102*, 99-113

Myrnes, B., Giercksky, K.-E. & Krokan, H. (1983) Interindividual variation in the activity of $O^6$-methyl guanine-DNA methyltransferase and uracil-DNA glycosylase in human organs. *Carcinogenesis*, *4*, 1565-1568

Nebert, D.W. & Atlas, S.A. (1978) The *Ah* locus: aromatic hydrocarbon responsiveness ... of mice and men. *Human Genet.*, *Suppl. 1*, 149-160

Oesch, F., Schmassmann, H. & Bentley, P. (1978) Specificity of human rat and mouse skin epoxide hydratase towards K-region epoxides of polycyclic hydrocarbons. *Biochem. Pharmacol.*, *27*, 17-20

Paigen, B., Gurtoo, H.L., Minowada, J., Houten, L., Vincent, R., Paigen, K., Parker, N.B., Ward, E. & Thompson-Hayner, N. (1977) Questionable relation of aryl hydrocarbon hydroxylase to lung cancer risk. *New Engl. J. Med.*, *297*, 346-350

Paigen, B., Ward, E., Steenland, K., Houten, L., Gurtoo, H.L. & Minowada, J. (1978) Aryl hydrocarbon hydroxylase in cultured lymphocytes of twins. *Am. J. human Genet.*, *30*, 561-571

Pasanen, M. & Pelkonen, O. (1981) Solubilization and partial purification of human placental cytochromes P-450. *Biochem. biophys. Res. Commun.*, *103*, 1310-1317

Pelkonen, O. & Nebert, D.W. (1982) Metabolism of polycyclic aromatic hydrocarbons: etiologic role in carcinogenesis. *Pharmacol. Rev.*, *34*, 189-222

Pelkonen, O. & Vahakangas, K. (1980) Metabolic activation and inactivation of chemical carcinogens. *J. Toxicol. environ. Health*, *6*, 989-999

Pelkonen, O., Vahakangas, K. & Nebert, D.W. (1980) Binding of polycyclic aromatic hydrocarbons to DNA: comparison with mutagenesis and tumorigenesis. *J. Toxicol. environ. Health*, *6*, 1009-1020

Pelkonen, O., Karki, N.T. & Tuimala, R. (1981) A relationship between cord blood and maternal blood lymphocytes and term placenta in the induction of aryl hydrocarbon hydroxylase activity. *Cancer Lett.*, *13*, 103-110

Perera, F.P. & Weinstein, I.B. (1982) Molecular epidemiology and carcinogen-DNA adduct detection: new approaches to studies of human cancer causation. *J. chronic Dis.*, *35*, 581-600

Perera, F.P., Poirier, M.C., Yuspa, S.H., Nakayama, J., Jaretzki, A., Curnen, M.M., Knowles, D.M. & Weinstein, I.B. (1982) A pilot project in molecular cancer epidemiology: determination of benzo(*a*)pyrene-DNA adducts in animal and human tissues by immunoassays. *Carcinogenesis*, *3*, 1405-1410

Pero, R.W., Miller, D.G., Lipkin, M., Markowitz, M., Gupta, S., Winawer, S.J., Enker, W. & Good, R. (1983) Reduced capacity for DNA repair synthesis in patients with or genetically prediscposed to colorectal cancer. *J. natl Cancer Inst.*, *70*, 867-875

Poirier, M.C. (1981) Antibodies to carcinogen-DNA adducts. *J. natl Cancer Inst.*, *67*, 515-519

Poirier, M.C., Santella, R.M., Weinstein, I.B., Grunberger, D. & Yuspa, S.H. (1980) Quantitation of benzo(*a*)pyrene-deoxyguanosine adducts by radioimmunoassay. *Cancer Res.*, *40*, 412-416

Poirier, M.D., Lippard, S.L., Zwelling, L.A., Ushay, H.M., Kerrigan, D., Thill, C.C., Santella, R.M., Grunberger, D. & Yuspa, S.H. (1982) Antibodies elicited against *cis*-diamminedichloroplatinum (II)-modified DNA are specific for *cis*-diamminedichloroplatinum (II)-DNA adducts formed *in vitro* and *in vivo*. *Proc. natl Acad. Sci. USA*, *79*, 6443-6447

Prough, R.A., Sipal, Z. & Jakobsson, S.W. (1977) Metabolism of benzo(*a*)pyrene by human lung microsomal fractions. *Life Sci.*, *21*, 1629-1636

Rahn, R.O., Chang, S.S., Holland, J.M., Stephens, T.J. & Smith, L.H. (1980) Binding of benzo(*a*)pyrene to epidermal DNA and RNA as detected by synchronous luminescence spectrometry at 77 K. *J. biochem. biophys. Meth.*, *3*, 285-291

Randerath, K., Reddy, M.V. & Gupta, R.C. (1981) $^{32}$P-labelling test for DNA damage. *Proc. natl Acad. Sci. USA*, *78*, 6126-6129

Reidenberg, M.M. & Drayer, D.E. (1978) Aromatic amines and hydrazines, drug acetylation and lupus erythematodes. *Human Genet.*, *Suppl. 1*, 57-63

Rudiger, H.W., Heisig, V. & Hain, E. (1980) Enhanced benzo[*a*]pyrene metabolism and formation of DNA adducts in monocytes of patients with lung cancer. *J. Cancer Res. clin. Oncol.*, *96*, 295-302

Sabadie, N., Richter-Reichhelm, H.B., Saracci, R., Mohr, U. & Bartsch, H. (1981) Interindividual differences in oxidative benzo[*a*]pyrene metabolism by normal and tumorous surgical lung specimens from 105 lung cancer patients. *Int. J. Cancer*, *27*, 417-425

Setlow, R.B. (1983) *Variation in DNA repair among humans*. In: Harris, C.C. & Autrup, H., eds, *Human Carcinogenesis*, New York, Academic Press, pp. 231-254

Shamsuddin, A.K.M., Sinopoli, N.T., Vahakangas, K., Hemminki, K., Boesch, R.R. & Harris, C.C. (1983) Identification of benzo[*a*]pyrene diol epoxide-DNA antigenicity in humans. (Abstract). *Fed. Proc.*, *42*, 1042

Shuster, S., Rawlins, M.D., Chapman, P.H. & Rogers, S. (1980) Decreased epidermal aryl hydrocarbon hydroxylase and localized pustular psoriasis. *Br. J. Dermatol.*, *103*, 23-26

Stoner, G.D., Harris, C.C., Autrup, H., Trump, B.F., Kingsbury, E.W. & Myers, G.A. (1978) Explant culture of human peripheral lung. I. Metabolism of benzo[*a*]pyrene. *Lab. Invest.*, *38*, 685-692

Stoner, G.D., Daniel, F.B., Schenck, K.M., Schut, H.A.J., Sandwisch, D.W. & Gohara, A.F. (1982) DNA binding and adduct formation of aflatoxin $B_1$ in cultured human and animal tracheobronchial and bladder tissues. *Carcinogenesis*, *3*, 1345-1348

Vahakangas, K., Pelkonen, O. & Sotaniemi, E. (1983) Cigarette smoking and drug metabolism. *Clin. Pharmacol. Ther.*, *33*, 375-380

Vaught, J.B., Gurtoo, H.L., Parker, N.B., LeBoeuf, R. & Doctor, G. (1979) Effects of smoking on benzo[*a*]pyrene metabolism by human placental microsomes. *Cancer Res.*, *39*, 3177-3183

Yamasaki, H., Huberman, E. & Sachs, L. (1977) Metabolism of the carcinogenic hydrocarbon benzo[*a*]pyrene in human fibroblast and epithelial cells. II. Differences in metabolism to water-soluble products and aryl hydrocarbon hydroxylase activity. *Int. J. Cancer*, *19*, 378-382

# GENETIC SUSCEPTIBILITY TO TOXIC SUBSTANCES AND ITS RELATIONSHIP TO CARCINOGENESIS

**J.Z. Hanke**

*Institute of Occupational Medicine, Lódź, Poland*

## SUMMARY

Although environment is a major factor that determines the incidence of cancer, in some cases heredity plays an important role. Some hereditary conditions cause a particular predisposition that depends upon host activation of chemical carcinogens. The metabolism of many carcinogens is related genetically to the *Ah* locus and, hypothetically, this gene can control their activation. Some inbred strains of mice and, probably, also some humans seem to possess a natural ability to metabolize chemical carcinogens, and therefore are more susceptible to cancer. The debrisoquine metabolic system, which is defective in 6-8% of the Caucasian population, is probably closely related to the *Ah* locus, and some relationship to cancer has been reported. *N*-Acetyl-transferase is known to be involved in acetylation, also, and thus in deactivation of the arylamines that are potent bladder carcinogens. Since the enzyme exhibits a distinct polymorphism, slow acetylation affects the susceptibility to cancer.

There are suggestions, also, that the genetically determined phenotype of glucose-6-phosphate dehydrogenase and $\alpha_1$-antitrypsin plays a role in individual susceptibility to cancer.

## DISCUSSION

Boveri (1914) was the first to suggest that cancer might be due to mutation in somatic cells. He based his idea on reports of abnormal mitoses in cancer cells and noted that cancer was associated with stimulated proliferation due to ageing, after exposure to certain environmental agents, and under certain hereditary conditions.

With respect to causation, human cancers may be divided, indeed, into four classes, according to the presence or absence of identifiable heritable (H) or environmental (E) factors: i.e., $H^-E^+$, $H^+E^-$, $H^+E^+$, $E^-E^-$. Any hypothesis about the incidence of cancer must take these classes into account, as well as the influence of age. Some hereditary conditions predispose to cancer due to a particular susceptibility to environmental agents. Two recessively inherited conditions in humans, xeroderma pigmentosum and atazia telangiectasia, present a predisposition to cancer associated with defective repair of DNA damage produced by radiation - in the former case ultra-violet, in the latter, ionizing radiation (Cleaver, 1968; Paterson *et al.*, 1976). Another condition that may operate in this fashion is Fanconi's anaemia.

The possibility that humans who are heterozygous for ataxia telangiectasia and Fanconi's anaemia may also be predisposed to cancer could assume considerable quantitative importance, since these heterozygotes comprise approximately 1% of the population (Swift, 1976). An intriguing kind of predisposition to cancer is the one that depends upon host activation of chemical carcinogens. It is clear, from carcinogenesis studies, that some chemicals are inactive in their natural state but are activated by enzymes present in animal tissues. Evidence is growing that metabolism to reactive intermediates, by cytochrome P-450-mediated monooxygenase, is a prerequisite for mutagenesis and carcinogenesis by many, if not all, polycyclic hydrocarbons (Heidelberger, 1975; Sims & Grover, 1974). These reactive intermediates probably bind covalently to numerous cellular macromolecules.

Nebert and Gielen (1972) presented evidence for a single gene difference between B6 and D2 inbred mouse strains in the induction of monooxygenase activity, aryl hydrocarbon hydroxylase (AHH) and cytochrome $P_1$-450 by treatment with methylcholanthrene (MC). In following years, using AHH activity to indicate the phenotype at the Ah locus, several laboratories found that about half, or slightly more than half, of all inbred mouse strains examined are responsive, while the remaining ones are nonresponsive (Kouri, 1976). Induction of AHH activity and cytochrome $P_1$-450 by MC is expressed almost exclusively as an autosomal dominant trait among offspring of the appropriate crosses of B6 and D2 inbred strains. Fibrosarcomas initiated by subcutaneously administered MC are associated with genetically mediated aromatic-hydrocarbon responsiveness among 14 inbred strains of mice, and the 'carcinogenic trait index' is over 42 in all responsive phenotype groups (Nebert *et al.*, 1978).

The Ah locus is highly complex; the simplest model that accommodates all existing data for the mouse comprises at least six alleles and two regulatory loci. It is now known (Haugen *et al.*, 1976) that the induction process involves, principally, de-novo protein synthesis rather than activation of preexisting moieties. The process by which the gene products of these Ah loci are regulatory is unknown, but it is possible that one of the gene products is the cytosol receptor, which is able to bind specific aromatic-hydrocarbon inducers (Nebert *et al.*, 1975).

It appears that mutation in nonresponsive mice involves a defective cytosol protein receptor, and that these mice do have the structural genes necessary for induction of AHH activity and its associated cytochromes by the very potent polycyclic aromatic inducers. It has become apparent in the past decade that different forms of P-450 generate different ratios of metabolites from the same substrate. Such differences in the metabolic profile of polycyclic hydrocarbons or other foreign compounds suggest, also, that differences may exist in the nature of the intermediates formed. Therefore, differences in the reactivity of these intermediates or products might result in marked dissimilarities in the toxicity or carcinogenicity of a given compound (Thorgeirsson & Nebert, 1977).

Hence, a responsive, but not a nonresponsive, polycyclic-hydrocarbon-treated mouse is subject to both quantitative and qualitative increases in the steady-state levels of certain reactive intermediates, due to both an increase in cytochrome $P_1$-450 content and an increased $P_1$-450:P-450 ratio in numerous tissues (Nebert *et al.*, 1977). The mouse genetic model of cytochrome $P_1$-450 inducibility has been extremely useful in elucidating the function of the mammalian monooxygenase system, especially with regard to the mechanisms of drug toxicity and chemical carcinogenesis.

Kellermann *et al.* (1973a,b) gave clinical geneticists and oncologists initial hope in modelling the process. The extent of AHH induction by MC in cultured mitogen-activated lymphocytes was examined in 353 healthy subjects. The distribution of inducibilities was trimodal, the groups being designated as 'low', 'intermediate' and 'high'. The data were consistent with the hypothesis of two alleles at a single locus.

Because of day-to-day variability in the lymphocyte-AHH assay, several laboratories found the assay and the trimodal distribution difficult to reproduce. In order to discover whether variation in the basal or the induced AHH activity ratio is genetically determined, Paigen *et al.* (1978) measured AHH in 48 pairs of twins. Both members of the twin pair were measured on the same day. Dizygotic twin pairs had higher variance than monozygotic twins, indicating that basal and induced AHH, as well as inducibility ratios, are all, to a large extent, genetically determined. These results confirm the earlier report of Kellermann *et al.* (1973a). Guirgis *et al.* (1976) claimed a correlation between high AHH inducibility and lung cancer. Laryngeal carcinoma, also, was reported to be associated with the high-inducibility phenotype. Trell *et al.* (1976) suggested that humans with intermediate and high AHH inducibility are at far greater risk of cancer than those with low AHH inducibility. Paigen *et al.* (1977) undertook a detailed reexamination of the role of AHH in relation to cancer risk. They tested lung-cancer patients to see whether the distribution of AHH inducibility was shifted toward the high end of the range, and found that approximately half of the 50 lung-cancer patients tested had low levels of AHH activity, much lower than was observed in the normal population.

Moreover, the enzyme level in a diseased subject may be altered by the disease, rather than by a condition preceding the disease state. Therefore, Paigen *et al.* (1977) decided to estimate the genetic state of AHH inducibility in lung-cancer patients indirectly, by measuring their first-degree relatives. The progeny should exhibit an AHH activity and inducibility halfway between that of the patient and that of the normal population, represented by the normal parent. A total of 57 couples was measured, and the distribution of AHH inducibility was the same for both the control population and the lung-cancer progeny. A chi-square analysis of the groups: low, intermediate and high, indicated no difference between the progeny and the control population. This finding disputes the highly significant difference in the distribution of AHH inducibility predicted by Kellermann *et al.* (1973a). Thus, Paigen *et al.* concluded that if lung-cancer patients do have a distribution of AHH inducibility that is different from that of the normal population, as reported, this difference is not a genetic one that is passed to their offspring. There is no simple explanation.

In trying to find a definitive answer to the question of whether the AHH genotype can ever be used to predict increased susceptibility to certain types of environmentally caused cancer, the original conclusions must be reiterated: that with respect to causation, human cancer may be divided into four classes according to the presence or absence of identifiable hereditary ($H^+$) or environmental ($E^+$) factors. In only one class, ($E^+H^+$), do both these factors play an important role.

The debrisoquine metabolic system (Idle & Smith, 1979) is probably closely related to the AHH system. Alicyclic oxidation of this antihypertensive drug is defective in about 6-8% of the Caucasian population, and the defect is a recessive trait that gives rise to a bimodal distribution of oxidative capacity. Since the defective hydroxylation is associated with a heightened sensitivity to the drug, adverse effects follow the normal therapeutic dose in deficient individuals. The deficiency in these individuals is not restricted to alicyclic oxidation; aromatic hydroxylation of quanosan and *ortho*-de-ethylation of phenacetin are also defective. The underlying biochemical basis for the polymorphism is not clear but may well reflect the relative amounts of the various cytochromes P-450 in the liver, or perhaps even some structural protein in the cytochrome matrix. From the occupational and environmental point of view, it is important to consider what other chemical conversions might be tied to the debrisoquine metabolic polymorphism. It has already been reported that Nigerians with liver cancer induced by aflatoxin exposure include a lower-than-control frequency of cases with poor metabolism of debrisoquine. These findings fit the hypothesis that poor metabolizers would be less capable of metabolic activation of the promutagen or precarcinogen, aflatoxin (Idle *et al.*, 1981).

Quite a different problem is posed by genetically determined differences in acetylation. Acetylation is a controlling factor in the metabolic rate of several drugs, since the acetylated drug is more easily excreted by the kidney than the free drug. In Western populations, about half the individuals are rapid acetylators and half are slow acetylators. Individuals can be reliably classified by comparing serum levels after a standard dose. For example, after a single dose of isoniazid, plasma concentrations have a bimodal distribution that contrasts markedly with the unimodal distribution shown after the administration of salicylate, which is not influenced by metabolic polymorphism. *N*-Acetyltransferase is known to be involved in acetylation, and thus deactivation, of arylamines that are potent bladder carcinogens. These compounds are among the most certain occupational carcinogens, e.g., β-naphthylamine, benzidine, 4-aminobiphenyl and 4-nitrobiphenyl, all of which have been used as industrial antioxidants. Lower *et al.* (1979) carried out an epidemiological study in Scandinavia to determine whether a difference in acetylating activity might contribute significantly to the risk of bladder cancer. Bladder cancers, of course, do not all represent occupational exposure, nor are all, or even most, necessarily due to arylamines. Thus, any effect of *N*-acetyltransferase is diluted by unrelated cases. Nevertheless, the study suggests that the urban Swedish population, presumably susceptible to occupational bladder cancer, has an excess of slow over fast acetylators, while the rural Swedish population of bladder-cancer patients is indistinguishable from controls.

However, the acetylation problem must be evaluated with great care. It is necessary to remember that some foreign compounds are not acetylated polymorphically, e.g., sulfanilamide, *para*-aminobenzoic acid and *para*-aminosalicylic acid, suggesting that other acetyltransferases may be present *in vivo* and involved in acetylation of these substrates. Some substances are acetylated by different acetyltransferases on different metabolic steps, as illustrated by procainamide metabolism. The parent compound is acetylated polymorphically, but the hydrolysis product, *para*-aminobenzoic acid, is acetylated monomorphically in the same individuals (Timbrell, 1982). Another example of genetically determined polymorphism is given by the differentiation of glucose-6-phosphate dehydrogenase (G-6-P D) activity in the general population and its influence on the cancer risk. At least 80 distinct, variant forms of G-6-P D have been identified (Motulsky *et al.*, 1971), some bound with low enzyme activity. Males are more likely to show enzymatic deficiency, because the gene determining the characteristic of G-6-P D is carried on the X chromosome. The X chromosome of males may be 'normal' or defective and the two male genotypes, 'reactor' or 'normal', are expected. Females may be classified in three groups, 'normal', 'intermediate' or 'reactor', depending on the presence of two normal X chromosomes, one normal and one defective, and two defective chromosomes, respectively. Most heterozygous females, however, show a measurable deficiency in the enzyme, although a few heterozygous females show quite a marked deficiency, as do homozygous men. An explanation for the variable expression in heterozygous women is found in the phenomenon of random X-chromosome deactivation.

Beaconsfield *et al.* (1965) postulated a correlation between the deficits of G-6-P D and cancer. The hypothesis was based on results gathered among Occidental Jews and was later supported by Naik and Anderson (1970) in Texas. Among American Negroes, a G-6-P D deficit was found in 7-17% of the population. Among the cancer patients, the deficit was found in 4.6% of men and 1.7% of women - only half the expected values. The difference was statistically significant, and cancer was more frequent among Negro women than men. Similar results were obtained by Sulis (1972) in Sardinia, where the G-6-P D deficit is very frequent (25-35%); among cancer patients, it is only 13%. The authors suggest that a block of glucose metabolism on the pentose shunt (where G-6-P D is the key enzyme) discriminates against the production of ribose and, as a result, against the synthesis of ribonucleic acid, the fundamental material of cell proliferation.

$\alpha_1$-Antitrypsin is a circulating protein comprising 90% of the $\alpha_1$-globulin fraction of plasma. It is now recognized that serum concentration of $\alpha_1$-antitrypsin is controlled by a pair of autosomal, codominant genes. As was first reported by Laurell and Eriksson (1963), individuals with severe $\alpha_1$-antitrypsin deficiency are predisposed to obstructive lung disease. These patients have the so-called protease inhibitor, (Pi) ZZ phenotype. Intermediate $\alpha_1$-antitrypsin deficiency is present in individuals heterozygous for Z and the normal M-gene (PiMZ phenotype), and also in individuals with other homozygous and heterozygous combinations of some of the known Pi alleles, more than ten in number. It is well documented, also (Berg & Eriksson, 1972; Palmer *et al.*, 1973), that the ZZ homozygote state is associated with cryptogenic hepatic fibrosis and cirrhosis in adults. Berg and Eriksson noted two hepatocellular carcinomas among 13 homozygous patients. The Z allele is associated with the presence of periodic acid-Schiff-positive, diastase-resistant, intracytoplasmic granules in the hepatic parenchyma (Gordon *et al.*, 1972). Reintoft and Hägerstrand (1979) found that of 56 patients with primary liver carcinoma, 10 showed nontumorous hepatocytes that contained diastase-resistant, periodic acid-Schiff-positive and $\alpha_1$-antitrypsin-positive globules. This rate of 18% among patients with liver carcinoma was compared with 6% in an unselected autopsy series.

It is interesting to note that the cancer-related urinary glycoprotein, EDC1, inhibits the section of trypsin and chymotrypsin on casein (Chawla *et al.*, 1978). Many tumours possess more proteolytic activity than their benign counterparts. Proteolytic activity associated with neoplasms may be responsible for the destruction of tissues surrounding tumours and may thereby contribute to invasive growth (Sylvén & Bois-Svensson, 1965). The problem of the relationship between $\alpha_1$-antitrypsin deficiency and cancer seems interesting but needs further investigation.

## REFERENCES


Beaconsfield, P., Rainsbury, R. & Kalton, G. (1965) Glucose-6-phosphate dehydrogenase deficiency and the incidence of cancer. *Oncologia*, *19*, 11-19

Berg, N.O. & Eriksson, S. (1972) Liver disease in adults with alpha$_1$-antitrypsin deficiency. *New Engl. J. Med.*, *287*, 1264-1267

Boveri, T.H. (1914) *Zur Frage der Entstehung maligner Tumoren*, Jena, Gustav Fischer

Chawla, R.K., Waldsworth, A.D. & Rudman, D. (1978) Antitryptic property of cancer-related glycoprotein EDC. *Cancer Res.*, *38*, 452-457

Cleaver, J.E. (1968) Defective repair replication of DNA in xeroderma pigmentosum. *Nature*, *218*, 652-656

Gordon, H.W., Dixon, J., Rogers, T.C. (1972) Alpha$_1$-antitrypsin (A1AT) accumulation in livers of emphysematous patients with A1AT deficiency. *Human Pathol.*, *3*, 361-370

Guirgis, H.A., Lynch, M.T., Mate, T., Harris, R.E., Wells, I., Caha, L., Anderson, J., Maloney, K. & Rankin, L. (1976) Aryl-hydrocarbon hydroxylase activity in lymphocytes from lung cancer patients and normal controls. *Oncology*, *33*, 105-109

Haugen, D.A., Coon, M.J. & Nebert, D.W. (1976) Induction of multiple forms of mouse liver cytochrome P-450. Evidence for genetically controlled *de novo* protein synthesis in response to treatment with beta-naphthyl flavone or phenobarbital. *J. biol. Chem.*, *251*, 1817-1827

Heidelberger, C. (1975) Chemical carcinogenesis. *Ann. Rev. Biochem.*, *44*, 79-121

Idle, J.R., Mahgoub, A., Sloan, T.P., Smith, R.L., Mbanefo, C.O. & Bababunmi, E.A. (1981) Some observations on the oxidation phenotype status of Nigerian patients presenting with cancer. *Cancer Lett.*, *11*, 331-338

Idle, J.R. & Smith, R.L. (1979) Polymorphism of oxidation at carbon centers of drugs and their clinical significance. *Drug Metab. Rev.*, *9*, 301-317

Kellermann, G., Luyten-Kellermann, M. & Shaw, C.R. (1973a) Genetic variation of aryl hydrocarbon hydroxylase in human lymphocytes. *Am. J. human Genet.*, *25*, 327-331

Kellermann, G., Shaw, C.R. & Luyten-Kellermann, M. (1973b) Aryl hydrocarbon hydroxylase inducibility and bronchogenic carcinoma. *New Engl. J. Med.*, *289*, 934-937

Kouri, R.E. (1976) *Relationship between levels of aryl hydrocarbon hydroxylase activity and susceptibility to 3-methylcholanthrene and benzo[a]pyrene-induced cancers in inbred strains of mice*. In: Freudenthal, R.E. & Jones, P.W., eds, *Carcinogenesis - A Comprehensive Survey*, Vol. 1, *Polynuclear Aromatic Hydrocarbons. Chemistry, Metabolism and Carcinogenesis*, New York, Raven Press, pp. 139-151

Laurell, C.B. & Eriksson, S. (1963) The electrophoretic $\alpha_1$-globulin pattern of serum in $\alpha$1-antitrypsin deficiency. *Scand. J. clin. Lab. Invest.*, *15*, 132-140

Lower, G.M., Jr, Nilsson, T., Nelson, C.E., Wolf, H., Gamsky, T.E. & Bryan, G.T. (1979) *N*-Acetyltransferase phenotype and risk in urinary bladder cancer: approaches in molecular epidemiology. Preliminary results in Sweden and Denmark. *Environ. Health Perspect.*, *29*, 71-79

Motulsky, A.G., Yoshida, A. & Stamatoyannopoulos, G. (1971) Variants of glucose-6-phosphate dehydrogenase. *Ann. N.Y. Acad. Sci.*, *179*, 636-643

Naik, S.N. & Anderson, D.E. (1970) G-6-P D deficiency and cancer. *Lancet*, *i*, 1060-1061

Nebert, D.W., Atlas, S.A., Guenter, T.H. & Kouri, R.E. (1978) *The Ah locus: genetic regulation of the enzymes which metabolize polycyclic hydrocarbons and the risk for cancer*. In: Gelboin, H.V. & Ts'o, P.O.P., eds, *Polycyclic Hydrocarbons and Cancer*, Vol. 2, *Molecular and Cell Biology*, New York, Academic Press, pp. 345-390

Nebert, D.W. & Gielen, J.E. (1972) Genetic regulation of aryl hydrocarbon hydroxylase induction in mouse. *Fed. Am. Soc. exp. Biol.*, *31*, 1315-1325

Nebert, D.W., Levitt, R.C., Jensen, N.M., Lambert, G.H. & Felton, J.S. (1977) Birth defects and aplastic anemia: differences in polycyclic hydrocarbon toxicity associated with Ah locus. *Arch. Toxicol.*, *39*, 109-132

Nebert, D.W., Robinson, J.R., Niwa, A., Kumaki, K. & Poland, A.P. (1975) Genetic expression of aryl hydrocarbon hydroxylase activity in the mouse. *J. cell. Physiol.*, *85*, 393-414

Paigen, B., Gurtoo, H.L., Minowada, J., Ward, E., Houten, L., Paigen, K., Reilly, A. & Vincent, R. (1978) *Genetics of aryl hydrocarbon hydroxylase in the human population and its relationship to lung cancer*. In: Gelboin, H.V. & Ts'o, P.O.P., eds, *Polycyclic Hydrocarbons and Cancer*, Vol. 2, *Molecular and Cell Biology*, New York, Academic Press, pp. 391-406

Paigen, B., Minowada, J., Gurtoo, H.L., Paigen, K., Parker, N.B., Ward, E., Thompson-Hayner, N., Bross, I.D.J., Bock, F. & Vincent, R. (1977) Distribution of aryl hydrocarbon hydroxylase inducibility in cultured human lymphocytes. *Cancer Res.*, *37*, 1829-1837

Palmer, P.E., Wolfe, H.J. & Gherardi, G.J. (1973) Hepatic changes in adult alpha$_1$-antitrypsin deficiency. *Gastroenterology*, *65*, 284-293

Paterson, M.C., Smith, B.P., Lohman, P.H.M., Anderson, A.K. & Fishman, L. (1976) Defective excision repair of $\gamma$-ray-damaged DNA in human (ataxia telangiectasis) fibroblasts. *Nature*, *260*, 444-447

Reintoft, I. & Hägerstrand, I.E. (1979) Does the Z gene variant of alpha$_1$-antitrypsin predispose to hepatic carcinoma? *Human Pathol.*, *10*, 419-424

Sims, P. & Grover, P.L. (1974) Epoxides in polycyclic aromatic hydrocarbon metabolism and carcinogenesis. *Adv. Cancer Res.*, *20*, 165-274

Sulis, E. (1972) G-6-P D deficiency and cancer. *Lancet*, *i*, 1185

Swift, M. (1976) *Cancer and genetics*. In: Bergsma, D., ed., *Birth Defects, The National Foundation of Dimes, Original Article Series,* Vol 12 (1), New York, Allan R. Liss

Sylvén, B. & Bois-Svensson (1965) On the chemical pathology of interstitial fluid. I. Proteolytic activities in transplanted mouse tumours. *Cancer Res.*, *25*, 458-468

Thorgeirsson, S.S. & Nebert, D.W. (1977) Genetic regulation of the metabolism of chemical carcinogens and other foreign compounds. *Adv. Cancer Res.*, *25*, 149-153

Timbrell, J.A. (1982) *Principles of Biochemical Toxicology*, London, Taylor & Francis

Trell, E., Korsgaard, R., Hood, S., Kitzing, P., Norden, G. & Simonssen, B.G. (1976) Aryl hydrocarbon hydroxylase inducibility and laryngeal carcinomas. *Lancet*, *ii*, 140

# COVALENT BINDING OF GENOTOXIC AGENTS TO PROTEINS AND NUCLEIC ACIDS

**L. Ehrenberg**

*Department of Radiation Biology, University of Stockholm, Sweden*

**SUMMARY**

To a large extent, initiators of current cancer incidence are unknown. This is due partly to lack of proper variables and low statistical power of epidemiological studies and to difficulties of risk estimation from experimental data. Considering these facts, as well as long latent times of genotoxic effects, monitoring systems aiming at risk prevention should: (1) respond soon after onset of exposure; (2) have sufficiently high power; (3) identify causative agents; and (4) permit risk quantitation. The determination of in-vivo adducts to DNA (the target in most genotoxic effects) and, especially, to blood proteins, fulfils these criteria, since the demonstration of protein adducts is a relevant measure of formation of the corresponding DNA adducts.

**DNA ADDUCT FORMATION IN GENOTOXICITY**

Following the discovery of Brookes and Lawley (1964) of correlations between carcinogenic and mutagenic potency of alkylating agents on one hand and their ability to bind to DNA on the other, overwhelming proof has been furnished that this is a general biological principle. From their own and other studies of experimental carcinogenesis, Miller and Miller (1977) concluded that most carcinogens/mutagens are electrophilic reagents or, in a majority of cases, give rise to electrophilic reagents through metabolism ('bioactivation') or through chemical change (Eq. 1).

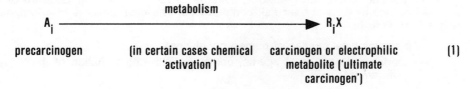

$$A_i \xrightarrow{\text{metabolism}} R_iX \qquad (1)$$

| precarcinogen | (in certain cases chemical 'activation') | carcinogen or electrophilic metabolite ('ultimate carcinogen') |
|---|---|---|

In cells, $R_iX$ react with nucleophilic atoms, $Y_j$, such as oxygen (O) and nitrogen (N) in nucleic acids and O, N and sulphur (S) in proteins.

$$R_iX + Y_j \longrightarrow R_iY_j \text{ adduct (e.g., alkylated product)} \quad (2)$$

Electrophiles in this sense comprise alkylating, acylating and arylating agents, nitrenium compounds, nitrite and other nitrosating agents, carbonyl compounds, metal ions (metal complexes) and free radicals (Ehrenberg & Osterman-Golkar, 1980) (only a few known mutagens do not belong to these groups, viz., non-reactive intercalating agents, base analogues).

The electrophilic reagent, $R_iX$, reacts randomly with all nucleophiles, with probabilities determined by reaction-kinetic rules, e.g., the reactivity of nucleophiles towards alkylating agents decreases in the order (Ross, 1962):

$$S > N > O \quad (3)$$

It is strongly indicated that, in general, dose-response relationships comprise a linear component, e.g.:

$$P(D) = a + bD + cD^2 + \ldots \quad (4)$$

which will predominate at low doses or levels (Ehrenberg *et al.*, 1983; von Bahr *et al.*, 1984). However, incomplete knowledge of repair kinetics at very low levels of DNA adducts leaves this question unresolved in some cases (Ehrenberg *et al.*, 1983).

A consequence of the random reaction pattern is that if only certain sites in DNA are critical in carcinogenesis or mutagenesis, the appearance of an electrophile in the DNA space will lead to chemical change at critical sites ('$DNA_c$'). This, together with the linearity of dose-response curves at low doses, led Ehrenberg and Osterman-Golkar (1980) to suggest that the reverse of the Miller and Miller theory is valid, viz., that any electrophile appearing in the DNA space of a cell is potentially mutagenic or carcinogenic.

Genotoxic effects are induced at very low doses or levels, apparently, at the present state-of-the-art, without any no-effect threshold (Ehrenberg *et al.*, 1983). Therefore, with low levels of potentially electrophilic chemicals in human environments (in a broad sense), the hazards of raised incidence of cancer and heritable damage, possibly also of embryonal damage, will predominate over those other kinds of toxicity that are mostly characterized by dose thresholds below which the risk is zero.

## NEEDS FOR IMPROVED METHODS FOR IDENTIFICATION AND QUANTITATION OF RISKS

A programme to prevent genotoxic hazards depends on the identification of causative agents and quantitation of their risk contributions. The main sources of information in this respect are data from:

- epidemiological studies of human populations,
- experimental studies of agents, combined with determination of their levels in the environment.

Epidemiological data indicate (Higginson & Muir, 1979) that some 70-90%, maybe more, of cancer cases in man are caused by environmental factors (in a broad sense), yet our knowledge of the identity of these factors is still fragmentary. Main factors so far known are tobacco smoking, alcohol and ultra-violet light (in skin cancer), and also small contributions from occupational exposure, including some ionizing radiation. The risk contribution of ionizing radiation to the general population has been estimated by risk coefficients determined from past exposure at relatively high doses. Most cases are referred to 'life-style' factors (Higginson & Muir, 1979), with indications of dietary factors (e.g., high fat intake, deficient intake of certain vitamins and selenium) and of factors pertaining to sexual habits and the reproduction pattern. These factors are supposed to act mainly *via* mechanisms that should be denoted promotion or cocarcinogenesis rather than initiation, the primary expected effect of genotoxic agents. Certain life-style factors can be changed only with great difficulty but, for preventive purposes, it is desirable to identify those environmental initiators that may be removable. However, very little is known in this respect. This situation may be related to the currently very scanty indications of mutagenicity in man.

Deficient knowledge of mechanisms (how important are endogenous processes?) and causative agents behind observed incidences of cancer and heritable diseases is, at least in part, a function of the long latent periods - which render the establishment of cause-effect relationships difficult - and of the low 'resolving power' (statistical power, Ehrenberg, 1977) of epidemiological as well as experimental studies. For these reasons, it is often difficult to detect a response that is less than 50-100% above the background incidence. Certainly, experimental studies may test defined factors at high doses, but a positive response in the laboratory organism must then be translated into risk in man, by extrapolation from the high experimental doses to the environmental dose or level and from the test system to man. There are many further factors that cannot be tested separately at high levels.

In this situation, it is of paramount importance to have 'early warning systems' that could:

(1) overcome the long latent times, and

(2) overcome the low statistical power of epidemiological and experimental studies.

For the purpose of prevention, the systems should further:

(3) identify causative agents.

Point (3) leads to the requirement that such systems should:

(4) permit quantitation of risk at a given dose.

## EARLY WARNING SYSTEMS

The extreme difficulty of the problem suggests that systems of this kind should comprise, concomitantly, all possible sources of information. The way the problem is treated, and the methods preferred, must take account of whether an investigation concerns:

- individuals at risk
- small groups with, e.g., occupational exposure, or
- large populations, such as the people of Finland.

Possible endpoints to be observed may be subdivided into (Bloom, (1981):

(1) reproductive disturbances (which fulfil criterion 1, above);

(2) genetic effects in somatic cells or gametes;

(3) biochemical and chemical endpoints.

This paper will not deal with endpoints (1) or (2). Although some of these endpoints [e.g., micronuclei (Jenssen & Ramel, 1976)] have a high statistical power, i.e., ability to detect a response to a low dose (criterion 2), these systems present the same difficulties that any epidemiological study presents with regard to separating of cause(s) of disease from confounding factors.

Biochemical and, especially, chemical analytical methods (Table 1) possess a statistical power many orders of magnitude higher than that of any biological endpoint. Since observations can be made very soon after the onset of exposure, they fulfil criteria (1) and (2) defined above. Those endpoints that involve structural and quantitative determination of (pre)mutagens or (pre)carcinogens in body fluids and excreta or their adducts to DNA in tissues will be able to identify risk factors (criterion 3) (see opening remarks of this paper). Finally, considering the relationship between degree of chemical change of DNA (dose of DNA adducts; Ehrenberg *et al.*, 1983) and response, or risk, the determination of DNA adducts is also able to permit risk quantitation (criterion 4).

**Table 1. Biochemical and chemical endpoints**

| | Test | Demonstration of |
|---|---|---|
| 1. | Demonstration of mutagens in body fluids | exposure |
| | 1.1. Chemical analysis | |
| | 1.2. Mutagenic activity of components in urine, etc. | also: metabolites with mutagenic activity |
| | 1.3. Metabolites in urine | |
| | (a) mercapturic acids, etc. | metabolic pathway |
| | (b) altered DNA bases | DNA damage<br>DNA repair |
| 2. | In-vivo 'dosimetry' of reactive compounds or reactive metabolites | |
| | 2.1. DNA adducts | DNA damage |
| | 2.2. Adducts to haemoglobin and other blood proteins | indirect measure |
| 3. | DNA-strand breaks | DNA damage<br>DNA repair |
| 4. | Determination of unscheduled DNA synthesis | |
| | (a) noninduced | repair of DNA damage |
| | (b) induced | repair capacity |

Since adduct formation is 'random' over cellular nucleophiles, protein adducts will be formed in the body concurrently with DNA adducts. Because modified amino acids are more easily determined than the corresponding changes in DNA, and proteins - e.g., haemoglobin and other blood proteins - are easily available in much larger quantities than DNA (a few mg in a 20-ml blood sample), haemoglobin adducts were proposed for indirect monitoring of DNA adducts and the ensuing risk (Osterman-Golkar *et al.*, 1976).

It was demonstrated in animal experiments that a number of compounds, both directly alkylating agents and compounds requiring bioactivation, gave a ratio:

$$\frac{\textbf{degree of protein alkylation (histidine adduct)}}{\textbf{degree of DNA alkylation (guanine-N-7 adduct)}} \qquad (5)$$

which is a approximately the same as the ratio of reaction rates with these macromolecules determined *in vitro*. This indicates that the dose in blood (erythrocytes) and in the DNA space of cells is approximately the same. When reactive metabolites are short-lived ($t_{1/2}$ less than about 1 min), as in the case of vinyl chloride or *N*-nitrosodimethylamine, haemoglobin alkylation gave a good estimate of liver DNA alkylation, whereas the degree of alkylation of DNA in more distal organs such as gonads was lower (Ehrenberg & Osterman-Golkar, 1980; Ehrenberg *et al.*, 1983). For such compounds, correction factors determined in animals have to be applied.

Hence, it may be concluded that determination of protein adducts is a relevant and quantifiable measure of DNA adducts.

Haemoglobin has a life-span of approximately four months in man. Since chemically changed molecules are not eliminated at low levels, a determination of haemoglobin adducts gives a value of the integrated dose (time-integral of level) during the months preceding sampling.

This long life-span of adducts contributes to the high resolution of the determination. Using gas chromatography-mass spectrometry, the method is, in principle, about one order of magnitude more sensitive than determination of the corresponding DNA adducts by immunochemical techniques. The sensitivity may be increased further but, in certain cases at least, it is counteracted by a background level of alkylation. In studies of histidine adducts in haemoglobin from personnel exposed to ethylene oxide, van Sittert *et al.* (1984) found that blood samples from unexposed persons also contained the same adducts, at variable levels. The existence of this background level, which has been verified in Swedish studies (Calleman *et al.*, unpublished data) and seen in laboratory animals, makes it impossible to identify those individuals who are exposed during working hours to less than about 3 ppm ethylene oxide. At these exposure levels, only mean values of exposed and control groups may be compared.

A determination of adducts by gas chromatography-mass spectrometry has the advantage over immunochemical methods, which require pre-preparation of specific antibodies, that adducts from a-priori unknown agents may be found, identified and quantitated. Consequently, a search can be made for unknown cancer initiators/mutagens that may be connected with food habits or may occur in other mixed exposures. In work with these aims, parallel studies of mercapturic acids and other metabolites in urine may be of importance in the determination of chemical structure. The amount of a particular mercapturic acid excreted during 24 h is likely to be some orders of magnitude larger than the amount of the corresponding adduct present in the haemoglobin in 10 ml blood. The usefulness of urine metabolites for risk quantitation is more doubtful. In any case, it will require calibration to the response at other endpoints.

## APPLICATIONS

The 'haemoglobin dosimetry' discussed in the previous section has been successfully applied in persons occupationally exposed to ethylene oxide (Calleman et al., 1978; van Sittert et al., 1984) and propylene oxide (Osterman-Golkar et al., 1984) (for the latter compound the background level of adducts is much lower than that of ethylene oxide).

From animal studies with vinyl chloride (Osterman-Golkar et al., 1977), it seems likely that the electrophilic metabolite is chloroethylene oxide, introducing the group 2-oxoethyl (I) onto nucleophilic atoms. This group was determined following reduction by $NaBH_4$ to 2-hydroxyethyl (II):

$$\underset{I}{\overset{O}{\underset{H}{C}}-CH_2-Y} \xrightarrow{NaBH_4} \underset{II}{HOCH_2-CH_2-Y} \qquad (6)$$

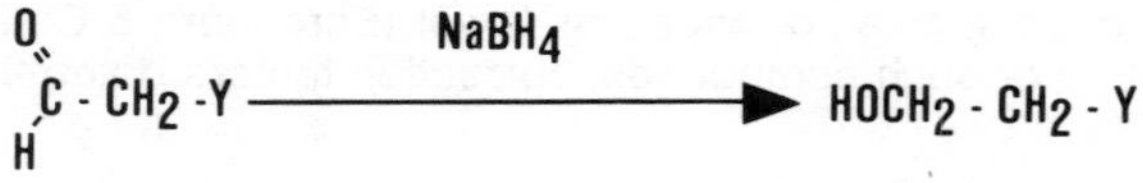

i.e., the same derivative that is obtained with ethylene oxide. In studies of exposed workers in the polyvinyl chloride industry, reduction of haemoglobin before analysis led to increased levels of II in control subjects also, i.e., indicating a background level of I-adducts.

Analytical errors render very uncertain the values determined as differences. In order to obtain a direct determination of I-adducts, the compound was reduced by $NaBH_4$ labelled with $^3H(T)$ or $^2H(D)$ (K. Svensson et al., in preparation).

$$\overset{O}{\underset{H}{C}}-CH_2-Y \xrightarrow{[^3H]\,NaBH_4} HOCH_2^*-CH_2-Y \qquad (7)$$

This gave a more reliable value of the level of I-adducts in vinyl chloride-exposed workers and, also, independent proof of the existence of background alkylation. The existence of background alkylation levels of certain chemical structures has implications beyond that of being a disturbance, or a source of error, in the monitoring of tissue doses in exposed persons. These levels might reflect exposure to mutagens or cancer initiators of exogenous or endogenous origin, with a causative role in current incidences of cancer and heritable diseases.

For this reason, it is important to clarify the origin of the observed alkylation levels. With simple alkyl groups such as those discussed here, several possibilities are open in each case. A good aid to identification of the ultimate electrophile is the fact that each species has a characteristic pattern of reaction rates, as determined by the 'substrate constant' (s), towards S (cysteine), N (histidine, N-terminal) and O (aspartic and glutamic acids), cf. Equation 3 and Segerbäck (1983).

In view of recent findings concerning the metabolism of urethane (Ribovich et al., 1982), this compound is a potential source of background 2-oxoethylations.

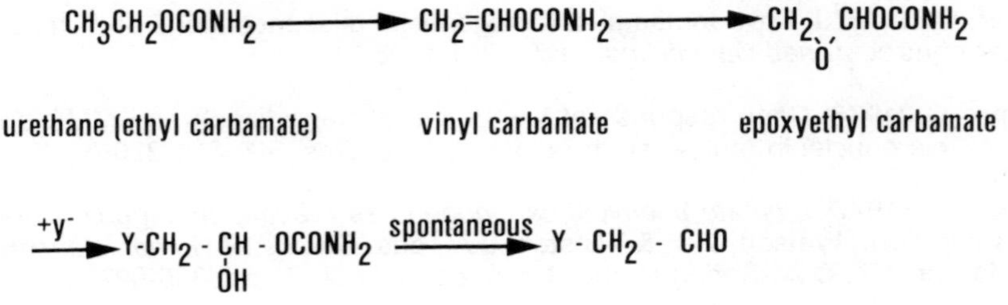


(8)

Since urethane is an efficient carcinogen, with widespread occurrence - it is formed in yeast fermentation and is, therefore, present in bread, beer and wine (Ough, 1976) - it is worth testing the hypothesis that urethane is an essential contributor to the largely unknown initiations, as discussed above, in current carcinogenesis.

## ACKNOWLEDGEMENT

This paper is based on work supported by the Swedish Work Environment Fund, the Swedish Board of Occupational Safety and Health and Shell International Research Maatschappij B.V.

## REFERENCES

von Bahr, B., Ehrenberg, L., Scalia-Tomba, G.-P. & Säfwenberg, J.-O. (1984) *Dose-response relationships of chemical carcinogens: Testing of mathematical models* (Swedish). Report to the Swedish Cancer Committee (in press)

Bloom, A.D., ed. (1981) *Guidelines for Studies of Human Populations Exposed to Mutagenic and Reproductive Hazards*, New York, March of Dimes Birth Defects Foundation, pp. 129-131

Brookes, P. & Lawley, P.D. (1964) Evidence for the binding of polynuclear aromatic hydrocarbons to the nucleic acid of mouse skin: relation between carcinogenic power of hydrocarbons and their binding to DNA. *Nature*, *202*, 781-784

Calleman, C.J., Ehrenberg, L., Jansson, B., Osterman-Golkar, S., Segerbäck, D., Svensson, K. & Wachtmeister, C.A. (1978) Monitoring and risk assessment by means of alkyl groups in hemoglobin in persons occupationally exposed to ethylene oxide. *J. environ. Pathol. Toxicol.*, *2*, 427-442

Ehrenberg, L. (1977) *Aspects of statistical inference in testing for genetic toxicity*. In: Kilbey, B.J., Legator, M., Nichols, W. & Ramel, C., eds, *Handbook of Mutagenicity Test Procedures*, Amsterdam, Elsevier, pp. 419-459

Ehrenberg, L. & Osterman-Golkar, S. (1980) Alkylation of macromolecules for detecting mutagenic agents. *Teratog. Carcinog. Mutag.*, *1*, 105-127

Ehrenberg, L., Moustacchi, E. & Osterman-Golkar, S. (1983) Dosimetry of genotoxic agents and dose-response relationships of their effects. *Mutat. Res.*, *123*, 121-182

Higginson, J. & Muir, C.S. (1979) Environmental carcinogenesis: misconceptions and limitations to cancer control. *J. natl Cancer Inst.*, *63*, 1291-1298

Jenssen, D. & Ramel, C. (1976) Dose response at low doses of X-irradiation and MMS on the induction of micronuclei in mouse erythroblasts. *Mutat. Res.*, *41*, 311-320

Miller, J.A. & Miller, E.C. (1977) *Ultimate chemical carcinogens as reactive mutagenic electrophiles*. In: Hiatt, H.H., Watson, J.D. & Winsten, J.A., eds, *Origins of Human Cancer*, Cold Spring Harbor, NY, Cold Spring Harbor Laboratory, Book B, pp. 605-627

Osterman-Golkar, S., Ehrenberg, L., Segerbäck, D. & Hällström, I. (1976) Evaluation of genetic risks of alkylating agents. II. Haemoglobin as a dose monitor. *Mutat. Res.*, *34*, 1-10

Osterman-Golkar, S., Hultmark, D., Segerbäck, D., Calleman, C.J., Göthe, R., Ehrenberg, L. & Wachtmeister, C.A. (1977) Alkylation of DNA and proteins in mice exposed to vinyl chooride. *Biochem. biophys. Res. Commun.*, *76*, 259-266

Osterman-Golkar, S., Bailey, E., Farmer, P.B., Gorf, S.M. & Lamb, J.H. (1984) Monitoring exposure to propylene oxide through the determination of hemoglobin alkylation. *Scand. J. Work environ. Health*, *10* (in press)

Ough, C.S. (1976) Ethyl carbamate in fermented beverages and foods. I. Naturally occurring ethyl carbamate. *J. Agric. Food Chem.*, *24*, 323-328

Ribovich, M.L., Miller, J.A., Miller, E.C. & Timmins, L.G. (1982) Labeled 1,$N^6$-ethenoadenosine and 3,$N^4$-ethenocytidine in hepatic RNA of mice given [ethyl-1,2-$^3$H or ethyl-1-$^{14}$C]ethyl carbamate (urethan). *Carcinogenesis*, *3*, 539-546

Ross, W.C.J. (1962) *Biological Alkylating Agents*, London, Butterworths

Segerbäck, D. (1983) Alkylation of DNA and hemoglobin in the mouse following exposure to ethene and ethene oxide. *Chem.-biol. Interact.*, *45*, 139-151

van Sittert, N.J., de Jong, G., Clare, M.G., Davis, R., Dean, B.J., Wren, L.J. & Wright, A.S. (1984) Cytogenetic, immunological and haematological effects in workers in an ethylene oxide manufacturing plant. *Br. J. ind. Med.* (in press)

# DOSIMETRY AND DOSE-RESPONSE RELATIONSHIPS

**H.-G. Neumann**

*Institute of Pharmacology and Toxicology, University of Würzburg, Federal Republic of Germany*

## INTRODUCTION

The shape of the dose-response relationship in carcinogenesis plays a significant role in the risk assessment of chemicals. Yet there seems to be continuing controversy about the issue. This paper is concerned with a particular aspect of biomonitoring genotoxic agents: the dose-dependence of metabolism and macromolecular binding. With this endpoint of response, instead of tumour formation, it should be easier to answer some of the questions about the application of biomonitoring to health surveillance of humans exposed to xenobiotics.

A basic question in this approach is whether, at low doses, a linear relationship exists between exposure and the tissue dose of potentially hazardous metabolites and cellular damage. Any deviations from linearity would constitute a basis for pharmacokinetic thresholds, and, also, would make it difficult to relate the analytical data obtained from measuring metabolites bound to macromolecules or excreted in the urine to the dose. Another question to be considered is the validity of determining protein-bound metabolites, those bound to haemoglobin, for example, for use as monitors for exposure and for damage of critical targets like DNA in liver or some extrahepatic tissue.

The discussion in this paper does not cover the problem of deviations from linear pharmacokinetics at high doses. It has become quite clear that numerous processes governing the fate of xenobiotics, such as absorption, liver uptake, metabolism, plasma protein binding and excretion, may become saturated for doses above $10^{-4}$ - $10^{-3}$ mol/kg in rodents and about a quarter of that in humans. The consequences may be either a larger than proportionate increase of reactive metabolites with dose or an increase that is less than proportionate. These effects may well have a bearing on extrapolations from high-dose experiments to a low-dose situation, but will be rarely encountered in biomonitoring.

## THE DOSE-DEPENDENCE OF MACROMOLECULAR BINDING

Several authors have studied the dose-dependence of macromolecular binding in liver, after oral administration of radioactively labelled carcinogens.

We have observed linear relationships over an extremely wide range of doses with *trans*-4-dimethylaminostilbene, a complete carcinogen for Zymbal's gland and an initiator for liver in rats (Gaugler & Neumann, 1979; Neumann, 1980). The lower end of this relationship is shown in Figure 1, the lowest dose being $5 \times 10^{-10}$ mol/kg. This is three orders of magnitude lower than the lowest daily dose with which Druckrey *et al.* (1963) were able to produce ear-duct tumours in rats. Within experimental error, there is a constant ratio in liver between binding to proteins, RNA and DNA.

**Fig. 1. Dose-response of *trans*-4-acetylaminostilbene in rats**

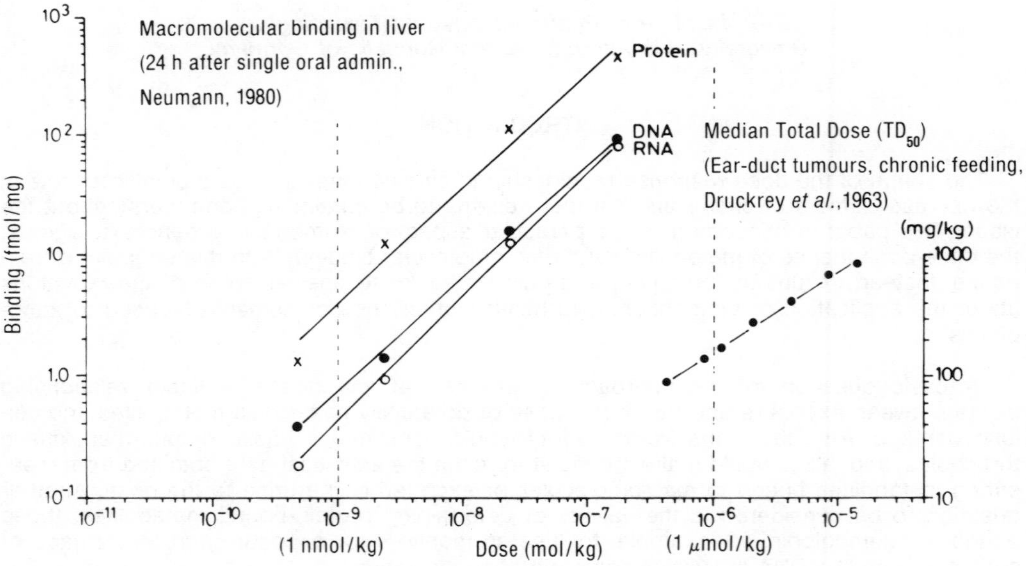


Linear relationships have been demonstrated also with subcarcinogenic doses to rats of the liver carcinogen, aflatoxin $B_1$ (Scott Appleton *et al.*, 1982). In this case, the lowest dose is $3 \times 10^{-11}$ mol/kg (Fig. 2). Again, the ratio of binding to different macromolecules remains constant with dose, but binding to proteins is lower than to DNA.

**Fig. 2. Dose-response of aflatoxin $B_1$ in rats**

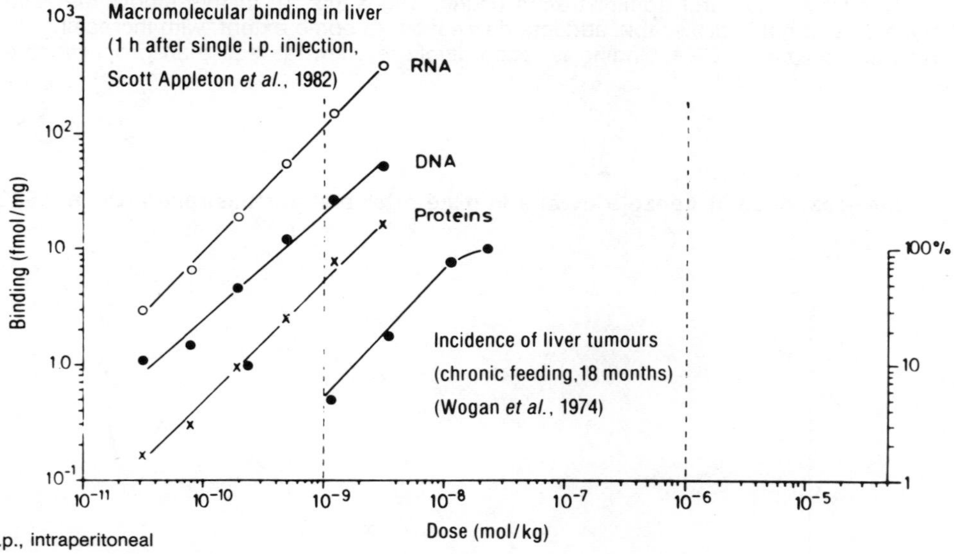


i.p., intraperitoneal

For *N*-nitrosodimethylamine, another liver carcinogen in rats, the lowest dose that produces tumours has been estimated at 1 × $10^{-6}$ mol/kg (Arai *et al.*, 1979; Crampton, 1980). The formation of DNA adducts is proportional to dose for doses that are two orders of magnitude smaller (Pegg & Perry, 1981; Fig. 3).

**Fig. 3. Dose-response of *N*-nitrosodimethylamine in rats**

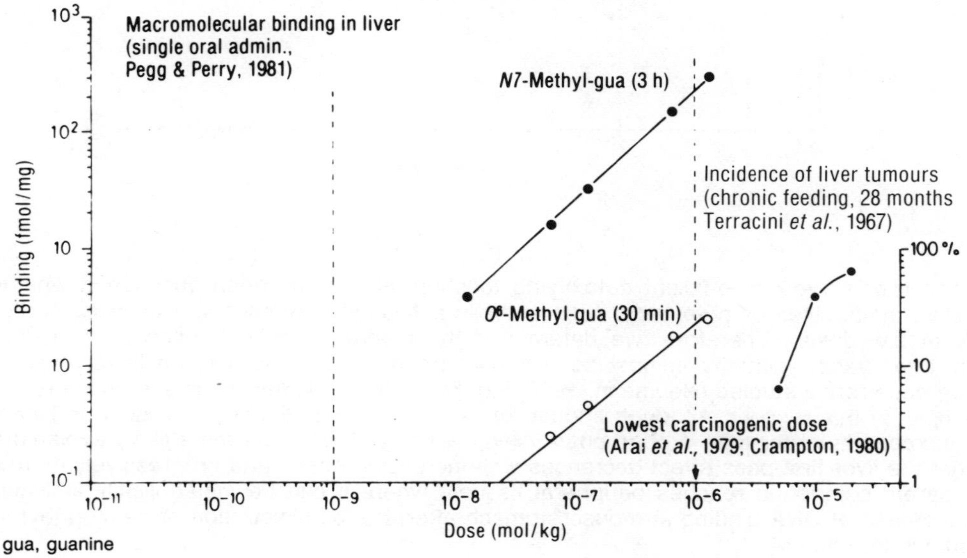


gua, guanine

Very recent results, also, have shown linear binding of benzo[*a*]pyrene metabolites to mouse liver DNA after oral administration (Dunn, 1983; Fig. 4). In this study, the ratio of hydrolysable and nonhydrolysable adducts decreases, to some extent, with increasing dose. The overall increase in DNA binding is, accordingly, somewhat less than proportionate to dose.

**Fig. 4. Dose-response of benzo[*a*]pyrene in mice after oral administration (Dunn, 1983)**

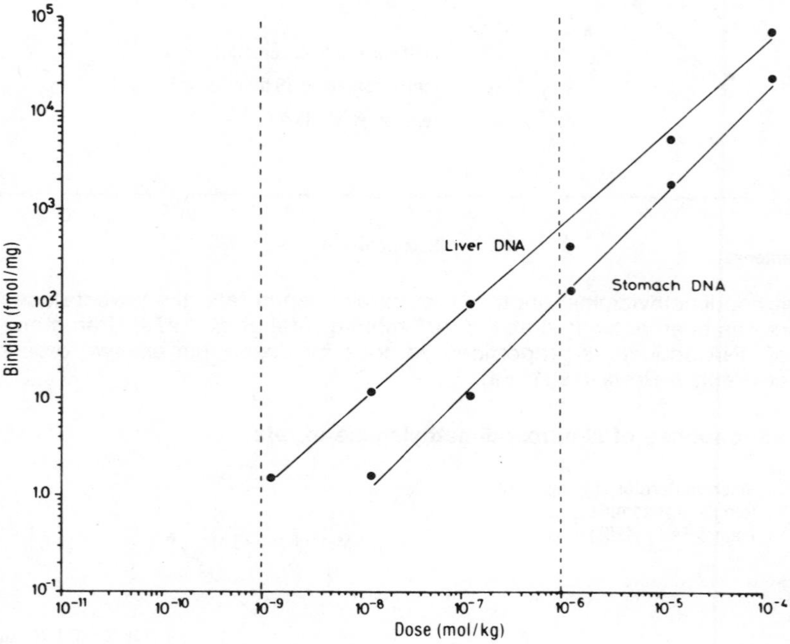


The liver serves an efficient detoxifying function, and it has been questioned whether reactive metabolites or precursors for metabolic activation may escape liver cells, particularly at low doses. Therefore, we determined the macromolecular binding in rat kidney, also, after *trans*-4-dimethylaminostilbene administration and found it to be linear over the entire dose range studied (Neumann, 1980; Fig. 5). *N*-Nitrosodimethylamine is an interesting example, in this respect. Although adduct formation increases linearly with dose in kidney, the increase is larger than proportionate (Pegg & Perry, 1981). The most likely explanation is that the liver first-pass effect decreases with increasing doses and progressively more of the parent compound reaches peripheral tissues, where it can be metabolically activated. the increase of DNA binding in mouse stomach after oral administration of benzo[*a*]pyrene is shown in Figure 4.

**Fig. 5. Dose-response of macromolecular binding in rat kidney**

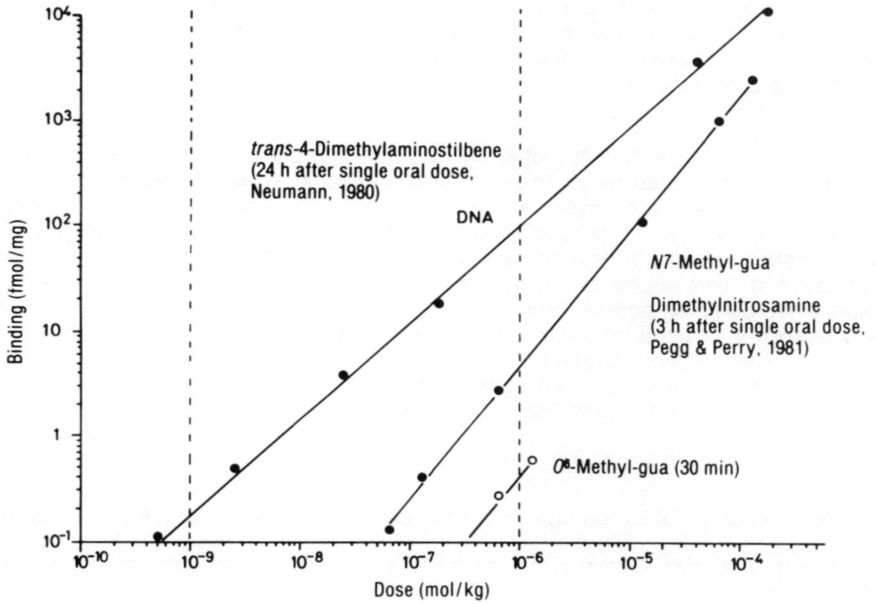


gua, guanine

All available data, therefore, indicate that deviations from linear pharmacokinetics do not exist down to doses of $10^{-9}$ - $10^{-10}$ mol/kg, i.e., absorption, metabolism, distribution and macromolecular binding follow first-order kinetics, even in this range (cf. Neumann, 1983a).

## BINDING TO HAEMOGLOBIN

It has been proposed that metabolites covalently bound to haemoglobin be used to monitor dose (Osterman-Golkar *et al.*, 1976; Neumann *et al.*, 1977; Wieland & Neumann, 1978). Immediately, the question arises as to whether reactive metabolites would be available at low doses, and what kind of a correlation exists between the reaction with a blood protein and tissue DNA. Some answers have been obtained with aromatic amines; haemoglobin-binding of alkylating agents is discussed by Farmer *et al.* (this volume) and Ehrenberg (this volume).

Common intermediates in the metabolism of most aromatic amines and aromatic nitro compounds are *N*-hydroxylamines and nitroso derivatives. Arylnitroso compounds react readily with sulfhydryl groups. The reaction with glutathione is rather complex and has been studied in some detail (Neumann *et al.*, 1977; Eyer, 1979; Dölle *et al.*, 1980; Eyer & Lierheimer, 1980). The primary reaction product either may be reduced by additional glutathione to the hydroxylamine or the amine, or it may rearrange to give a sulfinic acid amide. The composition of the final reaction products depends on the structure of the aryl moiety, pH and glutathione concentration. This reaction takes place in erythrocytes, also. *N*-Hydroxylamines and haemoglobin are cooxidized to yield the arylnitroso compound and methaemoglobin. Both reaction products can be reduced back, but part of the arylnitroso compound reacts with haemoglobin to give a sulfinic acid amide. This is stable *in vivo*, but can be hydrolysed under acidic conditions.

We have determined the haemoglobin-binding of a number of labelled aromatic amines in rats and have calculated an index for total binding (Table 1). In all cases, it was possible to remove most of the bound material (75-90%) under moderately acidic conditions. After the administration of *trans*-4-dimethylaminostilbene, the identified cleavage product was *trans*-4-aminostilbene; for benzidine, it was monoacetylbenzidine (Albrecht & Neumann, unpublished data). This supports the hypothesis that the sulfinic acid amide is, indeed, the predominant haemoglobin adduct. More haemoglobin adducts are produced by nitrobenzene than by acetanilide, indicating efficient in-vivo reduction of the nitro group. The data indicate, also, that haemoglobin binding may be used to monitor the absorption and reduction of nitro aromatics in general, as well as to demonstrate the reduction of azo derivatives by intestinal bacteria with subsequent absorption of the resulting arylamines.

**Table 1. Haemoglobin (Hb) binding-index (Albrecht & Neumann, unpublished data)**

| Compound[a] | Dose (μmol/kg) | Binding-index[b,c] |
|---|---|---|
| *Trans*-4-dimethylaminostilbene | 000.003-35 | 147 ± 9.6 (5)[c] |
| Nitrobenzene | 200 | 80 ± 2.7 (2) |
| Benzidine | 1.7 | 60 ± 1.8 (2) |
| 2-Acetylaminofluorene[d] | 93 | 21 ± 3.7 (3) |
| Acetanilid | 150 | 12 ± 1.0 (3) |
| Paracetamol | 6.9 | 0.42 ± 0.04 (2) |

[a] Single doses of labelled compounds were administered orally to rats

[b] Binding index $= \frac{\text{Binding (mmol/mol Hb)}}{\text{Dose (mmol/kg)}}$

[c] Average values from the number of animals given in brackets

[d] Ruthsatz and Neumann, unpublished data

Using *trans*-4-dimethylaminostilbene, we have studied the dose-dependence of the formation of the sulfinic acid amide. By a procedure that we propose for the biomonitoring of aromatic amines (Albrecht & Neumann, in preparation), we have isolated haemoglobin from blood samples of dosed animals, subjected it to acidic hydrolysis and determined *trans*-4-aminostilbene in the extracts. A perfectly linear correlation was obtained (Fig. 6). This confirms the results of previous experiments in which we determined total haemoglobin binding (Neumann *et al.*, 1977, 1980).

**Fig. 6. Dose-response of haemoglobin binding**

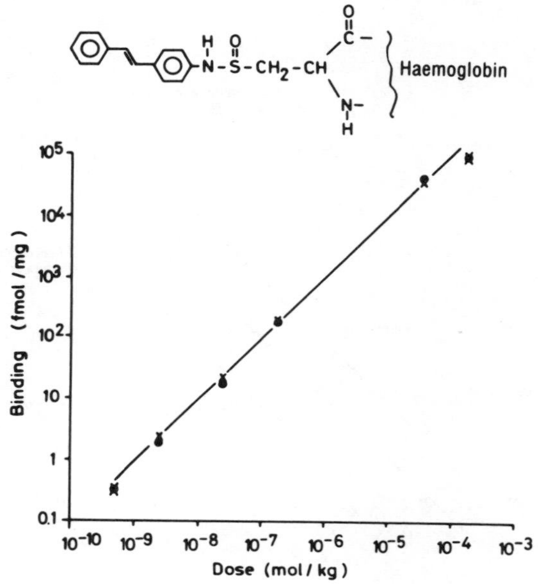


*trans*-4-Dimethylaminostilbene was administered orally to female Wistar rats. *trans*-4-Aminostilbene was determined after hydrolysis of haemoglobin. The structure represents the sulfinic acid amide which forms after the reaction of *trans*-4-nitrosostilbene with SH-groups. Values for two animals are given for each dose (Albrecht & Neumann, unpublished data)

With the aromatic amines, it is not assumed that the ultimate reactive metabolite which reacts with haemoglobin, i.e., a nitroso derivative, is identical to that which reacts with DNA. We have shown that the major adduct in liver contains an acetylaminobibenzyl moiety, which is attached to the C1 and $N^2$ position of guanine through its α and β-C atoms. This indicates that a hydroxamic acid ester is the ultimate reactive form (Zielinski & Neumann, 1983). Although this adduct is rather persistent *in vivo* (Baur & Neumann, 1980), we have not yet established whether it represents a critical lesion. *trans*-4-Nitrosostilbene, however, is the strongest mutagen of the *trans*-4-aminostilbene metabolites, and it has been proposed that an N-oxidation product, intermediate to the hydroxylamine and the nitroso derivative, might be the ultimate mutagen in the *Salmonella typhimurium* system (Glatt *et al.*, 1980; see also Neumann, 1983b). In this complex situation, which applies to other aromatic amines, also, the pharmacokinetic argument is particularly helpful. If apparent first-order kinetics prevail, as delineated from the dose-dependence of macromolecular binding, there should be a constant ratio between different metabolites and their reaction products. This is demonstrated in Figure 7. Binding of *trans*-4-aminostilbene metabolites to liver DNA and haemoglobin is proportional to dose over the entire dose range studied, except for the highest dose. This shows that there is a constant ratio between haemoglobin binding and DNA binding in the liver.

**Fig. 7. Dose-dependence of binding to liver DNA (•) and haemoglobin (o), 24 h after a single oral administration of *trans*-4-dimethylaminostilbene (—) or 2-acetylaminofluorene (—.—) to rats**

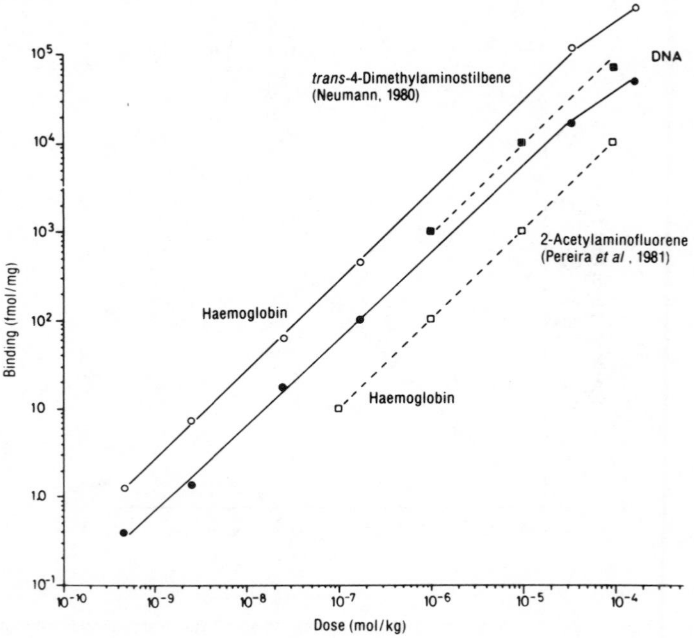


We assume that the values for haemoglobin binding of 2-acetylaminofluorene are too low, because the authors processed haemoglobin under acidic conditions, but this would not influence the linear correlation.

In contrast to the ratio for direct alkylating agents, which can be calculated from the differences in reaction rates with proteins and DNA, and which is reasonably constant for different tissues (Ehrenberg *et al.*, 1974), the ratio for aromatic amines is more variable. It depends on the rate constants of numerous interrelated metabolic pathways which, in addition, may be involved differently in different tissues. With 2-acetylaminofluorene, for instance, haemoglobin binding is considerably lower than with *trans*-4-dimethylaminostilbene, whereas DNA binding is more similar for these two compounds (Fig. 7). Thus, the ratio must be determined separately for each compound. It will vary from tissue to tissue and from species to species according to the metabolic situation. It should be emphasized, however, that haemoglobin binding is correlated to genotoxic risk.

## ACCUMULATION OF MACROMOLECULAR BINDING

Single exposures are rare events in normal life; the accumulation of macromolecular damage after repeated uptake of genotoxic compounds depends on the exposure interval and the elimination rates of the adducts. We administered 12 doses of *trans*-4-acetylaminostilbene within six weeks to rats. DNA binding increased steadily in all tissues studied. This indicates that the adducts are not readily repaired, and that the overall elimination time is considerably longer than the application interval of three to four days. A biological half-life,

$t_{1/2}$ = 22 d, has been found for DNA binding in liver, and this remained constant during the feeding period (Neumann, 1983c). Haemoglobin binding accumulated also, but, in accordance with the biological half-life $t_{1/2}$ = 14 d, reached a plateau after four weeks (Fig. 8). Due to these differences in elimination rates, the ratio of haemoglobin to DNA binding will depend on exposure time, to some extent.

**Fig. 8. Accumulation of macromolecular binding during repeated administration of *trans*-4-acetylaminostilbene (5 × $10^{-6}$ mol/kg) to rats (Hilpert & Neumann, unpublished data; cf. Neumann, 1983c)**

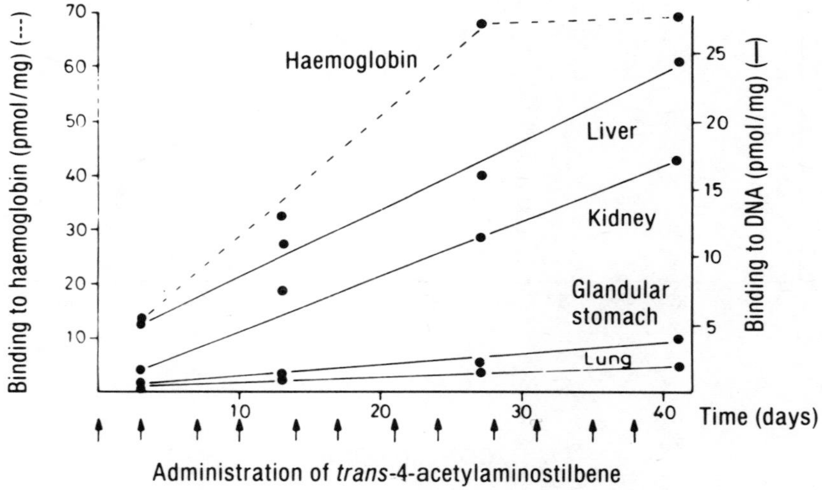


## ELIMINATION OF HAEMOGLOBIN BINDING

When we observed haemoglobin binding over a four-week period after a single administration of *trans*-4-dimethylaminostilbene, radioactivity decreased exponentially with time, i.e., linearly in a semi-logarithmic plot (Neumann, 1981). This posed an interesting problem. If the binding is stable *in vivo*, its decrease should be related to the life-span of erythrocytes. It should decrease linearly and approach zero after their mean lifetime. With benzene and benzo[*a*]pyrene, this has actually been observed (Calleman, personal communication). Therefore, we reexamined our finding with $^{14}C$-labelled benzidine, another aromatic amine. The compound was administered once to female rats, and haemoglobin binding was determined up to 70 days. The decline was clearly linear in a semi-logarithmic plot, from which a biological half-life of 11.5 days was determined (Fig. 9; Albrecht & Neumann, unpublished data). After 70 days, which is beyond the average lifetime of rat erythrocytes (63 days), there was still a significant radioactivity associated with the haemoglobin. This can be explained only by assuming that erythrocytes are labelled for an extended period of time after dosing, during which benzidine or monoacetylbenzidine is released from some deep compartment and is metabolically activated. For these processes, apparent first-order kinetics could be assumed, superimposed on the zero-order elimination of erythrocytes. Prior to these results, we believed that macromolecular binding is completed after one or two days, and that true elimination is measured afterwards. Obviously, this is not the case. For the use of haemoglobin as a dose monitor for aromatic amines, this means that each compound may have its own elimination rate and, therefore, its own accumulation characteristics.

**Fig. 9. Decrease of the haemoglobin binding index after a single oral administration of benzidine to rats (Albrecht & Neumann, unpublished data, mean ± SD of 3 experiments)**

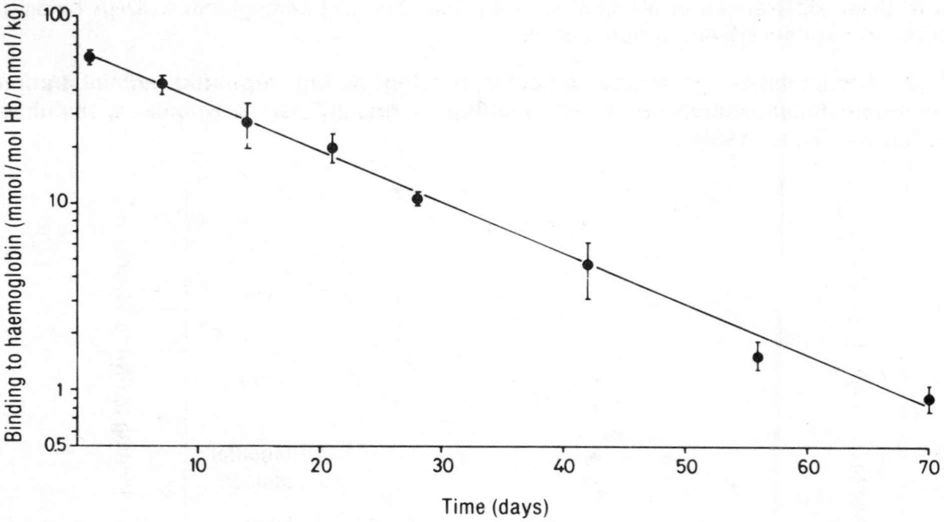


## CONCLUSIONS

A linear correlation has been established in several cases for the relationship: exposure - tissue dose - macromolecular binding. However, there may not be a corresponding linearity between macromolecular damage and tumour formation. The initial extent of DNA damage is modulated by repair and by persistence of the adducts formed. The resulting, integral, DNA damage still does not correlate with tumour formation. It may be related to the number of initiated cells, but when and where tumours arise depends on the promoting activity of the carcinogen itself in the broadest sense, and on secondary influences such as time and life-style, which may modulate later stages in carcinogenesis. It is barely possible to study the dose-response relationships for these endpoints, at present. Only when biochemical parameters can be related to late-stage effects, will it be possible to interpret this response in terms of dose and to understand better the further parameters that determine the shape of the overall dose-response curve.

## REFERENCES

Arai, M., Aoki, Y., Nakanishi, K., Miyata, Y., Mori, T. & Ito, N. (1979) Long-term experiment of maximal noncarcinogenic dose of dimethylnitrosamine for carcinogenesis in rats. *Gann*, *70*, 549-558

Baur, H. & Neumann, H.-G. (1980) Correlation of nucleic acid binding by metabolites of *trans*-4-aminostilbene derivatives with tissue-specific acute toxicity and carcinogenicity in rats. *Carcinogenesis*, *1*, 877-886

Crampton, R.F. (1980) Carcinogenic dose-related response to nitrosamines. *Oncology*, *37*, 251-254

Dölle, B., Töpner, W. & Neumann, H.-G. (1980) Reaction of arylnitroso compounds with mercaptans. *Xenobiotica*, *10*, 527-536

Druckrey, H., Schmähl, D. & Dischler, W. (1963) Dosage-effect relations in cancer induction with 4-dimethylamino-stilbene in rats. *Z. Krebsforsch.*, *65*, 272-288

Dunn, B.P. (1983) Wide range linear dose-response curve for DNA binding of orally administered benzo[*a*]pyrene in mice. *Cancer Res.*, *43*, 2654-2658

Ehrenberg, L., Hiesche, K.D., Osterman-Golkar, S. & Wennberg, I. (1974) Evaluation of genetic risks of alkylating agents: tissue dose in the mouse from air contaminated with ethylene oxide. *Mutat. Res.*, *24*, 83-103

Eyer, P. (1979) Reactions of nitrosobenzene with reduced glutathione. *Chem.-biol. Interact.*, *24*, 227-239

Eyer, P. & Lierheimer, E. (1980) Biotransformation of nitrosobenzene in the red cell and the role of glutathione. *Xenobiotica*, *10*, 517-526

Gaugler, B.J.M. & Neumann, H.-G. (1979) The binding of metabolites formed from aminostilbene derivatives to nucleic acids in the liver of rats. *Chem.-biol. Interact.*, *24*, 355-372

Glatt, H.R., Oesch, F. & Neumann, H.-G. (1980) Factors responsible for the metabolic formation and inactivation of bacterial mutagens from *trans*-4-acetylaminostilbene. *Mutat. Res.*, *73*, 237-250

Neumann, H.-G. (1980) Biochemical effects and early lesions in regard to dose-response studies. *Oncology*, *37*, 255-258

Neumann, H.-G. (1981) On the significance of metabolic activation and binding to nucleic acids of aminostilbene derivatives *in vivo*. *Natl Cancer Inst. Monogr.*, *58*, 165-171

Neumann, H.-G. (1983a) *The dose-dependence of DNA interactions of aminostilbene derivatives and other chemical carcinogens*. In: Hayes, A.W., Schnell, R.C. & Miya, T.S., eds, *Developments in the Science and Practice of Toxicology, ICT III*, Amsterdam, Elsevier Biomedical Press, pp. 135-144

Neumann, H.-G. (1983b) Role of extent and persistence of DNA modifications in chemical carcinogenesis by aromatic amines. *Recent Results Cancer Res.*, *84*, 77-89

Neumann, H.-G. (1983c) Role of tissue exposure and DNA lesions for organ-specific effects of carcinogenic *trans*-4-acetylaminostilbene in rats. *Environ. Health Perspect.*, *49*, 51-58

Neumann, H.-G., Metzler, M. & Töpner, W. (1977) Metabolic activation of diethylstilbestrol and aminostilbene derivatives. *Arch. Toxicol.*, *39*, 21-30

Osterman-Golkar, S., Ehrenberg, L., Segerbäck, D. & Hällström, I. (1976) Evaluation of genetic risks of alkylating agents. II. Hemoglobin as a dose monitor. *Mutat. Res.*, *34*, 1-10

Pegg, A.E. & Perry, W. (1981) Alkylation of nucleic acids and metabolism of small doses of dimethylnitrosamine in the rat. *Cancer Res.*, *41*, 3128-3132

Pereira, M.A.., Lin, L.-H.C. & Chang, L.W. (1981) Doses-dependency of 2-acetylaminofluorene binding to liver DNA and haemoglobin in mice and rats. *Toxicol. appl. Pharmacol.*, *60*, 472-478

Scott Appleton, B., Goetchius, M.P. & Campbell, T.C. (1982) Linear dose-response curve for the hepatic macromolecular binding of aflatoxin $B_1$ in rats at very low exposures. *Cancer Res.*, *42*, 3659-3662

Terracini, B., Magee, P.N. & Barnes, J.M. (1967) Hepatic pathology in rats on low dietary levels of dimethylnitrosamine. *Br. J. Cancer*, *21*, 559-565

Wieland, E. & Neumann, H.-G. (1978) Methemoglobin formation and binding to blood constituents as indicators for the formation, availability and reactivity of activated metabolites derived from *trans*-4-aminostilbene and related aromatic amines. *Arch. Toxicol.*, *40*, 17-35

Wogan, G.N., Paglialunga, S. & Newberne, P.M. (1974) Carcinogenic effects of low dietary levels of aflatoxin $B_1$ in rats. *Food Cosmet. Toxicol.*, *12*, 681-685

Zielinski, E. & Neumann, H.-G. (1983) The role of hydroxamic acid esters for the genotoxic effects of *trans*-4-acetylaminostilbene. *Eur. J. Cancer clin. Oncol.*, *19*, 1326

# KINETIC CONSIDERATIONS IN MONITORING EXPOSURE TO CHEMICALS


**A. Aitio**

*Department of Industrial Hygiene and Toxicology, Institute of Occupational Health, Helsinki, Finland*



## SUMMARY

The aim of biological exposure monitoring is to estimate the amount of the chemical absorbed in the body of each individual worker. As the concentration of any foreign chemical in body fluids is not constant, but shows an exposure-related fluctuation, this estimation may be successful only when based on knowledge of the behaviour of the chemical in the body, i.e., its kinetics.

Skin absorption is one of the reasons for performing biological monitoring of exposure. However, when the absorbed chemical is being monitored in the blood, skin absorption may be a source of considerable error since the concentration of the chemical in the blood specimen collected from the cubital vein does not represent the whole body but only the arm.

It should be borne in mind that the errors caused by kinetic variations are similar, independent of the type of measurement performed; thus, they pertain equally to the measurements of chemicals by traditional chemical analytical means and to measurements of the character of the chemical, e.g., mutagenicity.


## BASIC PHARMACOKINETIC CONCEPTS

Pharmacokinetics is a science that describes the behaviour of chemicals in the body. Its main application has been in pharmacology, where it is used to approximate the fate of a certain dose of a drug in the body, in order to find dosages that are not toxic, but are effective, that do not show too wide a variation in drug concentration, and do not lead to unacceptable drug accumulation.

In occupational toxicology, the use of kinetic data is, in a way, inverted; from concentrations of chemicals at a certain time-point, one tries to extrapolate either to total dose absorbed, or to the peak concentration, or to body burden of the chemical. In practice, most emphasis has to be placed on the kinetics of the disappearance of the chemical.

*Single dose*

When a dose of a chemical is absorbed in the body, its disappearance usually follows one of the model curves shown in Figure 1, where the scale is linear (A), or logarithmic (B). Curve X represents the (exceptional) case of disappearance by zero-order kinetics, i.e., the rate of disappearance remains constant irrespective of the concentration. This is what is called saturation kinetics; it is exemplified by ethyl alcohol, which is metabolized and cleared from the circulation at a constant speed. In principle, most chemicals would show this kind of disappearance kinetics, if a sufficiently high concentration were reached. However, for most chemicals in the occupational settings, saturating concentrations are not reached. It should be remembered that in the case of zero-order kinetics, no half-time may be given.

**Fig. 1. Disappearance from the blood of a chemical that follows zero-order kinetics (X); a chemical that follows first-order kinetics with a half-time of 0.7 days (Y1) or 7 days (Y2); and a chemical that is distributed in two compartments with approximate half-times of 0.7 and 7 days (Z)**

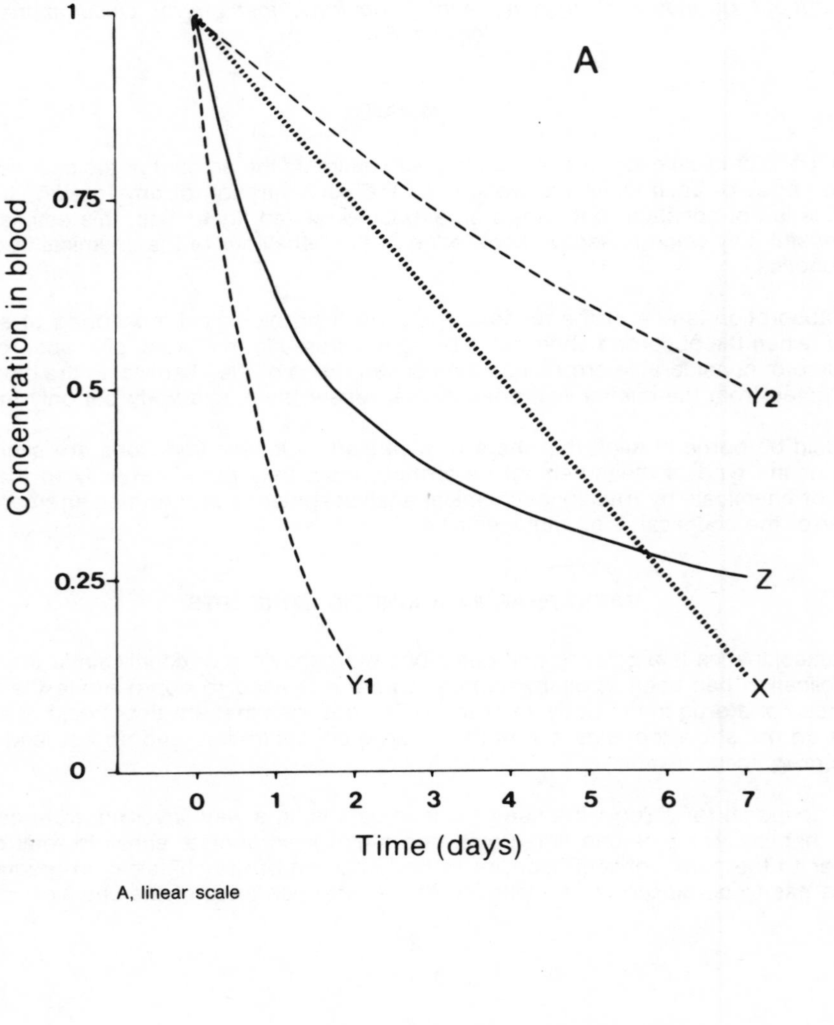


A, linear scale

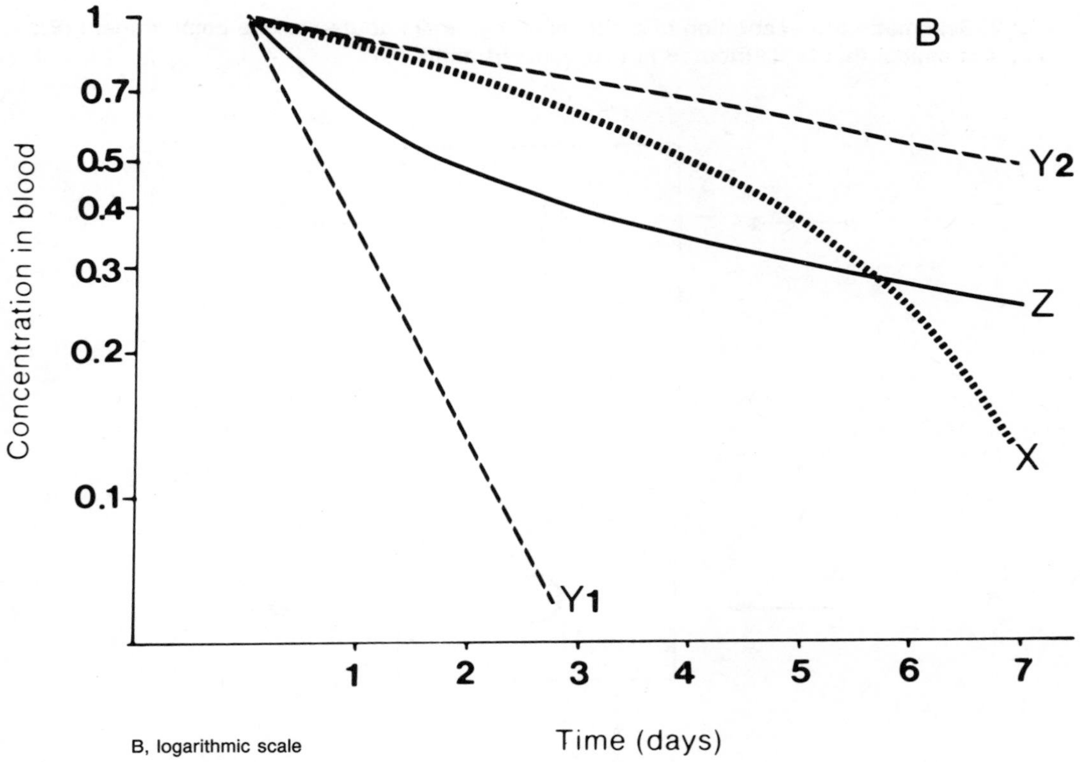


B, logarithmic scale

The curves Y1 and Y2 represent a rate of disappearance of the chemical that is directly proportional to the concentration, i.e., a constant fraction of the chemical disappears per unit time. In other words, the chemical shows first-order disappearance kinetics, the concentration at a specified time being, $C_t = C_0 \times e^{-kT}$, where $C_0$ is the concentration at time-point zero, T is the time elapsed and k the disappearance constant. From this equation one can derive a half-time - the time during which the concentration drops to half, $t_{1/2} = \ln 2/k$. From these equations one may, in principle, derive the concentration of the chemical at any point in time.

A prerequisite for the chemical to follow curve Y is that it be evenly distributed in the body (or that any tissue in the body show either a similar concentration, or no concentration at all). That is, the distribution of the chemical in the body may be described by the simple model, B, in Figure 2. This is very seldom the case, though. Generally, any chemical is distributed in the body in two or more more-or-less separate compartments, as illustrated in Figure 2 (C). Often, for example, the lipid solubility of the chemical forces it to become concentrated either in the body water or in tissues that are high in lipids. The two compartments (very often the kinetics may be described accurately enough by the rather simple, two-compartment model) show different half-times of the chemical, and these are then reflected, for example, in the concentrations of the chemical in the blood or urine. In this case, two - or more - consecutive half-times may be discerned in the concentrations measured, as illustrated by curve Z in Figure 1.

**Fig. 2. Schematic representation of a chemical that is distributed in one compartment (B); and a chemical that is distributed in two compartments (C)**

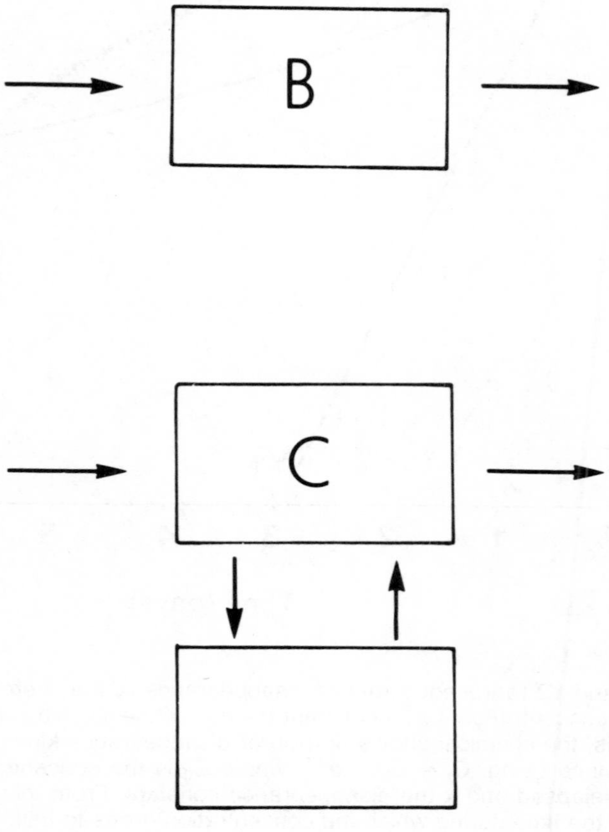


*Multiple doses*

Occupational exposure to chemicals is typically repetitive, day after day, week after week. In this situation, it is mainly the half-time of the chemical that determines the concentration changes of the chemical in body fluids, as exemplified in Figure 3.

**Fig. 3. Computer simulation of the concentration of a chemical in the blood after repetitive daily exposures**

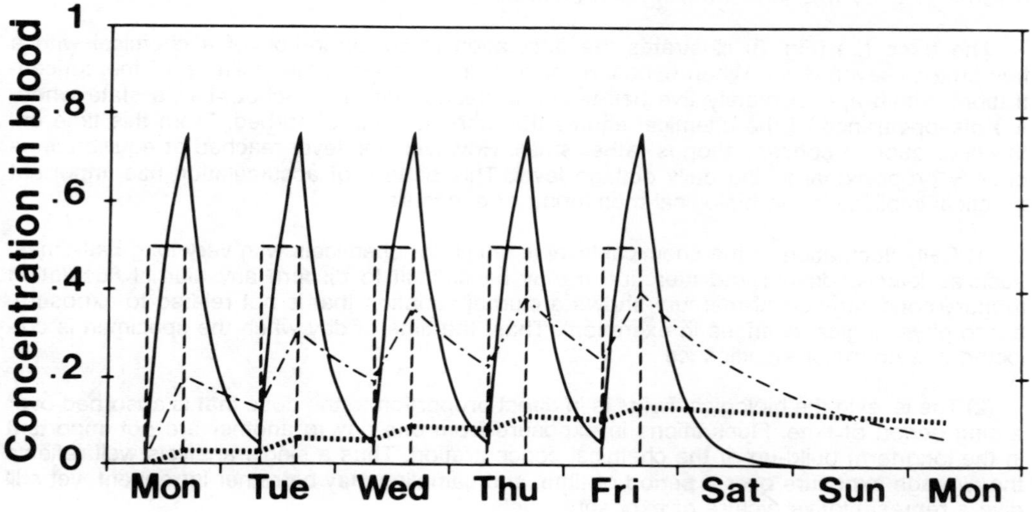


The bars denote exposure; the lines represent concentrations of chemicals with half-times of 4(—), 24 (—.—) and 168 h (...)

The continuous trace (Fig. 3) illustrates a chemical with a short half-time. Between exposures, the levels in body fluids return practically to zero, and no accumulation takes place, irrespective of the dose (as long as saturation is not reached). The daily changes in the concentration are very marked, from zero to a peak value (Fig. 3). The time of sampling is thus a major determinant of the concentration found. If the half-time is very short, it is practically impossible to evaluate anything but short-lived peak exposures; in these situations, the sampling must be performed immediately after the exposure. Should a chemical with this short half-time follow this excretion pattern all the way through, it would be impossible, by means of biological monitoring, to evaluate exposure over a working day, for example.

However, most chemicals are distributed in more than one compartment. For example, toluene shows at least three apparent half-times, viz., approximately 2 min, 1 h, and approximately 20 h. The last-mentioned half-time reflects the disappearance of toluene from body fat. As fat accumulation is a slow process - mainly limited by circulatory factors - and the release follows similar kinetics, this phase in the elimination may be related to a long-term exposure. In fact, fat functions as an integrator of blood-borne concentrations. As the half-time increases, after cessation of the exposure, strict timing of the sampling becomes less and less important. Thus, when sampling is done 15-20 h after the exposure, i.e., the morning after the exposure day and before the exposure of the sampling day starts, a fairly reliable picture may be obtained of the exposure on the day before.

For chemicals with short half-times, the concentrations measured from day to day are independent of each other. As it is rather common for exposure in the work place to vary considerably day by day, an estimation - even one well planned and executed - describes only one day at best. Thus, the half-time of the chemical has profound effects on sampling strategy. When the chemical shows a short half-time, a true picture of the exposure may be reached only by frequent sampling and analysis.

The trace (...) (Fig. 3) illustrates the fluctuation in concentration of a chemical with a half-time of seven days. When exposure begins, there is a gradual build-up of the concentration. Within approximately five half-times, a steady state is reached, i.e., a state where the disappearance of the chemical equals the amount being absorbed. From this time on, the fluctuation in concentration is rather small. However, the level reached at equilibrium is directly proportional to the daily dosage level. This pattern of accumulation has important practical implications in biological monitoring of exposure:

(1) Daily fluctuation of the chemical level is small; for chemicals with very long half-times, such as lead, cadmium and mercury, it may be difficult to discern any diurnal fluctuation (urinary concentration of mercury shows a diurnal variation that is not related to exposure, but to physiological changes in excretion). Thus, the time of day when the specimen is collected has no major significance.

(2) The level in the biological fluids is in direct proportion to the dose that is absorbed over a long period of time. Fluctuations in exposure from one day to another are not important in the long-term build-up of the chemical concentration. Thus a single analysis well reflects the average exposure over a period of time, and sampling may be rather infrequent, yet still give a representative picture of exposure levels.

*Skin absorption*

Many organic chemicals penetrate intact human skin easily. This has been shown, for example, for phenol, chlorinated phenols, several aromatic solvents, aromatic amines, organic nitrates and several different pesticides. In some instances, use of protective gloves is quite ineffective in preventing skin absorption. Skin absorption is one of the reasons for doing biological monitoring, as a complement to industrial hygiene measurements; the amount of a chemical absorbed through the skin need not bear any relationship with the concentration in the air. However, when the chemical itself is measured in circulating blood, skin absorption may cause a very marked error. Traditionally, the blood specimen is drawn from the cubital vein. If the forearms and hands have been in contact with a chemical that is absorbed through the skin, the measured concentration of that chemical represents only venous blood in the forearms. The concentration difference between the two hands may, under some circumstances, be 50-fold. As it is generally impossible to ensure that there is no skin exposure to a chemical that is absorbed through the skin, that chemical should not be measured in the blood at all. An alternative would be to measure a metabolite, in urine or blood, or the chemical itself in urine.

*Effect of meals*

Meals change the composition of the circulating blood. The most remarkable change is the increase in blood lipids. As many foreign chemicals in the work place are soluble in lipids, this meal-dependent increase in blood lipids changes the dissolution capacity of blood. It was shown, for example, that the venous blood:air concentration ratio of dichloromethane was directly related to the content of triglycerides in the blood, while that for *m*-xylene was higher after a meal. It is equally probable that lipid changes in the blood bring about changes in the blood:body fat distribution of chemicals. These changes may lead to erroneous interpretations of chemical concentrations in the blood. Therefore, optimally, specimen collection should be rigorously controlled with regard to meals. As the morning meal is usually less heavy, it seems advisable, if possible, to collect specimens in the morning.

*Kinetics of absorption*

The kinetics of absorption is very complicated for many occupational chemicals. The simplest case is that of many gaseous, organic solvents; they rapidly penetrate pulmonary membranes, and the absorption is a simple equilibrium reaction between inhaled air and blood. For aerosols, the situation is much more complicated, and depends on particle size, distribution of the chemical within the particle and the solubility of the chemical. It is important to note, also, that there are instances in which the amount of the chemical absorbed in the body proper is not important in relation to the toxic effect. It has been shown, for example, that after exposure to nickel derivatives in nickel refining, nasal mucosa retains slightly soluble nickel compounds for years. This nonabsorbed nickel causes the most severe nickel toxicity, the cancer. Thus the nickel that reaches the systemic circulation, and that which may be monitored in the blood or urine, is not meaningful at all.

Nickel represents a chemical with, apparently, very complex kinetics, also. This, however, is due to failure to speciate the nickel; one generally measures only total nickel, whereas the agent to which people are exposed may comprise several nickel derivatives with different solubilities and, therefore, different kinetic characteristics. In cases like this, one should be aware of the nature of the exposure, and interpret the results accordingly.

*Composite exposures*

Various nonspecific methods of monitoring exposure to carcinogenic and mutagenic agents, such as measurements of mutagenicity and thiol ethers, are discussed in this volume. One of the traditional arguments put forward in their favour is that they make it possible to monitor composite exposures, without actual knowledge of the chemical species concerned. It is implicit that such methods could measure even simultaneous exposures to a variety of mutagenic and chemical species, at a single measurement.

From what is stated above, concerning the importance of standardization of specimen collection, it is evident that for quantitative estimates of exposure, the kinetics of the phenomenon must be properly elucidated. Only when the kinetic features of the different mutagens to be monitored are similar enough, may quantitative monitoring of multiple chemical mutagens be accomplished by a single measurement.

# ASSESSING EXPOSURE OF INDIVIDUALS IN THE IDENTIFICATION OF DISEASE DETERMINANTS

**R. Saracci**

*International Agency for Research on Cancer, Lyon, France*

## BIOLOGICAL MARKERS OF EXPOSURE AND OF EARLY LESIONS

I interpret the title of my assignment in the framework of this seminar as meaning: what are the main advantages, disadvantages and problems of assessing exposure at the individual level, particularly by means of biological markers, rather than at the group level, in epidemiological research aimed at identifying chronic disease determinants and quantifying the size of their effects? This delineation leaves out - intentionally - the use of individual exposure monitoring in public health applications, e.g., to check compliance with existing control limits for pollutants, to establish new control limits or to monitor the health of individual workers. However, even public health applications may benefit indirectly from a discussion focussed on exposure assessment in epidemiological research.

There are two types of biological marker.

(1) *Markers of exposure* that is detectable, and possibly measurable on some continuous scale, by methods that are specific to single chemicals and lend themselves to a quantitation of exposure. This category includes a variety of techniques - physical, chemical, immunological - that measure foreign compounds and their metabolites in body fluids and tissues. If enough is known about kinetics, the measurements can lead to accurate assessments of the absorbed dose. This category also includes techniques for measuring the adducts of a foreign DNA-damaging compound with the DNA macromolecules or other macromolecules like the proteic moiety of haemoglobin, which may lead to an estimate of the dose available at critical sites. Where accurate calculations can be made, these methods represent, almost by definition, elective approaches for characterizing exposure to DNA-damaging agents.

(2) *Markers of early lesions* that are detectable by methods which are not chemical-specific, in the sense that they do not allow, when used alone, identification of the agent responsible for the lesion nor, *a fortiori*, quantification of the exposure of an individual to it. These markers encompass micronuclei, sister chromatid exchanges, chromosomal aberrations, unscheduled DNA synthesis, alkaline elution, lymphocyte mutation, sperm morphology, etc. Markers in this category are, in some sense, biologically significant in themselves and thus can be regarded as potential pathological endpoints. However, the key issue is, indeed, the biological level at which they are significant, the cell or tissue level or the level of the whole organism, i.e., in terms of health.

Certain markers, like mutagenic activity of, say, urine, are intermediate between category 1 and 2 but may be regarded preferably as category 2, as they reflect an early lesion-producing (e.g., mutagenic) ability and the effect is not specific to a single marker.

The use of markers of exposure and of early lesions leads to two types of epidemiological studies, each with its own potential advantages (Table 1). In the *first type*, one investigates the dependency relation of some late-occurring effect (like a birth defect or a cancer) on the marker. If the study uses markers of exposure that are both sensitive *and* specific for single DNA-damaging chemicals, it may contribute towards the establishment of causal relationships between exposure and effects. If, on the other hand, markers of early lesions are used, the situation is less clearcut. On the one hand, the advantage may lie in the fact that the marker (say, sister chromatid exchanges) may be a nonspecific signal that some unsuspected harmful exposure is, indeed, occurring. On the other hand, the advantage may be that the ultimate biological significance of the marker may be clarified by such a study.

**Table 1. Epidemiological research**

| | Exposure variables | Effect variables |
|---|---|---|
| 1. | Markers of exposure<br>Markers of early effects | Cancer, birth defects, etc. |
| 2. | Environmental measurements<br>Markers of exposure | Early effect markers<br>(e.g., sister chromatid exchanges; chromosomal aberrations, etc. |

The *second type* of epidemiological research explores the dependency relations of markers of early effects to the markers of exposure or to other measurements of exposure, such as the traditional questionnaire-based information or measurements in the external environment. If, by assumption or by independent demonstration, early effects are considered biologically relevant, this type of research has considerable potential. It replaces a long-term effect with a short-term one, usually an irreversible with a reversible one, thus allowing quicker successive studies, and also offers the possibility of experimental investigations in humans; further, by shortening the time interval between exposure and lesion, it leaves much less room for the operation of time-related confounding factors, permitting more direct inferences from the observed results.

## GROUP *VERSUS* INDIVIDUAL ASSESSMENT OF EXPOSURE

Both the advantages and the limitations of the types of studies just mentioned become more apparent in the light of two general questions that apply to all studies. Firstly, what are the advantages of measuring exposure at the individual level, as biological markers do, compared with measuring at the group level? Secondly, at the individual level of measurement, what are the advantages of biological markers over external environmental measurements or questionnaire assessment of exposure?

As to the first question, studies at the group level, i.e., so-called ecological studies, are often carried out in epidemiology. They prove relatively cheap most of the time, as they can capitalize on data collected, and already partly processed, for other purposes (e.g., administrative, economical, health care, etc.). A telling example, shown in Figure 1, is the now-classic relationship between tobacco consumption and lung-cancer death rates. The particular relationship shown in the figure refers to the 19 Italian regions. All it required was some hours spent digging out existing statistical data (Saracci, 1977) and making a few calculations. What may appear striking is that, from these data, it is possible to derive a neat exposure-response relationship, viz., estimates of the mortality ratios ('relative risk') of lung cancer as a function of daily cigarette consumption. As Table 2 shows, at each level of cigarette consumption there is less than 30% discrepancy between the estimated mortality ratios obtained from the Italian ecological data and from the American Cancer Society prospective study data (Hammond, 1966), the latter having involved the interview and follow-up of about one million people! The key point, however, is that one would hardly believe that the ratios from ecological data in the second column of Table 2 were a quantitative expression of a genuine causal link between tobacco and lung cancer without data from analytical studies (prospective or case-control) such as those in the third column. Indeed, evidence like that shown in Figure 1 was already available in the early 1950s and, being rightly regarded as merely suggestive, was chiefly used to prompt the inception of analytical studies. The basic reason for caution - sometimes pushed as far as scepticism - in interpreting ecological study results, is that there are no measures of exposure available at the individual, as opposed to the average group level, so that it is impossible to make a direct link between individuals who are exposed and those who die; it is quite possible that those who experience, say, lung cancer in a given geographical area are, in fact, not those exposed to an airborne pollutant in the same area. This type of occurrence is called 'aggregation-bias', which is characteristic of an ecological study. It was, for example, the inability to rule out confidently an aggregation bias, that led an International Agency for Research on Cancer Working Group (IARC, 1982) to regard the epidemiological evidence of carcinogenicity of aflatoxin $B_1$, solely based on ecological studies, as 'limited' rather than 'sufficient' (more recently some data from analytical investigations have become available (Bulatao-Jayme *et al.*, 1982); The aggregation bias may operate either - and more often - as a false positive association (i.e., an exposure/disease association which vanishes when exposure data at the individual level are used) or as a false negative association. Also, a false negative result may occur particularly when large groups are used (e.g., geographical areas with large populations) and/or the exposure is rare (e.g., an occupational exposure concentrated in few workplaces), by dilution of the few exposure-related cases in the majority of unrelated cases in the general population. Analyses of ecological data are almost unavoidable in epidemiological investigation of some problems, notably when an agent tends to be universally diffused, e.g., general air pollutants, drinking water pollutants, etc., where there may not be enough variation in exposure within a given population to be measurable with available methods and inter-population (group) comparisons may be a better alternative. As a rule, however, the intrinsic weaknesses of the ecological approach can be remedied only by individualizing the exposure assessment, i.e., performing it at the level of each individual in a study.

**Fig. 1. Cigarette consumption (1950-1951) and age-adjusted male lung-cancer death rates (1970-1971) in the 19 Italian regions**

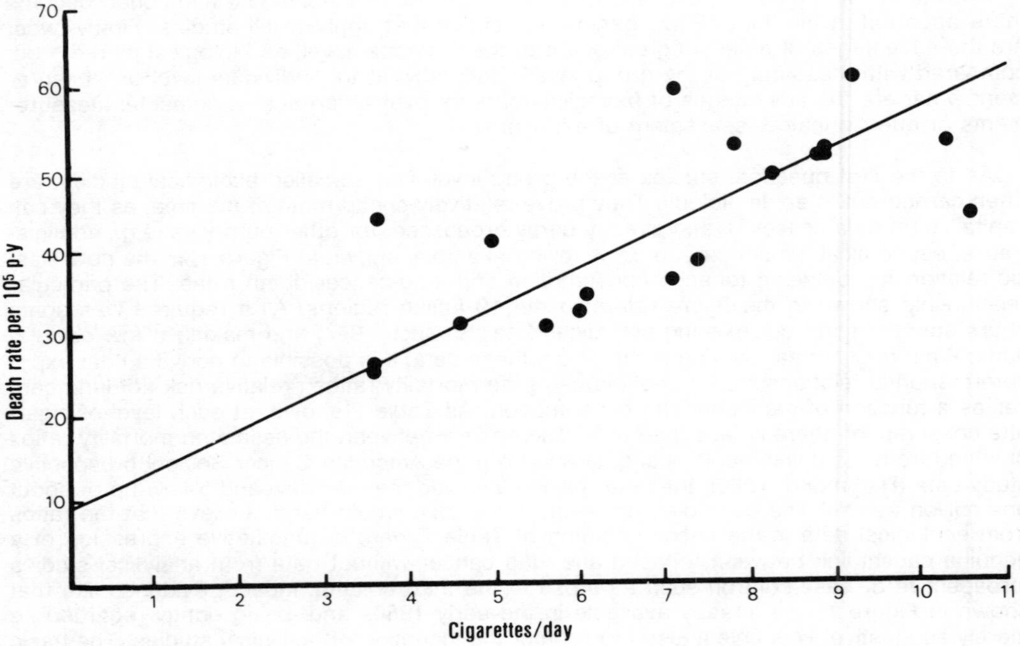


**Table 2. Mortality ratios of lung cancer as a function of smoking intensity**

| Cigarettes/day | Italian ecologic data analysis (males all ages) | American Cancer Society prospective study (males aged 35-84) |
|---|---|---|
| 0 | 1 | 1 |
| 1-9 | 3.6 | 4.6 |
| 10-19 | 8.8 | 7.5 |
| 20-39 | 16.7 | 13.1 |
| 40+ | 21.9 | 16.6 |

## INDIVIDUAL EXPOSURE ASSESSMENT: BIOLOGICAL *VERSUS* ENVIRONMENTAL MEASUREMENTS AND QUESTIONNAIRE INFORMATION

At the individual level, the merits of biological markers, either specific or nonspecific, must be compared with external exposure measurements or questionnaire-based information, as traditionally obtained in epidemiological studies.

One basic factor in any kind of marker comparison, and one which is often insufficiently stressed, is that one marker is preferable to another, other things being equal, if the relative risk associated with the first is found to be equal or higher than the relative risk associated with the second. To take a much-investigated domain, albeit outside the direct focus of genotoxic exposures, there is no doubt that pleural plaques are a marker of increased risk of pleural and bronchial neoplasias in the context of some occupational exposures to asbestos (Selikoff & Lee, 1978). However, the pertinent question is not whether subjects with plaques are at greater risk than subjects not exposed to asbestos, but whether they are at greater risk than other subjects apparently equally exposed to asbestos but not exhibiting any plaques. In other words, because plaques reflect either individual susceptibility or higher accumulated individual exposure, or both, is their presence a better predictor of increased risk than the fact that the worker has been operating for a number of years, say, in a dockyard? Notwithstanding the number of studies of asbestos workers, there is no clear answer to this question and the predictive significance of plaques is still to be established. The situation is not much better if one turns to another example - benzene exposure. It is true that pancytopenia, which is a relatively late effect, is usually accepted by haematologists as a marker of an increased risk of leukaemia (Goldstein, 1983), but when one moves to the more obviously interesting early effect marker of chromosomal aberrations, the position is once more unclear in quantitative terms. Do chromosomal aberrations mark an increased risk of leukaemia in the simple terms of level of exposure to benzene in a worker's environment, or in the even simpler terms that he has been for a number of years in a benzene-polluted environment and, related to this, do chromosomal aberrations appear at levels of exposure higher or lower than those for leukaemia? Such questions seem unanswerable at present, yet the research value of the biological markers depends largely, though not exclusively, on them.

Finally, a case which shows the advantage of a well-selected, sensitive and, at the same time, specific marker of exposure is that of primary hepatocellular carcinoma (again, outside genotoxic exposures, which do not offer too many examples at present). Here, the assessment of antecedent hepatitis, in particular hepatitis B infection, has been dramatically improved over other possible markers of exposure (e.g., clinical history) by the use of hepatitis B surface antigen and, more recently, by the antibodies to the core antigen. For example, Table 3 shows that a relative risk of 223 (1158.0/5.2) for primary hepatocellular carcinoma has been estimated in a prospective study in Taiwan (Beasley *et al.*, 1981), using hepatitis B surface antigen as a marker. A relative risk of this large size in a prospective study is most unlikely to be entirely explained by some biasing or confounding factor, and therefore suggests a causal role for the exposure under study, in this case, the one of which hepatitis B surface antigen is a marker, viz., the hepatitis B virus.

**Table 3. Deaths by cause and hepatitis B surface antigen (HBsAg) status on recruitment (modified from Beasley *et al.*, 1981)**

| HBsAg status on recruitment | Cause of death | | | Population at risk | PHC incidence[b] |
|---|---|---|---|---|---|
| | PHC[a] | Cirrhosis | Other | | |
| HBsAg positive | 40 | 17 | 48 | 3454 | 1158.0 |
| HBsAg negative | 1 | 2 | 199 | 19253 | 5.2 |

[a] Primary hepatocellular carcinoma

[b] Incidence of death from PHC per 100 000 during the time of the study

This leads to the consideration of a second important factor, that of uncontrolled confounding, and more generally of bias, which might affect a marker of risk and, in particular, a biological marker.

Measurements of biological markers can usually be freed from some important sources of bias, e.g., samples of lymphocytes can be tested for DNA adducts in the laboratory without knowledge of the exposure, nor of the disease status of the subjects from whom samples were taken. However, when the presence of a marker is tested in, say, lung-cancer cases and healthy controls, there is a possibility that metabolic derangements induced by the cancer may be a more important determinant of the marker's presence than the exposure itself.

Under these circumstances, an apparently cruder method of exposure assessment, recall through questionnaire, might provide information of greater validity than that of a biological test. In another respect, which is obvious but crucial, a recall method, particularly if supplemented by past environmental measurement may, in practice, work better than a biological marker. Currently available markers of exposure reflect exposure on a short-term scale (days, weeks or months) rather than on the long-term scale (years or quinquennia) which is usually relevant to the occurrence of such effects as cancers. The labile memory of a person questioned by an interviewer is better than the absence of elicitable memory of a blood sample, if we want to investigate cancer etiology now rather than waiting for 20 years in order to see the effect appear. There may be fortunate exceptions to this situation; some existing banks of biological material (mostly sera, not the best material for assessing DNA-damaging foreign compounds) may go some way towards meeting the need for testing biological markers, by the use of specimens collected many years ago from subjects who subsequently developed cancer.

Finally, one should not lose sight of the fact that biological marker associations with diseases are no less prone than questionnaire information to confounding effects by interfering variables. For the sake of example, consider that alcohol exposure is associated, most probably causally, with the occurrence of oesophageal cancer. However, alcohol drinkers show higher activities of some drug-metabolizing enzymes (Doull *et al.*, 1980). Therefore, any foreign compound, whether carcinogenic or not, that is handled by these same enzymes, and to which alcohol drinkers and nondrinkers are uniformly exposed, will produce metabolites which will be differentially distributed within the body fluids of drinkers and nondrinkers. The result will be, that if exposure to the xenobiotic is assessed through some external measurement, no association with oesophageal cancer will be found, but if exposure is assessed through levels of metabolites, a spurious association with oesophageal cancer will be found, purely by reason of the confounding metabolic effect of alcohol. I have not the experience in this field to assess whether situations of this type are common or not. However, taking into account the large number of factors (environmental and physiological) that are reported to influence handling of xenobiotics (Vesell, 1982), one may assume that perhaps more attention should be given to the play of these confounding factors when studies are designed. This question is all the more important, because published papers often give insufficient attention even to obvious confounders like age or sex in studies correlating exposure to biological markers and to early lesions, e.g., how can one interpret a 'statistically significant' difference from, says, 2% to 3% of chromosomal aberrations in subjects exposed and nonexposed to an agent, in the absence of a *tight* control over age (rather than just a coarse control, grouping subjects in large age intervals)?

## THE VALUE OF BIOLOGICAL MARKERS IN EPIDEMIOLOGY

Bringing together some common threads running through the arguments just sketched, a few points can be made in conclusion:

(1) General considerations, such as those just oulined, may serve as useful references in critically examining the value of biological markers for epidemiological studies. However, each marker must be considered in its own right, and the merits and demerits of one do not automatically transfer to another. Questions of acceptability, practicability and costs for large-scale studies need to be considered, as well as technical aspects. Markers of exposure in the strict sense, particularly those which, like haemoglobin adducts, may reflect integrated exposure to specific compounds over several weeks (Calleman, 1982), appear among the most attractive.

(2) As already emphasized by others (Fig. 2; Bridges, 1982) there is a need to calibrate progress from the exposure itself to exposure markers, early lesions and late lesions, both in animals and in man (in parallel), so that, hopefully, extrapolation from experimental systems to man may be made without too much uncertainty, and without having to await the full natural history of a disease in man before getting the necessary answers. Particular attention should be given to confounding factors in this type of calibration, and a certain amount of ingenuity, backed by a collaboration of epidemiologists, toxicologists and experimental scientists, will be required to identify the appropriate study designs, particularly in man.

**Fig. 2. Calibration of human response to DNA-damaging agents against experimental response (after Bridges, 1982)**

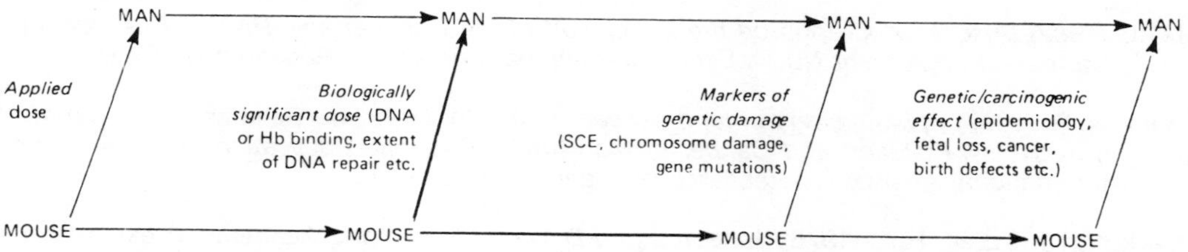


Hb, haemoglobin; SCE, sister chromatid exchanges

(3) For the time being, the use of markers in epidemiological investigation appears most useful and less debatable: (a) in those situations where external environmental measurements or questionnaire information is unable to define individual exposure, e.g., in problems of general environmental atmospheric pollution; (b) as a complement to environmental measurements and questionnaire information in selected situations.

Although better knowledge of the properties of biological markers is needed before they can be used as elective indicators of exposure in epidemiological studies, and although, somewhat circuitously, long-term epidemiological studies are needed to establish the ultimate value and health significance of such markers, the practical question is already open: do we wish to use some of these markers as partial substitutes for long-term effects and regulate public health action accordingly? As I said at the beginning, this question is outside my assignment, but I hope it will find an adequate place in the development of our discussions during this seminar.

## REFERENCES

Beasley, R.P., Hwang, L.Y., Lin, C.C. & Chien, C.S. (1981) Hepatocellular carcinoma and hepatitis B virus. A prospective study of 22 707 men in Taiwan. *Lancet*, *ii*, 1129-1133

Bridges, B.A. (1982) *Indicators of genotoxic exposures: status and prospects*. In: Bridges, B.A., Butterworth, B.E. & Weinstein, I.B., eds, *Indicators of Genotoxic Exposure (Banbury Report 13*), Cold Spring Harbor, NY, Cold Spring Harbor Laboratory, pp. 555-558

Bulatao-Jayme, J., Almero, E.M., Castro, A.C., Jardelezo, M.T.R. & Salarnat, L.A. (1982) A case-control dietary study of primary liver cancer risk from aflatoxin exposure. *Int. J. Epidemiol.*, *11*, 112-119

Calleman, C.J. (1982) In-vivo *dosimetry by means of alkylated hemoglobin - a tool in the design of tests of genotoxic effects*. In: Bridges, B.A., Butterworth, B.E. & Weinstein, I.B., eds, *Indicators of Genotoxic Exposure ( Banbury Report 13*), Cold Spring Harbor, NY, Cold Spring Harbor Laboratory, pp. 157-168

Doull, J., Klaassen, C.D. & Amdur, M.O. (1980) *Casarett and Doull's Toxicology*, 2nd ed., New York, MacMillan

Goldstein, B.O. (1983) Benzene is still with us. *Am. J. ind. Med.*, *4*, 585-587

Hammond, C.E. (1966) Smoking in relation to the death rates of one million men and women. *Natl Cancer Inst Monogr.*, *19*, pp. 127-204

IARC (1982) *IARC Monographs on the Evaluation of the Carcinogenic Risk of Chemicals to Humans*, *Supplement No. 4*, Lyon, International Agency for Research on Cancer

Saracci, R. (1977) *Epidemiology of lung cancer in Italy*. In: Mohr, U., Schmähl, D. & Tomatis, L., eds, *Air Pollution and Cancer in Man* (*IARC Scientific Publications No. 16*), Lyon, International Agency for Research on Cancer, pp. 205-215

Selikoff, I.J. & Lee, H.K. (1978) *Asbestos and Disease*, New York, Academic Press

Vesell, E.S. (1982) *Complex, dynamically interacting host factors that affect the disposition of drugs and carcinogens*. In: Bartsch, H. & Armstrong, B., eds, *Host Factors in Human Carcinogenesis* (*IARC Scientific Publications No. 39*), Lyon, International Agency for Research on Cancer, pp. 427-437

# MONITORING EXPOSURE TO 4-AMINOBIPHENYL *VIA* BLOOD PROTEIN ADDUCTS

**P.L. Skipper, L.C. Green, M.S. Bryant & S.R. Tannenbaum**

*Massachusetts Institute of Technology, Department of Nutrition and Food Science, Cambridge, MA, USA*

**F.F. Kadlubar**

*National Center for Toxicological Research, Jefferson, AR, USA*

## SUMMARY

The feasibility of monitoring exposure to 4-aminobiphenyl (4-ABP) was determined by measuring the formation of covalent blood protein adducts in rats following administration of the amine. A single major adduct was obtained from serum albumin, which was formed from 4-acetylaminobiphenyl and the single tryptophan residue of albumin. The adduct was isolated as part of a tetrapeptide, following Pronase digestion and reverse-phase high-performance liquid chromatography purification, and was identified by amino acid analysis, $^{1}$H-nuclear magnetic resonance and mass spectrometry. This adduct was cleared *in vivo* and showed a half-life of 4.7 days, essentially identical to that of albumin. Similarly, 4-ABP yielded a single major adduct with haemoglobin. This adduct showed a stability *in vivo* that was comparable to that of haemoglobin, and was shown to accumulate, with chronic dosing, to a level 30 times higher than that resulting from a single dose. Treatment of adducted haemoglobin, with 0.1 N hydrochloric acid in acetone, caused hydrolysis and release of 4-ABP. The formation of the haemoglobin adduct involves 4-nitrosobiphenyl as an intermediate product and, by implication, 4-hydroxyaminobiphenyl. Thus, the haemoglobin adduct is a measure of the fraction of 4-ABP that is *N*-hydroxylated directly, whereas the albumin adduct reflects that portion which is acetylated prior to *N*-hydroxylation. Approximately 0.02% of a dose of 4-ABP was bound to the tryptophan in albumin (dose of 2-80 mg/kg) and 5% was bound to haemoglobin as an acid-labile adduct (dose of 0.5-5000 μg/kg).

## INTRODUCTION

The concept, that accurate surveillance of exposure to carcinogens could be accomplished by measurement of the covalent adducts formed between the carcinogens and haemoglobin, was verified experimentally by studies on ethylene oxide and other alkylating agents (Ehrenberg & Osterman-Golkar, 1980). The prerequisites for dosimetry are that the adducts be stable and be formed in a dose-dependent fashion, and that sensitive analytical techniques be available for their quantification. These criteria are readily satisfied for the simple alkylating agents. The question of whether they can be met, also, for those carcinogens that are metabolically and chemically more complicated, is examined in this report, which details initial work aimed at the development of dosimetry for 4-aminobiphenyl.

4-Aminobiphenyl (4-ABP) is a recognized human bladder carcinogen (IARC, 1972). In experimental animals, however, it is known to initiate tumours in other organs as well (Clayson & Garner, 1976). It has been suggested that the target site is, at least partially, dependent on the extent to which the administered amine is acetylated (Kadlubar *et al.*, 1977; McQueen *et al.*, 1983). In the dog, which has almost no acetylase activity toward aromatic amines, 4-ABP is an effective bladder carcinogen. In species with active *N*-acetylation (rat, mouse, hamster), bladder tumours are rarely observed, although organs such as intestine, breast and liver are targets. In humans, aromatic amine acetylase activity is under genetic control, so that populations display two distinct phenotypes, the 'slow acetylator' and the 'fast acetylator'. Epidemiological studies have demonstrated a correlation between low acetylase activity and risk of bladder cancer, suggesting a role for aromatic amines in the etiology of this disease (Lower *et al.*, 1979; Cartwright *et al.*, 1982). Thus, dosimetry of 4-ABP becomes, at once, more complicated by the involvement of *N*-acetylation in the disposition of this carcinogen, and more rewarding because it might reveal insights into individual differences. In the present work, we show that it is possible, not only to quantify exposure to 4-aminobiphenyl, but also to measure the extent to which the compound is subject to the very important metabolic pathway of acetylation.

## MATERIALS AND METHODS

### *Animals and dosing*

[2,2'-$^3$H]-4-Aminobiphenyl (Midwest Research Institute, Kansas City, MO, USA) was diluted with unlabelled 4-ABP (Sigma, St Louis, MO, USA), to a specific activity of 0.75 mCi/mmol for the preparation of albumin adduct, or 30 mCi/mmol for all other studies; it was administered to young adult male Sprague-Dawley rats (Charles River, Wilmington, MA, USA), either as a solution in dimethyl sulfoxide by intraperitoneal injection (single-dose studies) or in corn oil by gavage (chronic study). For single-dose studies, blood was obtained by cardiac puncture. For multiple-dose studies, blood was obtained by puncture of a tail vein. Thirty animals were used to prepare albumin adduct and were given 4-ABP (100 mg/kg) as a solution in corn oil by gavage.

### *Isolation and fractionation of haemoglobin*

Blood was iced immediately after collection and centrifuged to separate red cells. After washing (3×) with saline, the cells were lysed with 3 vol distilled water. Four vol of 0.67 M phosphate (pH 6.5) were added to redissolve haemoglobin, and the haemolysate was then centrifuged at 25 000 × *g* at 4°C for 25 min to precipitate the cellular debris. The supernatant was dialysed against distilled water and then added dropwise to 20 vol of iced 1% concentrated hydrochloric acid in acetone. The precipitated globin was removed by centrifugation, and the acetone was evaporated. The residual aqueous solution was filtered to remove haem and further purified on a $C_{18}$ Sep-Pak (Waters Assoc., Milford, MA, USA). Three fractions were collected from the Sep-Pak: material that eluted with (1) 10 mmol/L potassium chloride buffer, pH 2.5, (2) buffer:methanol - 1:1, and (3) methanol. The 4-ABP elutes in fraction 2. That the source of the radioactivity in fraction 2 was 4-ABP was confirmed by high-performance liquid chromatography: over 90% of radioactivity cochromatographed with authentic 4-ABP.

*Albumin isolation and adduct purification*

Albumin was isolated from plasma by fractional precipitation with ammonium sulfate (Peters, 1962). It was dissolved in 0.05 M phosphate buffer (pH 8) and treated with Pronase (20 mg/g albumin) at 37°C in a shaking water-bath for 18 h. For clearance studies, this digest was injected directly onto a reverse-phase high-performance liquid chromatography column (Whatman ODS-3) for purification and quantification of the adduct by liquid scintillation counting. The column was eluted with a linear gradient of methanol in ammonium formate buffer (0.05 M, pH 6.4) over 40 min. In preparation for the clearance studies, the digest was subjected to an initial clean-up, which consisted of loading portions onto the high-performance liquid chromatography column and eluting successively with pure water and methanol, the latter containing all the radioactive components. The methanol eluants were pooled, concentrated and rechromatographed repeatedly, until five major products were obtained.

## RESULTS

Rats dosed with $^{3}$H-4-ABP accumulated a considerable fraction of the dose in their blood. Twenty-four hours after a single dose of 5 mg/kg, rats showed 8.0% of the administered radioactivity in the red blood cell compartment, and less than one-tenth of this in the plasma compartment. After dialysis of the washed, lysed, membrane-free red cells, 7.3% of the dose remained. Treatment with acidic acetone precipitated only 0.13% of the dose with the globin, while 6.6% of the dose was recovered in the acidic acetone solution. When this solution was evaporated and further purified on a $C_{18}$ Sep-Pak, 5.6% of the dose eluted in the 50:50 buffer:methanol wash. Chromatography of this eluate revealed that the major labelled product was $^{3}$H-4-ABP itself, accounting for 5% of the original dose. Since prolonged dialysis of the red cell lysate, and treatment of the haemolysate with acetone alone, in the absence of acid, both failed to liberate significant radioactivity, it appears that a form of 4-ABP had bound covalently to haemoglobin *in vivo*, and this adduct was acid-hydrolysed *in vitro*.

The radioactivity that was associated with albumin in the plasma compartment appeared in five separable products, following Pronase digestion of the albumin. The last of these to elute from the high-performance liquid chromatography column was the only one that contained 4-ABP covalently bound to an amino acid. It was identified as the tetrapeptide, $H_2$*N*-ala-try(-ABP)-ala-val, the carcinogen modification being 3-(tryptophan-*N*1-yl)-4-acetylaminobiphenyl. The identity was established by amino acid analysis, which revealed the presence of two equivalents of alanine and one of valine per $^{3}$H; by $^{1}$H-nuclear magnetic resonance spectroscopy in dimethyl sulfoxide-$d_6$ with a Bruker WM-500 spectrometer, which permitted a complete assignment of all expected resonances; by mass spectrometry (Kratos MS-50 with FAB source), which established the molecular weight as 654; and by ultra-violet spectroscopy, which established the absence of ionizable substituents on the chromophore. A second component was identified as 4-ABP by its retention time on high-performance liquid chromatography, and by isolation and gas chromatography, following derivatization with pentafluorobenzoyl chloride. A third component was identified as 4'-hydroxy-4-acetylaminobiphenyl, by the absence of amino acids as determined with an amino acid analyser; by $^1$H-nuclear magnetic resonance spectroscopy with a Bruker WH-270 spectrometer; by mass spectrometry with a Finnigan-MAT 4023 spectrometer, with a thermal in-beam desorption probe that produced a spectrum consisting of a molecular ion at m/z 227 and a base peak at m/z 185, corresponding to loss of ketene; and by ultra-violet spectroscopy which revealed the phenolic character of the compound. The remaining two compounds were not fully characterized, but it was inferred that they were not amino acid adducts by the results of $^1$H-nuclear magnetic resonance spectroscopy with a Bruker WM-500 spectrometer, and by their failure to yield detectable amino acids.

*In vivo*, the level of the acid-labile haemoglobin adduct formed was a linear function of the logarithm of the dose of 4-ABP, over the range of doses administered (Fig. 1). No detail-

**Fig. 1. In-vivo formation of the acid-sensitive haemoglobin-4-aminobiphenyl adduct as a function of single intraperitoneal doses of 4-aminobiphenyl**

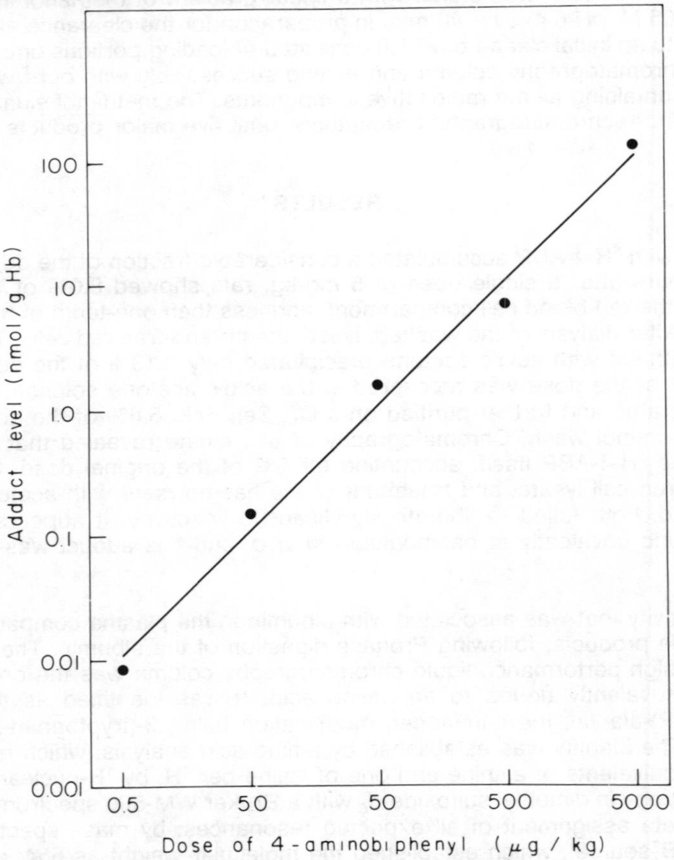


Adduct levels were determined after 24 h as described in the text. Hb, haemoglobin ed dose-response data were obtained for the albumin adduct formation but we observed, in the preparation of the adduct for identification, a yield of 0.02% based on estimated total albumin.

Chronic administration of 4-ABP led to an accumulation of radioactivity in the blood some 30-fold higher than that found after a single dose (Fig. 2). Upon cessation of dosing, the level of bound radioactivity decreased, initially, by 2.5% of its peak value per day, but after several weeks this rate had declined. About 60-65 days were required to clear all radioactivity from the blood.

**Fig. 2. Accumulation of haemoglobin adducts resulting from chronic administration of 4-aminobiphenyl**

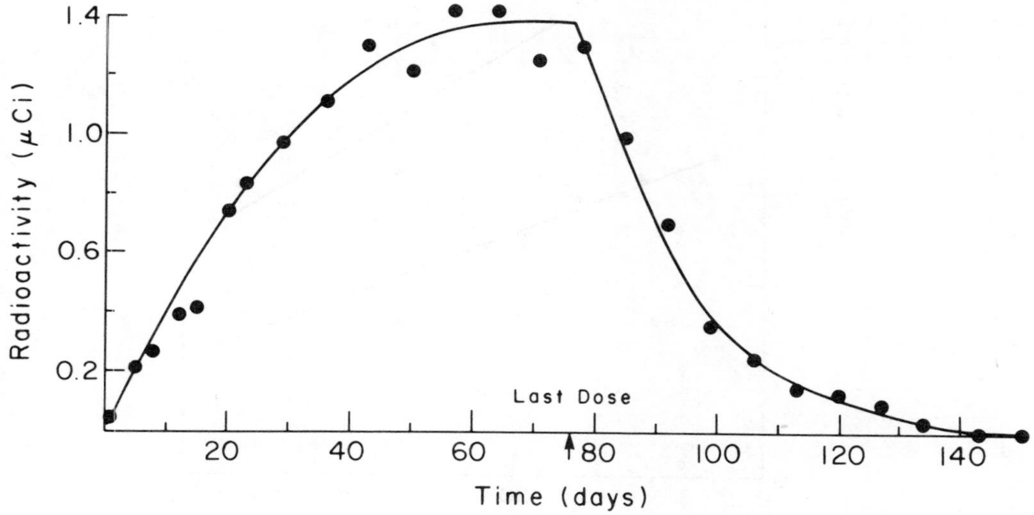


A dose of 2.05 μCi (11.4 μg) was administered on alternate days by gavage. Radioactivity is expressed as the total bound to all haemoglobin, based on a value of 6.4 ml blood/kg body weight.

The clearance of radioactivity associated with albumin was exponential (Fig. 3). The total activity clearance had a half-life of approximately 72 h, while clearance in the form of the one covalent adduct had a half-life of approximately 112 h. The latter value corresponds to 4.7 days, which is very close to reported values for the turnover of albumin in the rat ($t_{1/2}$ = 4 days).

**Fig. 3. Clearance of total $^3H$ (•) and 3-(tryptophan-*N*1-yl)-4-acetylaminobiphenyl (▲) from serum albumin, following administration of single doses of 4-aminobiphenyl**

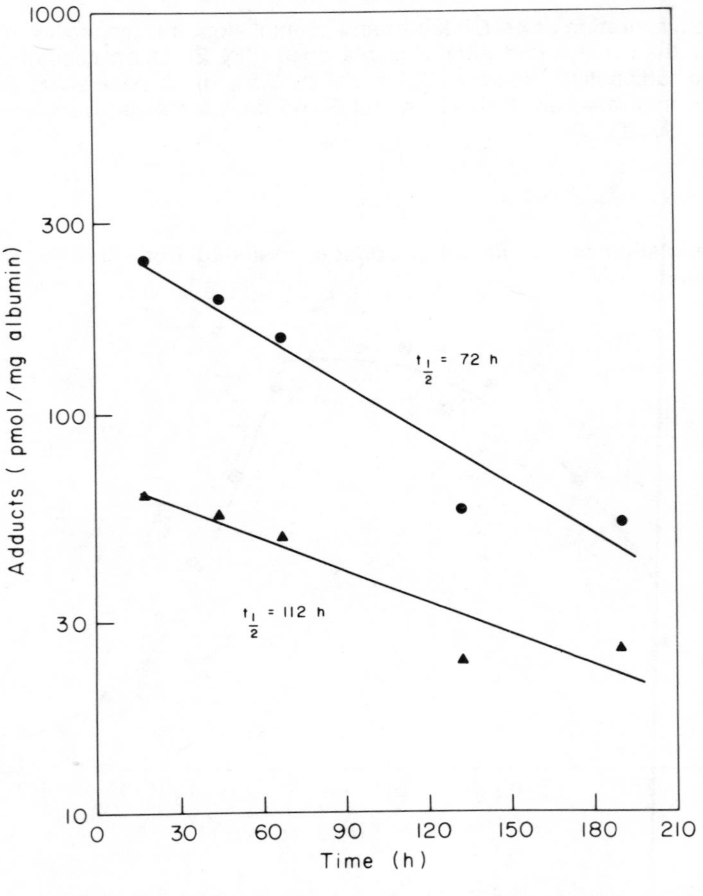


## DISCUSSION

Administration of 4-ABP to rats resulted in the formation of two distinct types of covalent adducts with blood proteins. In albumin, the site of attack is the single tryptophan residue, which, significantly, occurs in a sequence that is homologous in rat and man. The adduct is formed from acetylated 4-ABP. In haemoglobin, an adduct is formed that has not yet been identified with certainty. However, there is a considerable body of research on analogous aromatic amines, including aniline, aminofluorene and aminostilbene (Kiese & Taeger, 1976; Dölle *et al.*, 1980) as well as other, as yet unpublished data from our laboratories on the reaction of hydroxylaminobiphenyl, which make it almost certain that the adduct is a cysteine sulfinamide. The chemistry of the formation of this adduct is depicted in Figure 4. The intermediate compound, 4-nitrosobiphenyl, precludes the formation of any sulfinamide from acetylated 4-ABP. Thus, provided that other variables that affect the dose-response relationship of adduct formation are known, albumin adducts can be used as a dosimeter for the fraction of 4-ABP that undergoes *N*-acetylation, while haemoglobin adducts can be used to measure the remainder which is not acetylated.

**Fig. 4. Proposed scheme for formation *in vivo* and hydrolysis *in vitro* of the acid-sensitive haemoglobin-4-aminobiphenyl adduct**

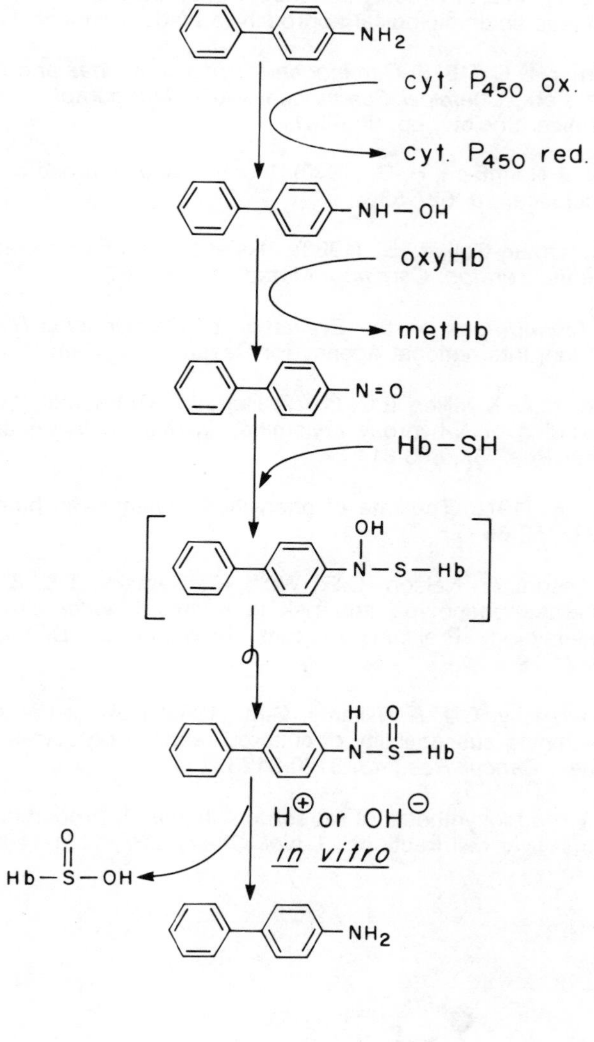


Hb, haemoglobin

**ACKNOWLEDGEMENTS**

This work was supported by Grant No. 5-P01-ES00597-13 from the National Institute of Environmental Health Sciences and Grant No. SIG-10-1 from the American Cancer Society.

**REFERENCES**

Cartwright, R.A., Glashan, R.W., Rogers, H.J., Ahmad, R.A., Barham-Hall, D., Higgins, E. & Kahn, M.A. (1982) Role of *N*-acetyltransferase phenotypes in bladder carcinogenesis: a pharmacogenetic epidemiological approach to bladder cancer. *Lancet*, *ii*, 842-846

Clayson, D.B. & Garner, R.C. (1976) *Carcinogenic aromatic amines and related compounds*. In: Searle, C.E., ed., *Chemical Carcinogens, ACS Monograph 173*, Washington, DC, American Chemical Society, pp. 366-461

Dölle, B., Töpner, W. & Neumann, H.-G. (1980) Reaction of arylnitroso compounds with mercaptans. *Xenobiotica*, *10*, 527-536

Ehrenberg, L. & Osterman-Golkar, S. (1980) Alkylation of macromolecules for detecting mutagenic agents. *Teratog. Carcinog. Mutag.*, *1*, 105-127

IARC (1972) *IARC Monographs on the Evaluation of Carcinogenic Risk of Chemicals to Man*, Vol. 1, Lyon, International Agency for Research on Cancer, pp. 74-79

Kadlubar, F.F., Miller, J.A. & Miller, E.C. (1977) Hepatic microsomal *N*-glucuronidation and nucleic acid binding of *N*-hydroxy arylamines in relation to urinary bladder carcinogenesis. *Cancer Res.*, *37*, 805-814

Kiese, M. & Taeger, K. (1976) The fate of phenylhydroxylamine in human red cells. *Arch. Pharmacol.*, *292*, 59-66

Lower, G.M., Jr, Nilsson, T., Nelson, C.E., Wolf, H., Gamsky, T.E. & Bryan, G.T. (1979) *N*-Acetyltransferase phenotype and risk in urinary bladder cancer: approaches in molecular epidemiology. Preliminary results in Sweden and Denmark. *Environ. Health Perspect.*, *29*, 71-79

McQueen, C.A., Maslansky, C.J. & Williams, G.M. (1983) Role of the acetylation polymorphism in determining susceptibility of cultured rabbit hepatocytes to DNA damage by aromatic amines. *Cancer Res.*, *43*, 3120-3123

Peters, T., Jr (1962) The biosynthesis of rat serum albumin. I. Properties of rat albumin and its occurrence in liver cell fractions. *J. biol. Chem.*, *237*, 1181-1185

## SESSION II

Chairman: A.E. Bennett
Rapporteur: P.H.M. Lohman

# BIOMONITORING OF CHEMICALS AND THEIR METABOLITES

**N.J. van Sittert**

*Health, Safety and Environment Division,*
*Occupational Health and Toxicology,*
*Biomedical Services,*
*Shell International Petroleum Maatschappij,*
*The Hague, The Netherlands*

## INTRODUCTION

Biomonitoring, or biological monitoring, can be defined as 'the measurement of internal exposure through analysis of a biological specimen', such as urine, blood, faeces, expired air (Zielhuis, 1978). The purpose of biological monitoring is to provide an estimate of the dose of a chemical at the level of the individual, the tissue or the target, thereby providing a better indication of received dose than, for example, measurements of the air concentration of a chemical in the breathing zone of the individual. Numerous published methods can be used to measure human exposure to chemicals, indirectly, by biological monitoring (WHO, 1982; Lauwerys, 1983). The substances monitored include industrial chemicals, agrochemicals and metals, most of which are not recognized as potential carcinogens or mutagens. With respect to the mutagens, however, relatively few methods have been published. This paper discusses the various possibilities for measurement of human exposure to potential carcinogens or mutagens by biological monitoring.

## METABOLISM OF CARCINOGENS AND MUTAGENS

Before biological monitoring can be applied in humans, the metabolism of the compound being monitored must be known. Thus, it is necessary to establish the nature and the kinetics of formation of the end products of metabolism in experimental species and, where possible, in humans.

Potential carcinogens and mutagens can be classified under two main headings:

(1) *Precursor carcinogens and mutagens* include known carcinogens and mutagens such as polycyclic aromatic and heterocyclic hydrocarbons, *N*-nitrosamines and *N*-nitrosamides, nitroaryl and furan compounds, aromatic and heterocyclic amines and azodyes, alkyltriazines and dialkylhydrazines and naturally occurring compounds such as aflatoxin and pyrrolizidine compounds. If precursor carcinogens and mutagens are not converted into ultimate carcinogens and mutagens in the organism, they are noncarcinogenic and nonmutagenic to that organism (Miller & Miller, 1976). Conversion may occur by spontaneous chemical reaction, i.e., a hydrolytic process, or by enzyme-mediated reactions. Most ultimate reactants are strong electrophiles, mainly alkylating and arylating species and, generally, have been found to be more effective than the parent compound in terms of carcinogenic and mutagenic potency. It has been proposed, also, that some ultimate reactants may have a free radical character (Cavalieri *et al.*, 1978).

(2) *Direct-acting carcinogens and mutagens* are ultimate mutagens and carcinogens *per se* and include, for example, alkylating and acylating agents.

*Reaction with cellular macromolecules*

Ultimate carcinogens and mutagens possess the intrinsic properties necessary for interaction with critical targets. For chemical mutagens, DNA has been established as the critical target (Drake & Baltz, 1976). Although less well understood than mutagenesis, interaction of the ultimate carcinogen with any or all of the information macromolecules, i.e., DNA, RNA and proteins, can be considered a critical lesion in the process of chemical carcinogenesis. These interactions are covalent bindings, that are formed by reaction of the electrophilic centres of ultimate mutagens or carcinogens with nucleophilic sites in nucleic acids (e.g., *N*-7 guanine, *N*-3 and *N*-1 adenine, *O*-6 guanine in DNA) and proteins (e.g., *N*-3 and *N*-1 histidine, *S*-cysteine, *N*-terminal valine). The nature of these covalent interactions depends on the chemical structure of the ultimate electrophilic reactant which, for precursor carcinogens and mutagens, is mainly determined by the action of mammalian enzyme systems. With exposure to direct-acting carcinogens and mutagens, the magnitude of key interactions depends on the efficiencies and capacities of deactivating enzymes. With exposure to precursor-carcinogens and mutagens, it is the balance between activating and deactivating enzymes that is important.

*Activation of precursor carcinogens and mutagens*

As discussed above, precursor carcinogens and mutagens can be converted into ultimate reactants, depending on the presence of activating enzymes. The reactions are, in most cases, oxygenation reactions, catalysed by membrane-bound microsomal oxygenases. Some of the reactive intermediates catalysed by the monooxygenase system are, for example, arylhydrocarbon epoxides, arylhalide epoxides, aliphatic halide epoxides and aliphatic epoxides.

By *N*-hydroxylation, arylamines and arylamides are metabolized into intermediates (proximate carcinogens), that may undergo further activation into ultimate carcinogens by conjugative or group transfer reactions with sulfate and glucuronic acid. Thus, highly polar conjugates are formed, that can be rapidly excreted *via* the bile and/or urine.

The metabolism of aliphatic nitrosamines involves a *C*-hydroxylation, and an unstable intermediate is produced, that yields a highly reactive carbonium ion.

*Deactivation of precursor carcinogens and mutagens*

Some enzyme-mediated reactions are directed towards deactivation of precursor carcinogens and mutagens. For example, important reductive reactions, such as the reduction of the azo group of aminoazo compounds, which yields two monocyclic amines, may be effective detoxification steps. Primary and secondary amines may undergo conjugation reactions to yield *N*-glucuronides. For certain aromatic compounds, e.g., chlorobenzene and nitrobenzene, direct hydroxylation (by insertion reaction into C-H bonds) has been shown to be involved in the metabolic pathway (Selander *et al.*, 1975).

Other enzyme reactions may include oxidative deamination, desulfuration and dechlorination. Also, mammalian tissues contain a large number of nonspecific esterases and amidases that can hydrolyse ester and amide linkages in potential carcinogens and mutagens.

*Deactivation of ultimate carcinogens and mutagens*

Mammalian tissues contain two enzyme systems, the *S*-glutathione transferases and the epoxide hydratases, that are especially efficient in scavenging electrophilic compounds, thereby protecting cellular macromolecules (DNA, RNA, protein) from attack by such agents. The *S*-glutathione transferases catalyse conjugation of the tripeptide, gluthatione (GSH), which contains a nucleophilic SH-group, with electrophilic reactants. The GSH conjugates are converted, by a series of enzymatic reactions, to cysteine conjugates and *N*-acetylcysteine conjugates (mercapturic acids), that are excreted in the urine or bile (Chasseaud, 1979). The electrophilic centres that are subject to attack by GSH include those in alkyl halides, alkene halides, arylhalides, aralkyl halides and the epoxides that are formed as intermediates during oxidative metabolic reactions. The epoxide hydratases catalyse the hydrolysis of electrophilic epoxides, thereby producing diols. These reactions generally result in deactivation of reactive epoxides.

Figure 1 shows the possible biochemical reactions that are involved in the metabolism of direct-acting and precursor carcinogens and mutagens.

## METABOLITES OF POTENTIAL INDUSTRIAL CARCINOGENS AND MUTAGENS

*Urinary metabolites*

A selection was made of 33 potential carcinogenic and/or mutagenic industrial chemicals, most of which are produced in quantities exceeding 10 000 tonnes per annum (Fishbein, 1979). The following classes of chemicals were included: alkylating agents (epoxides and aldehydes), alkylhalides, alkene halides, aralkylhalides, allyl derivatives, aromatic hydrocarbons, hydrazines and carbamates, aromatic amines and azodyes and cyclic ethers. Table 1 lists the urinary metabolites identified in experimental animals and, in some cases, in humans, after exposure to the various agents. Mercapturic acids or cysteine conjugates were detected in the urines of animals or humans after exposure to most of the chemicals in groups I-VII. However, none of the metabolism studies on the aromatic amines and azodyes (group VIII) reported the occurrence of mercapturic acids or cysteine conjugates, which may indicate that detoxification of active metabolites with GSH is not of great importance for these compounds. Furthermore, GSH conjugates were not detected in the urines of animals or humans after exposure to 1,4-dioxane.

**Fig. 1. Possible metabolic pathways of carcinogens and mutagens**

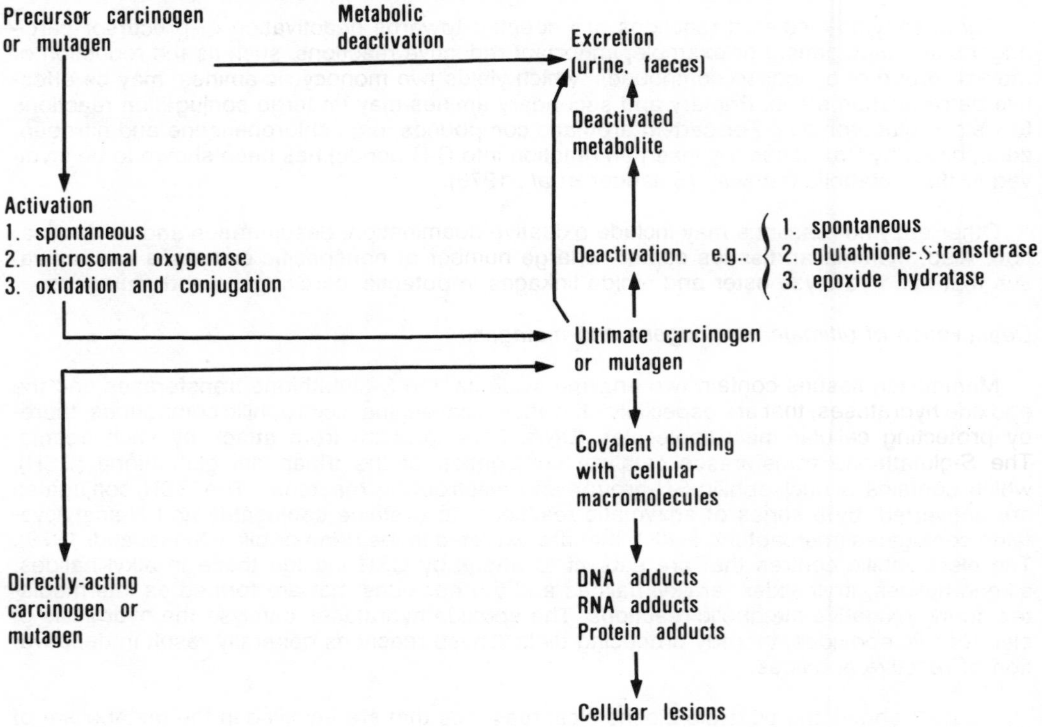


**Table 1. Urinary metabolites of potential industrial carcinogens and mutagens in experimental animals and humans**

| Class, chemical name, and structural formula | Species, dose and route of administration[a] | Urinary metabolite(s) (% of dose) | Reference |
|---|---|---|---|
| I Alkylating agents | | | |
| 1 Epoxides | | | |
| Ethylene oxide<br>$H_2C\text{—}CH_2$ (O) | Rat<br>2 mg/kg<br>i.p. | N-Acetyl-S-(2-hydroxyethyl) cysteine (33%)<br>S-(2-hydroxyethyl) cysteine (9%) | Jones & Wells (1981) |
| Propylene oxide<br>$H_3CHC\text{—}CH_2$ (O) | Rat<br>s.c. | N-Acetyl-S-(2-hydroxypropyl) cysteine | Barnsley (1966) |
| Epichlorohydrin<br>$ClH_2CHC\text{—}CH_2$ (O) | Rat<br>50 mg/kg<br>oral | N-Acetyl-S-(3 chloro-2-hydroxypropyl) cysteine (35%)<br>S-(2,3-dihydroxypropyl) cysteine (2%)<br>N-Acetyl-S-(2,3-dihydroxypropyl) cysteine (trace)<br>α-Chlorohydrin (4%) | Gingell et al. (1983) (unpublished) |

Table 1. (cont'd)

| Class, chemical name, and structural formula | Species, dose and route of administration[a] | Urinary metabolite(s) (% of dose) | Reference |
|---|---|---|---|
| Diglycidylether of bisphenol A (DGEBPA)[b] 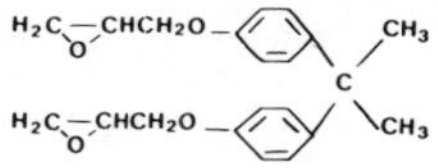 | Mouse<br>20-215 mg/kg<br>Oral | Bisdiol of DGEBPA (6%) | Climie et al. (1981) |
| | Rabbit; 5 and 50 mg/kg | Bisdiol of DGEBPA (10% at 5 mgm/kg<br>2% at 50 mg/kg) | Coveney (1983)<br>(unpublished) |
| Phenylglycidylether[b] 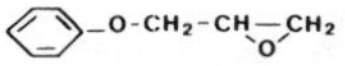 | Rat and rabbit<br>Oral | 2-Hydroxy-3-phenoxy propionic acid (94%)<br>N-Acetyl-S-(2-hydroxy-3-phenoxypropyl) cysteine (4%) | James et al. (1978) |
| 2. Aldehydes | | | |
| Formaldehyde<br>HCHO | Rat<br>40 mg/kg<br>i.p. | Formate<br>Formaldehyde<br>N,N-Diformylcysteine[c] | Mashford & Jones (1982) |
| Acetaldehyde<br>$CH_3CHO$ | Rat<br>6.2 mmol/kg<br>i.p. | Increase in alkali-hydrolysable thiols | Hemminki (1983) |
| Acrolein<br>$CH_2=CHCHO$ | Rat<br>s.c. | N-Acetyl-S-(3-hydroxypropyl) cysteine (10.5%) | Kaye (1973) |
| | Rat<br>10 mg/kg<br>Oral | N-Acetyl-S-(carboxyethyl) cysteine | Draminski et al. (1983) |
| II Alkylhalides | | | |
| Chloroform<br>$CHCL_3$ | Rat<br>NS | 2-Oxothiazolidine-4 carboxylic acid | Davidson et al. (1982) |
| Methylchloride<br>$CH_3Cl$ | Human<br>30-90 ppm<br>Inhalation | S-Methylcysteine | Van Doorn et al. (1980) |
| 1,2-Dichloroethane<br>$ClCH_2CH_2Cl$ | Rat<br>NS | N-Acetyl-S-(2-hydroxyethyl) cysteine<br>S-(2-hydroxyethyl) cysteine | Nachtomi et al. (1966) |
| 1,2-Dibromoethane<br>$BrCH_2CH_2Br$ | Rat<br>100 mg/kg<br>Oral | N-Acetyl-S-(2-hydroxyethyl) cysteine<br>S-(2-hydroxyethyl) cysteine | Nachtomi et al. (1966) |
| 1,2-Dichloropropane<br>$ClCH_2CHClCH_3$ | Rat<br>20 mg/kg<br>Oral | N-Acetyl-S-(2-hydroxypropyl) cysteine (25-35%)<br>N-Acetyl-S-(2,3-dihydroxypropyl) cysteine (minor)<br>β-Chlorolactate | Jones & Gibson (1980) |
| 1.2-Dibromo-3-chloropropane<br>$BrCH_2CHBrCH_2Cl$ | Rat<br>50 mg/kg<br>i.o. | S-(2,3-Dihydroxypropyl) cysteine ] mercapturic acids<br>1,3(Bis-cysteinyl) propan-2-ol ]<br>α-Chlorohydrin<br>α-Bromohydrin<br>Oxalic acid | Jones et al. (1979) |
| III Alkene halides | | | |
| Vinylchloride<br>$CH_2=CHCl$ | Rat<br>0.05-1.0-100 mg/kg<br>Oral | N-Acetyl-S-(2-hydroxyethyl) cysteine (18%)[d]<br>Thiodiglycolic acid (15%) | Watanabe et al. (1976) |

Table 1. (cont'd)

| Class, chemical name, and structural formula | Species, dose and route of administration[a] | Urinary metabolite(s) (% of dose) | Reference |
|---|---|---|---|
| Vinylidene chloride<br>$CH_2=CCl_2$ | Rat<br>1.0 and 50 mg/kg<br>Oral | N-Acetyl-S-(2-hydroxyethyl) cysteine<br>Thiodiglycolic acid ] (66%)[d] | McKenna et al. (1978) |
| | Rat<br>0.5-5.0 - mg/kg<br>Oral | N-Acetyl-S-(2-carboxymethyl) cysteine<br>Methylthio-acetylaminoethanol ] (48%)[d] | Reichert et al. (1979) |
| Chloroprene<br>$CH_2=CCl-CH=CH_2$ | Rat<br>50-200 mg/kg<br>Oral | Increase in thioethers | Summer & Greim (1980) |
| Allyl chloride<br>$CH_2=CH-CH_2Cl$ | Rat<br>400 mg/kg<br>s.c. | Allylmercapturic acid<br>N-Acetyl-S-(3-hydroxypropyl) cysteine | Kay et al. (1972) |
| Hexachlorobutadiene<br>$Cl_2C=CCl-CCl=CCl_2$ | Rat<br>1 and 50 mg/kg<br>Oral | 5 unidentified metabolites<br>(26% and 11%) | Reichert (1982) |
| 1,3-Dichloropropene[b]<br>$ClCH=CH-CH_2Cl$ | Rat<br>20 mg/kg<br>Oral | N-Acetyl-S-(3-chloroprop-2-enyl) cysteine (82%) | Climie et al. (1979) |
| IV Aralkylhalides | | | |
| Benzylchloride<br>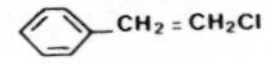 | Rabbit<br>200 mg/kg<br>Oral | N-Acetyl-S-(benzyl) cysteine (49%) | Bray et al. (1958) |
| V Allyl derivatives | | | |
| Amides | | | |
| Acrylonitrile<br>$CH_2=CHCN$ | Rat<br>30 mg/kg<br>Oral | N-Acetyl-S-(2-cyanoethyl) cysteine (25%)<br>4-Acetyl-3-carboxy-5-tetrahydro-1,2-2H-thiazine (25%) | Langvardt et al. (1979) |
| Acrylamide<br>$CH_2=CH-C(=O)NH_2$ | Rats<br>NS | N-Acetyl-S-(3-amino-3-oxopropyl) cysteine (70%) | Langvardt et al. (1979) |
| VI Aromatic hydrocarbons | | | |
| Benzene<br> | Human<br>NS | Phenol (major) | Lauwerys (1979) |
| | Rat<br>NS | Phenylmercapturic acid (minor)<br>Catechol (minor) | |
| Styrene<br>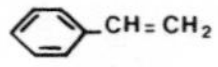 | Human | Mandelic acid<br>Phenylglyoxylic acid | Engström et al. (1978) |
| | Rat<br>250 mg/kg<br>i.p.<br>(3 weeks) | Mandelic acid<br>Phenylglyoxilic acid<br>Hippuric acid<br>N-acetyl-S-(1-phenyl-2-hydroxyethyl) cysteine (7%)<br>N-acetyl-S-(2-phenyl-2-hydroxyethyl) cysteine (3.6%)<br>N-acetyl-S-phenacylcysteine (0.1%) | Seutter-Berlage et al. (1978) |

Table 1. (cont'd)

| Class, chemical name, and structural formula | Species, dose and route of administration[a] | Urinary metabolite(s) (% of dose) | Reference |
|---|---|---|---|
| VII <u>Hydrazines</u> | | | |
| <u>Carbamates</u> | | | |
| Hydrazine<br>$NH_2NH_2$ | Rat<br>30 mg/kg<br>i.p. | Hydrazine (30%)<br>Acid hydrolysable derivative of hydrazine (25%) | Springer <u>et al.</u> (1981) |
| Urethane<br>(Ethylcarbamate)<br>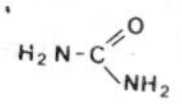 | Human<br>30 mg/kg<br>Oral | Ethyl mercapturic acid (0.13%)<br><u>N</u>-Acetyl-<u>S</u>-carbethoxycysteine (0.9%)<br>Acetyl-<u>N</u>-hydroxyurethane (0.6%) | Boyland & Nery (1965) |
| VIII <u>Aromatic amines and azodyes</u> | | | |
| Benzidine<br>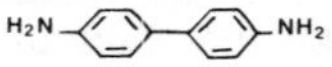 | Human | Unchanged benzidine<br>3,3'-Dihydroxybenzidine<br>Mono- and diacetylbenzidine<br>3-Hydroxy-benzidine<br><u>N</u>-Hydroxy-<u>N</u>-acetyl aminobenzidine | Fishbein (1979) |
| 3,3'-Dichlorobenzidine<br>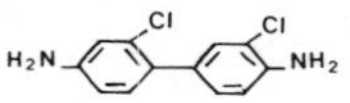 | Monkey<br>0.2 mg/kg<br>i.v. | Unchanged 3,3'-dichloro benzidine | Kellner <u>et al.</u> (1973) |
| 4,4'-Methylene-bis-(2-chloroaniline)<br>(MOCA)<br>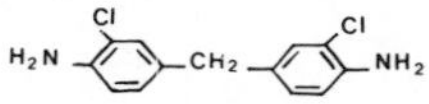 | Rat<br>1-13-100 mg/kg<br>i.p. | Unchanged MOCA (0.5%)<br>Conjugated MOCA (1.2%)<br>Conjugated polar metabolites (13%) | Farmer <u>et al.</u> (1981) |
| 4 Aminoazobenzene<br>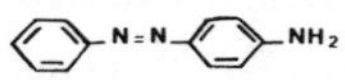 | Rat<br>NS | Conjugated <u>N</u>-hydroxy-<u>N</u>-acetyl aminoazobenzene | Sato <u>et al.</u> (1966) |
| IX <u>Cyclic ethers</u> | | | |
| 1,4-Dioxane<br>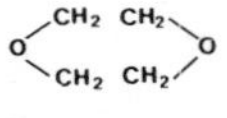 | Rat<br>10 mg/kg<br>i.v. | β-Hydroxyethoxyacetic acid (92%)[d] | Braun & Young (1977) |
| | Rat<br>1.4 g/kg<br>Oral | 2-Hydroxyethoxyacetic acid -lactone | Woo <u>et al.</u> (1977) |

[a] i.p., intraperitoneal; s.c., subcutaneous; NS, not specified; i.v., intravenous

[b] Industrial production less than 10 000 tons per annum

[c] Tentative urinary metabolite

[d] Dose-dependent excretion of urinary metabolites

In the urines of animals exposed to epoxides, diols were excreted, also, as metabolites, demonstrating the importance of epoxide hydratase as a detoxifying enzyme for this class of chemicals.

The relationship between exposure levels and the amount of excreted urinary metabolite has been studied with some chemicals, e.g., vinyl chloride, vinylidene chloride, 1,4-dioxane (Watanabe *et al.*, 1976; Young *et al.*, 1976; McKenna *et al.*, 1978). The fate of these chemicals was shown to be dose-dependent in experimental animals. At the highest exposure levels, there was a deviation in the linearity between dose and the amount of excreted urinary metabolites.

In conclusion, metabolism studies on most of the chemicals in Table 1 showed the presence of specific urinary metabolites that were excreted in reasonably high amounts. Such metabolites, therefore, may be used for biological monitoring of human exposure to the respective chemicals.

### *DNA and protein adducts*

Table 2 lists those chemicals in Table 1 that have been tested for covalent binding with cellular macromolecules (DNA and protein) in experimental animals. For some chemicals, the ability to form covalent binding with haemoglobin was studied, because the measurement of haemoglobin adducts might be a future method of biomonitoring human exposure to potential carcinogenic and mutagenic chemicals.

**Table 2. In-vivo covalent binding of potential industrial mutagens and carcinogens with DNA and protein, in experimental animals and humans**

| Class and chemical name | Species | Binding with DNA (reference) | Binding with protein (reference) |
|---|---|---|---|
| I Alkylating agents | | | |
| 1 Epoxides | | | |
| Ethylene oxide | Rat<br>Human | 7-*N*-(2'-hydroxyethyl) deoxyguanosine (Ehrenberg *et al.* (1974) | $N_3$-(2'-hydroxyethyl) histidine in haemoglobin (Osterman-Golkar *et al.* (1976; Calleman *et al.* 1978) |
| Propylene oxide | Rat | | $N_3$-(2'-hydroxypropyl) histidine in haemoglobin (Farmer *et al.*, 1982) |
| Butylene oxide | Rat | – (Paul & Pavelka 1971) | |
| 2. Aldehydes | | | |
| Acrolein | Rat | + (Munsch *et al.*, 1974) | |
| II Alkylhalides | | | |
| Chloroform | Rat | + (Stott *et al.*, 1981) | + (haemoglobin) (Pereira & Chang, 1982) |

**Table 2 (contd)**

| Class and chemical name | Species | Binding with DNA (reference) | Binding with protein (reference) |
|---|---|---|---|
| 1,2-Dibromoethane | Rat | + (Hill et al., 1978) | |
| 1,2-Dibromo-3-chloro-propene | Rat | + (Lipscomb et al., 1977) | + (Lipscomb et al., 1977) |
| III Alkene halides | | | |
| Vinylchloride | Rat | 7-N-(2'-oxoethyl) deoxyguanosine (Laib et al., 1981)<br>1,N[6]-Ethenodeoxyadenosine<br>3, N[4]-ethenodeoxycytidine (Green & Hathway, 1978) | + (Bolt et al., 1976) |
| Vinylidenechloride | Rat | | + (McKenna et al., 1978) |
| IV Aralkylhalides | | | |
| Benzylchloride | Mouse | 7N-benzyldeoxyguanosine (Walles, 1981) | + (haemoglobin) (Walles, 1981) |
| V Aromatic hydrocarbons | | | |
| Benzene | Rat | + (Lutz & Schlatter, 1977) | + (haemoglobin) (Pereira & Chang, 1981) |
| Styrene | Rat | | + (Savolainen & Vaino, 1977) |
| VI Hydrazines, carbamates | | | |
| Urethane | Rat | + (Pound et al., 1976) | + (Prodi et al., 1970) |
| VIII Aromatic amines and azodyes | | | |
| Benzidine | Rat | N-(Deoxyguanosin-8-yl)-N'-acetyl-benzidine (Martin et al., 1983) | + (haemoglobin) (Pereira & Chang, 1981) |
| 4-Aminoazobenzene | Rat | + (Sonnebichler & Reichert, 1978) | |

The formation of specific adducts with cellular DNA, of the chemicals in Table 2, has been studied in animals exposed to ethylene oxide, vinyl chloride, benzyl chloride and benzidine. The formation of specific haemoglobin adducts has been studied in animals exposed to ethylene oxide and propylene oxide; these compounds have been studied, also, to determine the relationship between the amount of DNA or haemoglobin adduct and exposure levels.

By determining the amount of 7-*N*-(2'-hydroxyethyl)guanine in DNA, Wright (1983) found, except for testicular DNA, a uniform level of alkylation of DNA, in a wide range of tissues, in rats that had received inhalation exposure (6 h) to $^{14}C$ ethylene oxide at air concentrations of 1, 10 and 33 ppm. Osterman-Golkar *et al.* (1976) and Calleman *et al.* (1978) reported that, in rats and humans exposed to ethylene oxide, the amount of $N_3$-(2'-hydroxyethyl)histidine in haemoglobin was reasonably proportional to the dose received. Farmer *et al.* (1982) found a linear relationship between exposure to propylene oxide over the dose range 0-2000 ppm and the amount of $N_3$-(2'-hydroxypropyl)histidine in haemoglobin of rats.

## BIOLOGICAL MONITORING OF POTENTIAL INDUSTRIAL CARCINOGENS AND MUTAGENS

As discussed, biological monitoring by quantitative measurement of urinary metabolites could, in principle, be applied for most potential carcinogens and mutagens listed in Table 1. Ideally, however, exposure to a chemical should be measured at the level of the critical cellular target (target dose). The target dose is defined as the concentration of the ultimate carcinogen or mutagen at the critical target in a given time (Ehrenberg *et al.*, 1974). Since DNA has been established as the critical cellular target for mutagens and, probably, for carcinogens, the most relevant approach would be to determine the DNA dose. This may be achieved by determining the amount of DNA adduct that is formed by reaction of the carcinogens or mutagens with cellular DNA. However, measurement of DNA adducts may be difficult in humans, because the availability of DNA is restricted in accessible samples (e.g., blood) and because many DNA adducts are unstable, e.g., measurement of 3-alkyladenine would be useful only in studies of recent exposure. Osterman-Golkar *et al.* (1976) and Ehrenberg and Osterman-Golkar (1980) suggested that the assay of stable adducts, that are formed with blood proteins such as haemoglobin, may be a suitable method to estimate the target dose (DNA dose) in humans exposed to carcinogens or mutagens. However, before this method can be used with confidence, the proportional relationships between haemoglobin adducts and DNA adducts must be shown to be approximately constant, at several exposure levels, in at least two experimental species. This will be the case, particularly, for compounds requiring metabolic activation (Wright, 1981).

### *Measurement of urinary metabolites*

*Selective methods*: Analytical methods have been developed to determine urinary metabolite concentrations for the following compounds in Table 1: the epoxide, diglycidylether of bisphenol A; the chlorinated hydrocarbons, methyl chloride, vinyl chloride and 1,3-dichloropropene; the aromatic hydrocarbons, benzene and styrene; the allyl derivative, acrylonitrile; the aromatic amines, benzidine, 3,3'-dichlorobenzidine and 4,4'-methylene-bis(2-chloroaniline) and the cyclic ether, 1,4-dioxane.

Table 3 lists the concentrations of urinary metabolites in humans occupationally exposed to the above chemicals and, when available, corresponding air concentrations. Background levels of urinary metabolites in supposedly nonexposed subjects are shown in the table, also. Relatively high background levels were found for urinary *S*-methylcysteine, phenol and thiodiglycolic acid, the metabolites of methyl chloride, benzene and vinyl chloride, respectively. Therefore, exposure to these chemicals at levels below 1-5 ppm might produce urinary concentrations of their respective metabolites within the background range. This may limit the application of biological monitoring for the measurement of low individual exposure to these compounds. By sensitive analytical methods combined with absence of significant background levels in controls, low exposure levels can be measured for aromatic amines, acrylonitrile, the diglycidylether of bisphenol A, 1,4-dioxane and styrene.

**Table 3. Levels of urinary metabolites in occupationally exposed and nonexposed persons**

| Chemical | Urinary metabolite | Method[a] | Sensitivity | Urinary concentration in occupationally exposed persons (corresponding air concentrations in brackets) | Urinary concentration in nonexposed persons | Reference |
|---|---|---|---|---|---|---|
| Diglycidylether of bisphenol A (DGEBPA) | Bisdiol of DGEBPA | HPLC | 0.02 mg/l | n.d. | < 0.02 mg/l | Eadsforth (1983) (unpublished) |
| Methylchloride | *S*-Methylcysteine | GLC | not published | 50-700 μmol $CH_3SH$/mmol creatinine[b] (30-90 ppm; 8 h) | 0-90 μmol $CH_3SH$/mmol creatinine | Van Doorn *et al.* (1980) |
| Vinylchloride | Thiodiglycolic acid | GLC/MS | 0.05 mg/l | 0.3-4.0 mg/l (0.14-7.0 ppm; 8 h) | <0.05-1.3 mg/l | Müller *et al.* (1978)<br>Müller *et al.* (1979) |
| 1,3-Dichloropropene | *N*-Acetyl-*S*-(3-chloro-prop-2-enyl) cysteine | GLC | 0.3 mg/l | n.d. | n.d. | Eadsforth (1982) (unpublished) |
| Acrylonitrile | Acrylonitrile | GLC | 5 μg/l | 105 μg/l[b] (3.7 ppm; 8 h)<br>10.9 μg/l[b] (0.17 ppm; 8 h) | <5 μg/l | Sakurai *et al.* (1978) |
| Benzene | Phenol | GLC | 1 mg/l | 45-50 mg/l[b] (10 ppm; 8 h) | <20 mg/l | Lauwerys (1979) |
| Styrene | Mandelic acid | GLC | 1 mg/l | 700 mg/l[b] (31 ppm, 8 h)<br>1200 mg/l[b] (55 ppm; 8 h)<br>1600 mg/l[b] (74 ppm; 8 h) | <5 mg/l | Engström *et al.* (1978) |
| Benzidine | Benzidine<br>*N*-Acetyl benzidine<br>*N,N'*-Diacetyl-benzidine | GLC<br>GLC<br>GLC | 1 μg/l<br>2 μg/l<br>0.2 μg/l | n.d. | <1 μg/l<br><2 μg/l<br><0.2 μg/l | Nony & Bowman (1980) |
| 3,3'-Dichloro-benzidine (DCB) | DCB<br>*N*-Acetyl DCB | GLC/MS<br>GLC/MS | 0.2 μg/l<br>0.2 μg/l | | <0.2 μg/l<br><0.2 μg/l | Hurst *et al.* (1981) |
| 4,4-Methylene-bis(2-chloroaniline) (MOCA) | MOCA | GLC | 40 μg/l<br>1 μg/l | 250 μg/l (<0.02 mg/m³)<br>n.d. | <40 μg/l<br>n.d. | Linch *et al.* (1971)<br>van Roosmalen & Klein (1979) |
| 1,4-Dioxane | β-hydroxyethoxy-acetic acid | GLC/MS | 0.1 mg/l | 50 ± 25 mg/l (1.6 ± 0.7 ppm; 8 h) | n.d. | Young *et al.* (1976)<br>Braun (1977) |

[a] n.d., not determined; GLC, gas liquid chromatography; MS, mass spectrometry; HPLC, high-performance liquid chromatography

[b] Samples taken at the end of shift

*Nonselective methods*: Nonselective tests for detection of exposure to carcinogens and mutagens include (1) the determination of urinary thioethers and (2) the determination of mutagens in urine.

(1) A number of carcinogenic and mutagenic compounds have been examined in experimental animals, using the thioether assay for the presence of urinary thioethers (van Doorn *et al.*, 1982). A series of known carcinogens, e.g., aromatic amines and azodyes, did not produce increased urinary thioether levels. This finding agrees with the reported absence of specific mercapturic acids or cysteine conjugates in the urines of animals exposed to such compounds (Table 1). However, some compounds that lack carcinogenic or mutagenic properties produced increased urinary thioether levels, e.g., benzyl alcohol, biphenyl, mesitylene, toluene and *o*-xylene.

In the industrial situation, application of the thioether test has been somewhat limited. An increase in urinary thioethers has been demonstrated in workers in certain segments of the chemical industry, including rubber workers (Seutter-Berlage *et al.*, 1977; Vainio *et al.*,

1978). However, the increase was generally not more than twice the background values. It has been shown that smoking is a confounding factor and that the excretion of thioethers was related to the number of cigarettes consumed per day. Urinary thioether concentrations in smokers were correlated with the results of mutagenicity assays in urine (van Doorn *et al.*, 1979).

Currently, the main value of the thioether test for industry may be that it is able to indicate overexposure to unsuspected electrophilic compounds.

(2) As discussed previously, chemical substances metabolized by animals or humans may be excreted in the urine as conjugated products (e.g., arylamines). Mutagenic substances in the urine can be detected by microbial indicators. Enzymes such as β-glucuronidase and arylsulfatase must be added to the test system, to release the chemical from its conjugated form. Furthermore, activating systems can be used, such as the conventionally used microsomal oxidizing S-9 mix or the liver cytosol fraction. Enhanced mutagenic activity has been shown in urines of workers in the rubber industry (Falck *et al.*, 1980), of workers exposed to epichlorohydrin concentrations of over 25 ppm (Kilian *et al.*, 1978) and of workers exposed to petroleum coke and pitch (Pasquini *et al.*, 1982).

Smoking appeared to be an important confounding factor in most studies of occupationally exposed workers.

### *DNA and protein adducts*

*Alkylation of haemoglobin*: The application of measurements of haemoglobin adducts in humans has been limited, so far, to the epoxides, ethylene oxide and propylene oxide. 'Cold' analytical methods, i.e., methods that do not rely on measurement of radiolabelled material, are required to measure the extent of alkylation in human haemoglobin. Gas chromatography/mass spectrometry techniques have been developed for quantitative determination of *S*-methylcysteine, and of $N_3$-(2'-hydroxyethyl)- and $N_3$-(2'-hydroxypropyl)histidine residues in haemoglobin (Calleman *et al.*, 1978; Farmer *et al.*, 1982; van Sittert *et al.*, 1984). Such methods tend to be rather laborious, but the easier immunochemical techniques, some of which are being used to assess carcinogen-DNA adducts in biological specimens (Montesano *et al.*, 1982), are not yet available to assess haemoglobin adducts. Table 4 reviews the alkylation to specific amino acids in haemoglobin that has been found in persons occupationally exposed to ethylene oxide and propylene oxide, and in persons believed to have had no such occupational exposure. With regard to ethylene oxide, variable and relatively high background levels of $N_3$-(2'-hydroxyethyl)histidine residues in haemoglobin were found in the control group. Therefore, alkylations due to occupational exposure to ethylene oxide at low levels (below 1 ppm) could not be identified on an individual basis (van Sittert *et al.*, 1984). Calleman *et al.* (1978) reported $N_3$-(2'-hydroxyethyl)histidine levels in haemoglobin that were beyond the background range, in humans who were exposed to ethylene oxide in air at concentrations above 30 ppm. Osterman-Golkar (1983) reported much lower background levels of $N_3$-(2'-hydroxypropyl)histidine residues, but very high background levels of *S*-methylcysteine residues were found in human haemoglobin (Bailey *et al.*, 1981). Such background alkylations may have clear toxicological and methodological implications; methods are currently being developed to determine whether there are corresponding background levels in DNA. Some laboratories, also, are attempting to establish the origin of background alkyl residues in haemoglobin. This research may lead to the detection of unsuspected exposure to potential carcinogenic or mutagenic compounds which, if undetected, could confound the results of epidemiological and other studies.

**Table 4. Degree of alkylation of amino acids in haemoglobin in occupationally exposed and nonexposed individuals**

| Chemical | Haemoglobin adduct | Method[a] | Sensitivity of method (nmol/g globin) | Levels in occupationally exposed persons (nmol/g globin) (corresponding air concentrations in brackets) | Levels in non-exposed persons (nmol/g globin) | Reference |
|---|---|---|---|---|---|---|
| Ethylene oxide | N,-(2'-hydroxyethyl) histidine | GLC/MS | 0.05 | 0.4–14 (<1->30 ppm) | <0.05 | Calleman *et al.* (1978) |
| Ethylene oxide | N,-(2'-hydroxyethyl) histidine | GLC/MS | | | 0.17–1.5 | Osterman-Golkar (1983) |
| Ethylene oxide | N,-(2'-hydroxyethyl) histidine | GLC/MS | 0.02 | <0.02–9.7 (0.05–8 ppm; 8 h) | <0.02–4.7 | van Sittert *et al.* (1984) |
| Propylene oxide (PO) | N,-(2'-hydroxypropyl) histidine | GLC/MS | | 0.9–1.2 (low levels of PO) 6–10 (average level 10 ppm) | <0.1–0.4 | Osterman-Golkar (1984) |
| Methylating agents | S-Methylcysteine | GLC/MS | | | 15.4 ± 1.8 | Baily *et al.* (1981) |

[a] GLC, gas liquid chromatography; MS, mass spectrometry

*Alkylation of nucleic acids*: It would be useful if 'cold' analytical techniques could be used to determine the extent of DNA adducts that are formed by the reaction of electrophiles with cellular DNA. Immunological techniques have been developed to assess specific, nonradiolabelled DNA adducts in experimental animals after exposure to known carcinogens such as acetylaminofluorene and *N*-nitroso compounds (Montesano *et al.*, 1982). At present, however, no similar analytical method is available to quantitate the DNA adducts of potential industrial carcinogens and mutagens (Table 2).

An alternative method of measuring the formation of DNA adducts might be to determine the amount of alkylated nucleic acid bases that is excreted in urine (Montesano *et al.*, 1982). For example, 3-methyladenine which, apparently, is not a normal urinary constituent, was excreted in the urine of rats exposed to dimethylsulfate and *N*-nitrosodimethylamine (Löfroth *et al.*, 1974; Shaikh *et al.*, 1980). However, in order to measure low industrial exposure to potential carcinogens and mutagens by this method, more sensitive, possibly immunological, techniques might need to be developed.

## CONCLUDING REMARKS

At present, four methodologies are used to monitor human exposure to potential industrial carcinogens and mutagens by determination of chemicals or their metabolites in body fluids. These comprise (1) measurement of urinary metabolites, (2) measurement of alkylation of haemoglobin, (3) urinary mutagenicity assays and (4) the thioether assay.

Measurement of urinary metabolites shows the proportion of the ultimate carcinogen or mutagen that is excreted in the urine unchanged, or after detoxification by metabolic or chemical reactions. It does not measure the proportion that evades detoxification and reacts with cellular macromolecules. Thus, measurement of urinary metabolites gives an indirect estimate of the carcinogenic or mutagenic dose in the whole body, but the fraction that bonds to cellular macromolecules (e.g., DNA) is not quantified, and the method is of limited value in calculating risk estimates. However, firstly, because no other specific methodology is available for biomonitoring exposure to the majority of potential industrial carcinogens or mutagens (Table 1) and, secondly, because this method gives valuable information about recent exposure (most of these compounds are rapidly metabolized and excreted in experimental species), measurement of urinary metabolites should be recommended in biological monitoring programmes. Metabolic pathways have been studied for most chemicals listed in Table 1; therefore, the development of practical analytical techniques to determine urinary metabolites is possible, in principle. Methods developed so far show that the sensitivity of analytical techniques is satisfactory, and that exposure to 0.1 ppm or less of a compound is detectable. However, for low occupational exposure to some alkyl groups of simple structure, e.g., methyl- and ethylhalides, vinylhalides, etc., the value of urinary monitoring is reduced by the presence of relatively high background values of urinary metabolites in supposedly nonexposed individuals. Furthermore, urinary metabolites of simple alkyl groups are not specific for a single chemical. For example, thiodiglycolic acid is a common urinary metabolite of vinylhalides and ethylhalides.

By determining the amount of haemoglobin alkylation, the target dose (DNA dose) of the chemical can be estimated. This technique, therefore, might be more meaningful for calculating risk estimates. Compounds like ethylene oxide, benzyl chloride and, probably, propylene oxide, are distributed evenly throughout the body in experimental animals, giving approximately the same dose in different organs (Walles, 1981; Wright, 1983). Therefore, with these chemicals, it can be reasonably assumed that the haemoglobin dose will approximate the DNA dose. However, particularly for compounds that need metabolic activation, validation studies should be carried out, to investigate the proportional relationship between haemoglobin and DNA alkylation in different species at several exposure levels. For exposure monitoring, measurement of the degree of haemoglobin alkylation is particularly useful in determining the average dose that is received over a longer period, i.e., the life-time of the erythrocyte. This method has been used to monitor occupational exposure to ethylene and propylene oxide. The analytical technique for detection of haemoglobin alkylation is very sensitive; in principle, it is possible to detect exposure to a few parts per billion of a particular compound. However, the presence of background haemoglobin alkylations (methyl groups in cysteine; 2-hydroxyethyl and 2-hydroxypropyl groups in histidine) limits the possibility of monitoring low occupational exposure to methylating agents, ethylene oxide and, to a lesser degree, propylene oxide in individuals. Binding to haemoglobin has been demonstrated, in experimental animals exposed to a range of potential industrial carcinogens and mutagens (Table 2). However, much research is still needed, including the development of less laborious, possibly immunochemical techniques, before measurement of haemoglobin alkylation can be used routinely as a dose monitor for exposure to these compounds.

The urinary mutagenicity assay is a sensitive but nonspecific method, which is able to monitor, qualitatively, exposure to certain mutagenic substances or their active metabolites that are excreted in urine.

The urinary thioether test is an insensitive, nonspecific method and may be useful only to monitor overexposure to (unknown) electrophilic substances.

## REFERENCES

Bailey, E., Connors, T.A., Farmer, P.B., Gorf, S.M. & Rickard, J. (1981) Methylation of cysteine in haemoglobin following exposure to methylating agents. *Cancer Res.*, *41*, 2514-2517

Barnsley, E.A. (1966) The formation of 2-hydroxypropylmercapturic acid from 1-halogenopropanes in the rat. *Biochem. J.*, *100*, 362-372

Bolt, H.M., Kappus, H., Kaufmann, R., Appel, K.E., Buchter, A. & Bolt, W. (1976) *Metabolism of $^{14}$C-vinyl chloride* in vitro *and* in vivo. In: Rosenfeld, C. & Davis, W., eds, *Environmental Pollution and Carcinogenic Risks* (*IARC Scientific Publications No. 13*; *INSERM Symposia Series Vol. 52*), Lyon, International Agency for Research on Cancer, pp. 151-164

Boyland, E. & Nery, R. (1965) The metabolism of urethane and related compounds. *Biochem. J.*, *94*, 198-208

Braun, W.H. (1977) Rapid method for the simultaneous determination of 1,4-dioxan and its major metabolite, β-hydroxyethoxyacetic acid, concentrations in plasma and urine. *J. Chromatogr.*, *133*, 263-266

Braun, W.H. & Young, J.D. (1977) Identification of β-hydroxy ethoxyacetic acid as the major urinary metabolite of 1,4-dioxane in the rat. *Toxicol. appl. Pharmacol.*, *39*, 33-38

Bray, H.G., James, S.P. & Thorpe, W.V. (1958) Metabolism of some ω-halogenoalkylbenzenes and related alcohols in the rabbit. *Biochem. J.*, *70*, 570-579

Calleman, C.J., Ehrenberg, L., Jansson, B., Osterman-Golkar, S., Segerbäck, D., Svensson, K. & Wachtmeister, C.A. (1978) Monitoring and risk assessment by means of alkyl groups in hemoglobin in persons occupationally exposed to ethylene oxide. *J. environ. Pathol. Toxicol.*, *2*, 427-442

Cavalieri, E., Roth, R., Grandjean, C., Althoff, J., Patil, K., Liakus, S. & March, S. (1978) Carcinogenicity and metabolic profiles of 6-substituted benzo[*a*]pyrene derivatives on mouse skin. *Chem.-biol. Interact.*, *22*, 35-51

Chasseaud, L.F. (1979) The role of glutathione and glutathione *S*-transferases in the metabolism of chemical carcinogens and other electrophilic agents. *Adv. Cancer Res.*, *29*, 175-274

Climie, I.J.G., Hutson, D.H., Morrison, B.J. & Stoydin, G. (1979) Glutathione conjugation in the detoxification of (Z)-1,3-dichloropropene (a component of the nematocide D-D) in the rat. *Xenobiotica*, *9*, 149-156

Climie, I.J.G., Hutson, D.H. & Stoydin, G. (1981) Metabolism of the epoxy resin component 2,2-bis (4-(2,3-epoxypropoxy)phenyl) propane, the diglycidyl ether of bisphenol A (DGEBPA) in the mouse. Part II. Identification of metabolites in urine and faeces following a single oral dose of $^{14}$C-DGEBPA. *Xenobiotica*, *11*, 401-424

Davidson, I.W.F., Summer, D.D. & Parker, J.C. (1982) Chloroform: a review of its metabolism, teratogenic, mutagenic, and carcinogenic potential. *Drug. Chem. Toxicol.*, *5*, 1-87

van Doorn, R., Bos, R.P., Leijdekkers, C.M., Wagenaar-Zegers, M.A.P., Theuws, J.L.G. & Henderson, P.T. (1979) Thioether concentration and mutagenicity of urine from cigarette smokers. *Int. Arch. occup. environ. Health*, *43*, 159-166

van Doorn, R., Borm, P.T.A., Leijdekkers, C.M., Henderson, P.T., Reuvers, J. & van Bergen, T.J. (1980) Detection and identification of *S*-methylcysteine in urine of workers exposed to methylchloride. *Int. Arch. occup. environ. Health*, *46*, 99-109 van Doorn, R., Bos, R.P., Brouns, R.M.E. & Henderson, P.T. (1982) Is de bepaling van thioethers in urinemonsters bruikbaar bij de biologische monitoring van genotoxische belasting door de arbeidsomgeving. *Tijdschrift Soc. Geneesk.*, *60*, 30-34

Drake, J.W. & Baltz, R.H. (1976) The biochemistry of mutagenesis. *Ann. Rev. Biochem.*, *45*, 11-37

Draminski, W., Eder, E. & Henschler, D. (1983) A new pathway of acrolein metabolism in rats. *Arch. Toxicol.*, *52*, 243-247

Ehrenberg, L. & Osterman-Golkar, S. (1980) Alkylation of macromolecules for detecting mutagenic agents. *Teratog. Carcinog. Mutag.*, *1*, 105-127

Ehrenberg, L., Hiesche, K.D., Osterman-Golkar, S. & Wennberg, I. (1974) Evaluation of genetic risks of alkylating agents: tissue dose in the mouse from air contaminated with ethylene oxide. *Mutat. Res.*, *24*, 83-103

Engström, K., Haerkoenen, H., Pekari, K. & Rantanen, J. (1978) Evaluation of occupational styrene exposure by ambient air and urine analysis. *Scand. J. Work environ. Health*, *4*, 121-123

Falck, K., Sorsa, M. & Vainio, H. (1980) Mutagenicity in urine of workers in rubber industry. *Mutat. Res.*, *79*, 45-52

Farmer, P.B., Rickard, J. & Robertson, S. (1981) The metabolism and distribution of 4,4'-methylene-bis (2-chloroaniline) (MBOCA) in rats. *J. appl. Toxicol.*, *1*, 317-322

Farmer, P.B., Gorf, S.M. & Bailey, E. (1982) Determination of hydroxypropylhistidine in haemoglobin as a measure of exposure to propylene oxide using high resolution gas chromatography mass spectrometry. *Biomed. mass Spectrom.*, *9*, 69-71

Fishbein, L. (1979) *Potential Industrial Carcinogens and Mutagens* (*Studies in Environmental Science 4*), Amsterdam, Elsevier Scientific Publishing Company

Green, T. & Hathway, D.E. (1978) Interactions of vinylchloride with rat-liver DNA *in vivo*. *Chem.-biol. Interact.*, *22*, 211-224

Hemminki, K. (1983) Urinary sulfur containing metabolites after administration of ethanol, acetaldehyde and formaldehyde to rats. *Toxicol. Lett.*, *11*, 1-6

Hill, D.L., Shih, T.W., Johnston, T.P. & Struck, R.F. (1978) Macromolecular binding and metabolism of the carcinogen 1,2-dibromoethane. *Cancer Res.*, *38*, 2438-2442

Hurst, R.E., Settine, R.L., Fish, F. & Roberts, E.C. (1981) Analysis of urine for parts-per-trillion levels of aromatic diamines with capillary chromatography and selected-ion monitoring mass spectrometry. *Anal. Chem.*, *53*, 2175-2179

James, S.P., Pheasant, A.E. & Solheim, E. (1978) Metabolites of 1,2-epoxy-3-phenoxy and 1,2-epoxy-3-(p-nitro-phenoxy) propane. *Xenobiotica*, *8*, 219-228

Jones, A.R. & Gibson, J. (1980) 1,2-Dichloropropane: metabolism and fate in the rat. *Xenobiotica*, *10*, 835-846

Jones, A.R. & Wells, G. (1981) The comparative metabolism of 2-bromoethanol and ethylene oxide in the rat. *Xenobiotica*, *11*, 763-770

Jones, A.R., Fakhouri, G. & Gadiel, P. (1979) The metabolism of the soil fumigant 1,2-dibromo-3-chloropropane in the rat. *Experientia*, *35*, 1432-1434

Kaye, C.M., Clapp, J.J. & Young, L. (1972) The metabolic formation of mercapturic acids from allyl halides. *Xenobiotica*, *2*, 129-139

Kaye, C.M. (1973) Biosynthesis of mercapturic acids from allylalcohol, allyl esters and acrolein. *Biochem. J.*, *134*, 1093-1101

Kellner, H.M., Christ, O.E. & Lotzesch, K. (1973) Animal studies on the kinetics of benzidine and 3,3-dichlorobenzidine. *Arch. Toxicol.*, *31*, 61-79

Kilian, D.J., Pullin, T.G., Connor, T.H., Legator, M.S. & Edwards, H.N. (1978) Mutagenicity of epichlorohydrin in the bacterial assay system: evaluation by direct in-vitro activity and in-vivo activity of urine from exposed humans and mice. *Mutat. Res.*, *53*, 72

Laib, R.J., Gwinner, L.M. & Bolt, H.M. (1981) DNA alkylation by vinylchloride metabolites: etheno derivatives or 7-alkylation of guanine. *Chem.-biol. Interact.*, *37*, 219-231

Langvardt, P.W., Putzig, C.L., Young, J.D. & Braun, W.H. (1979) Isolation and identification of urinary metabolites of vinyl-type compounds: application to metabolites of acrylonitrile and acrylamide. *Toxicol. appl. Pharmacol.*, *48*, A161

Lauwerys, R. (1979) *Human Biological Monitoring of Industrial Chemicals. 1. Benzene* (*Document EUR 6570*), Luxembourg, Health and Safety Directorate, Commission of the European Communities

Lauwerys, R.R. (1983) *Industrial Chemical Exposure: Guidelines for Biological Monitoring*, Davis, CA, Biomedical Publications

Lipscomb, T.P., Shih, T.W. & Hill, D.L. (1977) Distribution and metabolism of (1,3-$^{14}$C) 1,2-dibromo-3-chloropropane in rats. *Pharmacologist*, *19*, 168

Löfroth, G., Osterman-Golkar, S. & Wennerberg, R. (1974) Urinary excretion of methylated purines following inhalation of dimethylsulphate. *Experientia*, *30*, 641-642

Lutz, W.K. & Schlatter, C. (1977) Mechanism of the carcinogenic action of benzene: irreversible binding to rat liver DNA. *Chem.-biol. Interact.*, *18*, 241-245

Linch, A.L., O'Connor, G.B., Barnes, J.R., Killian, A.S. & Neeld, W.E. (1971) Methylene-bis-ortho-chloroaniline (MOCA): evaluation of hazards and exposure control. *Am. ind. Hyg. Assoc. J.*, *32*, 802-819

Martin, C.N., Beland, F.A., Kennelly, J.C. & Kadlubar, F.F. (1983) Binding of benzidine, *N*-acetylbenzidene, *N,N'*-diacetylbenzidene and Direct Blue 6 to rat liver DNA. *Environ. Health Perspect.*, *49*, 101-106

Mashford, P.H. & Jones, A.R. (1982) Formaldehyde metabolism by the rat: a re-appraisal. *Xenobiotica*, *12*, 119-129

McKenna, M.J., Zempel, J.A., Madrid, E.O., Braun, W.H. & Gehring, P.J. (1978) Metabolism and pharmacokinetic profile of vinylidene chloride in rats following oral administration. *Toxicol. appl. Pharmacol.*, *45*, 821-835

Miller, E.C. & Miller, J.A. (1976) *The metabolism of chemical carcinogens to reactive electrophiles and their possible mechanisms of action in carcinogenesis*. In: Searle, C.S., ed., *Chemical Carcinogens* (*ACS Monograph 173*), Washington DC, American Chemical Society, pp. 737-762

Montesano, R., Rajewsky, M.F., Pegg, A.E. & Miller, E. (1982) Development and possible use of immunological techniques to detect individual exposure to carcinogens: International Agency for Research on Cancer/International Programme on Chemical Safety Working Group Report. *Cancer Res.*, *42*, 5236-5239

Müller, G., Norpoth, K., Kusters, E., Herweg, K. & Versin, E. (1978) Determination of thiodiglycolic acid in urine specimens of vinylchloride exposed workers. *Int. Arch. occup. environ. Health*, *41*, 199-205

Müller, G., Norpoth, K. & Wickramasinghe, R.H. (1979) An analytical method, using GC-MS, for the quantitative determination of urinary thiodiglycolic acid. *Int. Arch. occup. environ. Health*, *44*, 185-191

Munsch, N., Merano, E. & Frayssinet, C. (1974) Incorporation d'acroleine (H-3) dans le foie du rat. *Biochimie*, *56*, 1433-1436

Nachtomi, E.E., Alumot, E. & Bondi, A. (1966) The metabolism of ethylene dibromide in the rat. I. Identification of detoxification products in urine. *Isr. J. Chem.*, *4*, 239-246

Nony, C.R. & Bowman, M.C. (1980) Trace analysis of potentially carcinogenic metabolites of an azodye and pigment in hamster and human urine as determined by two chromatographic procedures. *J. Chromatogr. Sci.*, *18*, 64-74

Osterman-Golkar, S. (1983) *Tissue doses in man: implications in risk assessment*. In: Hayes, A.W., Schmell, R.C. & Miya, T.S., eds, *Developments in the Science and Practice of Toxicology, ICT*, Amsterdam, Elsevier Biomedical Press

Osterman-Golkar, S., Ehrenberg, L., Segerbäck, D. & Hällström, I. (1976) Evaluation of genetic risks of alkylating agents. II. Haemoglobin as a dose monitor. *Mutat. Res.*, *34*, 1-10

Pasquini, R., Monarca, S., Sforzolini, G., Conti, R. & Fagioli, F. (1982) Mutagens in urine of carbon electrode workers. *Int. Arch. occup. environ. Health*, *50*, 387-395

Paul, J.S. & Pavelka, M.A. (1971) Covalent binding of a carcinogenic epoxide to DNA *in vivo*. *Fed. Proc.*, *30*, 448

Pereira, M.A. & Chang, L.W. (1981) Binding of chemical carcinogens and mutagens to rat hemoglobin. *Chem.-biol. Interact.*, *33*, 301-305

Pereira, M.A. & Chang, L.W. (1982) Binding of chloroform to mouse and rat hemoglobin. *Chem.-biol. Interact.*, *39*, 89-99

Pound, A.W., Franke, F. & Lawson, T.A. (1976) The binding of ethyl carbamate to DNA of mouse liver *in vivo*: the nature of the bound molecule and the site of binding. *Chem.-biol. Interact.*, *14*, 149-163

Prodi, G., Rocchi, P. & Grilli, S. (1970) In-vivo reaction of urethane with nucleic acids and proteins. *Cancer Res.*, *30*, 2887-2892

Reichert, D. (1982) Distribution, elimination and metabolism of hexa chloro-1,3-butadiene in rats. *Naunyn-Schmiedeberg's Arch. Pharmacol.* (*Suppl.*), *319*, R17

Reichert, D., Werner, H.W., Metzler, M. & Henschler, D. (1979) Molecular mechanism of 1,1-dichloroethylene toxicity: excreted metabolites reveal different pathways of reactive intermediates. *Arch. Toxicol.*, *42*, 159-169

van Roosmalen, P.B. & Klein, A.L. (1979) An improved method for the determination of 4,4'-methylene bis (2-chloroaniline) (MOCA) in urine. *Am. ind. Hyg. Assoc. J.*, *40*, 106-109

Sakurai, H., Onodera, M., Utsunomiya, T., Minakuchi, H., Iwai, H. & Matsumura, H. (1978) Health effects of acrylonitrile in acrylic fibre factories. *Br. J. ind. Med.*, *35*, 219-222

Sato, K., Poirier, L.A., Miller, J.A. & Miller, E.C. (1966) Studies on the N-hydroxylation and carcinogenicity of 4-aminoazobenzene and related compounds. *Cancer Res.*, *26*, 1678-1687

Savolainen, H. & Vainio, H. (1977) Organ distribution and nervous system binding of styrene and styrene oxide. *Toxicology*, *8*, 135-141

Selander, H.G., Jerina, D.M. & Daly, J.W. (1975) Metabolism of chlorobenzene with hepatic microsomes and solubilised cytochrome P-450 systems. *Arch. Biochem. Biophys.*, *168*, 309-321

Seutter-Berlage, F., van Dorp, H., Kosse, H.G.J. & Henderson, P.T. (1977) Urinary mercapturic acid excretion as a biological parameter of exposure to alkylating agents. *Int. Arch. occup. environ. Health*, *39*, 45-51

Seutter-Berlage, F., Delbressine, L.P.C., Smeets, F.L.M. & Ketelaars, H.C.J. (1978) Identification of three sulphur-containing urinary metabolites of styrene in the rat. *Xenobiotica*, *8*, 413-418

Shaikh, B., Huang Shing-Kwan, S. & Pontzer, N.J. (1980) Urinary excretion of methylated purines and 1-methylnicotinamide following administration of methylating carcinogens. *Chem.-biol. Interact.*, *30*, 253-256

van Sittert, N.J., de Jong, G., Clare, M.G., Davis, R., Dean, B.J., Wren, L.J. & Wright, A.S. (1984) Cytogenetic, immunological and haematological effects in workers in an ethylene oxide manufacturing plant. *Br. J. ind. Med.* (in press)

Sonnebichler, J. & Reichert, F. (1978) Wechselwirkung von p-Dimethyl aminoazobenzol mit Nichthistoprotein aus Rattenleberchromatin. *Z. Krebsforsch.*, *91*, 55-61

Springer, D.L., Krivak, B.M., Broderick, D.J., Reed, D.J. & Dost, F.N. (1981) Metabolic fate of hydrazine. *J. Toxicol. environ. Health*, *8*, 21-29

Stott, W.T., Reitz, R.H., Schumann, A.M. & Watanabe, P.G. (1981) Genetic and nongenetic events in neoplasia. *Food Cosmet. Toxicol.*, *19*, 567-576

Summer, K.H. & Greim, H. (1980) Detoxification of chloroprene (2-chloro-1,3-butadiene) with glutathione in the rat. *Biochem. biophys. Res. Commun.*, *96*, 566-573

Vainio, H., Savolainen, H. & Kilpikari, I. (1978) Urinary thioether of employees of a chemical plant. *Br. J. ind. Med.*, *35*, 232-234

Walles, S.A. (1981) Reactions of benzyl chloride with haemoglobin and DNA in various organs in mice. *Toxicol. Lett.*, *9*, 379-387

Watanabe, P.G., McGowan, G.R. & Gehring, P.J. (1976) Fate of [$^{14}$C]vinyl chloride after single oral administration in rats. *Toxicol. appl. Pharmacol.*, *36*, 339-352

Woo, Y.T., Arcos, J.C., Argus, M.F., Griffin, G.W. & Nishiyana, K. (1977) Metabolism *in vivo* of dioxane: identification of p-dioxane-2-one as the major urinary metabolite. *Biochem. Pharmacol.*, *6*, 1535-1538

WHO (1982) *Field Surveys of Exposure to Pesticides* (*VBC/82.1*), Geneva, World Health Organization

Wright, A.S. (1981) *New strategies in biochemical studies for pesticide toxicity*. In: Bandall, S.K., Macro, T.G., Golberg, L. & Leng, M.L., eds, *The Pesticide Chemist and Modern Toxicology* (*ACS Symposium Series No. 160*), Washington DC, American Chemical Society, pp. 285-304

Wright, A.S. (1983) *Molecular dosimetry techniques in human risk assessment: an industrial perspective*. In: Hayes, A.W., Schmell, R.C. & Miya, T.S., eds, *Developments in the Science and Practice of Toxicology, ICT*, Amsterdam, Elsevier Biomedical Press, pp. 311-318

Young, J.D., Braun, W.H., Gehring, P.J., Horwath, B.S. & Damiel, R.L. (1976) 1,4-Dioxane and β-hydroxyethoxyacetic acid excretion in urine of humans exposed to dioxane vapours. *Toxicol. appl. Pharmacol.*, *38*, 643-646

Zielhuis, R.L. (1978) Biological monitoring. *Scand. J. Work environ. Health*, *4*, 1-18

# EXCRETION OF THIOETHERS IN URINE AFTER EXPOSURE TO ELECTROPHILIC CHEMICALS

**P.T. Henderson[1], R. van Doorn[2], C.-M. Leijdekkers[1] & R.P. Bos[1]**

[1]*Institute of Pharmacology, Toxicology Unit, University of Nijmegen, The Netherlands*

[2]*Occupational Health Service 'Midden-Ijssel' Deventer, The Netherlands*

## SUMMARY

Electrophilic agents - a class of chemicals that includes most genotoxic compounds - can be inactivated by reaction with glutathione or other SH-bearing molecules. The conjugates so formed often appear in the urine as mercapturic acids or other thioether products.

This paper critically reviews the suitability of the urinary thioether assay as a method for the detection of exposure to electrophilic agents or their precursors.

In practice, the greatest value of the thioether assay appears to lie in its signal function. This is demonstrated for cigarette smokers and industrial workers involved in chemical waste incineration. Whenever increased thioether excretion is observed, it is likely to be due to exposure to one or more suspect compounds. However, when the thioether concentration ranges within the limits of the normal value, one must not conclude that there is no, or negligible, exposure.

More specific applications of the assay of thio compounds in urine allow development of selective methods that may be useful for biological monitoring.

## INTRODUCTION

It is well known that reactive, intermediary metabolites are generated during enzymatic oxidation of certain chemicals, and that, owing to their electrophilic properties, such toxic intermediates (epoxides, carbonium ions, etc.) can bind covalently to proteins, RNA and/or

DNA, by reaction with nucleophilic SH-, $NH_2$- or OH-groups. The subsequent structural modification of vital cellular macromolecules, often indicated by the term 'chemical lesion', is the first step towards a toxic effect. Compounds that are commonly metabolized *via* the formation of electrophilic intermediates are sometimes called potentially alkylating substances. This class of compounds includes most mutagenic and carcinogenic substances and other substances that can cause severe tissue damage.

Whether exposure to an alkylating substance leads to an initiation of toxic effects depends on several factors. Not all electrophilic metabolites formed during biotransformation will bind covalently to cellular macromolecules. In the cell, there are several detoxification mechanisms that are able to deactivate reactive electrophiles. In this respect, binding to glutathione is of particular importance.

Glutathione is a tripeptide, consisting of the amino acids glycine, cysteine and glutamine, and is present in high concentrations in many cells. An important property of glutathione is that it has a nucleophilic SH-group. In theory all electrophilic agents conjugate with glutathione, but the extent to which this occurs will depend upon the electrophilic reactivity and upon the catalyzing action of glutathione S-transferase (Chasseaud, 1979).

As long as a century ago, Baumann and Preusse (1879) and Jaffe (1879) demonstrated that certain sulfur-containing compounds appeared in the urine of dogs after administration of halogenated aromatic compounds like bromobenzene. These compounds were identified as *N*-acetylcysteine conjugates and were called mercapturic acids. In fact, premercapturic acids, those labile thioethers that form mercapturic acids upon acidification of aqueous solutions, are the real metabolites of aromatic hydrocarbons in urine (Gillham & Young, 1968). The mechanism of mercapturic-acid formation was not conclusively established until many years after the first discovery of mercapturic acids in urine. It was shown that glutathione was the major, and probably the only source of the cysteine moiety (Barnes *et al.*, 1959; Bray *et al.*, 1959). Glutathione conjugates are converted to cysteine conjugates in two enzymatic reactions: removal of the glutamyl by $\gamma$-glutamyl transferase and removal of the glycyl moiety by cysteinylglycinase. These enzymes are present in many tissues. Often, cysteine conjugates are *N*-acetylated by an *N*-acetyltransferase and the mercapturic acids are excreted subsequently in urine (Moldèus *et al.*, 1978).

The reaction of electrophilic intermediary metabolites with glutathione can prevent covalent binding of the electrophiles to cellular macromolecules. In this way, glutathione conjugation represents a protective mechanism (e.g., Fig. 1).

Although glutathione conjugation generally results in the formation of thioethers that are less toxic than their parent electrophiles, there are a few exceptions where glutathione conjugates have mutagenic properties (Rannug *et al.*, 1978; Bladeren *et al.*, 1979).

Glutathione conjugation results in the formation of thioethers that are excreted as glutathione conjugates in bile and as cysteine conjugates, premercapturic acids, mercapturic acids and other thioethers in bile and urine. When, somewhere in the organism, electrophiles react with glutathione, mercapturic acids and other thioethers are expected to appear in the urine. Most mercapturic acids that have been identified as urinary metabolites originate from xenobiotic compounds, but some endogenous electrophiles are also excreted as mercapturic acids (Chasseaud, 1976, 1979). As a consequence, the question arises whether the synthesis of mercapturic acids, which in turn may lead to elevated urinary excretion of these detoxification products, can be considered a parameter of exposure to electrophilic compounds.

**Fig. 1. Metabolic scheme of bromobenzene**

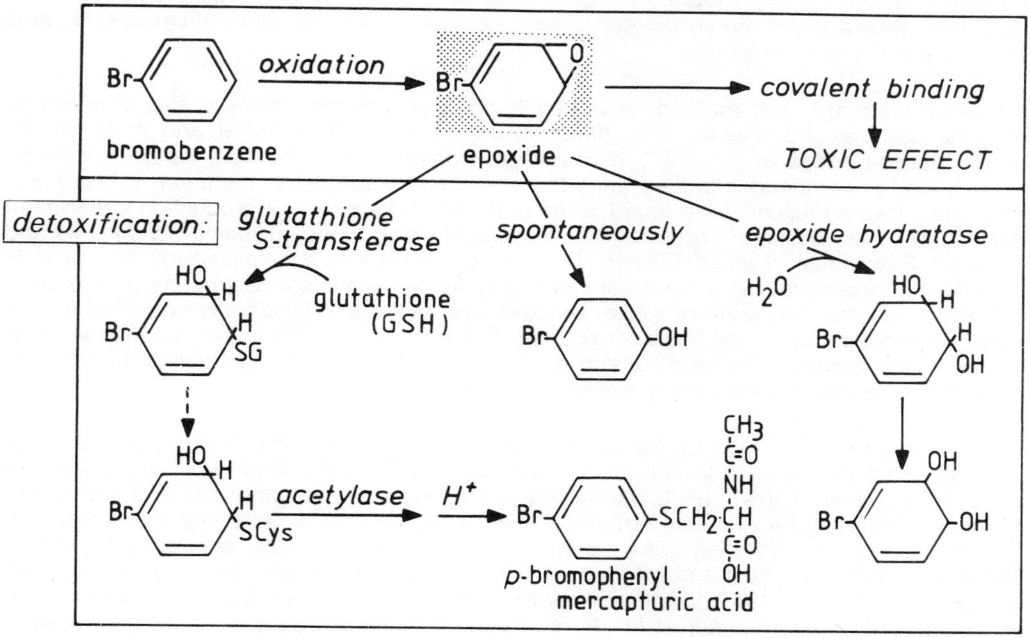


The reaction of 3,4-bromobenzene oxide (epoxide) with glutathione leads to the formation of a mercapturic acid and protects the cell from toxic effects due to covalent binding (adapted from Gillette, 1973).

A number of reports have appeared concerning the detection of hazardous exposure by nonselective measurements of urinary thioether concentrations (Savolainen *et al.*, 1977; Seutter-Berlage *et al.*, 1977; Pentz, 1978; Vainio *et al.*, 1978; Pentz, 1979; Summer *et al.*, 1979a,b; van Doorn *et al.*, 1979, 1980a,b, 1981a,b; Kilpikari & Savolainen, 1982; Heinonen *et al.*, 1983).

The present paper attempts a critical review of the possibilities and limitations of the so-called thioether, or mercapturic acid, assay as a method for detecting exposure to electrophilic agents and their precursors.

## METHODS

Several procedures have been described for nonselective detection of thio compounds (including mercapturic acids) in urine samples of persons exposed to electrophilic compounds (Seutter-Berlage *et al.*, 1977; Vainio *et al.*, 1978; van Doorn *et al.*, 1979; Summer *et al.*, 1979b). By application of these procedures, a large group of divalent sulfur compounds can be detected. In some methods, a distinction is made between different forms in which the urinary thio compounds occur: mercaptans (R-SH, also called thiols), disulfides (R-S-S-R') or thioethers (R-S-R').

The simplest assay procedure includes alkaline hydrolysis of thio compounds. The concentration of the thiols so formed is determined by the method of Ellman (1959), which is based on generation of the yellow dye, 2-nitro-benzoic acid, from dithiobisnitrobenzoic acid in the presence of thiol-groups.

Some years ago, this method was used to compare concentrations of thio compounds in urine samples from workers in different departments of chemical and metal industries (Seutter-Berlage *et al.*, 1977). On average, workers in the chemical industry excreted more thioether compounds in their urine than did their colleagues in the metal industry. For example, elevated values were found in the urine of workers from a production department where there was possible exposure to a mixture of compounds, including acrylonitrile and biphenyl. Savolainen *et al.* (1977) and Vainio *et al.* (1978) also determined thio compounds by a similar procedure. At a chemical plant, they found higher concentrations of thio compounds in the morning urine of rubber workers and radial tyre builders over that of clerks, plastic monomer mixers and footwear preparers. Pentz (1978, 1979) measured the excretion of bound thio-groups in hospital patients and in healthy volunteers and studied the influence of certain diseases on the urinary excretion of thio conjugates.

In all these studies, high background values and large variations in individual values of nonexposed persons were found, even though corrections were made for the presence of pre-existing urinary thiols. This is due mainly to high urinary concentrations of the disulfide, cysteine, which can be eliminated from the final SH content by the following modification of the thioether assay. Acidified urine samples are extracted with ethyl acetate, the extracts are evaporated to dryness and the residue is taken up in water and subsequently hydrolysed (van Doorn *et al.*, 1981a). An overview and some details of this procedure are presented in Figure 2. In order to adjust for differences in the concentration of urine, the thioether concentration is usually expressed as mmol SH per mol creatinine. The procedure is quite selective for mercapturic acids and other thioethers which are extracted into ethyl acetate. Generally very low recoveries are found for cysteine conjugates. The possibility of interference in the assay by sulfur-containing drugs (like penicillin antibiotics) must also be considered.

## FACTORS INFLUENCING THIOETHER EXCRETION

It is important to note that precursors of the excreted thiocompounds do not necessarily possess carcinogenic properties. In animal experiments, administration of notoriously genotoxic compounds (e.g., benzidine and several other carcinogenic arylamines), on the one hand, did not lead to increased thioether concentrations in urine. On the other hand, several chemicals, for which neither carcinogenic nor mutagenic effects are known, were found to be substantially converted into thioethers. Examples are bromobenzene, phorone and *o*-xylene. The apparently electrophilic nature of such compounds or of their reactive metabolites, however, makes them suspect and would justify genotoxic re-evaluation.

The question of whether the thioether assay can help to assess harmful human exposure to chemicals (e.g., workers occupationally exposed to electrophilic substances) leads to a second question: is there a quantitative relationship between the measured thioether values in urine and the degree of exposure? To answer this question, one has to know which are the exogenous and endogenous factors that determine the amount of the thioethers excreted into the urine of an individual after exposure to electrophilic substances.

**Fig. 2. Scheme for nonselective determination of thioethers in urine (van Doorn *et al.*, 1981a)**

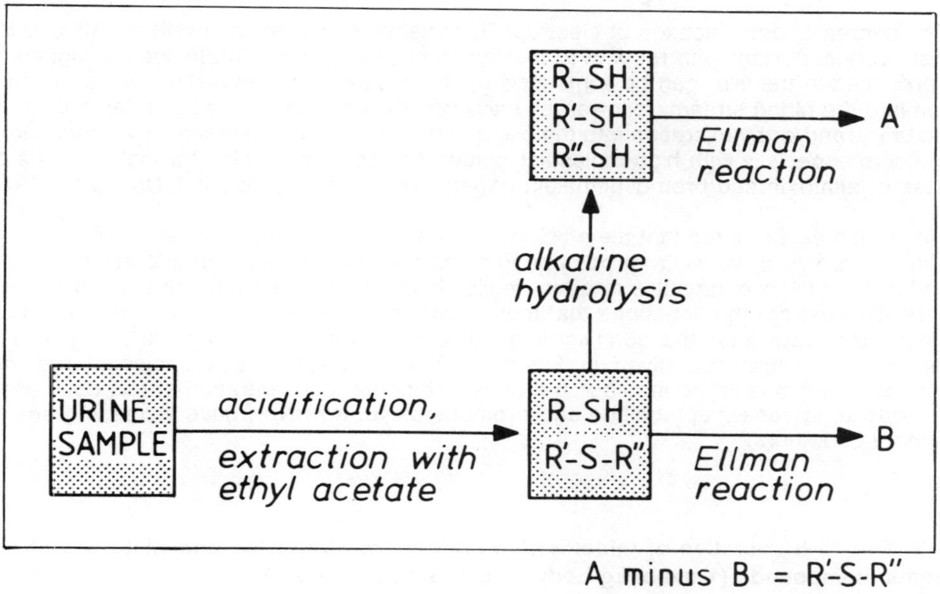


*Extraction*: Thawed urine samples were centrifuged for 5 min at 3000 g before processing. Aliquot samples of 5.0 ml of clear urine were transferred to glass-stoppered tubes and the pH was adjusted to 1.5-2.0 with 4 N hydrochloric acid. After the addition of 8.0 ml of ethyl acetate, the layers were shaken vigorously for 15 min by a shaking apparatus, then separated by centrifugation at 3000 × *g* for 5 min. The ethyl acetate layer was removed and the extraction procedure was repeated with another 8.0 ml of ethyl acetate. The collected ethyl acetate layers were evaporated to dryness using a rotary evaporator and the residue was taken up in 2.0 ml of distilled water.

*Hydrolysis*: Alkaline hydrolysis was performed on 1.0 ml samples in brown-glass screw-capped tubes by the addition of 0.5 ml of 4 N sodium hydroxide, followed by saturation with nitrogen (bubbling for 15 sec), then by subjection of the closed tubes to a boiling waterbath for 50 min. The tubes were cooled in ice for 10 min. Under mixing, 0.5 ml of 4 N hydrochloric acid was added.

*Ellman reaction*: Exactly 5 min later, the SH-concentration was determined according to Ellman (1959) with slight modifications. A 0.25-ml aliquot sample of the aqueous solution was added to a freshly made mixture consisting of 2.0 ml of 0.5 M phosphate buffer (pH of 7.1) and 0.3 ml of a 5,5'-dithiobis-(2-nitrobenzoic acid) solution (0.4 mg 5,5'-dithiobis-(2-nitrobenzoic acid) per ml of 1% sodium citrate solution). Absorbances were read at 412 nm on a Pye Unicam SP 1750 spectrophotometer. Corrections were made for the contribution of the extract and of the 5,5'-dithiobis-(2-nitrobenzoic acid) solution to the absorbance. The SH concentration was calculated from the corrected absorbance and the molar absorbance of the reference compound, *N*-acetyl-*L*-cysteine, in the SH determination, i.e., A-B = R'-S-R".

The dose is of primary importance. Although thioether production generally increases with the dose, after a large dose, the glutathione content of cells may be depleted, and any further increase in the dose may not give a proportional increase in thioether production. In humans, such a large exposure seldom occurs except in acute intoxications. The mode of exposure, also, may influence the biological fate of the chemical concerned. A substance taken up by inhalation during an 8-h exposure period is often metabolized differently from the same amount ingested or absorbed instantaneously. Further, endogenous factors play an important role in the amount of thioether finally excreted into the urine. These factors include the fraction of the absorbed amount which is metabolized into reactive intermediates and the degree to which these electrophiles are detoxified. Volatile compounds are often exhaled largely unchanged and are thus less liable to metabolic conversion.

Antagonistic and synergistic effects may appear, however, for example due to inhibition and induction of biotransformational enzymes (Kilpikari & Savolainen, 1982).

The degree of detoxification of electrophilic metabolites by reaction with glutathione is, of course, very important with respect to the thioether production. Glutathione conjugates that are produced in the liver can be eliminated by the organism in two distinct ways: by mobilization into the blood stream, followed by excretion as mercapturic acid or related thioethers in the urine, and/or by excretion into the bile and elimination *via* the faeces. Excretion into the bile occurs especially with high-molecular glutathione conjugates like, for instance, the conjugates of aflatoxin and bromosulfthalein (Whelan *et al.*, 1970; Degen & Neumann, 1978).

Finally, it must be noted that the efficiency of the thioether assay (the recoveries of extraction and hydrolysis) varies from compound to compound. For mercapturic acids, however, the influence of these variables is often negligible compared with the influence of biological factors. To illustrate the variations that may occur, the thioether assay was applied to urine samples from rats after the administration of some aromatic compounds in equal doses of each. The results are given in Figure 3. The thioether values showed large variations; paracetamol failed to increase the urinary thioether concentration at this dose. Biphenyl, *o*-xylene, styrene, naphthalene and bromobenzene, however, gave increasing amounts of urinary thioethers.

**Fig. 3. The 24-h excretion of thioethers in urine by rats after the administration of some aromatic compounds (1 mmol/kg body weight; intraperitoneal)**

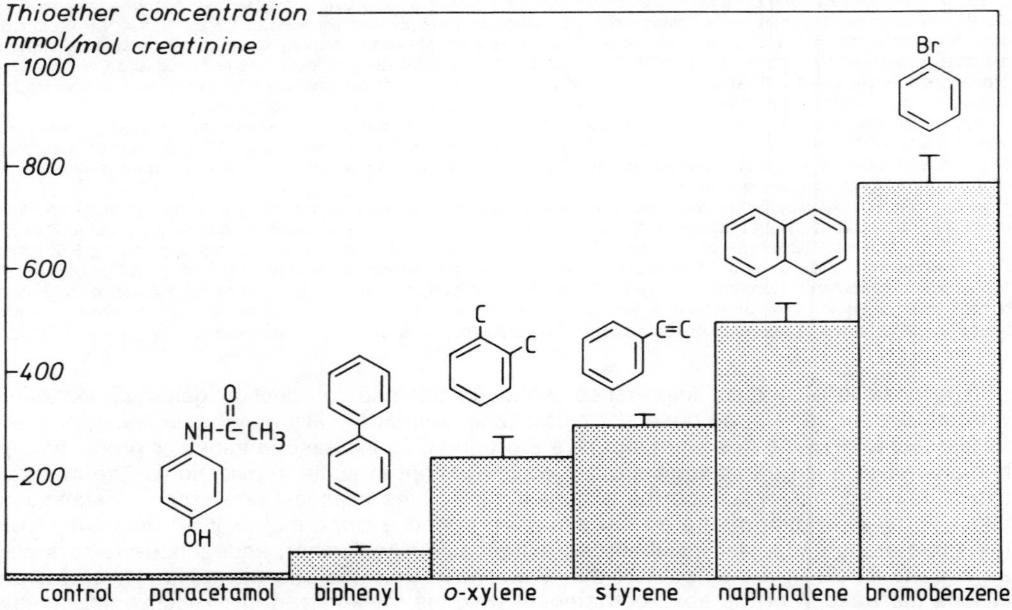


Note the large variations in the urinary thioether values.

## PRACTICAL SIGNIFICANCE: SIGNAL FUNCTION

An attempt was made to find a dose-response relationship in exposed persons. Human volunteers inhaled different amounts of a mixture of potentially carcinogenic substances, viz., cigarette smoke, and urinary thioether concentrations were assayed (van Doorn *et al.*, 1979). Cigarette smoke contains several carcinogens, e.g., benzopyrenes, nitrosamines, vinyl chloride and numerous suspect pyrolysis products of amino acids. It is, therefore, not surprising that smoking caused an increase in the urinary excretion of thioethers (Fig. 4). For 20 days, thioether concentrations were measured in afternoon urine samples from a smoker who consumed a gradually increasing number of cigarettes daily. It is interesting that the thioether excretion followed the smoking pattern. In addition, a good correlation was found between the presence of mutagenic substances in urine and the number of cigarettes consumed, demonstrating that not all harmful components of cigarette smoke are effectively detoxified. Recently, Heinonen *et al.* (1983) found no differences in urinary excretion of thioethers between smokers of low-tar and of medium-tar cigarettes. This suggests that the tar content of cigarettes should not be regarded as the decisive factor in increased thioether production.

**Fig. 4. Thioether concentrations (mmol SH per mol creatinine) (□) and mutagenic activity (revertants/$10^6$ bacteria per mmol creatinine) (▤) in the urine of a cigarette smoker**

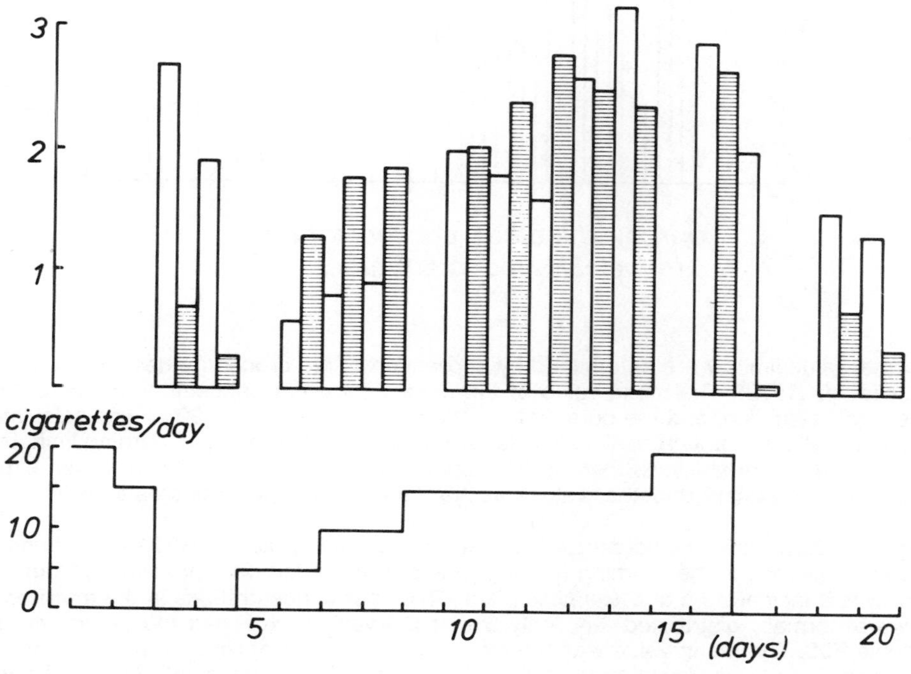


In the thioether determination, a correction was made for the presence of disulfides by using sodium borohydride (from van Doorn *et al.*, 1979). Note that the values are considerably lower than values found by application of the method described in Figure 2.

Due to its nonselective nature, the sensitivity of the thioether assay is limited by a relatively high 'nonexposure' level. The presence of a background was shown in urine samples of nonexposed people (Fig. 5), which were collected at random during the day and included samples obtained from smokers.

**Fig. 5. Distribution of thioether concentrations in a group of 196 urine samples obtained from nonexposed people (including cigarette smokers)**

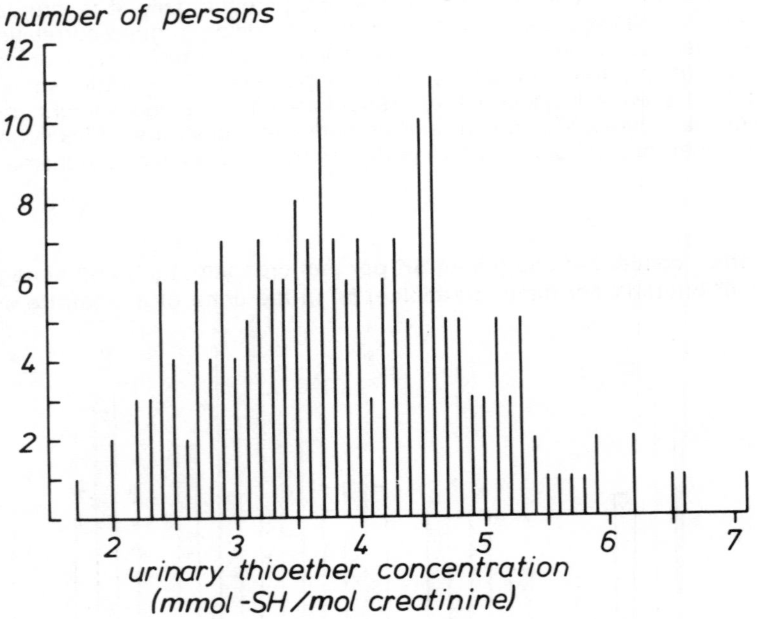


Normal value limits were determined by application of the procedure described by Rümke and Bezemer (1972). The mean value of the urinary thioether concentration was 3.8 mmol SH per mol creatinine, and the outer limit of the percentile P95 ($\gamma = 90\%$) was calculated to be 5.9 mmol SH per mol creatinine. The latter should be regarded as a warning limit; by this method of statistical analysis, there is 90% confidence that, at most, 5% of a non-exposed population have urinary thioether values above 5.9 mmol SH per mol creatinine.

Figure 6 shows how the percentile limit can be applied in practice. Sixty-seven urine samples were collected in the morning (pre-work) and in the afternoon (post-work) from operators of waste incinerators at a chemical plant. Thioether concentrations in the morning samples were normally distributed, and only 3% of the values exceeded the outer limit of the percentile P95. The mean value was 3.9 mmol SH per mol creatinine. In the afternoon samples, however, a large deviation from the normal distribution was seen; 24% of the values exceeded the outer limit of the percentile P95 and the mean value increased to 5.1 mmol SH per mol creatinine. No differences were found in morning and afternoon urine samples from a control group of workers in another department of the same plant.

**Fig. 6. Distribution of thioether concentrations in urine from operators of chemical waste incinerators**

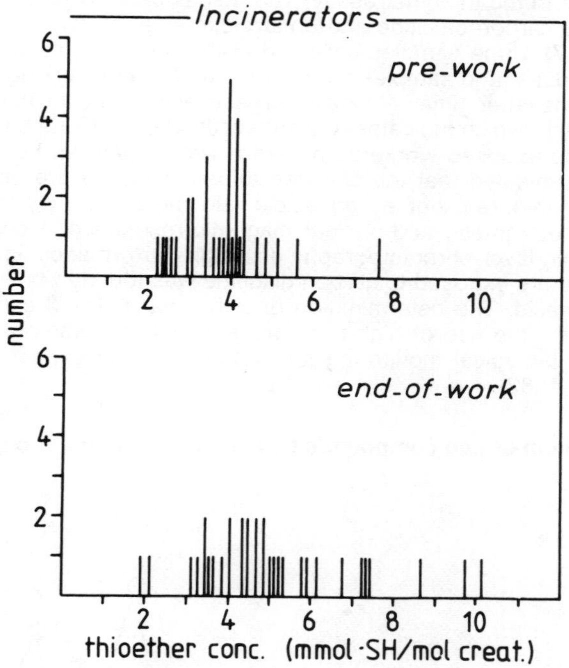


Urine samples were collected at the beginning (upper part of the figure) and at the end (lower part) of the working day (from van Doorn *et al.,* 1981a).

Obviously, in tracing the exposure of people to (mixtures of) electrophilic compounds, the nonselective thioether assay can have an important function. When the compounds involved are unknown, the results of the assay can be interpreted only in a qualitative sense. In many instances, however, when an assessment of risk or of the degree of exposure is desired, qualitative results are insufficient, and suspected compounds must be identified in order to make quantitative verdicts possible.

## SPECIFIC APPLICATIONS

When the electrophilic compound that causes an enhanced thioether excretion is known, and interference by other substances can be excluded, a more specific application of the thioether assay may be possible. Inclusion of the assay in a biological monitoring programme is then conceivable.

The following example, involving exposure to carbon disulfide in a viscose rayon industry, may illustrate this. Measurements were taken of the urinary excretion of thio compounds by persons exposed occupationally to carbon disulfide; the ambient concentration in the production department was about 7 ppm. Pre- and post-work urine samples were collected for 7 days and subjected to the thioether assay. The results demonstrate a positive relationship between exposure to carbon disulfide and urinary excretion of thio compounds (see results for one worker, Fig. 7). Urine samples collected at the beginning of the working day (10:00 pm) showed lower values, and samples taken at the end of the working day (6:00 am) showed higher values, time after time. A small increase was found in the pre-work values in the course of the week, which indicates a slight accumulation. Urine samples obtained from other carbon disulfide-exposed workers showed a similar pattern, but control samples did not. Further studies revealed that the elevated thioether values are not caused by carbon disulfide itself or by thiourea, but by an acidic thio metabolite. By the use of gas-chromatography-mass-spectrometry and nuclear magnetic resonance, a compound isolated (by solvent extraction, thin-layer chromatography and high-performance liquid chromatography) from the urine of workers exposed to carbon disulfide was found to be identical to 2-thiothiazolidine-4-carboxylic acid. The determination of 2-thiothiazolidine-4-carboxylic acid in urine of viscose workers, by the use of high-performance liquid chromatography, has promising prospects in routine biological monitoring for carbon disulfide exposure (van Doorn *et al.*, 1981c; Rosier *et al.*, 1982).

**Fig. 7. Urinary excretion of thio compounds by a spinner during a 7-day period (night shift)**

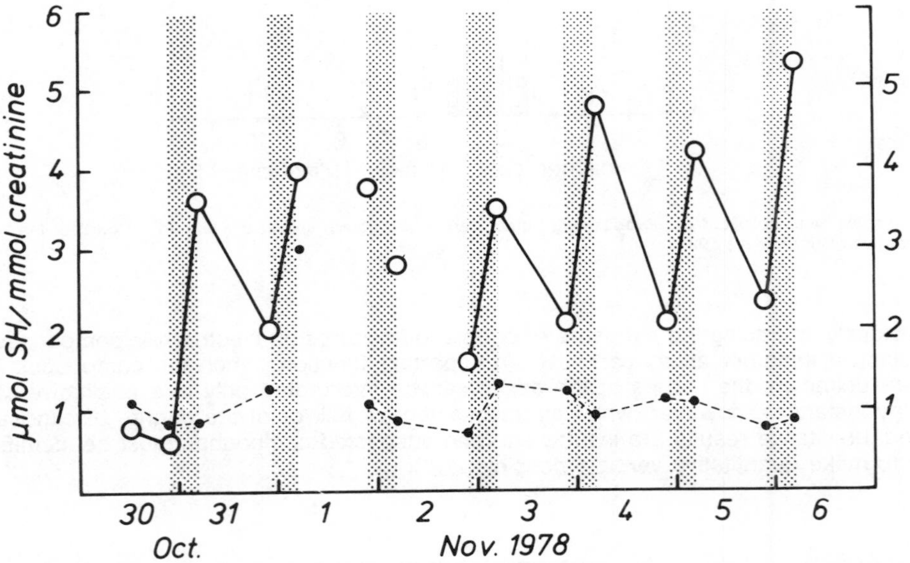


The subject was a nonsmoker, who worked in a viscose plant where he was exposed to about 20 mg carbon disulfide/m$^3$. (•), disulfides and thiols; (o), other thio compounds (including thioethers and thioesters). The dotted areas correspond to the work periods of the day (from van Doorn *et al.*, 1981b).

A second example, dealing with the detection of methyl chloride exposure, demonstrates the limited possibilities of the nonselective thioether assay. Exposure of a group of workers to methyl chloride in the ambient air at concentrations of up to 90 ppm could not be detected by assaying urine samples collected at various times for the presence of thioethers. Since relatively low concentrations of mercapturic acids would be readily detected, this means that methyl mercapturic acid is not a major metabolite of methyl chloride. Using a detection method based on alkaline hydrolysis of the urine samples with subsequent detection of methyl mercaptan in the 'head-space' of the solutions, large increases in the excretion of methyl thio compounds were found in the urine of some workers in the group exposed to methyl chloride (Fig. 8). This metabolite (or metabolites) has not so far been identified.

**Fig. 8. Concentrations of thioether (mmol SH/mol creatinine) (A) and methyl thio compounds (mmol $CH_3SH$/mol creatinine) (B) in the urine of industrial workers exposed to methyl chloride in comparison with concentrations in the urine of control subjects (from van Doorn *et al.*, 1980a)**

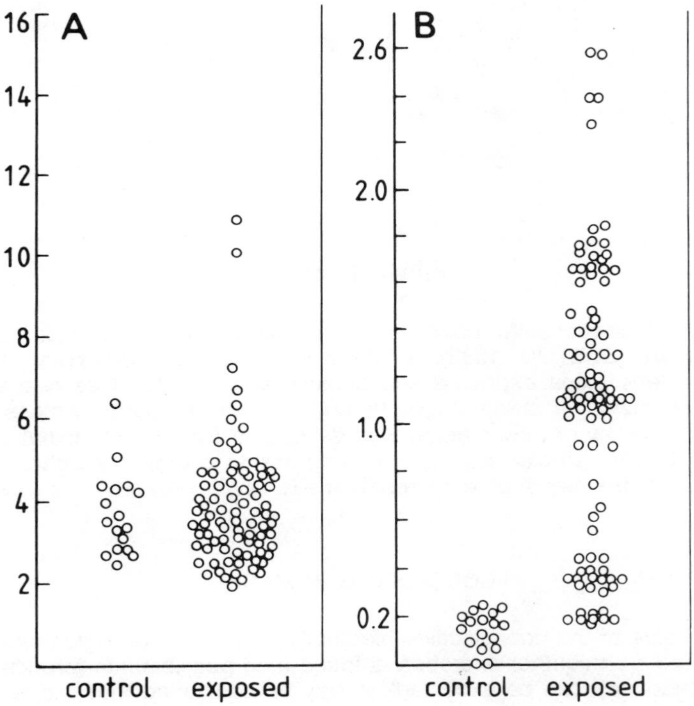


These examples show that applications of the nonselective assay of urinary thio-compounds can aid development of more selective methods that may be useful in biological monitoring programmes.

A similar procedure is followed in the attempt to develop a method for detection of detoxification products in urine after exposure to ethylene oxide. At present, there is no reliable biological method to monitor exposure to this compound. Current investigations show that inhalation exposure of rats to ethylene oxide (50-250 ppm, for 4 h) results in a dose-dependent increase in urinary thioether excretion (Fig. 9). Assuming that human biotransformation of ethylene oxide is comparable with that in the rat, only very high exposures are likely to be detected with the thioether assay. Thus, for biological monitoring of occupational exposure to ethylene oxide, selective methods allowing more sensitive detection of the thioether metabolites are needed.

**Fig. 9. The 24-h excretion of thioethers in urine by rats (adult males weighing about 250 g) after inhalation exposure to various concentrations of ethylene oxide for 4 h**

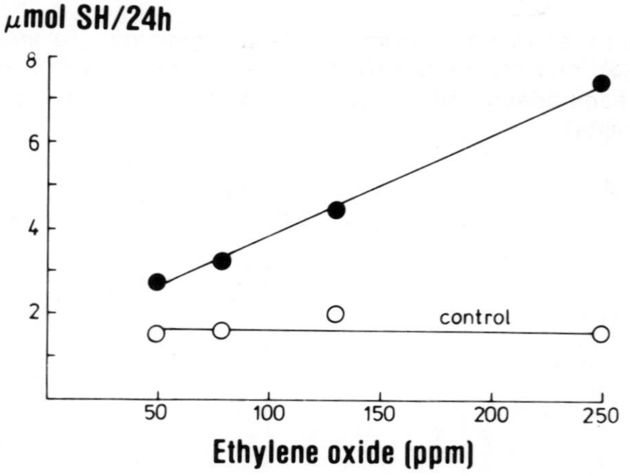


The existence of several sulfur-containing metabolites of ethylene oxide has been proposed in the literature (ECETOC, 1982). A thioether compound was isolated from the urine of rats after ethylene oxide exposure and subsequently identified as *N*-acetyl-*S*-carboxymethylcysteine by thin-layer chromatography and gas chromatography-mass spectrometry. A selective gas chromatography method was developed for the determination of this compound in urine. Further studies are now in progress to discover whether this metabolite is also excreted in the urine of exposed humans, e.g., workers in a hospital sterilization plant.

## CONCLUDING REMARKS

The primary value of the nonselective thioether assay lies in its signal function: when an increase in the urinary thioether excretion is found (and possible interference by other exogenous thio compounds can be excluded), it may be concluded that exposure and absorption of one or more electrophilic substances has occurred.

A simple calculation shows that, after a worker has remained for 8 h in a workroom environment where the air is contaminated with an electrophilic substance at a concentration of 1 ppm, the thioether concentration in the urine will increase by 25 mmol SH per mol creatinine at most. This level would occur only when, after complete absorption of the compound, 100% was converted to thioether and excreted in urine. In reality, less will be absorbed, a smaller fraction will be detoxified by conjugation with SH-bearing molecules, and only a part of the conjugates will be excreted as thioethers in urine. At present, little is known about the conversion of electrophilic compounds to urinary mercapturic acids and other thioethers in humans. Therefore, when elevated thioether levels are measured in human urine only qualitative interpretations are possible for most cases.

In summary, when the nonselective thioether assay is used and values are found within the limits of the normal distribution, one cannot conclude that no, or negligible, exposure has occurred. If the measurements have been taken correctly, the absence of an increase in thioether excretion may be a consequence of one or more of the following factors: (1) The wrong time for sampling was chosen. (2) The chemical(s) in question were absorbed, but only a minute part of the absorbed amount was excreted in urine as thioether. (3) There has been absorption but no detoxification, or incomplete detoxification, has occurred.

For case 3, especially, it is important to combine the thioether assay with other nonspecific exposure assays.

## ACKNOWLEDGEMENTS

Financial support was obtained from the Praeventiefonds and from the Labour Inspectorate, Dutch Ministry of Social Affairs.

## REFERENCES

Barnes, M.M., James, S.P. & Wood, P.B. (1959) The formation of mercapturic acids. 1. Formation of mercapturic acid and the level of glutathione in tissues. *Biochem. J.*, *71*, 680-690

Baumann, E. & Preusse, C. (1879) Über Bromphenylmercaptursäure. *Ber. dtsch. chem. Ges.*, *12*, 806-810

van Bladeren, P.J., van der Gen, A., Breimer, D.D. & Mohn, G.R. (1979) Stereoselective activation of vicinal dihalogen compounds to mutagens by glutathione conjugation. *Biochem. Pharmacol.*, *28*, 2521-2524

Bray, H.G., Franklin, T.J. & James, S.P. (1959) The formation of mercapturic acids. 2. The possible role of glutathione. *Biochem. J.*, *77*, 690-696

Chasseaud, L.F. (1976) *Conjugation with glutathione and mercapturic acid excretion.* In: Arias, I.M. & Jakoby, W.B., eds, *Glutathione, Metabolism and Function*, New York, Raven Press, pp. 77-114

Chasseaud, L.F. (1979) The role of glutathione and glutathione S-transferases in the metabolism of chemical carcinogens and other electrophilic agents. *Adv. Cancer Res.*, *29*, 175-275

Degen, G.H. & Neumann, H.G. (1978) The major metabolite of aflatoxin $B_1$ in the rat is a glutathione conjugate. *Chem.-biol. Interact.*, *22*, 239-255

van Doorn, R., Bos, R.P., Leijdekkers, C.-M., Wagenaars-Zegers, M.A.P., Theuws, J.L.G. & Henderson, P.T. (1979) Thioether concentration and mutagenicity of urine from cigarette smokers. *Int. Arch. occup. environ. Health*, *43*, 159-166 van Doorn, R., Borm, P.J.A., Leijdekkers, C.-M., Henderson, P.T., Reuvers, J. & van Bergen, T. (1980a) Detection and identification of *S*-methylcysteine in urine of workers exposed to methylchloride. *Int. Arch. occup. environ. Health*, *46*, 99-109

van Doorn, R., Bos, R.P., Brouns, R.M.E., Leijdekkers, C.-M. & Henderson, P.T. (1980b) Effect of toluene and xylenes on liver glutathione and their urinary excretion as mercapturic acids in the rat. *Arch. Toxicol.*, *43*, 293-304

van Doorn, R., Bos, R.P., Brouns, R.M.E., Leijdekkers, C.-M. & Henderson, P.T. (1981a) Enhanced excretion of thioethers in urine of operators of chemical waste incinerators. *Br. J. ind. Med.*, *38*, 187-190

van Doorn, R., Leijdekkers, C.-M., Henderson, P.T., Vanhoorne, M. & Vertin, P.G. (1981b) Determination of thio compounds in urine of workers exposed to carbon disulfide. *Arch. environ. Health*, *36*, 289-298

van Doorn, R., Delbressine, L.P.C., Leijdekkers, C.-M., Vertin, P.G. & Henderson, P.T. (1981c) Identification and determination of 2-thiothiazolidine-4-carboxylic acid in urine of workers exposed to carbon disulfide. *Arch. Toxicol.*, *47*, 51-58

ECETOC (1982) *Toxicity of Ethylene Oxide and its Relevance to Man* (*Technical Report No. 5*), Brussels

Ellman, G.L. (1959) Tissue sulfhydryl groups. *Arch. Biochem. Biophys.*, *82*, 70-77

Gillham, B. & Young, L. (1968) Biochemical studies of toxic agents. The isolation of premercapturic acids from the urine of animals dosed with chlorobenzene and bromobenzene. *Biochem. J.*, *109*, 143-147

Gillette, J.R. (1973) *Factors that affect the covalent binding and toxicity of drugs*. In: Acheson, G.H., ed., *Proceedings of the 5th International Congress on Pharmacology*, Vol. 2, *Toxicological Problems*, Basel, Karger, pp. 187-202

Heinonen, T., Kytöniemi, V., Sorsa, M. & Vainio, H. (1983) Urinary excretion of thioethers among low-tar and medium-tar cigarette smokers. *Int. Arch. occup. environ. Health*, *52*, 11-16

Jaffe, M. (1879) Über die nach Einführung von Brombenzol und Chlorbenzol im Organismus entstehende schwefelhaltige Säuren. *Ber. dtsch. chem. Ges.*, *12*, 1092-1098

Kilpikari, I. & Savolainen, H. (1982) Increased urinary excretion of thioether in new rubber workers. *Br. J. ind. Med.*, *39*, 401-403

Moldèus, P., Jones, D.P., Ormstadt, K. & Orrenius, S. (1978) Formation and metabolism of a glutathione-S-conjugate in isolated rat liver and kidney cells. *Biochem. biophys. Res. Commun.*, *83*, 195-200

Pentz, R. (1978) Influence of diet and xenobiotics on the urinary excretion of bound thiol groups in man and rats. *Naunyn-Schmiedenberg's Arch. Pharmacol., Suppl. 302*, R20

Pentz, R. (1979) Urinary excretion of bound thiol groups in healthy and diseased man and rats. *Naunyn-Schmiedenberg's Arch. Pharmacol., Suppl. 308*, R51

Rannug, U., Sundvall, A. & Ramel, C. (1978) The mutagenic effect of 1,2-dichloroethane on *Salmonella typhimurium*. I. Activation through conjugation with glutathione *in vitro*. *Chem.-biol. Interact.*, *20*, 1-16

Rosier, J., Vanhoorne, M., Grosjean, R., van de Walle, E., Billemont, G. & van Peteghem, C. (1982) Preliminary evaluation of urinary 2-thiothiazolidine-4-carboxylic acid (TTCA) levels as a test for exposure to carbon disulfide. *Int. Arch. occup. environ. Health*, *51*, 159-167

Rümke, C.L. & Bezemer, P.D. (1972) Methoden voor de bepaling van normale waarden. *Ned. Tijdschr. Geneesk.*, *116*, 1559-1568

Savolainen, H., Vainio, H. & Kilpikari, I. (1977) Urinary thioether determination as a biological exposure test. *Acta pharmacol. toxicol.*, *41* (*Suppl. IV*), 33

Seutter-Berlage, F., van Dorp, H.L., Kosse, H.G.J. & Henderson, P.T. (1977) Urinary mercapturic acid excretion as a biological parameter of exposure to alkylating agents. *Int. Arch. occup. environ. Health*, *39*, 45-51

Summer, K.H., Rozman, K., Coulston, F. & Greim, H. (1979a) Species differences in the excretion of mercapturic acids between rats and chimpanzees dosed with naphthalene and diethylmaleate. *Toxicol. appl. Pharmacol.*, *48*, A160

Summer, K.H., Rozman, K., Coulston, F. & Greim, H. (1979b) Urinary excretion of mercapturic acids in chimpanzees and rats. *Toxicol. appl. Pharmacol.*, *50*, 207-212

Vainio, H., Savolainen, H. & Kilpikari, I. (1978) Urinary thioether of employees of a chemical plant. *Br. J. ind. Med.*, *35*, 232-234

Whelan, G., Hoch, J. & Combes, B. (1970) A direct assessment of the importance of conjugation for biliary transport of sulfobromophtalein sodium. *J. Lab. clin. Med.*, *75*, 542-557

# USE OF ALKYLATED PROTEINS IN THE MONITORING OF EXPOSURE TO ALKYLATING AGENTS

**P.B. Farmer, E. Bailey & J.B. Campbell**

*MRC Toxicology Unit, MRC Laboratories,*
*Woodmansterne Road, Carshalton,*
*Surrey SM5 4EF, UK*

## SUMMARY

Measurement of reaction products of alkylating agents with proteins has been used as a monitor of in-vivo exposure to over 30 such compounds. Doses in animals exposed to directly-acting alkylating agents (e.g., methyl methanesulfonate, ethylene oxide) are directly related to the production of alkylated amino acids in haemoglobin. The erythrocyte dose of alkylating agent, calculated from the extent of haemoglobin alkylation, is in some cases related to liver and extra-hepatic DNA doses; thus, detection of alkylation of haemoglobin may be taken as an indication of a reaction at the carcinogenic target site.

## INTRODUCTION

Interaction between alkylating agents and proteins was postulated 60 years ago, when Cashmore and McCombie (1923) suggested that the vesicant action of mustard gas (bis[2-chloroethyl)sulfide) and related compounds might be due to a reaction with 'the amino acids present in the skin'. Since then, the chemistry of alkylating agent-protein interactions has been studied in considerable detail. Much information has been derived from work on enzymes, where many of the protein modification reagents that are used for active site investigations or in structure determinations are alkylating agents. In 1974, Ehrenberg *et al.* suggested the possibility of using the reaction products of alkylating agents with proteins as an indicator of in-vivo exposure to these compounds. Haemoglobin has been favoured as the target protein owing to its ready availability from erythrocytes and its long lifetime. Also, with the alkylating agents ethylene oxide, *N*-nitrosodimethylamine (Osterman-Golkar *et al.*, 1976), methyl methanesulfonate (Segerbäck *et al.*, 1978) and benzo[*a*]pyrene (Calleman, 1982), it has been shown that the alkylated protein is stable throughout the life-span of the erythrocytes. Although haemoglobin has been the most widely used protein for exposure monitoring, it may not be the most suitable target to detect the formation of highly reactive metabolites produced in the liver. Attention has been directed recently towards the use of serum albumin (Hemminki & Savolainen, 1980), which has a shorter life-time than haemoglobin but is synthesized in the liver and, therefore, may be in closer contact with active metabolites.

A summary of some alkylating agents for which haemoglobin or albumin binding has been studied is given in Table 1. Both directly-acting and metabolically activated, carcinogenic, alkylating agents are well represented on the list, although it is noteworthy that noncarcinogenic control substances have not been well studied. Thus, although the results to date suggest that genotoxic alkylating compounds are universally capable of reacting with haemoglobin, the reverse argument - that nongenotoxic compounds do not react with haemoglobin - has not been verified.

**Table 1. Protein binding of alkylating agents**

| Compound | Protein | Species | Analytical method[a] | Reference |
|---|---|---|---|---|
| Methyl methanesulfonate | Hb | mouse | A,B | Segerbäck *et al.* (1978) |
| | Hb | rat | C | Farmer *et al.* (1980) |
| | | | | Bailey *et al.* (1981) |
| | Hb | rat | A | Pereira & Chang (1981)[b] |
| *N*-Nitrosodimethylamine | Hb | mouse | A | Osterman-Golkar *et al.* (1976) |
| | Serum | rat | A | Hemminki & Savolainen (1980) |
| | Hb | rat | C | Bailey *et al.* (1981) |
| | Hb | rat | A | Pereira & Chang (1981) |
| | Erythrocyte[c] | human | A | Kim *et al.* (1981) |
| Methyl bromide | Hb | mouse | A | Djalali-Behzad *et al.* (1981) |
| Methyl chloride | Erythrocyte, plasma[d] | human | A | Redford-Ellis & Gowenlock (1971) |
| Dichlorvos | Hb | mouse | A | Segerbäck & Ehrenberg (1981) |
| Ethylene oxide | Hb | mouse | A | Osterman-Golkar *et al.* (1976) |
| | Hb | human | C,D | Calleman *et al.* (1978) |
| | Hb | rat | A,C,D | Osterman-Golkar *et al.* (1984) |
| | Hb | mouse | A | Segerbäck (1983) |
| Propylene oxide | Hb | rat | C | Farmer *et al.* (1982) |
| Vinyl chloride | Hb | mouse | A | Osterman-Golkar *et al.* (1977) |
| Ethylene | Hb | mouse | A,C | Ehrenberg *et al.* (1977) |
| | | | A | Osterman-Golkar & Ehrenberg (1982) |
| | Hb | mouse | | Segerbäck (1983) |
| Benzo[*a*]pyrene | Hb | mouse | A | Calleman (1982) |
| | Hb | rat | A | Pereira & Chang (1981) |
| Chloroform | Hb | rat, mouse | A | Pereira & Chang (1981, 1982a,b) |
| 2-Acetylaminofluorene | Hb | rat | A | Pereira & Chang (1981, 1982b) |
| | Hb | rat, mouse | A | Pereira *et al.* (1981) |
| | Serum | rat | A | Hemminki & Savolainen (1980) |
| Benzyl chloride | Hb | mouse | A | Walles (1981) |
| Aflatoxin $B_1$ | Hb | rat | A | Pereira & Chang (1981) |
| *trans*-Dimethylaminostilbene | Hb, plasma | rat | A | Neumann (1980) |
| *trans*-4-Aminostilbene | Hb, plasma | rat | A | Wieland & Neumann (1978) |
| 4-Aminobiphenyl | Hb, albumin | rat | A[e] | Tannenbaum *et al.* (1983) |

[a] Analytical methods: A, radiochemical; B, gas chromatography; C, gas chromatography-mass spectrometry; D, amino acid analysis

[b] Other compounds studied for haemoglobin (Hb) binding in the rat, by Pereira & Chang using radiochemical methods, were *N*-nitrosomethylurea, *N*-nitrosoethylurea, *N*-methyl-*N*'-nitro-*N*-nitrosoguanidine, 3-methylcholanthrene, *N*-nitrosodiethylamine, benzene, 7,12-dimethylbenz[*a*]anthracene, aniline, benzidine, carbon tetrachloride, naphthalene, phenol.

[c] In-vitro experiment using human erythrocytes

[d] In-vitro experiment using human erythrocytes and plasma

[e] Structure of adduct determined by mass, ultra-violet and fluorescence spectra

## EXPERIMENTAL APPROACHES TO THE DETERMINATION OF ALKYLATED PROTEINS

As indicated in Table 1, radiochemical methods for detecting alkylating agent-protein adducts have been most widely used. Such methods have the advantage of extremely high sensitivity, limited only by the specific activity of the alkylating agent used. One possible drawback in the use of labelled compounds is that alkylation reactions may not be distinguishable from biosynthetic incorporation of the label into the protein. Thus, for example, the metabolism of *N*-nitrosodimethylamine produces a one-carbon fragment which is readily incorporated into other molecules by biosynthetic pathways (Magee & Farber, 1962; Magee & Hultin, 1962; Osterman-Golkar *et al.*, 1976). Ideally, therefore, radiochemical approaches should involve determination of the label in an alkylated amino acid isolated from the protein, rather than of that in the intact protein.

Radiochemical methods of the type listed in Table 1 are clearly unsuitable for monitoring human exposure to nonradioactive alkylating agents. The table includes several examples of the use of chromatographic methods, which are able to measure such exposure. The use of a high-sensitivity detection system, such as a mass spectrometer, combined with a high-resolution gas-chromatography system (GC-MS), is necessary to analyse successfully the alkylated material present in human samples, which are commonly in the concentration range 0-10 pmol/mg protein. The determination of alkylated amino acids in proteins using GC-MS is limited to studies of alkylating agents of low molecular weight. Up to the present, there are no published reports on GC-MS studies of alkylated amino acids with alkyl groups of chain length greater than $C_3$.

Despite the fact that methods based on GC-MS are sometimes time-consuming and tedious (and, consequently, are not ideally suited to routine human monitoring), they have been used successfully in a wide variety of animal experiments (Ehrenberg *et al.*, 1977; Farmer *et al.*, 1980; Bailey *et al.*, 1981; Farmer *et al.*, 1982; Osterman-Golkar *et al.*, 1984) and for limited human studies (Calleman *et al.*, 1978).

An example of a GC-MS assay that is currently in use in our laboratory is illustrated in the schematic diagram of Figure 1. *N*-3'-(2-Hydroxyethyl)histidine is one of the modified amino acids produced in proteins following their exposure to ethylene oxide (Fig. 2). A method of quantifying this amino acid may be suitable for monitoring occupational exposure to ethylene oxide or to its metabolic precursor, ethylene. As indicated in Figure 1, an internal standard labelled with a stable isotope (Campbell, 1983) is required for mass spectrometric determination of the alkylated amino acid. For such syntheses, it is important to ensure that the isotope is well incorporated in the standard, so that no contribution is made by the standard to the unlabelled material being analysed. Additionally, the isotopic label must be stable to the chemically rigorous isolation and derivatization procedures used for the amino acid. These conditions were satisfied by $d_4$-hydroxyethylhistidine, which contained no detectable nondeuterated material and which preserved its $d_4$ label intact throughout the isolation procedure.

An analogous analytical approach is used to analyse *N*-3'-(2-hydroxypropyl)histidine in order to monitor exposure to propylene oxide (Farmer *et al.*, 1982). In this case, a pentadeuterated internal standard is used (Campbell, 1983). Exposure to methylating agents may be similarly monitored by quantitative determination by GC-MS of *S*-methylcysteine, using either a trideuterated internal standard (Farmer *et al.*, 1980) or an enantiomeric internal standard (Bailey *et al.*, 1980).

**Fig. 1. Quantitation of *N*-3'-(2-hydroxyethyl)histidine in haemoglobin**

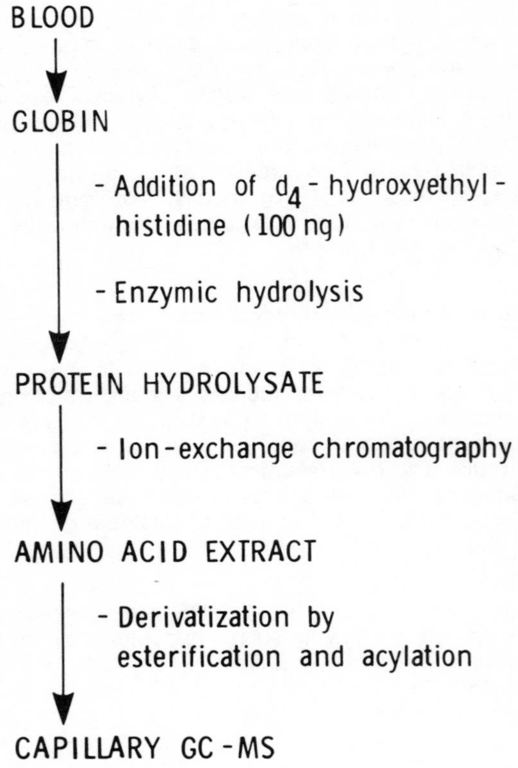


**Fig. 2. Formation of *N*-3'-(2-hydroxyethyl)histidine following interaction of ethylene oxide with histidine in a protein**

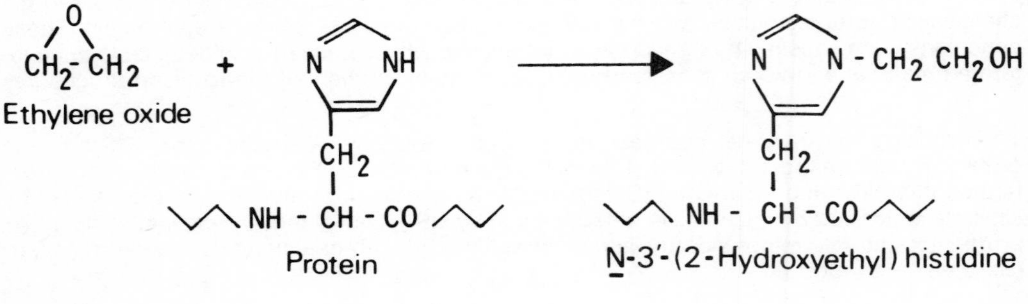

## DOSE-RESPONSE RELATIONSHIPS FOR ALKYLATED AMINO ACIDS IN PROTEINS

In order to use determinations of protein alkylation as a monitoring procedure, there must be a reliable dose-response relationship between the exposure dose of alkylating agent and the production of alkylated amino acid. This requirement appears to have been satisfied for the exposures studied to date although, naturally, all the experiments were carried out in animals rather than in man. Thus, GC-MS determination of the production of *S*-methylcysteine in rat haemoglobin, following intraperitoneal treatment with methyl methanesulfonate, showed that the alkylated amino acid was linearly related to dose (Bailey *et al.*, 1981). Radiochemical experiments by Segerbäck *et al.* (1978) also showed that the degree of alkylation of haemoglobin in mice was linearly dependent on the quantity of methyl methanesulfonate injected.

For ethylene oxide, an almost linear relationship was observed between the production of *N*-3'-(2-hydroxyethyl)histidine in haemoglobin of rats and the dose of ethylene oxide (Fig. 3). In this case, the alkylating agent was administered by inhalation of doses of 0-100 ppm (30 h/week for 2 yr) (Osterman-Golkar *et al.*, 1984).

**Fig. 3. Production of *N*-3'(2-hydroxyethyl)histidine (HOEtHis) in rat haemoglobin (Hb) following exposure by inhalation to ethylene oxide (Osterman-Golkar *et al.*, 1984)**

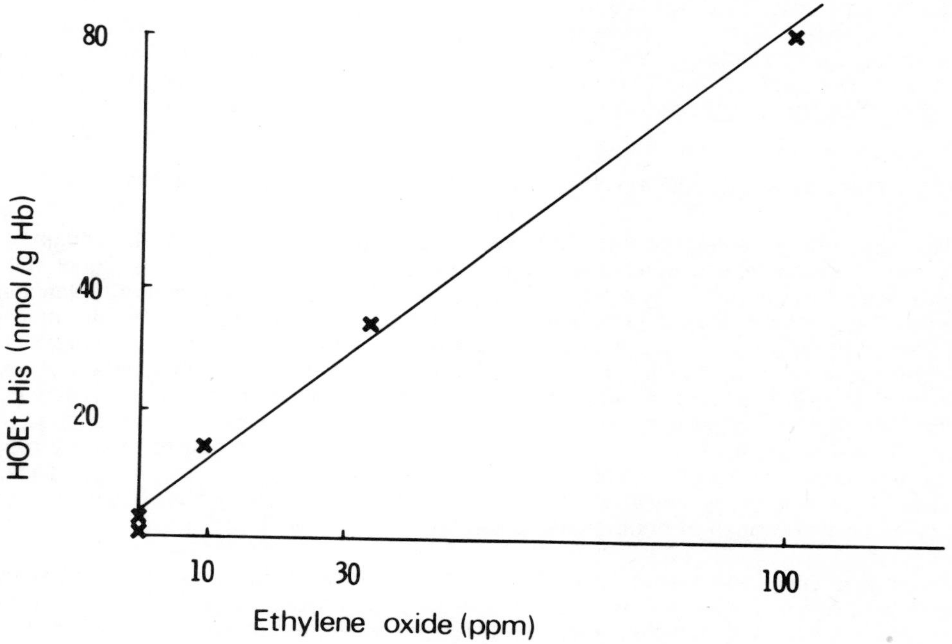


Other agents studied for dose-response relationships include the following: *trans*-4-dimethylaminostilbene (Neumann, 1980), with which total binding to rat haemoglobin increased linearly over a $10^5$-fold dose range; chloroform (Pereira & Chang, 1982a), which showed a linear dependence of binding to rat haemoglobin after oral administration of 0.1 to 100 μmol/kg; 2-acetylaminofluorene (Pereira *et al.*, 1981), which showed linear binding from 0.1 to 100 μmol/kg in rat and mouse haemoglobin. Only for *N*-nitrosodimethylamine (Bailey *et al.*, 1981) was the dose-response nonlinear over an intraperitoneal dose range of 0-400 μmol/kg in rats. This is currently under investigation.

## BACKGROUND LEVELS OF ALKYLATED AMINO ACIDS

For both *S*-methylcysteine (Bailey *et al.*, 1981; Farmer, 1982) and *N*-3'-(2-hydroxyethyl)histidine (Osterman-Golkar *et al.*, 1984), background levels of alkylated amino acids have been observed. The presence of these background levels means that low degrees of exposure cannot be easily monitored. The source of the alkylated amino acids is unknown, as yet. Although the presence of a methylated amino acid is perhaps unsurprising (owing to the abundance of other methylated amino acids present *in vivo*), detection of the hydroxyethylated histidine was unexpected. If it is proved conclusively that this alkylation is not caused by an experimental artefact, other sources for the alkyl group must be considered, e.g., exposures to ethylene, ethylene oxide, ethylene chlorohydrin, dichloroethane or ethanolamine (coupled with in-vivo nitrosation).

The background levels of *S*-methylcysteine are species-dependent (Bailey *et al.*, 1981; Farmer, 1982) and range from 6-480 nmol/g. For hydroxyethylhistidine, only two species, rat and human, have been studied in detail, and in these species, the amounts of alkylated amino acid appear to be comparable (average, about 2 nmol/g haemoglobin), although a considerable range of values was observed (Osterman-Golkar *et al.*, 1984; Osterman-Golkar & Farmer, unpublished data). Animals and humans show little or no background of *N*-3'-(2-hydroxypropyl)histidine, used for monitoring exposure to propylene oxide.

## DOES THE PRESENCE OF ALKYLATED PROTEINS INDICATE A GENOTOXIC RISK ?

The dose-response relationships described above show that alkylated haemoglobin is produced following in-vivo exposure to alkylating agents. The amount of product will be directly related to the erythrocyte dose of active alkylating agent. The observed relationship between haemoglobin alkylation and the exposure dose in animal experiments indicates that the erythrocyte dose and the exposure dose are directly related, also, in these cases. However, observation of haemoglobin alkylation may be considered an indication of genotoxic risk only if it has been demonstrated that such alkylations correlate with reactions at the target DNA-site, i.e., that the erythrocyte dose is directly related to the target dose (Fig. 4). In an extreme situation, the active alkylating agent would be distributed evenly throughout all the organs of the body and the target dose would equal the erythrocyte dose. In practice, of course, many alkylating agents do not distribute evenly, either because of their short lifetime or because of compartmentalization.

Amounts of DNA- and haemoglobin-binding products have been compared for a variety of radioactive alkylating agents. For vinyl chloride (Osterman-Golkar *et al.*, 1977) and *N*-nitrosodimethylamine (Osterman-Golkar *et al.*, 1976), on the one hand, which give short-lived metabolites, the doses received by organs far from the liver are considerably reduced. Ethylene oxide, on the other hand, appears to be distributed fairly uniformly. In the rat, alkylation of guanine-*N*-7 in liver-DNA and testis-DNA by ethylene oxide was about 150% and 50%, respectively, of the value expected from the degree of alkylation of haemoglobin (Osterman-Golkar *et al.*, 1984). Similarly, in the mouse (Segerbäck, 1983), the extent of guanine

**Fig. 4. Proposed mechanism of covalent binding of alkylating agents *in vivo***

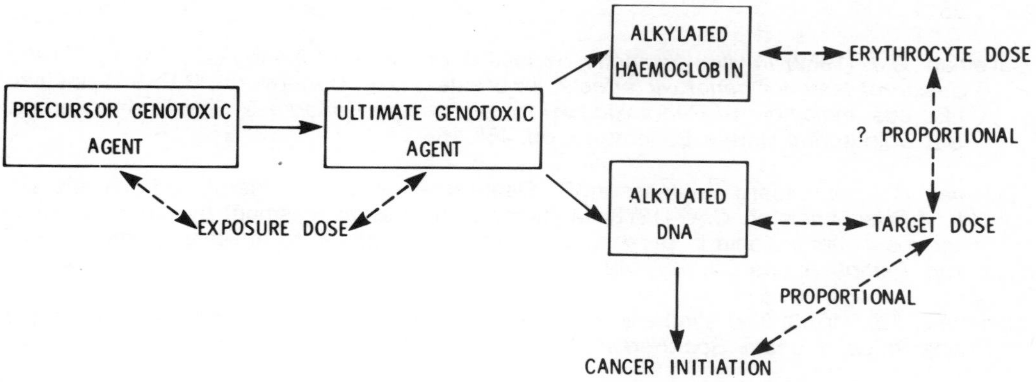


alkylation in liver, spleen and testis did not deviate more than two-fold from the amount expected from haemoglobin alkylation. Thus, the degree of alkylation of DNA can be estimated approximately from the erythrocyte dose for ethylene oxide.

Further support for the use of haemoglobin (or other protein) alkylation as an indication of DNA-binding has come from Neumann (1980), who has shown that the dose-response of *trans*-4-dimethylaminostilbene binding to plasma proteins and to haemoglobin is proportional to the dose-response for binding to liver-DNA. Similarly, Pereira *et al.* (1981) have illustrated that the dose-response curves for 2-acetylaminofluorene binding to haemoglobin and rat liver-DNA are closely related over a large dose range.

Thus, in *some* instances, it seems possible to monitor DNA-binding by determination of protein binding and, therefore, detection of the latter may be taken as an indication of a genotoxic risk. In this respect, the presence of background levels of alkylation in humans (see above) could be considered an indication of an existing risk, although it is of unknown magnitude as yet. One of the most important questions remaining to be answered is how the risk in humans is *quantitatively* related to the extent of DNA-binding. This information will be accessible only when the chemical nature of the target site and the interactions of the alkylated target with cellular repair systems are understood in detail.

## ACKNOWLEDGEMENTS

The authors gratefully acknowledge the technical help of Ms S.M. Gorf and J.H. Lamb in the execution of the experimental work described in this paper.

## REFERENCES

Bailey, E., Farmer, P.B. & Lamb, J.H. (1980) The enantiomer as internal standard for the quantitation of the alkylated amino acid *S*-methyl-*L*-cysteine in haemoglobin by gas chromatography-chemical ionisation mass spectrometry with single ion detection. *J. Chromatogr.*, *200*, 145-152

Bailey, E., Connors, T.A., Farmer, P.B., Gorf, S.M. & Rickard, J. (1981) Methylation of cysteine in haemoglobin following exposure to methylating agents. *Cancer Res.*, *41*, 2514-2517

Calleman, C.J. (1982) *In-vivo dosimetry by means of alkylated hemoglobin - a tool in the design of tests for genotoxic effects*. In: Bridges, B.A., Butterworth, B.E. & Weinstein, I.B., eds, *Indicators of Genotoxic Exposure* (*Banbury Report 13*), Cold Spring Harbor, NY, Cold spring Harbor Laboratory, pp. 157-168

Calleman, C.J., Ehrenberg, L., Jansson, B., Osterman-Golkar, S., Segerbäck, D., Svensson, K. & Wachtmeister, C.A. (1978) Monitoring and risk assessment by means of alkyl groups in hemoglobin in persons occupationally exposed to ethylene oxide. *J. environ. Pathol. Toxicol.*, *2*, 427-442

Campbell, J.B. (1983) The synthesis of *N*(Γ)-(2-hydroxyethyl)histidine and their deuterated analogues. *J. chem. Soc. Perkin's Trans.*, *1*, 1213-1217

Cashmore, A.E. & McCombie, H. (1923) The interaction of ββ'-dichlorodiethylsulphide, sulphoxide and sulphone with glycine ester and with potassium phthalimide. *J. chem. Soc.*, *123*, 2884-2890

Djalali-Behzad, G., Hussain, S., Osterman-Golkar, S. & Segerbäck, D. (1981) Estimation of genetic risks of alkylating agents. VI. Exposure of mice and bacteria to methyl bromide. *Mutat. Res.*, *84*, 1-9

Ehrenberg, L., Hiesche, K.D., Osterman-Golkar, S. & Wennberg, I. (1974) Evaluation of genetic risks of alkylating agents: tissue dose in the mouse from air contaminated with ethylene oxide. *Mutat. Res.*, *24*, 83-103

Ehrenberg, L., Osterman-Golkar, S., Segerbäck, D., Svensson, K. & Calleman, C.J. (1977) Evaluation of genetic risks of alkylating agents. III. Alkylation of haemoglobin after metabolic conversion of ethene to ethene oxide *in vivo*. *Mutat. Res.*, *45*, 175-184

Farmer, P.B. (1982) *The occurrence of* S-*methylcysteine in the hemoglobin of normal untreated animals*. In: Bridges, B.A., Butterworth, B.E. & Weinstein, I.B., eds, *Indicators of Genotoxic Exposure* (*Banbury Report 13*), Cold Spring Harbor, NY, Cold Spring Harbor Laboratory, pp. 169-175

Farmer, P.B., Bailey, E., Lamb, J.H. & Connors, T.A. (1980) Approach to the quantitation of alkylated amino acids in haemoglobin by gas chromatography-mass spectrometry. *Biomed. mass Spectrom.*, *7*, 41-46

Farmer, P.B., Gorf, S.M. & Bailey, E. (1982) Determination of hydroxypropylhistidine in haemoglobin as a measure of exposure to propylene oxide, using high resolution gas chromatography mass spectrometry. *Biomed. mass Spectrom.*, *9*, 69-71

Hemminki, K. & Savolainen, H. (1980) Alkylation of rat serum proteins by dimethylnitrosamine and acetylamino fluorene. *Toxicol. Lett.*, *6*, 433-437

Kim, S., Paik, W.K., Choi, J., Lotlikar, P.D. & Magee, P.N. (1981) Microsome-dependent methylation of erythrocyte proteins by dimethylnitrosamine. *Carcinogenesis*, *2*, 179-182

Magee, P.J. & Farber, E. (1962) Toxic liver injury and carcinogenesis. Methylation of rat liver nucleic acids by dimethylnitrosamine *in vivo*. *Biochem. J.*, *83*, 114-124

Magee, P.J. & Hultin, T. (1962) Toxic liver injury and carcinogenesis. Methylation of proteins of rat liver slices by dimethylnitrosamine *in vitro*. *Biochem. J.*, *83*, 106-114

Neumann, H.-G. (1980) Dose-relationship in the primary lesion of strong electrophilic carcinogens. *Arch. Toxicol.*, *Suppl. 3*, 69-77

Osterman-Golkar, S. & Ehrenberg, L. (1982) Covalent binding of reactive intermediates to hemoglobin as an approach for determining the metabolic activation of chemicals-ethylene. *Drug Metab. Rev.*, *13*, 647-660

Osterman-Golkar, S., Ehrenberg, L., Segerbäck, D. & Hällström, I. (1976) Evaluation of genetic risks of alkylating agents. II. Haemoglobin as a dose monitor. *Mutat. Res.*, *34*, 1-10

Osterman-Golkar, S., Hultmark, D., Segerbäck, D., Calleman, C.J. & Gothe, R. (1977) Alkylation of DNA and proteins in mice exposed to vinyl chloride. *Biochem. biophys. Res. Commun.*, *76*, 259-266

Osterman-Golkar, S., Farmer, P.B., Segerbäck, D., Bailey, E., Calleman, C.J., Svensson, K. & Ehrenberg, L. (1984) Dosimetry of ethylene oxide in the rat by quantitation of alkylated histidine in hemoglobin. *Teratog. Carcinog. Mutag.*, *3*, 395-405

Pereira, M.A. & Chang, L.W. (1981) Binding of chemical carcinogens and mutagens to rat hemoglobin. *Chem.-biol. Interact.*, *33*, 301-305

Pereira, M.A. & Chang, L.W. (1982a) Binding of chloroform to mouse and rat hemoglobin. *Chem.-biol. Interact.*, *39*, 89-99

Pereira, M.A. & Chang, L.W. (1982b) *Hemoglobin binding as a dose monitor for chemical carcinogens*. In: Bridges, B.A., Butterworth, B.E. & Weinstein, I.B., eds, *Indicators of Genotoxic Exposure* (*Banbury Report 13*), Cold Spring Harbor, NY, Cold Spring Harbor Laboratory, pp. 177-187

Pereira, M.A., Lin, L.-H.C. & Chang, L.W. (1981) Dose dependency of 2-acetylaminofluorene binding to liver DNA and hemoglobin in mice and rats. *Toxicol. appl. Pharmacol.*, *60*, 472-478

Redford-Ellis, M. & Gowenlock, A.H. (1971) Studies on the reaction of chloromethane with human blood. *Acta pharmacol. toxicol.*, *30*, 36-48

Segerbäck, D. (1983) Alkylation of DNA and hemoglobin in the mouse following exposure to ethene and ethene oxide. *Chem.-biol. Interact.*, *45*, 139-151

Segerbäck, D. & Ehrenberg, L. (1981) Alkylating properties of dichlorvos (DDVP). *Acta pharmacol. toxicol.*, *49*, (*Suppl. V*), 56-66

Segerbäck, D., Calleman, C.J., Ehrenberg, L., Löfroth, G. & Osterman-Golkar, S. (1978) Evaluation of genetic risks of alkylating agents. IV. Quantitative determination of alkylated amino acids in haemoglobin as a measure of the dose after treatment of mice with methyl methanesulphonate. *Mutat. Res.*, *49*, 71-82

Tannenbaum, S., Skipper, P.L., Green, L.C., Obiedzinski, M.W. & Kadlubar, F. (1983) Blood protein adducts as monitors of exposure to 4-aminobiphenyl. *Proc. Am. Assoc. Cancer Res.*, *24*, 69

Walles, S.A. (1981) Reaction of benzyl chloride with haemoglobin and DNA in various organs in mice. *Toxicol. Lett.*, *9*, 379-387

Wieland, E. & Neumann, H.-G. (1978) Methemoglobin formation and binding to blood constituents as indicators for the formation, availability and reactivity of activated metabolites derived from *trans*-4-aminostilbene and related aromatic amines. *Arch. Toxicol.*, *40*, 17-35

# IMMUNOLOGICAL METHODS FOR DETECTION OF CARCINOGEN-DNA ADDUCTS

**J. Adamkiewicz, P. Nehls & M.F. Rajewsky**

*Institut für Zellbiologie (Tumorforschung), Universität Essen (GH) D-4300 Essen 1, FRG*

**SUMMARY**

Considerable advances have been made during recent years, with regard to the detection and quantification of carcinogen- or mutagen-induced, structural modifications in the DNA of mammalian cells, by the introduction of immunoanalytical methods, in particular in conjunction with monoclonal antibodies (Mab). Antibodies are characterized by an outstanding capacity for the specific recognition of subtle alterations of molecular structure. They can, therefore, be used as sensitive detection probes in assays for DNA modifications caused by low levels of DNA-reactive (e.g., environmental) agents. Depending on the purpose of analysis, various types of immunoassays can be performed. The competitive radioimmunoassay (RIA) represents a routinely applicable, reproducible and sensitive assay for the detection of defined carcinogen-DNA adducts in hydrolysates of cellular DNA, in body fluids or in urine. Depending on their particular design, enzyme immunoassays (EIA) may have exceptionally low detection limits, due to the enzymatic amplification of the measured radioactivity or colour intensity. Similarly, recently established immuno-slot-blot (ISB) techniques are also characterized by very high sensitivity. Immunocytological assays (ICA) use Mab in conjunction with electronically intensified immunofluorescence for detection of modified DNA components in the nuclei of individual cells. Finally, single modified deoxynucleosides can be detected and localized in individual DNA molecules by immuno-electron microscopy (IEM).

**INTRODUCTION**

A variety of environmental mutagens and carcinogens, chemotherapeutic agents, as well as ultra-violet light and ionizing radiation, cause structural alterations of cellular macromolecules. With respect to carcinogenesis, genomic deoxyribonucleic acid (DNA) appears to be the most critical target (Grover, 1979). Besides point mutations, a variety of other genetic mechanisms based on carcinogen-induced DNA modifications have been considered that may lead to malignant transformation (Rajewsky, 1983). With regard to the cellular uptake, bioactivation and biological effects of DNA-reactive agents, numerous studies, both *in vivo* and *in vitro*, have shown not only considerable interspecies and interindividual variations, but also differences between tissues and cell types, cell cycle and differentiation stages, as well as sex-dependent differences (Bartsch & Armstrong, 1982; Graf & Jaenisch, 1982; Rajewsky, 1983). In order to correlate the dose of a given agent with its biological effects (e.g., cytotoxicity, mutation or the induction of cancer), the overall *exposure dose* is not a satisfactory parameter. Instead, the *effective dose* should be considered, i.e., the degree of interaction with and damage to critical cellular target molecules (e.g., DNA).

Highly sensitive methods are required for the dosimetry of specific structural modifications resulting from the reaction of carcinogens and mutagens with DNA. With the exception of recently developed 'postlabelling' techniques (Randerath *et al.*, 1981; Gupta *et al.*, 1982; Randerath, this volume), radiochromatographic methods (Baird, 1979) are not only restricted in their applicability to the use of [$^{3}$H]- or [$^{14}$C]-labelled DNA-reactive agents, but also limited in sensitivity by the specific radioactivity of these agents that can be achieved and, consequently, by the large amounts of DNA (cells) required for analysis. Immunoanalytical procedures using polyclonal or monoclonal antibodies (Mab) specifically directed against structurally modified components of DNA have opened new possibilities for detection and quantitation of the 'fingerprints' of nonradioactive (e.g., environmental) agents in small DNA samples, and in individual cells or DNA molecules (Rajewsky *et al.*, 1980; Müller & Rajewsky, 1981; Adamkiewicz *et al.*, 1982; Müller *et al.*, 1982). In this paper, we shall discuss the analytical potential of Mab that specifically recognize deoxynucleosides carrying small alkyl residues (Rajewsky *et al.*, 1980; Adamkiewicz *et al.*, 1982). This class of DNA modifications typically results from the reaction of cellular DNA with carcinogenic and mutagenic *N*-nitroso compounds (Lawley, 1976; Pegg, 1977; Grover, 1979; Kohn, 1979; O'Connor *et al.*, 1979; Singer, 1979; Rajewsky, 1980). The principles, and the advantages and disadvantages, of different types of immunological assays for alkylated DNA constituents (i.e., competitive radioimmunoassay, RIA; enzyme immunoassay, EIA; immunocytological assay, ICA; immuno-slot-blot, ISB; and immuno-electron microscopy, IEM) are described and their detection limits and sensitivities are compared.

## IMMUNOANALYSIS OF ALKYL-DEOXYNUCLEOSIDES IN CELLULAR DNA EXPOSED TO *N*-NITROSO COMPOUNDS

### *Antibodies specific for alkyl-deoxynucleosides*

Precise structural characterization of DNA adducts is a prerequisite for establishing specific immunoassays. Although this condition is not yet fulfilled in the case of many chemical agents known to react with DNA, various DNA modifications have been identified, particularly resulting from exposure of cells to alkylating *N*-nitroso compounds (alkylnitrosamines, alkylnitro-nitrosoguanidines, alkylnitrosoureas), *N*-acetoxy-*N*-2-acetylaminofluorene, benzo[*a*]pyrene, and aflatoxin $B_1$. Antisera or Mab have been raised against several reaction products of these chemicals with DNA and used in appropriate immunoassays (reviewed by Müller & Rajewsky, 1981; Poirier, 1981; Müller *et al.*, 1982). Table 1 lists a number of specific DNA modifications for which immunoassays have been described. Due to the rapidly increasing number of antibodies against modified DNA components, this table is, however, not complete. The following sections preferentially deal with antibodies directed against $O^6$-ethyl-2'-deoxyguanosine ($O^6$-EtdGuo), one of the alkyl-deoxynucleosides produced in DNA exposed to ethylating agents such as *N*-ethyl-*N*-nitrosourea, *N*-nitrosodiethylamine, ethylmethanesulfonate, diethylsulfate, or *N*-ethyl-*N*'-nitro-*N*-nitrosoguanidine. *O*$^6$-Alkyl-deoxyguanosine is of particular interest in terms of DNA repair, as well as mutagenesis and carcinogenesis (Rajewsky *et al.*, 1977; Rajewsky, 1980, 1983), and has been chosen by our group as a model DNA modification for establishing and standardizing immunoanalytical methodology. Antisera and a collection of Mab have been raised against *O*$^6$-EtdGuo (Müller & Rajewsky, 1978, 1980; Rajewsky *et al.*, 1980; Adamkiewicz *et al.*, 1982; J. Adamkiewicz & M.F. Rajewsky, in preparation) and used in various immunological assays. This permits a comparison of these assays, using a precisely defined DNA modification as the antigen, and the same set of well-characterized antibodies.

**Table 1. Structurally modified DNA components for which immunoassays have been described**

| Structurally modified DNA component | Reference |
|---|---|
| *O*$^6$-Methyl-2'-deoxyguanosine | Wild *et al.* (1983) |
| *O*$^6$-Butyl-2'-deoxyguanosine<br>*O*$^4$-Butyl-2'-deoxythymidine<br>*O*$^2$-Butyl-2'-deoxythymidine | Saffhill *et al.* (1982) |
| *O*$^6$-Ethyl-2'-deoxyguanosine<br>*O*$^6$-Isopropyl-2'-deoxyguanosine | Rajewsky *et al.* (1980) |
| *O*$^4$-Ethyl-2'-deoxythymidine | Adamkiewicz *et al.* (1982) |
| 8-Acetoxy-*N*-2-acetylaminofluorene- 2'-deoxyguanosine | Kriek *et al.* (1981); |
| 8-Acetoxy-aminofluorene-2'-deoxyguanosine | Poirier *et al.* (1979) |
| Benzo[*a*]pyrene-modified DNA | Poirier *et al.* (1980) |
| Aflatoxin $B_1$-modified DNA | Haugen *et al.* (1981) |
| Ultra-violet-modified DNA | Strickland & Boyle (1981) |

### *Properties of antibodies specific for* O$^6$*-EtdGuo*

Modified nucleosides are haptens, and require coupling to a carrier protein (e.g., keyhole limpet haemocyanin) to elicit an immunological response (Müller & Rajewsky, 1981). Of the various known carcinogen or mutagen DNA adducts, those with smaller modifications such as deoxynucleosides carrying small alkyl residues (e.g., methyl or ethyl groups linked covalently to an oxygen or nitrogen atom) may be expected to be less well recognized by antibodies than large 'bulky' adducts such as those resulting from the interaction of DNA with polycyclic hydrocarbons. However, this assumption is not supported by the experimental evidence that has accumulated during the past years. Indeed, a series of Mab with very high affinity and specificity for alkyldeoxynucleosides has been developed (Table 1). The binding characteristics of these Mab have been worked out by determination of their cross-reactivities with other modified or naturally occurring DNA constituents, and Mab with very low degrees of cross-reactivity have been obtained (Rajewsky, 1980; Adamkiewicz *et al.*, 1982). As an example, the cross-reactivities of a panel of Mab raised against *O*$^6$-EtdGuo (Adamkiewicz *et al.*, 1982) with the corresponding unmodified 2'-deoxyguanosine (dGuo) are shown in Figure 1. With the best of these Mab, dGuo at concentrations $>10^7$-fold the concentration of *O*$^6$-EtdGuo (i.e., up to $3 \times 10^{-3}$M dGuo, equivalent to the dGuo content of 2.3 mg of hydrolysed DNA/ml in the RIA sample) does not interfere with the quantitation of *O*$^6$-EtdGuo in the competitive RIA. These data clearly show that antibodies are indeed exceptionally sensitive reagents for the specific detection of small alterations of nucleoside structure.

**Fig. 1. Cross-reactivities with 2'-deoxyguanosine of a series of mouse × mouse (EM) and rat × rat (ER) hybridoma-secreted monoclonal antibodies (Mab) specific for *O*⁶-ethyl-2'-deoxyguanosine (*O*⁶-EtdGuo)**

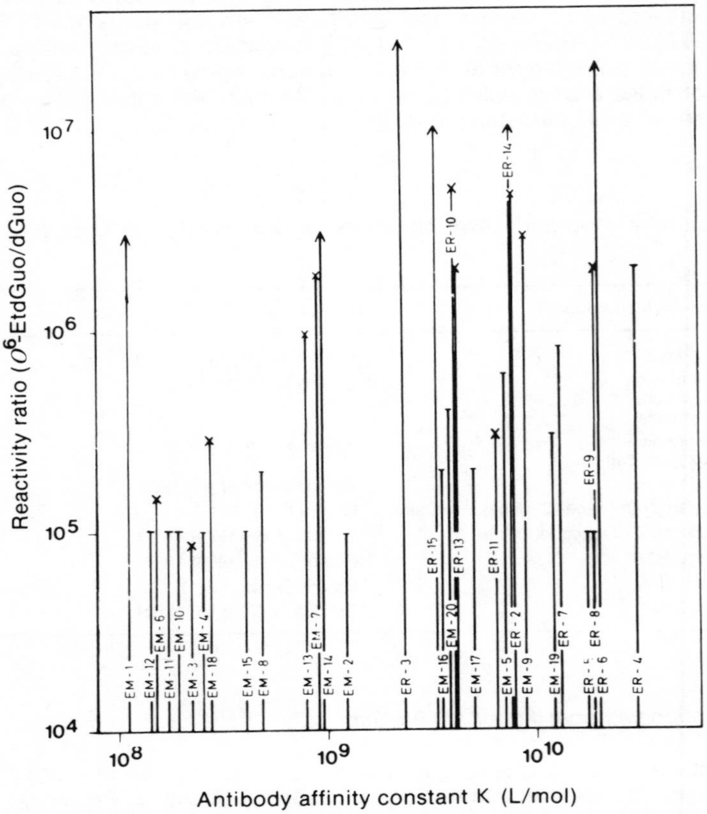


Cross-reactivities are expressed as reactivity ratios indicating the amounts of dGuo required for 50% (T), 10% (⤒), or <10% (↑) inhibition of tracer-antibody binding (ITAB) in the competitive radioimmunoassay (RIA) as multiples of the corresponding amounts of *O*⁶-EtdGuo. Antibody affinity constants were calculated as described by Müller (1980).

*Competitive radioimmunoassay (RIA)*

The basic principle of the competitive RIA is shown in Figure 2. A constant amount of radioactively labelled antigen (tracer), low enough to permit a sensitive assay, but high enough to minimize the time required for accurate liquid scintillation counting, is mixed in the assay tubes with a constant amount of antibody sufficient to bind 50% of the tracer. At this tracer-antibody ratio, small changes in the concentration of either antibody or antigen significantly affect tracer-antibody binding, as can be judged from the typical sigmoidal antibody titration curve shown in Figure 3. The slope of this curve is steepest at 50% tracer-antibody binding. Upon addition of nonradioactive antigen (inhibitor, Fig. 2) - either a standard dilution series of known concentration or a (DNA) sample with an unknown content of antigen - a fraction of bound tracer depending in size on the inhibitor concentration is displaced from the antibodies by the inhibitor. In Figure 2, tracer-antibody binding is inhibited by

**Fig. 2. Principle of the competitive radioimmunoassay**

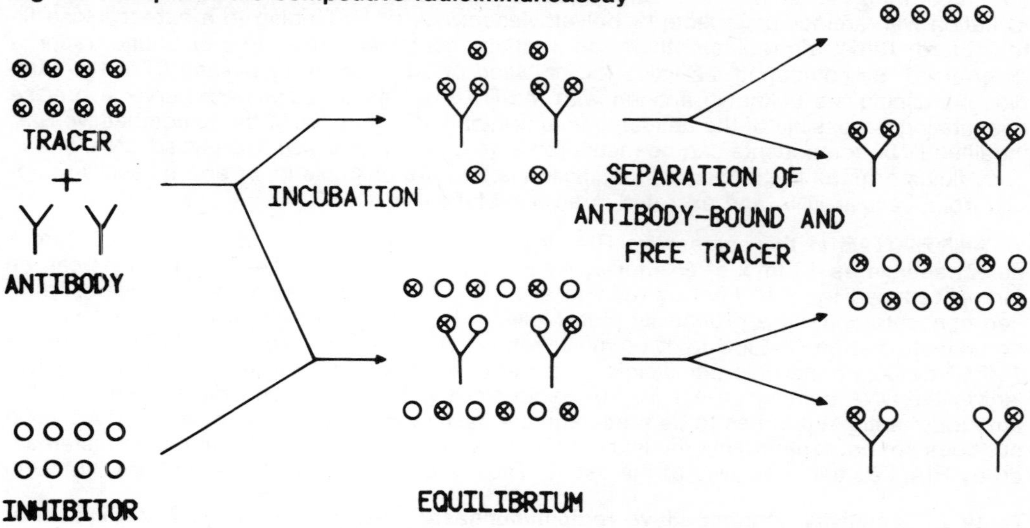


**Fig. 3. Typical titration curve of the supernatant of a Mab-secreting hybridoma cell culture**

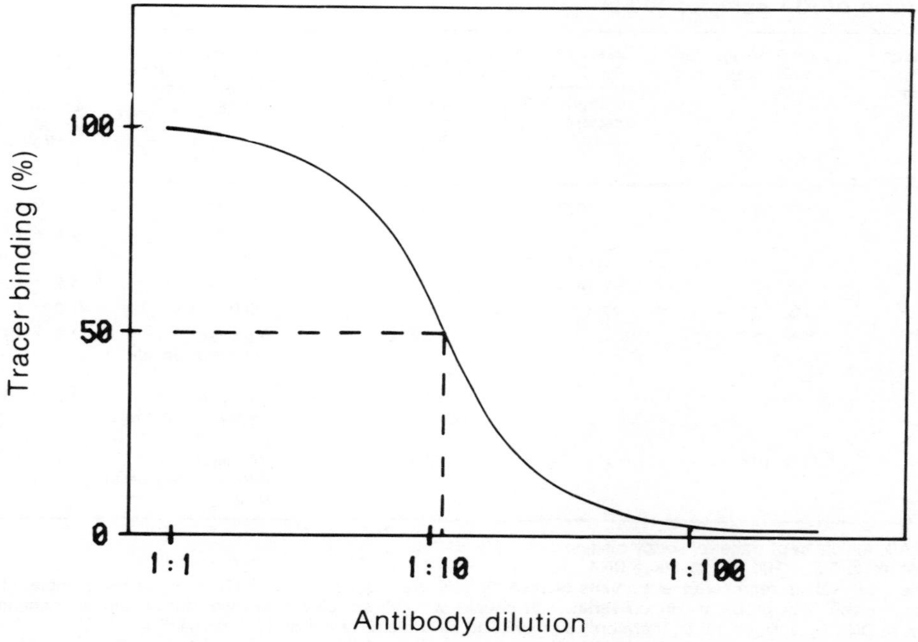


A constant amount of radioactive tracer war incubated with varying amounts of antibody. After removal of antibody-bound tracer by ammonium sulfate precipitation, the percentage of antibody-bound tracer was calculated.

50%. Following the removal of antibody-bound tracer from the RIA samples, usually by precipitation with ammonium sulfate or polyethyleneglycol, or by binding to a nitrocellulose filter (Chard, 1982), the radioactivity in the fractions containing either free or bound tracer is determined. By comparing the value for inhibition of tracer-antibody binding (ITAB) in samples containing the unknown antigen with the ITAB values of a standard curve, a precise quantitation is possible of the antigen concentration in the sample. With the competitive RIA, modified DNA components can be measured in (e.g., enzymatically) hydrolysed DNA, and in body fluids or urine in the case of modifications that are unstable in, or enzymatically removed from, cellular DNA and excreted in an unmetabolized form.

At 50% ITAB in the competitive RIA, the best anti-$O^6$-EtdGuo Mab shown in Figure 1 detect as little as 40 fmol of $O^6$-EtdGuo in a 100-μl RIA sample. The detection limit can be further reduced to ~ 10 fmol by reading at 20% ITAB (when ITAB is plotted against antigen concentrations in a probability plot; Müller & Rajewsky, 1980), so that $O^6$-EtdGuo can be quantified at an $O^6$-EtdGuo:dGuo molar ratio in DNA of ~ $2 \times 10^{-7}$ (i.e., the equivalent of 700 $O^6$-EtdGuo molecules per diploid genome) in a hydrolysate of 100 μg of DNA (equivalent to the DNA content of ~ $1.6 \times 10^7$ diploid cells). This detection limit is defined by the amount of antigen that has to be present in the assay tube to give a measurable ITAB value but does not necessarily limit the level of modifications in cellular DNA that can be determined by RIA, i.e., the sensitivity of the assay. Table 2 demonstrates that even the $O^6$-EtdGuo

**Table 2. Sensitivity of competitive radioimmunoassay (RIA) for $O^6$-ethyl-2'-deoxyguanosine ($O^6$-EtdGuo)**
***Preconditions*: 1) Antibody affinity constant ($K_A$): $10^{10}$ $M^{-1}$**
**2) Tracer concentration: $3 \times 10^{-10}$ M (20 - 30 Ci/mmol)**
**3) Volume of RIA sample: 100 μl**

| Amount of $O^6$-EtdGuo required for 50 % ITAB[a] (fmol) | $O^6$-EtdGuo: dGuo molar ratio in DNA | Amount of hydrolysed DNA required for 50 % ITAB in one RIA sample[b,c] | | Concentration of hydrolysed DNA in RIA sample (μg/ml) | Number of diploid cells corresponding to required amount of hydrolysed DNA[e] |
|---|---|---|---|---|---|
| | | without HPLC[d] separation prior to RIA (μg) | with HPLC[d] separation prior to RIA (mg) | | |
| 50 | $1 \times 10^{-2}$ | 0.0077 | - | 0.154 | $1.2 \times 10^3$ |
| 50 | $1 \times 10^{-3}$ | 0.0774 | - | 1.54 | $1.2 \times 10^4$ |
| 50 | $1 \times 10^{-4}$ | 0.774 | - | 15.4 | $1.2 \times 10^5$ |
| 50 | $1 \times 10^{-5}$ | 7.74 | - | 154 | $1.2 \times 10^6$ |
| 50 | $1 \times 10^{-6}$ | 77.4 | - | 1540 | $1.2 \times 10^7$ |
| 50 | $1 \times 10^{-7}$ | - | 0.774 | modified deoxynucleoside(s) only | $1.2 \times 10^8$ |
| 50 | $1 \times 10^{-8}$ | - | 7.74 | modified deoxynucleoside(s) only | $1.2 \times 10^9$ |
| 50 | $1 \times 10^{-9}$ | - | 77.4 | modified deoxynucleoside(s) only | $1.2 \times 10^{10}$ |

[a] ITAB, Inhibition of tracer-antibody binding
[b] Assuming $6.5 \times 10^{-4}$ mol dGMP/g DNA
[c] The given values represent the amounts of DNA hydrolysate required for 50% ITAB in one RIA sample. In practice, for a given DNA probe, a dilution series with double or triple samples is assayed for $O^6$-EtdGuo content. The amount of DNA hydrolysate must, therefore, be multiplied by a factor of 8.5 or 12.5, respectively.
[d] Separation of $O^6$-EtdGuo from other unmodified and modified deoxynucleosides by high-performance liquid chromatography (HPLC) (high DNA-hydrolysate concentrations) ~ 100% recovery of inhibitor assumed
[e] Assuming $6.7 \times 10^{-12}$ g DNA per diploid cell; values not corrected for DNA loss during isolation

content of DNA samples with $O^6$-EtdGuo:dGuo molar ratios $<10^{-7}$ can be determined by competitive RIA. In this case, the ($O^6$-EtdGuo)-containing fraction is separated from the remaining constituents of a DNA hydrolysate (by high-performance liquid chromatography) and concentrated by evaporation (Speed Vac Concentrator, Savant Instruments, Hicksville, NY, USA). Under these conditions, the sensitivity of the competitive RIA is limited only by the total amount of DNA available for analysis. Contamination with exogenous $O^6$-EtdGuo or other cross-reactants during the separation and concentration steps may be a source of error in the measurements but can, of course, be avoided by using clean working conditions.

*Enzyme immunoassay (EIA)*

The basic principle of the EIA is the coupling of either the antibody or the antigen to an enzyme, and quantitation of the immunological reaction *via* the formation of a product resulting from the conversion of a suitable substrate by the enzyme. A variety of EIAs and EIA modifications have been described (reviewed, e.g., by Harris *et al.*, 1982). A simple version is illustrated in Figure 4. In this case, equal amounts of the antigen are bound to a solid phase (e.g., plastic surface of microtitre plates, filters). After incubation with the first, antigen-specific antibody and the removal of unbound antibodies by a washing step, a second, enzyme-labelled antibody specific for the first antibody is added. After washing, the enzymatic activity (formation of a quantifiable product in the sample) is quantitated and compared with the activity of standard samples containing a known amount of antigen. The enzymatic activity in the sample is proportional to the amount of the second antibody, the first antibody and the amount of antigen recognized by the first antibody. Alternatively, the EIA can be performed as a direct assay, with the enzyme conjugated directly to the first antibody, or as a multi-step assay with the second antibody serving as a bridge to allow the binding of an enzyme complex to amplify the signal. The most commonly used enzymes are alkaline phosphatase and peroxidase; however, the use of β-galactosidase, glucose oxidase or catalase, has also been described. Depending on the product generated by these enzymes, either visible colour, fluorescence or radioactivity is measured (Harris *et al.*, 1982). The characteristic advantage of the EIA over other immunoanalytical procedures lies in its capability to intensify a weak signal (i.e., a low amount of enzyme in the reaction tube due to a correspondingly low amount of antigen) by allowing a prolonged reaction time of the enzyme. Thus, EIAs with very low detection limits (<1 fmol of DNA adduct) have been described by Harris *et al.* (1982; ultrasensitive enzymatic radioimmunoassay) and by Van der Laken *et al.* (1982; high sensitivity enzyme-linked immunosorbent assay). In spite of their apparent superiority over the RIA, the practical application of EIAs for ultrasensitive determinations of carcinogen-DNA adducts has, in the past, been hampered by an only moderate reproducibility and accuracy. This was due mainly to the fact that homogenous binding of the DNA to all wells of the microtitre plates used as the solid phase (a necessary prerequisite for the comparison of standards and samples) often could not be achieved. In addition, even low levels of nonspecific antibody binding often increased the background in these ultrasensitive EIAs. However, these technical difficulties can be solved and have already been overcome by some investigators (e.g., Kriek *et al.*, this volume).

*Immuno-slot-blot (ISB)*

A special type of solid-phase immunoassay is represented by a recently established immuno-slot-blot procedure (ISB) (Nehls *et al.*, 1984a). In principle similar to a non-competitive EIA, in this assay, DNA containing a specific modified nucleoside (e.g., $O^6$-EtdGuo) is bound to a nitrocellulose filter and incubated with a Mab specific for this modified DNA

**Fig. 4. Principle of the enzyme immunoassay**

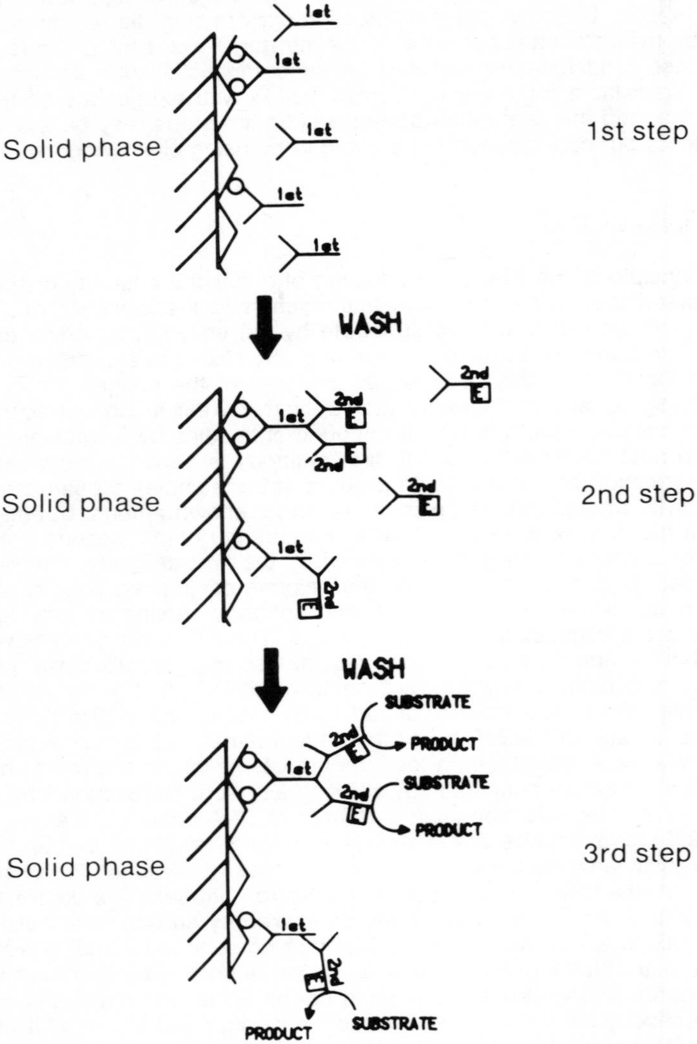


(⌇) DNA containing a modified nucleoside that can be detected by specific antibodies (Y). The second antibody is labelled with an enzyme (E).

component. The binding of the antibody is detected by a second, enzyme-labelled antibody directed against the first antibody, or by a multi-step procedure, as mentioned previously. In the latter modification, the ISB, like the EIA, profits from the enzymatic intensification of weak signals. In Figure 5, $O^6$-EtdGuo is detected in DNA by ISB with the use of a biotinylated second antibody specific for the first Mab and a complex of biotinylated alkaline phosphatase and avidin (Leary *et al.*, 1983). The applied Mab, which is characterized by a detec-

**Fig. 5. Immuno-slot-blot (ISB) of *O*⁶-ethyl-2'-deoxyguanosine (*O*⁶-EtdGuo)-containing DNA**

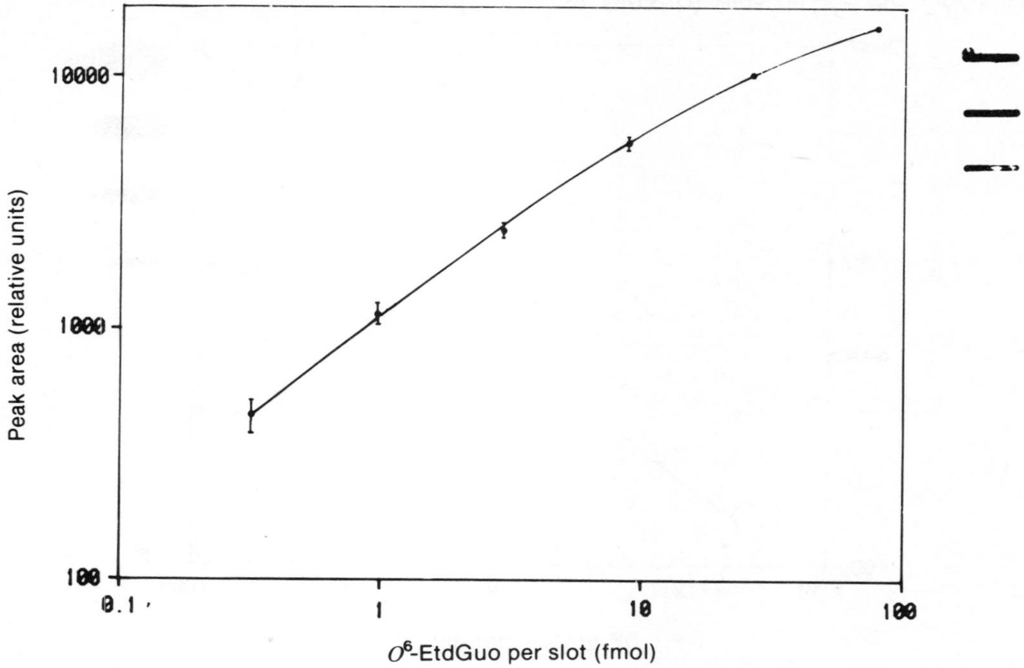


1.5 μg of a single-stranded rat DNA containing 78 fmol (slot 1), 26 fmol (slot 2), 8.7 fmol (slot 3), 2.9 fmol (slot 4), 0.96 fmol (slot 5), 0.32 fmol (slot 6), and 0 fmol (slots 7 and 8) of *O*⁶-EtdGuo were blotted onto a nitrocellulose filter (Schleicher & Schuell, Dassel, FRG). After successive incubation with a specific anti-*O*⁶-EtdGuo Mab, a second biotinylated antibody specific for the first antibody, and a complex of biotinylated alkaline phosphatase and avidin, enzymatic activity was visualized according to Leary *et al.* (1983). Colour intensities were evaluated by densitometry.

tion limit of 40 fmol of *O*⁶-EtdGuo at 50% ITAB in the competitive RIA (see above), detects ⩾ 0.3 fmol of *O*⁶-EtdGuo in a 3-μg sample of DNA bound to a nitrocellulose filter, thus allowing the quantitation of *O*⁶-EtdGuo at *O*⁶-EtdGuo:dGuo molar ratios in DNA of $\geqslant 2 \times 10^{-7}$. Alternatively, the second antibody can be labelled with $^{125}$iodine and its binding detected by autoradiography. In this case, weak signals can be intensified by increasing exposure time. In comparison to the competitive RIA, the ISB requires significantly smaller amounts of DNA (i.e., cells, biopsy specimens) for high-sensitivity analysis. Thus, in a 140-pg sample of DNA containing *O*⁶-EtdGuo at an *O*⁶-EtdGuo:dGuo molar ratio of $4 \times 10^{-4}$, 0.036 fmol of *O*⁶-EtdGuo can still be detected by ISB using a $^{125}$I-labelled second antibody and an exposure time of 26 h (Fig. 6).

### *Immunocytological assay (ICA)*

When immunocytological staining procedures are used, the antigen can be detected directly and quantified approximately in tissues or in individual cells, without a requirement for its isolation prior to analysis. For reasons described below, ICAs (as well as the IEM pro-

**Fig. 6. Immuno-slot-blot (ISB) analysis of different amounts of ethylated DNA containing *O*⁶-ethyl-2'-deoxyguanosine (*O*⁶-EtdGuo)**

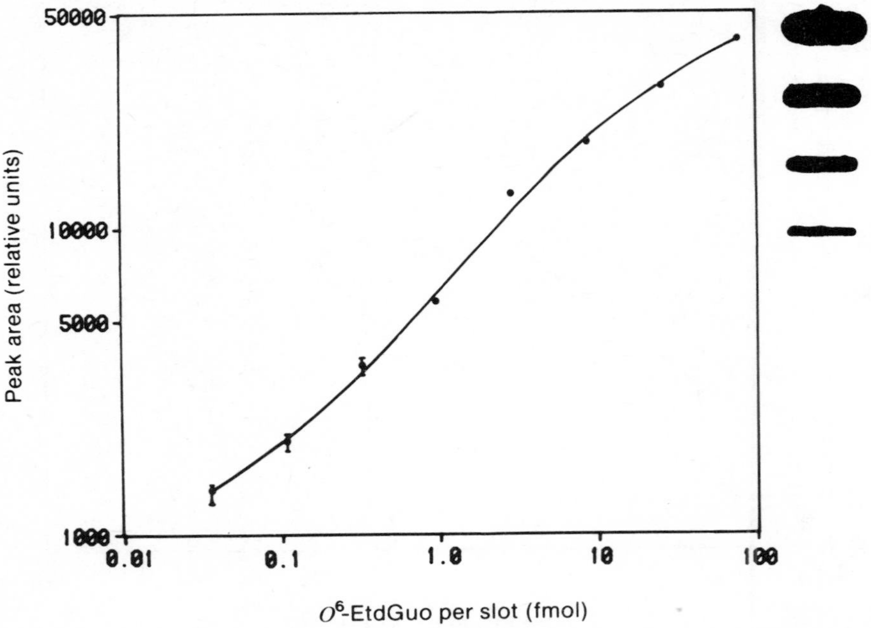


In descending order, amounts of 300, 100, 33.3, 11.1, 3.7, 1.2, 0.41, and 0.14 ng of DNA ($O^6$-EtdGuo:dGuo molar ratio of 4 x $10^{-4}$) per slot, corresponding to 78, 26, 8.7, 2.9, 0.96, 0.32, 0.11, and 0.036 fmol, respectively, of $O^6$-EtdGuo per slot, were blotted onto a nitrocellulose filter following heat denaturation of the DNA. After binding of a specific anti-($O^6$-EtdGuo)-Mab (first antibody), a $^{125}$I-labelled, second antibody was allowed to react with the first antibody. The filter was exposed to Kodak X-Omat film for 26 h, the optical density of the film was recorded by densitometry, and the integral values for the peak areas (corresponding to the slot position) were determined.

cedures discussed in the subsequent section) are the domain of Mabs. A central prerequisite for ICAs is the accessibility of the antigen for the specific antibody. In the case of carcinogen- or mutagen-DNA adducts, the cell (cytocentrifuged cell samples, cell smears, frozen or paraffin-embedded sections, squash preparations of tissue fragments, etc.) must be fixed in order to immobilize the antigens, to preserve the morphology of the cells or tissues, and to allow free access of the antibody into and out of the cell nucleus and cytoplasm. The antigen is visualized either by labelling the first, specific antibody (direct assay), or by a second, labelled antibody (indirect assay) specific for the first antibody, or by labelling one component in a multi-step procedure with a fluorochrome or an enzyme. As in the case of the EIA described in a previous section, a variety of modifications is possible and has been reported (for an overview see Bullock & Petrusz, 1982). Thus, peroxidase, alkaline phosphatase and β-galactosidase have been used in combination with substrates yielding precipitates. Bound fluorochrome-labelled antibodies can be visualized directly by fluorescence microscopy.

Necessary prerequisites for the immunostaining of $O^6$-EtdGuo in the DNA of individual cells are the fixation of cells, RNase-treatment to remove ($O^6$-ethylguanosine)-containing RNA (a potential cross-reactant) and in-situ denaturation of cellular DNA. The complex struc-

**Fig. 7.** Immunofluorescence staining of $O^6$-ethyl-2'-deoxyguanosine ($O^6$-EtdGuo) in nuclear DNA of cultured brain tumour cells of rat (a) untreated or (b) treated with *N*-ethyl-*N*-nitrosourea to give an $O^6$-EtdGuo:dGuo molar ratio in DNA of $\sim 8 \times 10^{-6}$

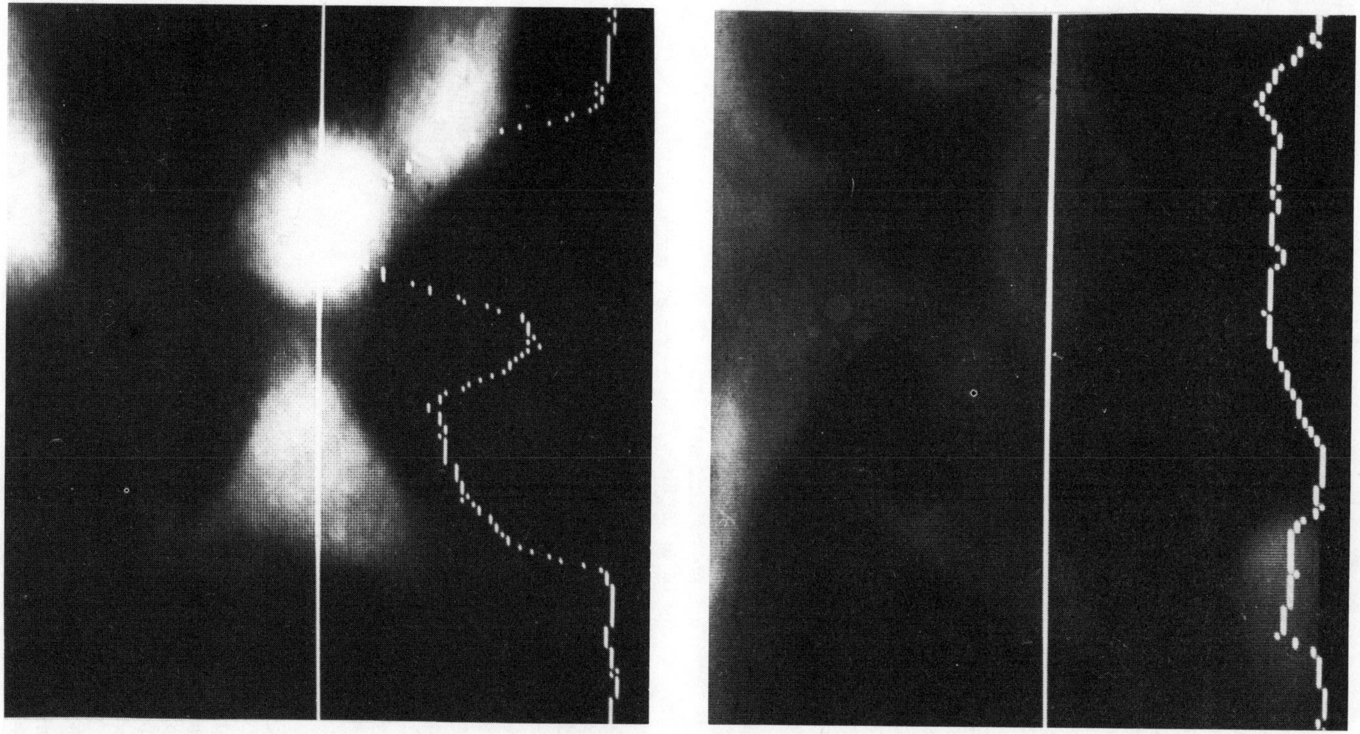

Light intensities along a single TV-line (blanked white) are simultaneously displayed by the image analysis system (dotted curve).

ture of cells, and the drastic procedure required to render the antigenic structures in DNA accessible to the antibody, represent potential sources of artefacts and interferences with the assay. Among the critical factors are the stability of the antigen during the immunostaining procedure, masking of the antigen by the process of fixation, nonspecific binding of antibody to cellular structures, 'specific' binding of antibody to cellular epitopes not identical with the desired antibody recognition site on $O^6$-EtdGuo, and autofluorescence of the cells. Most of these problems can be overcome by selecting the most suitable Mabs from a larger collection and by systematic optimization of the staining procedure (Adamkiewicz *et al.*, 1983; J. Adamkiewicz & M.F. Rajewsky, in preparation). For quantitation of $O^6$-EtdGuo in the nuclei of individual cells, the signals produced by fluorochrome-labelled antibodies are amplified by electronic image-intensification combined with a sensitive television camera, and evaluated with the aid of a microprocessor-based image storage and analysis system (Adamkiewicz *et al.*, 1983; Fig. 7). The sensitivity of the ICA for $O^6$-EtdGuo presently allows the detection of ~700 $O^6$-EtdGuo residues per diploid genome, corresponding to an $O^6$-EtdGuo:dGuo molar ratio of $\sim 3 \times 10^{-7}$. The detection and quantitation of carcinogen-DNA adducts in individual cells isolated from tissues or body fluids, or in cells within tissues containing multiple, phenotypically distinct cell subpopulations, is particularly useful for the monitoring of exposure, and for the characterization of intercellular and interindividual variations of cellular repair capacity.

*Immuno-electron microscopy (IEM)*

In combination with transmission electron microscopy, Mab permit the direct visualization of specific carcinogen-modified sites in individual DNA molecules (Nehls *et al.*, 1983; de Murcia *et al.*, 1979, polyclonal antibodies). Thus, $O^6$-EtdGuo can be localized in double-stranded DNA by the binding of an antibody molecule directed against this ethylation product. When a protein-free DNA spreading technique is used, antibody binding sites can be clearly identified without requirement for a second antibody carrying an electron-dense label. This method has already been successfully applied in studying the distribution of $O^6$-EtdGuo in DNA exposed to ethylnitrosourea *in vitro* (P. Nehls, E. Spiess, E. Weber & M.F. Rajewsky, submitted for publication) or *in vivo* (Nehls *et al.*, 1984b). As judged from investigations on a plasmid DNA of known nucleotide sequence, bound antibody molecules can be localized on DNA with a resolution of ~ 30 base pairs (P. Nehls, E. Spiess & M.F. Rajewsky, in preparation). Prior to the processing of DNA samples for IEM, unbound antibody molecules must be separated and removed from the DNA-antibody complexes. This can be done by gel filtration, ion exchange chromatography or by centrifugation. Depending on the separation method and the antibody used, up to 60% of the $O^6$-EtdGuo molecules present in double-stranded DNA can be visualized by IEM.

## CONCLUSIONS

From the data shown in the preceding sections it is clear that antibodies, especially Mab, can be applied in adequate assays as very specific and sensitive reagents for detection and quantitation of low amounts of modifications induced in DNA by DNA-reactive agents, even in the presence of a large excess of unmodified DNA constituents. It should, however, not be overlooked that the production of Mab specific for carcinogen-DNA adducts involves a number of difficult (and time-consuming) steps. Thus, the structure of the antigens must be determined, the corresponding haptens must be synthesized and coupled to carrier proteins for immunization, and Mab have to be developed and characterized. Furthermore, the antigens must be rather stable in the immunized animal in order to induce a specific immune response. Recent advances in the in-vitro immunization methodology (reviewed by Reading, 1982) may facilitate parts of these procedures considerably, by reducing the influence of the immunological phenomena of tolerance and suppression and by shortening the time required until clones of antibody-producing hybridomas can be grown.

Antibodies can be applied in different immunological assays, depending on the characteristics of the antigen and on the aims of the respective study. Based on measurements using the same set of Mab specific for $O^6$-EtdGuo, we have, in the present paper, compared various immunoanalytical procedures with respect to their principles and their sensitivity and specificity. The competitive RIA is a reproducible, routinely applicable assay requiring the antibody, the radioactively labelled antigen (tracer) and a liquid scintillation counter. Using high-performance liquid chromatography, a particular modified nucleoside can be isolated from the bulk DNA prior to analysis by RIA. This permits measurements of very low levels of specific modified deoxynucleosides. For example, 10 $O^6$-EtdGuo molecules per diploid genome can be quantitated in 4 mg of ethylated DNA using a Mab that detects 10 fmol of $O^6$-EtdGuo in a competitive RIA (see section on competitive RIA). Immunoassays such as the EIA or the ISB, where the measured signal (i.e., colour development or silver grains on a sensitive X-ray film) is amplified with the time of incubation or exposure, have significantly lower detection limits than the competitive RIA and, therefore, require considerably less DNA. In addition, the synthesis of a radioactive antigen required as tracer in the RIA is no longer necessary. It should, however, be noted that these noncompetitive immunoassays do not offer the possibility of isolating carcinogen-modified DNA components from bulk DNA prior to analysis (as in the case of the RIA). EIA and ISB thus have obvious advantages over the conventional RIA although more practical experience will be needed to convert these assays into routinely applicable laboratory procedures.

ICA requires more sophisticated equipment and experience than do RIA, EIA or ISB. At the level of the light microscope, the detection limit of ICA is defined by the amount of a specific DNA modification per cell nucleus detectable by immunostaining. The detection limit thus also defines the sensitivity of the assay, because (contrary to RIA and EIA) the measured signal cannot be increased by applying a higher amount of sample. In practice, the detection limit of ICA is dependent on the background ('noise') level of the sample, which, in turn, depends on the individual Mab selected for analysis, the immunostaining procedure chosen and the properties of the electronic equipment used for image intensification.

By ICA, about 700 $O^6$-EtdGuo molecules per diploid genome can presently be detected above background, using electronically intensified direct immunofluorescence and a Mab selected for very low nonspecific background binding from a large collection of Mab (Adamkiewicz *et al.*, 1983; J. Adamkiewicz & M.F. Rajewsky, in preparation). The specific advantage of ICA lies in the fact that individual cells can be analyzed (e.g., cell smears, cell cultures, biopsy specimens). Finally, at the level of the transmission electron microscope, IEM opens the possibility for analysis of carcinogen- and mutagen-induced, specific structural alterations of DNA at the level of individual DNA molecules, e.g., in gene sequences of known nucleotide sequence.

## ACKNOWLEDGEMENTS

Research was supported by the Deutsche Forschungsgemeinschaft (SFB 102/A9), the Commission for the European Communities (ENV-544-D[B]), and by the Wilhelm und Maria Meyenburg Stiftung.

## REFERENCES

Adamkiewicz, J., Drosdziok, W., Eberhardt, W., Langenberg, U. & Rajewsky, M.F. (1982) *High-affinity monoclonal antibodies specific for DNA components structurally modified by alkylating agents.* In: Bridges, B.A., Butterworth, B.E. & Weinstein, I.B., eds, *Indicators of Genotoxic Exposure* (*Banbury Report 13*), Cold Spring Harbor, NY, Cold Spring Harbor Laboratory, pp. 265-276

Adamkiewicz, J., Ahrens, O., Huh, N. & Rajewsky, M.F. (1983) Quantitation of alkyl-deoxynucleosides in the DNA of individual cells by high-affinity monoclonal antibodies and electronically intensified, direct immunofluorescence. *J. Cancer Res. clin. Oncol.*, *105*, A15

Baird, W.M. (1979) *The use of radioactive carcinogens to detect DNA modification.* In: Grover, P.L., ed., *Chemical Carcinogens and DNA*, Boca Raton, FL, CRC Press, pp. 59-83

Bartsch, H. & Armstrong, B., eds (1982) *Host Factors in Human Carcinogenesis* (*IARC Scientific Publications No. 39*), Lyon, International Agency for Research on Cancer

Bullock, G.R. & Petrusz, P., eds (1982) *Techniques in Immunocytochemistry*, Vol. 1, London, Academic Press

Chard, T. (1982) *An Introduction to Radioimmunoassay and Related Techniques*, 2nd ed., Amsterdam, Elsevier Biomedical Press

Graf, T. & Jaenisch, R., eds (1982) *Tumorviruses, Neoplastic Transformation and Differentiation. Current Topics in Microbiology and Immunology 101*, Berlin, Springer-Verlag

Grover, P.L., ed. (1979) *Chemical Carcinogens and DNA*, Boca Raton, FL, CRC Press

Gupta, R.C., Reddy, M.V. & Randerath, K. (1982) $^{32}$P-Postlabeling analysis of non-radioactive aromatic carcinogen-DNA adducts. *Carcinogenesis*, *3*, 1081-1092

Harris, C.C., Yolken, R.H. & Hsu, I.C. (1982) Enzyme immunoassays: applications in cancer research. *Meth. Cancer Res.*, *20*, 213-243

Haugen, Å., Groopman, J.D., Hsu, I.C., Goodrich, G.R., Wogan, G.N. & Harris, C.C. (1981) Monoclonal antibody to aflatoxin $B_1$-modified DNA detected by enzyme immunoassay. *Proc. natl Acad. Sci. USA*, *78*, 4124-4127

Kohn, K.W. (1979) DNA as a target in cancer chemotherapy: measurements of macromolecular DNA damage produced in mammalian cells by anticancer agents and carcinogens. *Meth. Cancer Res.*, *16*, 291-345

Kriek, E., van der Laken, C.J. & Welling, M. (1981) *Immunological detection and quantification of the reaction products of 2-acetylaminofluorene with guanine in DNA.* In: Bartsch, H. & Armstrong, B., eds, *Host Factors in Human Carcinogenesis (IARC Scientific Publications No. 39*), Lyon, International Agency for Research on Cancer, pp. 541-550 van der Laken, C.J., Hagenaars, A.M., Hermsen, G., Kriek, E., Keupers, A.J., Nagel, J., Scherer, E. & Welling, M. (1982) Measurement of *O*$^6$-ethyldeoxy-guanosine and *N*-(deoxyguanosine-8-yl)-*N*-acetyl-2-aminofluorene in DNA by high-sensitive enzyme immunoassays. *Carcinogenesis*, *3*, 569-572

Lawley, P.D. (1976) *Carcinogenesis by alkylating agents*. In: Searle, C.E., ed., *Chemical Carcinogens* (*ACS Monograph No. 173*), Washington DC, American Chemical Society, pp. 83-244

Leary, J.J., Brigati, D.J. & Ward, D.C. (1983) Rapid and sensitive colorimetric method for visualizing biotin-labeled DNA probes hybridized to DNA or RNA immobilized on nitrocellulose: bio-blots. *Proc. natl Acad. Sci. USA*, *80*, 4045-4049

Müller, R. (1980) Calculation of average antibody affinity in anti-hapten sera from data obtained by competitive radioimmunoassay. *J. immunol. Meth.*, *34*, 345-352

Müller, R. & Rajewsky, M.F. (1978) Sensitive radioimmunoassay for detection of O$^6$-ethyldeoxyguanosine in DNA exposed to the carcinogen ethylnitrosourea *in vivo* or *in vitro*. *Z. Naturforsch.*, *33c*, 897-901

Müller, R. & Rajewsky, M.F. (1980) Immunological quantification by high-affinity antibodies of *O*$^6$-ethyldeoxyguanosine in DNA exposed to *N*-ethyl-*N*-nitrosourea. *Cancer Res.*, *40*, 887-896

Müller, R. & Rajewsky, M.F. (1981) Antibodies specific for DNA components structurally modified by chemical carcinogens. *J. Cancer Res. clin. Oncol.*, *102*, 99-113

Müller, R., Adamkiewicz, J. & Rajewsky, M.F. (1982) *Immunological detection and quantification of carcinogen-modified DNA components*. In: Bartsch, H. & Armstrong, B., eds, *Host Factors in Human Carcinogenesis* (*IARC Scientific Publications No. 39*), Lyon, International Agency for Research on Cancer, pp. 463-479

de Murcia, G., Lang, M.-C.E., Freund, A.-M., Fuchs, R.P.P., Daune, M.P., Sage, E. & Leng, M. (1979) Electron microscopic visualization of *N*-acetoxy-*N*-2-acetylaminofluorene binding sites in ColE1 DNA by means of specific antibodies. *Proc. natl Acad. Sci. USA*, *76*, 6076-6080

Nehls, P., Adamkiewicz, J. & Rajewsky, M.F. (1984a) Immuno-slot-blot: a highly sensitive immunoassay for the quantitation of carcinogen-modified nucleosides in DNA. *J. Cancer Res. clin. Oncol.*, *108* (in press)

Nehls, P., Rajewsky, M.F., Spiess, E. & Werner, D. (1984b) Highly sensitive sites for guanine-O$^6$ ethylation in rat brain DNA exposed to *N*-ethyl-*N*-nitrosourea *in vivo*. *EMBO J.*, *3*, 327-332

Nehls, P., Spiess, E. & Rajewsky, M.F. (1983) Visualization of *O*$^6$-ethylguanine in ethylnitrosourea-treated DNA by immune electron microscopy. *J. Cancer Res. clin. Oncol.*, *105*, A23

O'Connor, P.J., Saffhill, R. & Margison, G.P. (1979) N-*Nitroso compounds: biochemical mechanisms of action*. In: Emmelot, P. & Kriek, E., eds, *Environmental Carcinogenesis*, Amsterdam, Elsevier Biomedical Press, pp. 73-96

Pegg, A.E. (1977) Formation and metabolism of alkylated nucleosides: Possible role in carcinogenesis by nitroso compounds and alkylating agents. *Adv. Cancer Res.*, *25*, 195-269

Poirier, M.C. (1981) Antibodies to carcinogen-DNA adducts. *J. natl Cancer Inst.*, *67*, 515-519

Poirier, M.C., Dubin, M.A. & Yuspa, S.H. (1979) Formation and removal of specific acetylaminofluorene-DNA adducts in mouse and human cells measured by radioimmunoassay. *Cancer Res.*, *39*, 1377-1381

Poirier, M.C., Santella, R., Weinstein, I.B., Grunberger, D. & Yuspa, S.H. (1980) Quantitation of benzo[a]pyrene-deoxyguanosine adducts by radioimmunoassay. *Cancer Res.*, *40*, 412-416

Rajewsky, M.F. (1980) *Specificity of DNA damage in chemical carcinogenesis*. In: Montesano, R., Bartsch, H. & Tomatis, L., eds, *Molecular and Cellular Aspects of Carcinogen Screening Tests* (*IARC Scientific Publications No. 27*), Lyon, International Agency for Research on Cancer, pp. 41-54

Rajewsky, M.F. (1983) Structural modifications and repair of DNA in neuro-oncogenesis by *N*-ethyl-*N*-nitrosourea. *Recent Results Cancer Res.*, *84*, 63-76

Rajewsky, M.F., Augenlicht, L.H., Biessmann, H., Goth, R., Hülser, D.F., Laerum, O.D. & Lomakina, L.Y. (1977) *Nervous system-specific carcinogenesis by ethylnitrosourea in the rat: molecular and cellular mechanisms*. In: Hiatt, H.H., Watson, J.D. & Winsten, J.A., eds, *Origins of Human Cancer, Book B: Mechanisms of Carcinogenesis*, Cold Spring Harbor, NY, Cold Spring Harbor Laboratory, pp. 709-726

Rajewsky, M.F., Müller, R., Adamkiewicz, J. & Drosdziok, W. (1980) *Immunological detection and quantification of DNA components structurally modified by alkylating carcinogens (ethylnitrosourea)*. In: Pullman, B., Ts'o, P.O.P. & Gelboin, H., eds, *Carcinogenesis: Fundamental Mechanisms and Environmental Effects*, Dordrecht, Reidel, pp. 207-218

Randerath, K., Reddy, M.V. & Gupta, R.C. (1981) $^{32}$P-labeling test for DNA damage. *Proc. natl Acad. Sci. USA*, *78*, 6126-6129

Reading, C.L. (1982) Theory and methods for immunization in culture and monoclonal antibody production. *J. immunol. Meth.*, *53*, 261-291

Saffhill, R., Strickland, P.T. & Boyle, J.M. (1982) Sensitive radioimmunoassay for $O^6$-*n*-butyldeoxyguanosine, $O^2$-*n*-butylthymidine and $O^4$-*n*-butylthymidine. *Carcinogenesis*, *3*, 547-552

Singer, B. (1979) *N*-Nitroso alkylating agents: formation and persistence of alkyl derivatives in mammalian nucleic acids as contributing factors in carcinogenesis. *J. natl Cancer Inst.*, *62*, 1329-1339

Strickland, P.T. & Boyle, J.M. (1981) Characterization of two monoclonal antibodies specific for dimerised and non-dimerised adjacent thymidines in single stranded DNA. *Photochem. Photobiol.*, *34*, 595-601

Wild, C.P., Smart, G., Saffhill, R. & Boyle, J.H. (1983) Radioimmunoassay of $O^6$-methyldeoxyguanosine in DNA of cells alkylated *in vitro* and *in vivo*. *Carcinogenesis*, *4*, 1605-1609

# BIOCHEMICAL (POSTLABELLING) METHODS FOR ANALYSIS OF CARCINOGEN-DNA ADDUCTS

**K. Randerath, E. Randerath, H.P. Agrawal & M.V. Reddy**

*Department of Pharmacology Baylor College of Medicine Houston, Texas 77030, USA*

## SUMMARY

Radioactive carcinogens have provided most of our present knowledge about the interactions between carcinogens and components of biological systems. The requirement of radioactive carcinogens restricts carcinogen-DNA binding studies to chemicals that are readily available in isotopically labelled form, i.e., a minute fraction of all potentially mutagenic or carcinogenic chemicals. To extend the scope of carcinogen-DNA binding studies, an alternative method, which does not require radioactive test chemicals, has been developed. In this approach, radioactivity ($^{32}P$) is incorporated into DNA constituents by polynucleotide kinase-catalysed ($^{32}P$)-phosphate transfer from ($\gamma$-$^{32}P$)ATP *after* exposure of the DNA, *in vitro* or *in vivo*, to a nonradioactive, covalently binding chemical; alteration of DNA nucleotides is shown by the appearance of extra spots on autoradiograms from thin-layer chromatograms of digests of the chemically modified DNA. Adduct levels are quantitated by scintillation counting. The sensitivity of the technique depends, to some extent, on the chemical structure of the adducts, in that greater sensitivity is achieved if adducts can be separated, as a class, from the normal nucleotides. An estimated 80% of all carcinogens can be separated in this way, giving rise to bulky and/or aromatic substituents in DNA. Under present conditions, one such adduct in $10^9$ - $10^{10}$ normal nucleotides can be detected. A total of 41 compounds has been studied, so far. Binding to DNA of rodent liver and skin was readily detected by the $^{32}P$-postlabelling assay for all known carcinogens among these compounds, and adducts were detected in DNA from tissues of smokers.

## INTRODUCTION

Any chemical that is capable of forming covalent bonds with DNA of somatic and reproductive mammalian cells *in vivo* is a potential mutagen, carcinogen and teratogen. Since such genotoxic chemicals may be of natural or man-made origin, exposure to them cannot be completely eliminated. In view of the dangers posed by such covalently binding chemicals to present and future human generations, human contact with them must be minimized. Methods that allow detection and quantitation of the DNA-binding potential of chemicals, directly *in vivo*, should have considerable value in the detection of gene-altering chemicals in the environment, provided that they can be applied to a large number of chemicals of diverse structure and that the sensitivity is sufficient to detect low DNA-binding activities. Because covalent DNA binding of chemicals *in vivo* may range from one adduct in $10^3$ normal nucleotides to one adduct in $10^9$ - $10^{10}$ nucleotides, such methods, ideally, should be capable of detecting extremely low binding, of the order of a single adduct per diploid mammalian genome containing about $1.2 \times 10^{10}$ DNA nucleotides.

As shown by Lutz (1979, 1982), the covalent binding index, defined as μmol of chemical bound per mol of DNA nucleotide/mmol of chemical administered per kg body weight of animal, exhibits a good quantitative correlation with the hepatocarcinogenic potency of chemicals of diverse structure. Therefore, methods for the analysis of adducts in DNA should not only detect but also quantitate DNA-binding activity. This paper reviews recent efforts to develop such an ultrasensitive method, involving $^{32}$P-postlabelling of adducts (Randerath *et al.*, 1981; Gupta *et al.*, 1982; Randerath *et al.*, 1983; Reddy *et al.*, 1984), and describes the results obtained by applying the method to a number of genotoxic chemicals of diverse structure.

## MATERIALS AND METHODS

The sources of materials used in the $^{32}$P-postlabelling procedure, as well as safety precautions and chromatographic and autoradiographic procedures, have been reported previously (Gupta *et al.*, 1982). ($\gamma$-$^{32}$P)ATP was synthesized in the laboratory as described (Gupta *et al.*, 1982). To remove background material from maps of aromatic carcinogen-DNA adducts, polyethyleneimine-cellulose thin layers were given a final development in 0.35 M magnesium chloride before autoradiography (Reddy *et al.*, 1984). Screen-intensified autoradiography was performed at -80°C. For in-vivo modification of DNA, female BALB/c or CD-1 mice (25 g) and male Sprague-Dawley or Fischer rats (200 g) were maintained on standard laboratory diet and water *ad libitum*. A list of chemicals used for animal treatments is given in Table 2. For adduct detection in mouse skin DNA, the backs of mice were shaved with clippers three days prior to treatment, which was performed by topical application of four doses, of 1.2 μmol each, of test compounds in 200 μl of solvent. Compounds were dissolved in acetone, except for azodyes, which were dissolved in acetone/water (7:3, v/v). The treatments were at 0 h, 24 h, 48 h and 72 h for arylamines and derivatives, azo compounds and nitro compounds, and at 0 h, 6 h, 30 h, and 54 h for polycyclic aromatic hydrocarbons. Control mice were given 200 μl of solvent alone. DNA was isolated 24 h after the last treatment. For adduct detection in mouse liver DNA, animals were given a single intraperitoneal dose (150 mg/kg) of methylating agent in 0.2 ml of 0.9% sodium chloride; control mice received 0.9% sodium chloride alone. DNA was extracted 3 h after administration. Alkenylbenzenes were given to mice by intraperitoneal injection of compound (400 mg/kg) in 0.1 ml tricaprylin, and liver DNA was isolated 24 h after treatment; control mice received tricaprylin alone. For adduct detection in rat liver DNA, animals were given a single intraperitoneal dose of test chemical (40 mg/kg) in 0.3 ml of dimethyl sulfoxide; control rats received vehicle alone. DNA was isolated 4 h after administration. The treatment with mycotoxins is outlined in the legend to Figure 7.

$^{32}$P-Postlabelling analysis of DNA adducts was performed as described previously (Gupta *et al.*, 1982; Reddy *et al.*, 1984), using 1-2 μg of DNA for enzymatic digestion and 0.15-0.3 μg of DNA nucleotides for $^{32}$P-labelling. For estimation of adduct levels, spots were excised from replicate maps and counted by the Cerenkov assay (Gupta *et al.*, 1982). Appropriate blank areas of the chromatogram were assayed also, and their count rates subtracted from the sample count rates. For aromatic adducts, 150-460 μCi of $^{32}$P-labelled digest were applied to the thin-layer chromatograms for counting of adducts and 0.25 μCi was applied in order to assay normal nucleotides. Results were calculated from:

$$\text{relative adduct labelling} = \frac{\text{cpm in adducts}}{(\text{cpm in normal nucleotides}) \times \text{dilution factor}}$$

Under conditions of excess ATP (Gupta *et al.*, 1982; Reddy *et al.*, 1984), relative adduct labelling, as defined above, accurately reflects the adduct levels in the samples analysed; a relative adduct labelling value of $10^{-7}$ corresponds to a DNA modification level of about 0.3 pmol of adduct/mg of DNA.

## RESULTS

If the $^{32}$P-postlabelling scheme is to test for the capacity of chemicals to bind to DNA, it should be applicable to most or all covalently binding chemicals; thus, a major question that had to be answered in the initial phase of method development was whether many adducts of diverse structure were amenable to $^{32}$P-postlabelling by T4 polynucleotide kinase-catalysed phosphorylation. So far, 41 chemicals have been studied, comprising arylamines and derivatives, azo compounds, nitroaromatics, polycyclic aromatic hydrocarbons, alkenylbenzenes, mycotoxins and methylating agents. In every case, $^{32}$P-labelling of carcinogen-DNA derivatives could be readily detected. A total of 174 different adducts was detected as $^{32}$P-labelled derivatives on maps from the various in-vivo modified DNA samples. Thus, the $^{32}$P-postlabelling method was found applicable to a very large number of chemically diverse adducts. Most of the compounds studied are listed in Table 2.

The $^{32}$P-postlabelling procedure comprises four consecutive steps (Fig. 1): (1) Digestion of modified DNA to 3'-mononucleotides; (2) incorporation of $^{32}$P into the 3'-mononucleotides; (3) removal of normal nucleotides; and (4) thin-layer chromatographic separation and autoradiography of $^{32}$P-labelled adduct nucleotides. As illustrated in Figure 2, the latter compounds are (5'-$^{32}$P)deoxyribonucleoside 3',5'-bisphosphates. In addition to PEI-cellulose thin-layer chromatography, reversed-phase thin-layer chromatography on octadecylsilane layers was able to separate the normal nucleotides from the adduct nucleotides (Reddy *et al.*, 1984). To remove the labelled normal nucleotides and resolve the ($^{32}$P)adducts, a four-directional anion-exchange polyethyleneimine-cellulose thin-layer chromatography system was developed (Fig. 3). In this procedure, freshly prepared labelled digest (14-18 μCi/μl) is applied slowly to the origin (Fig. 3, OR), located close to the centre of the thin-layer sheet (20 × 20 cm). Development is begun immediately after sample application, without drying of the origin area. Conditions for the various developments are indicated by the legend to Figure 3 and are detailed in Gupta *et al.* (1982) and Reddy *et al.* (1984). This procedure has been applied successfully to aromatic carcinogens with two to six aromatic rings; for less aromatic carcinogens (such as alkenylbenzenes, sterigmatocystin, and aflatoxin $B_1$), the normal nucleotides were best removed from the adducts by octadecylsilane reversed-phase thin-layer chromatography (Reddy *et al.*, 1984).

**Fig. 1. $^{32}P$-Postlabelling assay of carcinogen-DNA adducts involves four steps: digestion of DNA to mononucleotides, $^{32}P$-labelling of digestion products, removal of $^{32}P$-labelled nonadduct components and fingerprinting**

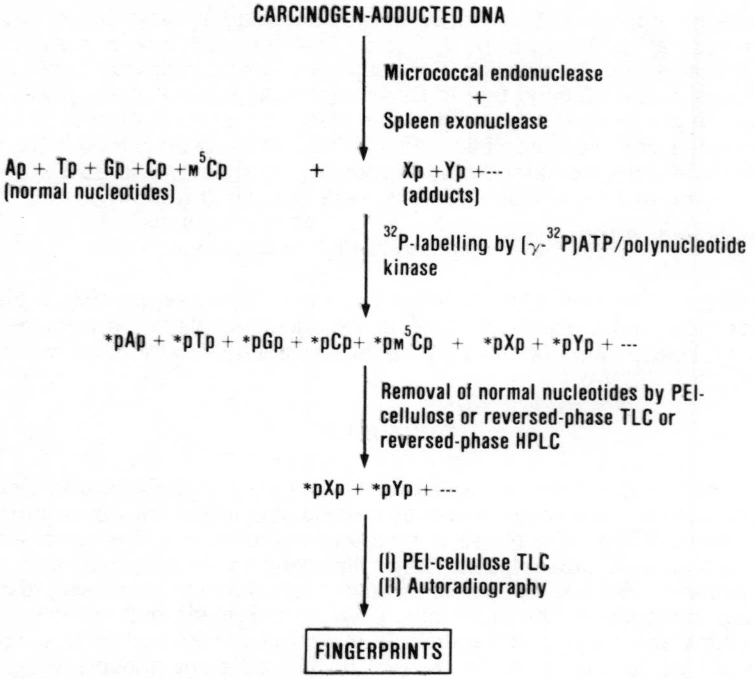


Prefix d for deoxyribo- was omitted for the nucleotides (from Gupta *et al.*, 1982, with permission). TLC, thin-layer chromatography; HPLC, high-performance liquid chromatography

**Fig. 2. Structure of 5'-$^{32}P$-labelled 3',5'-bisphosphate of *N*-acetyl-*N*-(deoxyguanosine-8-yl)-2-aminofluorene**

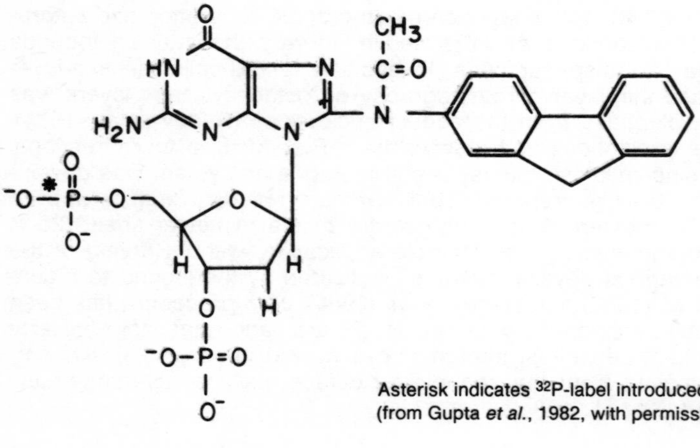

Asterisk indicates $^{32}P$-label introduced by enzymatic ($^{32}P$)phosphate transfer (from Gupta *et al.*, 1982, with permission).

**Fig. 3. Diagram of four-directional polyethyleneimine-cellulose thin-layer chromatography for the separation of $^{32}P$-labelled carcinogen-DNA adducts**

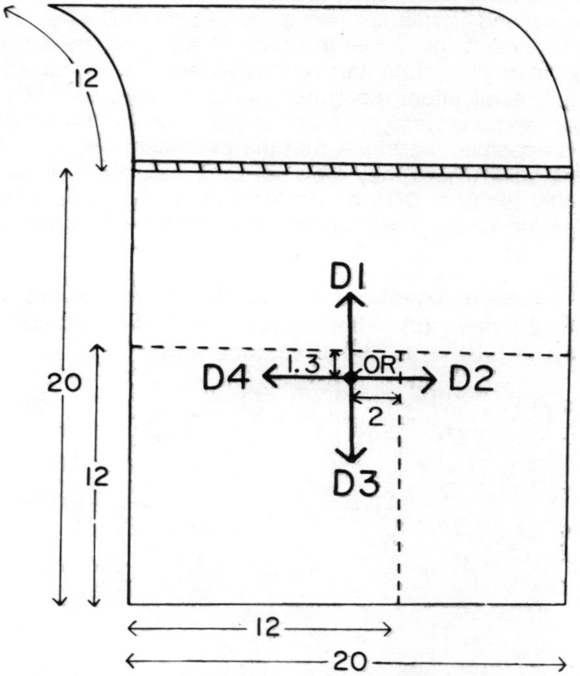


A Whatman 1 wick was attached to the top of the polyethyleneimine-cellulose thin layer, and $^{32}P$-labelled DNA digest was applied at the origin (OR). Developments in direction 1 (D1) and D2 were in 1.1 M lithium chloride and 2.5 M ammonium formate (pH 3.5), respectively. These developments served to remove $^{32}P$-labelled normal DNA nucleotides, $^{32}P_i$, and contaminants, while the adducts were retained at OR. The adducts were separated, subsequently, by development in D3 and D4 with 3 M lithium formate, 7-8.5 M urea (pH 3.5), and 0.8 M lithium chloride, 0.5 M Tris-hydrochloric acid, 7-8.5 M urea (pH 8.0), respectively (from Gupta *et al.*, 1982, with permission).

The lower level of adduct detection was found to be one adduct in (3.5-6) $\times$ $10^7$ nucleotides, for the standard four-directional polyethyleneimine-cellulose thin-layer chromatographic procedure, while one adduct in about $10^8$ nucleotides could be detected if normal nucleotides were removed from the adduct nucleotides by reversed-phase thin-layer chromatography, the increase in sensitivity being due to reduced background radioactivity (Reddy *et al.*, 1984). The technique involves simultaneous labelling of normal and adduct nucleotides, which enables accurate quantitation of DNA adduct levels. It is possible, also, to isolate the adducts first and then label them in the virtual absence of normal nucleotides, a technique that increases the sensitivity of detection to one adduct in about $10^{10}$ normal nucleotides, but produces a loss of quantitation accuracy compared to the standard technique (K. Randerath & E. Randerath, unpublished data).

Some examples of application of the postlabelling method are given here; other examples have been described by Gupta *et al.* (1982), Randerath *et al.* (1983) and Reddy *et al.* (1984). The method was applied to skin DNA from mice treated topically with the polycyclic aromatic hydrocarbons, benzo[*a*]pyrene, 3-methylcholanthrene and 7,12-dimethylbenz[*a*]anthracene, respectively. As shown by autoradiography (Fig. 4), a large number of

$^{32}P$-labelled 7,12-dimethylbenz[*a*]anthracene-DNA adducts was detected in digests of DNA obtained from mouse skin at several time points after carcinogen treatment. Four of these adducts (spots 3, 4, 5, and 6) were highly persistent (Fig. 4c,d). These adducts were detected in mouse epidermis and dermis as late as seven months after application of a single dose of 1.2 μmol 7,12-dimethylbenz[*a*]anthracene to skin. As shown in Figure 5, substantial removal of 7,12-dimethylbenz[*a*]anthracene-DNA adducts occurred during the first two weeks after carcinogen application; thereafter, adducts underwent little change. Analogous results were reported for benzo[*a*]pyrene and 3-methylcholanthrene-DNA adducts (Randerath *et al.*, 1983). It is possible, therefore, that the persistent adducts occupy specific genomic sites in quiescent cells where they may not be amenable to repair, because of localized conformational alterations of DNA or shielding by associated proteins. Table 1 summarizes our data (Randerath *et al.*, 1983) on the formation and persistence of these adducts in mouse skin.

**Fig. 4. Maps of $^{32}P$-Labelled digests of mouse skin DNA isolated at 1 day (a), 6 days (b), 14 days (c) and 28 days (d), after topical application of four 1.2 μmol doses of 7,12-dimethylbenz[*a*]anthracene, in 200 μl acetone each**

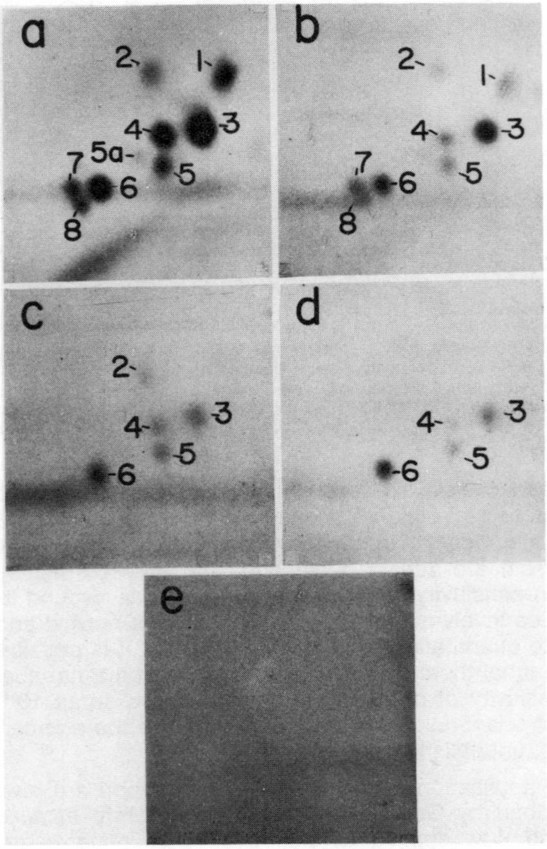


(e), digest of DNA from control mice that had received acetone only. Separation was by the standard procedure (Fig. 3) and spots were visualized by autoradiography (E. Randerath, H.P. Agrawal & K. Randerath, unpublished data)

**Fig. 5. Time course of relative adduct labelling (RAL = $cpm_{adducts}/cpm_{total}$) observed in the experiment illustrated in figure 4**

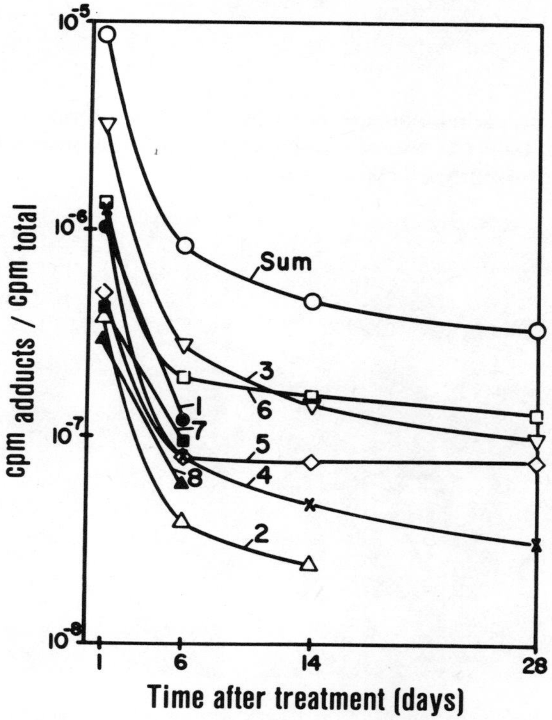


A relative adduct labelling value of $10^{-6}$ corresponds to an adduct level of 3 fmol/μg of DNA. Assays were performed by Cerenkov counting of excised spots. Relative standard deviations were < ±15% (E. Randerath, H.P. Agrawal & K. Randerath, unpublished data).

**Table 1. Total adduct levels in mouse skin DNA at various time points after polycyclic aromatic hydrocarbon treatment, expressed as: X = number of nucleotides per single adduct or Y = number of adducts per genome[a]**

| Days after treatment[b] | BP[c] | | DMBA[d] | | MC[e] | |
|---|---|---|---|---|---|---|
| | X | Y | X | Y | X | Y |
| 1 | $6.2 \times 10^4$ | $1.9 \times 10^5$ | $6.0 \times 10^4$ | $2.0 \times 10^5$ | $1.3 \times 10^5$ | $9.2 \times 10^4$ |
| 6 | $3.8 \times 10^5$ | $3.2 \times 10^4$ | $5.0 \times 10^5$ | $2.4 \times 10^4$ | $4.2 \times 10^5$ | $2.9 \times 10^4$ |
| 14 | $1.4 \times 10^6$ | $8.6 \times 10^3$ | $1.1 \times 10^6$ | $1.1 \times 10^4$ | $1.4 \times 10^6$ | $8.6 \times 10^3$ |
| 28 | $2.7 \times 10^6$ | $4.4 \times 10^3$ | $1.4 \times 10^6$ | $8.6 \times 10^3$ | $1.5 \times 10^6$ | $8.0 \times 10^3$ |

[a] Assuming that the genome of a diploid mammalian cell contains $1.2 \times 10^{10}$ nucleotides

[b] Mice were treated with four 1.2-μmol doses of polycyclic aromatic hydrocarbon, as described in the text

[c] Benzo[*a*]pyrene

[d] 7,12-Dimethylbenz[*a*]anthracene

[e] 3-Methylcholanthrene

Adducts obtained from mouse skin DNA after topical application of the known skin carcinogen, 4-nitroquinoline-1-oxide, also exhibited a typical fingerprint (Fig. 6). Eight adducts were found, and the total level of modification one day after carcinogen application was one adduct in about $7.4 \times 10^5$ nucleotides (Reddy *et al.*, 1984).

**Fig. 6. Standard 4-D polyethyleneimine-cellulose thin-layer maps of $^{32}$P-labelled digests of control mouse skin DNA (A) and of mouse skin DNA 1 day after the last application of four 1.2-μmol doses of 4-nitroquinoline-1-oxide (B)**

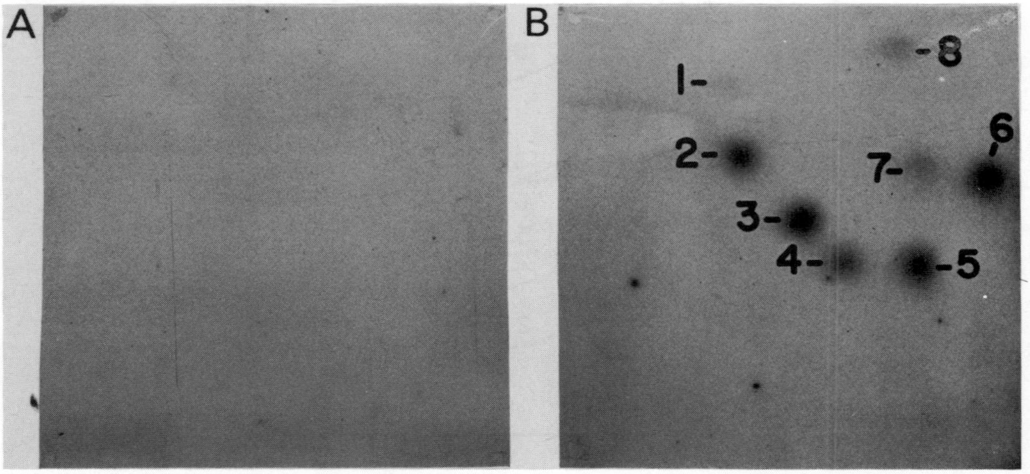


Separation was by the standard procedure (Fig. 3) and spots were visualized by autoradiography (M.V. Reddy, R.C. Gupta, E. Randerath & K. Randerath, submitted for publication)

Figure 7 illustrates the application of the $^{32}$P-postlabelling method to rat liver DNA, after exposure *in vivo* to the mycotoxins, aflatoxin $B_1$ and sterigmatocystin, respectively. These compounds are known potent hepatocarcinogens (Wogan, 1973; Purchase & van der Watt, 1970), and in-vivo DNA binding of aflatoxin $B_1$ has been studied extensively (Essigmann *et al.*, 1982), but such data were not available for sterigmatocystin-DNA binding *in vivo*. Our studies on these mycotoxins revealed that the adducts seen on $^{32}$P-labelled fingerprints (Fig. 7) were mostly oligo- (di- and tri-)nucleotides, because enzymatic digestion, under our standard conditions (Gupta *et al.*, 1982), of DNA containing such adducts led to the formation of oligonucleotides (mostly di- and trinucleotides), rather than the usual mononucleotide adducts (M.V. Reddy & K. Randerath, unpublished data). Sterigmatocystin-DNA adducts were detectable in rat liver DNA as late as 3.5 months after injection of a single dose (9 mg/kg) of sterigmatocystin to male Fischer rats (M.V. Reddy & K. Randerath, unpublished data). When equal doses were compared, sterigmatocystin led to an approximately 10-fold lower level of DNA modification than did aflatoxin $B_1$, in accord with the nearly 10-fold lower carcinogenic potency of the former compound compared to the latter (Purchase & van der Watt, 1970).

**Fig. 7. Polyethyleneimine-cellulose thin-layer maps of aflatoxin $B_1$ ($AFB_1$)-DNA (A and B) and sterigmatocystin (ST)-DNA (C and D) adducts**

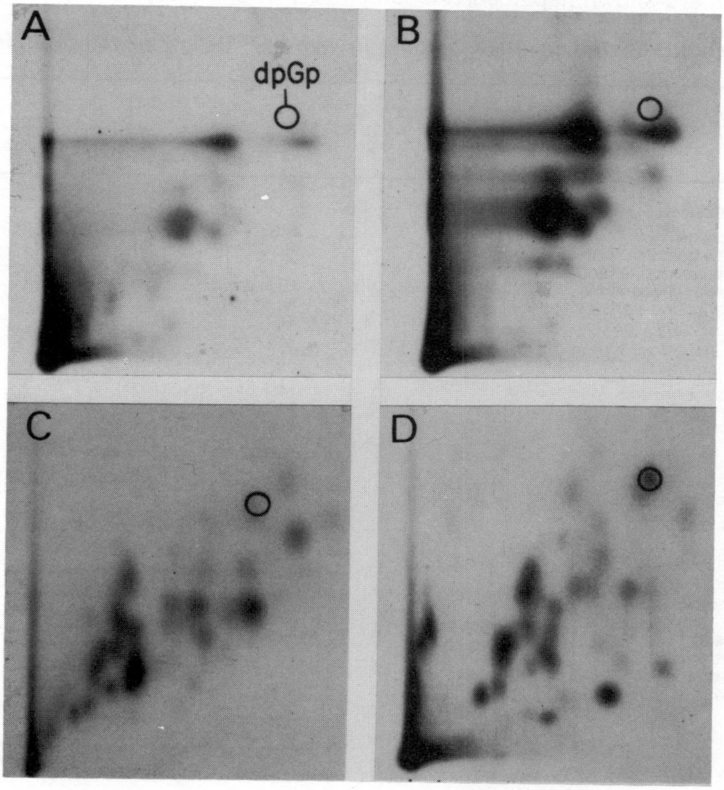


A, 0.125 mg/kg $AFB_1$; B, 1 mg/kg $AFB_1$; D, 1 mg/kg ST. Panels A, B and D were obtained from DNA modified *in vivo*; panel C was from DNA modified *in vitro* with ST, in the presence of m-chloroperoxybenzoic acid, under conditions similar to those used by Martin and Garner (1977) for chemical modification of DNA with aflatoxin $B_1$. For the in-vivo samples, rat liver DNA was isolated 2 h after intraperitoneal injection of the indicated dose of carcinogen, dissolved in 0.2 ml dimethylsulfoxide. Adduct purification was by octadecylsilane reversed-phase thin-layer chromatography. Final mapping was by two-dimensional polyethyleneimine-cellulose thin-layer chromatography in 3 M ammonium formate (pH 3.5) and 1.15 M Tris-HCl (pH 8.0) (M.V. Reddy & K. Randerath, unpublished data).

Table 2 lists most of the compounds that we have studied so far, together with their sources, the tissues investigated, the number of adducts detected for each compound and an estimation of total adduct level for each compound. Except for anthracene, pyrene and perylene, all the compounds gave rise to $^{32}P$-labelled adducts at the levels indicated. It seems important that, within the group of polycyclic aromatic hydrocarbons, a good correlation was observed between the carcinogenic potency of individual compounds for mouse skin (Dipple, 1976; Phillips *et al.*, 1979) and their binding levels to mouse skin DNA. In particular, we failed to detect DNA binding of the noncarcinogens, anthracene, pyrene and perylene (at a detection sensitivity of one adduct in about $5 \times 10^7$ nucleotides), while the strong carcinogens, benzo[*a*]pyrene, 3-methylcholanthrene and 7,12-dimethylbenz[*a*]anthracene, showed the highest levels of DNA binding. The $^{32}P$-postlabelling assay also showed

DNA binding of the alkenylbenzenes, safrole, estragole and methyleugenol, *in vivo* (M.V. Reddy, D.H. Phillips, R.E. Haglund & K. Randerath, unpublished data). These naturally occurring compounds are known hepatocarcinogens (Miller *et al.*, 1983).

**Table 2. Compounds tested for DNA binding *in vivo* by $^{32}$P-postlabelling analysis**

| Compounds | Source[a] | Tissue[b] | DNA adducts | |
|---|---|---|---|---|
| | | | No. | Levels[d] |
| *Arylamines and derivatives* | | | | |
| 2-Acetylaminofluorene | S | MS | 6 | ++ |
| *N*-Hydroxy-2-acetylaminofluorene | N | RL | 16 | +++ |
| *N*-Hydroxy-2-acetylaminophenanthrene | N | RL | 10 | +++ |
| *N*-Hydroxy-4-acetylaminobiphenyl | N | RL | 10 | +++ |
| *N*-Hydroxy-4-acetylamino-*trans*-stilbene | N | RL | 9 | +++ |
| 4-Aminobiphenyl | S | MS | 1 | + |
| Benzidine | S | MS | 3 | + |
| *Azo compounds* | | | | |
| 4-Dimethylaminoazobenzene | N | MS | 2 | + |
| Congo red | S | MS | 2 | + |
| Evan's blue | S | MS | 1 | + |
| *Nitro compounds* | | | | |
| 4-Nitroquinoline-1-oxide | S | MS | 8 | ++ |
| 2,6-Dinitrotoluene | S | MS | 3 | + |
| *Polycyclic aromatic hydrocarbons* | | | | |
| Benzo[*a*]pyrene | S | MS | 5 | +++ |
| | | RL | 2 | ++ |
| 7,12-Dimethylbenz[*a*]anthracene | S | MS | 8 | +++ |
| 3-Methylcholanthrene | S | MS | 13 | +++ |
| Benzo[*e*]pyrene | S | MS | 5 | + |
| 1,2-Benzanthracene | S | MS | 2 | + |
| 1,2,3,4-Dibenzanthracene | S | MS | 6 | + |
| 1,2,5,6-Dibenzanthracene | S | MS | 3 | ++ |
| 1,12-Benzoperylene | S | MS | 2 | ++ |
| Chrysene | A | MS | 2 | ++ |
| Anthracene | S | MS | ND[e] | - |
| Pyrene | S | MS | ND[e] | - |
| Perylene | S | MS | ND[e] | - |
| *Alkenylbenzenes* | | | | |
| Safrole | A | ML | 5 | +++ |
| Estragole | A | ML | 5 | +++ |
| Methyleugenol | K | ML | 4 | +++ |
| *Methylating agents* | | | | |
| *N,N*-Nitrosodimethylamine | A | ML | 5 | ++++ |
| 1,2-Dimethylhydrazine | S | ML | 5 | ++++ |
| *N*-Methyl-*N*-nitrosourea | A | ML | 5 | ++++ |
| Streptozotocin | U | ML | 5 | ++++ |
| *Mycotoxins* | | | | |
| Aflatoxin $B_1$ | S | RL | 9[f] | ++++ |
| Sterigmatocystin | S | RL | 15[f] | +++ |

[a] S, Sigma Chemical Company, St Louis, MO; N, National Cancer Institute Chemical Carcinogen Reference Standard Repository; A, Aldrich Chemical Company, Kalamazoo, MI; U, Upjohn Company, Milwaukee, WI; K, ICN K & K Laboratories, Plainview, NY

[b] MS, mouse skin; ML, mouse liver; RL, rat liver

[c] These numbers reflect the total number of adducts detected, including those requiring very prolonged exposure before they could be detected. Approximate limit of detection was one adduct in $5 \times 10^7$ - $10^8$ nucleotides.

[d] Total adduct levels: +, one adduct in $>10^7$ nucleotides; ++, one adduct in $5 \times 10^5$ - $10^7$ nucleotides; +++, one adduct in $10^4$ - $5 \times 10^5$ nucleotides; ++++, one adduct in $<10^4$ nucleotides

[e] Not detected

[f] These adducts were oligonucleotides containing covalently bound carcinogen (see text)

**Fig. 8. Schematic polyethyleneimine-cellulose fingerprints of $^{32}$P-labelled adducts formed after in-vivo modification of DNAs with the indicated aromatic carcinogens**

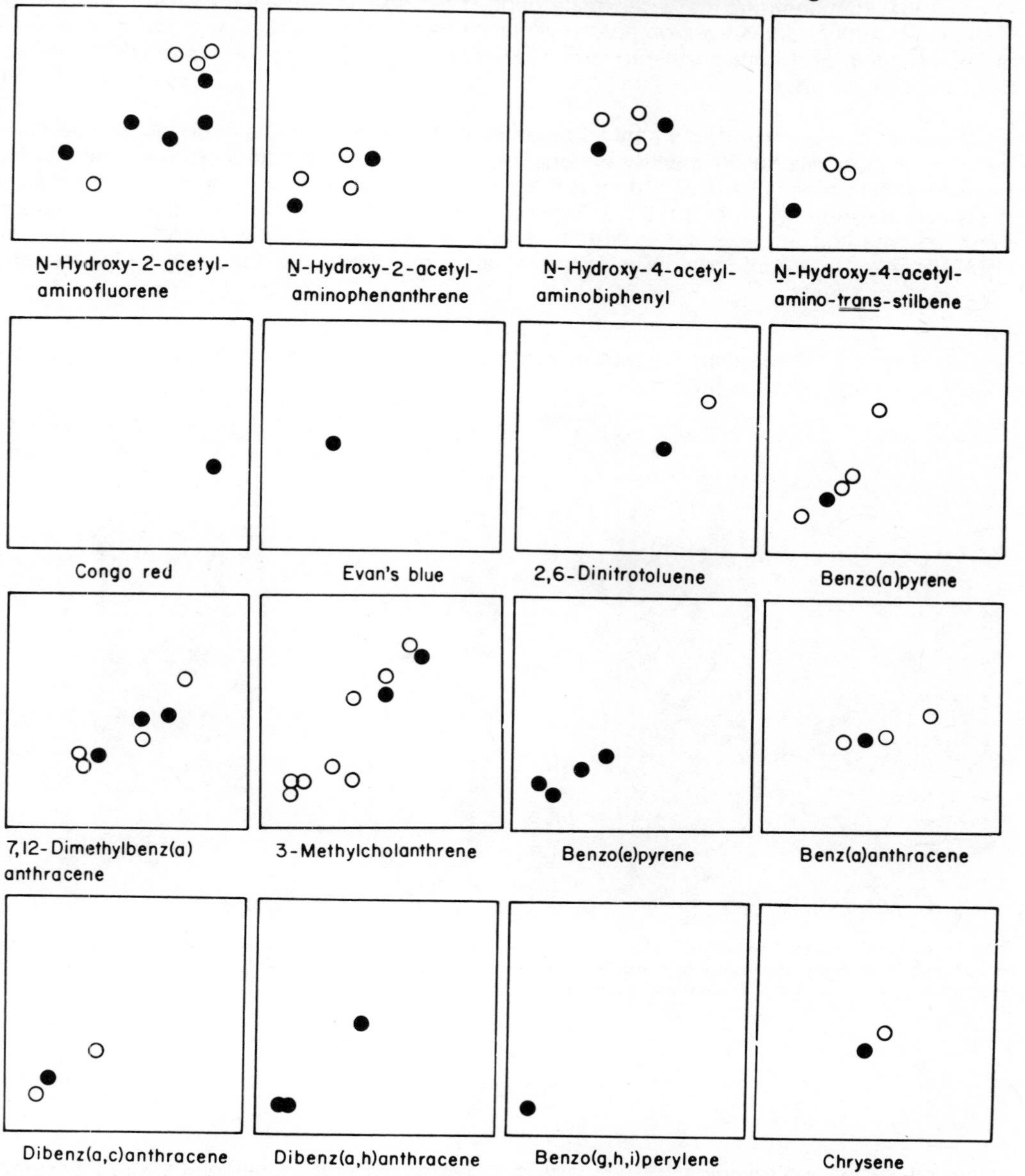


Tissues examined are listed in Table 2; benzo[*a*]pyrene adducts were from mouse skin DNA. The standard procedure for adduct separation (Fig. 3) was used. Urea concentration in D3 (bottom to top) and D4 (left to right) solvents was 8.5 M for $^{32}$P-labelled adducts induced by polycyclic aromatic hydrocarbons, and 7 M for the rest. Solid circles represent major adducts. Very weak adducts are not shown (M.V. Reddy, R.C. Gupta, E. Randerath & K. Randerath, submitted for publication)

Figure 8 shows schematic fingerprints for a series of aromatic carcinogen-DNA adducts. The examples in this paper and elsewhere (Gupta *et al.*, 1982; Randerath *et al.*, 1983; Reddy *et al.*, 1984) demonstrate that each compound gives rise to a characteristic fingerprint of $^{32}$P-labelled adduct derivatives on polyethyleneimine-cellulose maps. Therefore, on the basis of such fingerprints, it is possible to identify the carcinogen to which the particular DNA has been exposed *in vivo*.

The postlabelling assay was applied recently to DNA from humans exposed to potentially genotoxic chemicals. In preliminary experiments, carcinogen-DNA adducts were detected in buccal mucosa brushings (B.P. Dunn, H.F. Stich, H.P. Agrawal & K. Randerath, unpublished data) and placentas (E. Randerath, R. Everson & R. Santella, unpublished data) of smokers. Thus, the method appears to be directly applicable to the detection of adducts in human DNA and may, therefore, contribute to the monitoring of human exposure to mutagenic and carcinogenic agents.

**Fig. 9. Maps of $^{32}$P-labelled digests of oral mucosal DNA from a nonsmoker (left) and a heavy smoker/drinker (right)**

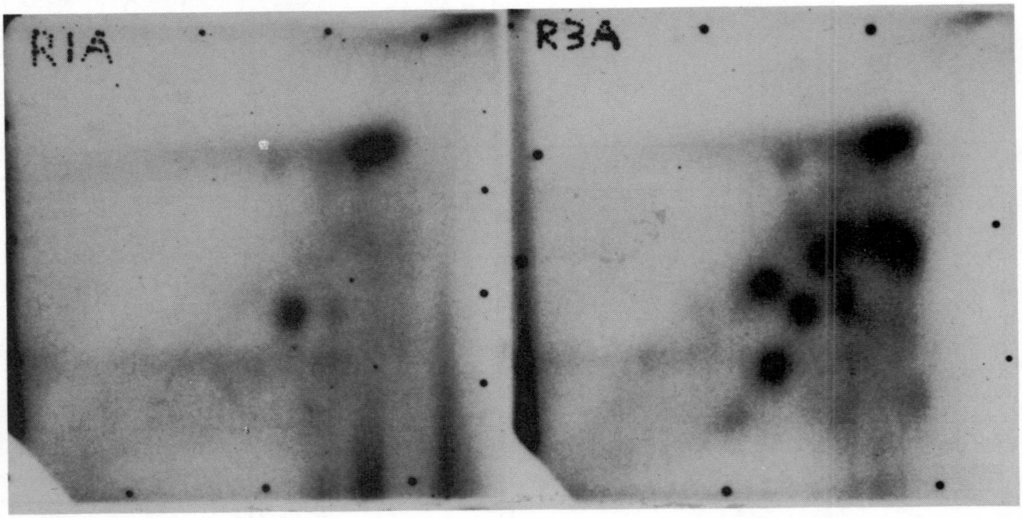


The $^{32}$P-labelled samples (derived from about 0.7 μg DNA each) were first purified by reversed-phase thin-layer chromatography, then mapped on polyethyleneimine-cellulose thin layers according to the standard procedure (B.P. Dunn, H.F. Stich, H.P. Agrawal & K. Randerath, unpublished data)

## DISCUSSION

In this paper, we review a $^{32}$P-postlabelling method, recently developed in our laboratory, for the analysis of carcinogen-nucleic acid adducts. While we have focussed mainly on analysis of carcinogen-DNA adducts, the method is also applicable to carcinogen-RNA adducts. The salient features of the new $^{32}$P-postlabelling test for covalent DNA binding of chemicals can be summarized as follows:

(1) The method enables detection of minute amounts of adduct formed by the reaction of DNA with *nonradioactive* chemicals. In principle, at least, DNA binding of any chemical can be assayed.

(2) Since radioactive carcinogens are not required for in-vivo studies, the assay is relatively inexpensive.

(3) The method is applicable both to individual compounds and to mixtures.

(4) The chemical identity of carcinogens and/or adducts need not be known for DNA binding to be detected.

(5) If the fingerprint patterns obtained with pure carcinogens or carcinogenic mixtures are known, then these fingerprints enable identification of the carcinogen or carcinogen mixture to which the test DNA was originally exposed.

(6) Small (μg) amounts of DNA are required for analysis; thus, the method can be applied to mg amounts of biological material.

(7) The method is highly sensitive; for aromatic carcinogens, a few adducts per mammalian genome can be detected in a cell population or a tissue.

(8) The method gives accurate quantitation.

(9) The method is potentially useful for investigating repair, removal and loss of adducts from cell or tissue DNA; the effects of anticarcinogens, chemopreventive agents, and/or metabolic inhibitors on adduct formation and persistence can be studied, also.

(10) The method is applicable to DNA from humans exposed to mutagens or carcinogens, and is, therefore, a tool for monitoring human exposure to these compounds.

The results reviewed in this paper and by others (reviewed by Hemminki, 1983) suggest that chemicals that are capable of forming covalent bonds with DNA in mammalian tissues are likely to be carcinogenic. The $^{32}P$-postlabelling assay is able to assess whether covalent bond formation between a chemical and DNA *in vivo*, by itself, indicates carcinogenic potential of the particular compound. Obviously, more compounds must be investigated before a definite answer can be given to this question. Ashby (1983) recommended that a substantial proportion of the resources that are currently earmarked for chronic carcinogenicity bioassays, might be better employed on short-term in-vivo evaluations in rodents of chemicals known to be genotoxic from various in-vitro tests. On the basis of our results, the $^{32}P$-postlabelling assay is a suitable short-term test for the detection of carcinogen-DNA adducts in animal tissues, i.e., an assay for potential carcinogenicity of chemicals.

This paper has shown, also, that the $^{32}P$-postlabelling technique is able to assess an important property of chemical carcinogens, that of the formation of persistent DNA adducts, which may play a crucial role in carcinogenesis. When the test is used to detect covalent binding of chemicals to DNA, therefore, an evaluation of adduct persistence should be included, also. Compounds such as 7,12-dimethylbenz[*a*]anthracene or sterigmatocystin, which give rise in animals to highly persistent, essentially irreparable adducts, induce irreversible toxic effects in the genetic material of mammals *in vivo*; human exposure to such compounds must be minimized by appropriate regulatory measures. Since the $^{32}P$-postlabelling assay is capable of determining the extent of covalent DNA binding of chemicals as well as adduct persistence, it may become an important tool for risk assessment of genotoxic chemicals.

## ACKNOWLEDGEMENTS

Our work on the development and application of the $^{32}P$-postlabelling method of adduct analysis was supported by USPHS grants CA 25590, CA 32157 and CA 10893 (P6), awarded by the National Cancer Institute. We wish to thank a large number of colleagues and friends - too numerous to be named here - who have greatly contributed to this work by encouraging us in the development of the method and/or by providing samples used in the standardization of experimental conditions. The contributions of Dr Ramesh C. Gupta to the development of the method are also gratefully acknowledged.

## REFERENCES

Ashby, J. (1983) The unique role of rodents in the detection of possible human carcinogens and mutagens. *Mutat. Res.*, *115*, 177-213

Dipple, A. (1976) *Polynuclear aromatic carcinogens*. In: Searle, C.E., ed., *Chemical Carcinogens*, Washington DC, American Chemical Society, pp. 245-314

Essigmann, J.M., Croy, R.G., Bennett, R.A. & Wogan, G.N. (1982) Metabolic activation of aflatoxin $B_1$: patterns of DNA adduct formation, removal, and excretion in relation to carcinogenesis. *Drug Metab. Rev.*, *13*, 581-602

Gupta, R.C., Reddy, M.V. & Randerath, K. (1982) $^{32}P$-postlabeling analysis of non-radioactive aromatic carcinogen-DNA adducts. *Carcinogenesis*, *3*, 1081-1092

Hemminki, K. (1983) Nucleic acid adducts of chemical carcinogens and mutagens. *Arch. Toxicol.*, *52*, 249-285

Lutz, W.K. (1979) In-vivo covalent binding of organic chemicals to DNA as a quantitative indicator in the process of chemical carcinogenesis. *Mutat. Res.*, *65*, 289-356

Lutz, W.K. (1982) *The covalent binding index - DNA binding* in vivo *as a quantitative indicator for genotoxicity*. In: Bridges, B.A., Butterworth, B.E. & Weinstein, I.B., eds, *Indicators of Genotoxic Exposure* (*Banbury Report 13*), Cold Spring Harbor, NY, Cold Spring Harbor Laboratory, pp. 189-202

Martin, C.N. & Garner, R.C. (1977) Aflatoxin B-oxide generated by chemical or enzymic oxidation of aflatoxin $B_1$ causes guanine substitution in nucleic acids. *Nature*, *267*, 863-865

Miller, E.C., Swanson, A.B., Phillips, D.H., Fletcher, T.L., Liem, A. & Miller, J.A. (1983) Structure-activity studies of the carcinogenicities in the mouse and rat of some naturally occurring and synthetic alkenylbenzene derivatives related to safrole and estragole. *Cancer Res.*, *43*, 1124-1134

Phillips, D.H., Grover, P.L. & Sims, P. (1979) A quantitative determination of the covalent binding of a series of polycyclic hydrocarbons to DNA in mouse skin. *Int. J. Cancer*, *23*, 201-208

Purchase, I.F.H. & van der Watt, J.J. (1970) Carcinogenicity of sterigmatocystin. *Food Cosmet. Toxicol.*, *8*, 289-295

Randerath, K., Reddy, M.V. & Gupta, R.C. (1981) $^{32}$P-labelling test for DNA damage. *Proc. natl Acad. Sci. USA*, *78*, 6126-6129

Randerath, E., Agrawal, H.P., Reddy, M.V. & Randerath, K. (1983) Highly persistent polycyclic aromatic hydrocarbon-DNA adducts in mouse skin: detection by $^{32}$P-postlabeling analysis. *Cancer Lett.*, *19*, 231-239

Reddy, M.V., Gupta, R.C., Randerath, E. & Randerath, K. (1984) $^{32}$P-Postlabeling test for covalent DNA binding of chemicals *in vivo*: application to a variety of aromatic carcinogens and methylating agents. *Carcinogenesis*, *5*, 231-243

Wogan, G.N. (1973) Aflatoxin carcinogenesis. *Meth. Cancer Res.*, *7*, 309-344

# MONITORING ENDOGENOUS NITROSAMINE FORMATION IN MAN

**H. Ohshima & H. Bartsch**
*Unit of Environmental Carcinogens and Host Factors,*
*International Agency for Research on Cancer,*
*Lyon, France*

## SUMMARY

Results from animal experiments and studies in human subjects indicated that the amount of nitrosoproline (NPRO) excreted in the 24-h urine, following ingestion of precursors, is an index for the rate of endogenous nitrosation; this method was found to be sensitive, reproducible and could be satisfactorily applied to human subjects in clinical and field studies.

*N*-Nitrosothiazolidine 4-carboxylic acid (NTCA) and its 2-methyl derivative (NMTCA) were also identified in the urine of human subjects. As the respective amino precursors (thiazolidine 4-carboxylic acids) can be formed by reaction of formaldehyde or acetaldehyde with cysteine, measurement of NTCA and NMTCA in urine may provide a further index for endogenous nitrosation in the human body and may also allow monitoring of exposure of human subjects to aldehydes, nitrate and nitrite.

The yield of nitroso compounds formed endogenously in the human body was shown to be linked to the intake of precursors, but several inhibitors and catalysts, either as pure substances or occurring in complex mixtures, were shown to modify the nitrosation reaction *in vivo*. In particular, ingestion of ascorbic acid after nitrate-rich meals was efficient in lowering human exposure to endogenously formed *N*-nitroso compounds.

A dose-response relationship was established for the formation of NPRO in rats *in vivo*, after concurrent administration of various concentrations of the precursors, L-proline and sodium nitrite. The logarithm of the amount of NPRO formed was found to be proportional to the logarithm of the product of the proline dose and the square of the nitrite dose. On the basis of these results, a kinetic model was formulated allowing the estimation of the daily precursor dose quantity, ([amine][nitrite]$^2$), required to give 50% tumour incidence in rats after two years of feeding. The potential application of this model, for the estimation of carcinogenic risk from endogenously formed *N*-nitrosamines in humans, is discussed.

Our results demonstrate unequivocally the endogenous formation of *N*-nitroso compounds in the human body, the significance of which in human carcinogenesis remains to be established.

## INTRODUCTION

We have developed a simple and sensitive method (*N*-nitrosoproline test) for quantitative estimation of endogenous nitrosation (Ohshima & Bartsch, 1981a; Ohshima *et al.*, 1982a,b). It is based on the findings that certain *N*-nitrosamino acids, such as *N*-nitrosoproline (NPRO), following ingestion of precursors (proline and $NO_3$ or $NO_2$), are excreted unchanged in the urine and faeces (Dailey *et al.*, 1975; Chu & Magee, 1981; Ohshima *et al.*, 1982b).

More recently, *N*-nitrosothiazolidine 4-carboxylic acid (NTCA) and *N*-nitroso-2-methylthiazolidine 4-carboxylic acid (NMTCA) were isolated and identified in the urine of human subjects for the first time (Ohshima *et al.*, 1983a; Tsuda *et al.*, 1983). The origin and the biological significance of these two compounds is not known. However, the easily nitrosatable amino precursors, thiazolidine 4-carboxylic acid and its 2-methyl derivative, may readily be formed by reaction of formaldehyde or acetaldehyde with L-cysteine, respectively; therefore, measurement of NTCA and NMTCA in urine may allow monitoring of exposure of human subjects to precursors like formaldehyde, acetaldehyde, nitrate and nitrite and, also, thus provide a further index for endogenous nitrosation in the human body.

Our method of measuring nitrosated amino acids in urine has allowed study of the kinetics and factors affecting nitrosation *in vivo* in human subjects and in animals after administration of a nitrosatable amino acid and nitrosating agents. Moreover, clinical and field studies in human subjects could be implemented, in order to evaluate, finally, the possible role of endogenous *N*-nitroso compounds in humans related to cancer at specific sites such as the stomach and the oesophagus (Bartsch *et al.*, 1983). A brief summary of the results obtained to date is presented, including the description of a kinetic model that predicts carcinogenic effects caused by endogenous nitrosation.

## MATERIALS AND METHODS

### *Collection of urine samples and analysis of* N-*nitroso compounds*

In one experiment (Fig. 1) 24-h urine samples were collected from one human volunteer over a 30-day period; from day 16 to 30, the diet was supplemented with 100 mg ascorbic acid three times/day after each meal. Urine samples were spiked with *N*-nitrosopipecolic acid as an internal standard, and analyses for NPRO and *N*-nitrososarcosine were carried out according to a published procedure (Ohshima & Bartsch, 1981a; Ohshima *et al.*, 1982b), after the compounds were converted to their methyl esters by diazomethane. A modified solvent extraction was used to analyse NTCA and NMTCA in human urine (Ohshima *et al.*, 1984). Samples were analysed on a tracer 550 gas chromatograph, which was interfaced to a Thermal Energy Analyzer (TEA 502).

**Fig. 1. The excretion of background levels of nitrosamino acids in the urine of a human subject and the effect of ascorbic acid intake (mean values)**

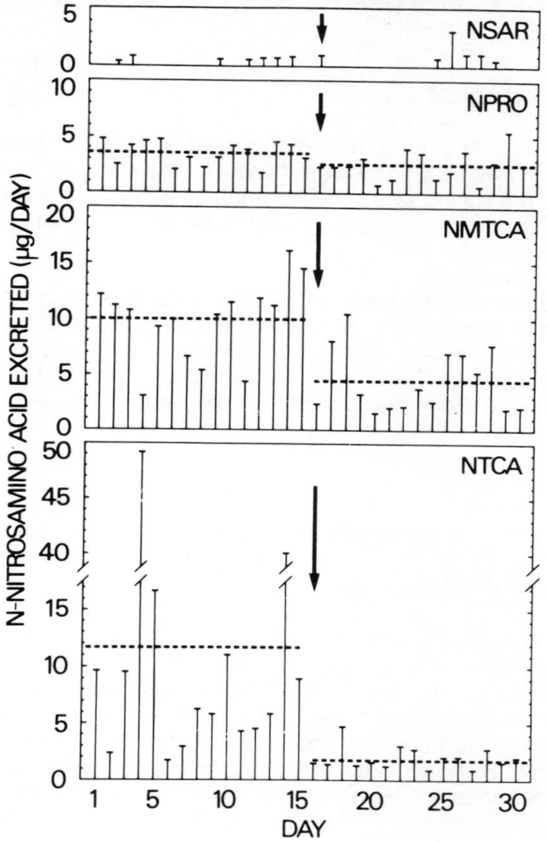


Arrows indicate the start of ascorbic acid intake. NSAR, *N*-nitrososarcosine; NPRO, *N*-nitrosoproline; NMTCA, *N*-nitroso-2-methylthiazolidine 4-carboxylic acid; NTCA, *N*-nitrosothiazolidine 4-carboxylic acid

*The NPRO test* (Ohshima & Bartsch, 1981a; Bartsch *et al.*, 1983)

Human subjects are given: (1) 200 ml beetroot juice (containing 260 mg nitrate); (2) 30 min later, 10 ml of L-proline solution (500 mg); (3) the subjects fast for a further 2 h (during urine collection, foodstuffs rich in nitrate and cured meat, smoked fish and beer are avoided); (4) 24-h urine samples are collected in plastic bottles containing 10 g sodium hydroxide; (5) 100-ml urine aliquots are stored at -20°C prior to analysis; no artefactual formation or degradation of NPRO, *N*-nitrososarcosine or nitrate/nitrite was shown to occur for three months.

The application of the NPRO method to human subjects, to the best of our knowledge, does not involve a health risk, for the following reasons: when ingesting beetroot juice (as a source of nitrate) and proline, (1) the experiments involve only an increased intake of commonly occurring food ingredients at dose ranges that are considered normal daily intake; (2) the absence of carcinogenic and mutagenic effects of NPRO is established; and (3) the natural occurrence of low levels of NPRO in many foodstuffs and in the urine of human subjects is known.

*Formation of NTCA and NMTCA in rats*

Male BD VI rats (250 $\pm$ 25 g) were fasted overnight, but received water *ad libitum*. Excretion rates of NTCA and NMTCA were determined in experiments in which the compound (dissolved in 1 ml of 0.9% saline at a concentration of 100 μg/ml) was given to rats by stomach tube. The urine and faeces were collected separately in plastic tubes containing 5 ml of sodium hydroxide for 24 h.

Endogenous formation of NTCA and NMTCA in rats was studied according to a similar procedure previously reported for that of NPRO (Ohshima *et al.*, 1982b). The precursor compounds were freshly prepared, using distilled water, and given to rats by stomach tube at concentrations given in Table 5 (final volume, 2 ml/rat). The amounts of NTCA and NMTCA formed *in vivo* in rats were estimated from those excreted in the 24-h urine.

## RESULTS AND DISCUSSION

*Kinetics of endogenous NPRO formation*

Kinetic studies were carried out on the formation of NPRO *in vivo* in a human volunteer who had ingested vegetable juice (as a source of nitrate) and 30 min later an aqueous solution of proline (see Materials and methods). Urinary excretion of NPRO was monitored as an index of endogenous nitrosation (Ohshima & Bartsch, 1981a). The formation of NPRO *in vivo* was found to be proportional to the proline dose but strongly dependent on nitrate intake; when more than 260 mg was consumed, excreted NPRO increased exponentially with the dose of nitrate ingested. Formation of NPRO in healthy human subjects was ~ 23 μg/24 h per person, corresponding to 0.002 and 0.004% of the ingested amounts of nitrate (325 mg) and proline (500 mg), respectively.

Endogenous formation of NPRO was observed in a human subject who ingested pickled vegetables (Chinese cabbage) as a source of nitrate and nitrite, followed 1 h later by a dose of proline (Ohshima & Bartsch, 1982). *N*-Nitroso compounds can thus be formed in the body after simultaneous ingestion of nitrate (nitrite)-containing vegetables and nitrosatable precursors, which may be present in foodstuffs or in the stomach.

In humans, in addition to nitrosated amino acids, formation and detection of *N*-nitrosopiperazine in the urine (Bellander *et al.*, 1984) and of *N*-nitrosodimethylamine from aminopyrine (Spiegelhalder & Preussmann, 1984) has been described.

*A kinetic model for tumour induction in rats after feeding of nitrite and amine*

A dose-response study on NPRO formation in rats *in vivo* was carried out after concurrent administration of proline (in the diet) and nitrite (in the drinking water) (Ohshima *et al.*, 1983b). In keeping with the kinetics for nitrosation of proline *in vitro*, the total amount of

NPRO excreted in the urine and faeces was found to be proportional to the dose of proline and to the square of nitrite dose, as described by Equation 1:

$$\log [\text{NPRO}] = 1.394 + 0.439 \log ([\text{proline}][\text{nitrite}]^2) \qquad \text{(Eq. 1)}$$

the amount of NPRO was expressed in nmol per rat per day, and the amounts of proline and nitrite in μmol per rat per day.

Tumour induction by concurrent administration of nitrite and a nitrosatable amine can be regarded as depending mainly (but not exclusively) on: (1) the rate of endogenous nitrosation of the amino compound, and (2) the carcinogenic activity of the *N*-nitrosamine formed *in vivo*. On the basis of results obtained from our dose-response study (Eq. 1) and from available data on nitrosation kinetics and carcinogenic potency of selected nitrosamines in the literature, a kinetic model (Eq. 2) was formulated to estimate the daily precursor doses ([amine][nitrite]$^2$) required to give 50% tumour incidence in rats after two years of the feeding.

$$\log [PD_{50}] = 2.406 + \log [TD_{50}]/0.439 - \log [k_{n,a}/k_{n,p}] \qquad \text{(Eq. 2)}$$

where $TD_{50}$ is the dose of nitrosamine (nmol per kg body weight per day) that will give 50% tumour yield after two years' exposure, while $PD_{50}$ is the precursor dose quantity, [amine][nitrite]$^2$ in μmol$^3$ per kg body weight per day, that will induce tumours in 50% of the animals after two years' feeding; $k_{n,a}$ and $k_{n,p}$ are the pH-dependent, nitrosation-rate constants for amine and proline, respectively.

It has been reported previously (Table 1) that simultaneous administration of amine and nitrite can induce tumours in rats. Therefore, the daily precursor doses from these long-term carcinogenicity experiments could be compared with the $PD_{50}$ values calculated by Equation 2. We assumed that in the previous experiments, the average body weight of the rats was 250 g, with a daily intake of 25 g food and 25 ml water. Only when the precursor dose quantities for the earlier reports were almost equal to or exceeded the calculated $PD_{50}$ values, were tumours induced in a high incidence (Table 1). Thus, the calculated $PD_{50}$ values are compatible with the results of previous carcinogenicity experiments, confirming the usefulness of this kinetic model for quantitative risk estimation in experimental animals.

The kinetic model derived from the study on rats may be used to predict nitrosamine formation in humans. The experimental data in our previous paper (Ohshima & Bartsch, 1981a), in which NPRO formation was quantitatively monitored in a human volunteer after ingestion of beetroot juice (as a source of nitrate) and proline, were compared with the values for NPRO calculated by the kinetic model. The amount of NPRO formed in a typical experiment after ingestion of 260 mg nitrate and 500 mg proline was 9.6 ± 1.4 μg (mean ± SD, n = 4) per 24 h per person, which is compatible with the amount of 8.6 μg per 24 h calculated by Equation 1, assuming that 1% of the total nitrate ingested in beetroot juice can be converted to nitrite *in vivo* (Spiegelhalder *et al.*, 1976; Tannenbaum *et al.*, 1976). At higher doses of nitrate intake, Equation 1 tended to underestimate the amount of NPRO actually formed. Thus, the kinetic model may predict formation of nitrosamines.

Provided that the following assumptions are valid, this model may be used for risk estimations for the formation of *N*-nitrosodimethylamine (NDMA) and *N*-nitrosopyrrolidine (NPYR) in humans: (1) the rate of nitrosation in the human body is similar to that in rats, although it

**Table 1. Comparison of precursor doses used in previous carcinogenicity experiments with those calculated from our kinetic model (Eq. 2) (table reproduced from Ohshima *et al.*, 1983b, with permission)**

| Amine (%) | $NaNO_2$ (%) | Precursor dose quantity[a] [amine] $[nitrite]^2$ ($\mu mol^3$/kg bw/day) | Duration of treatment | Tumour incidence[b] (%) | Calculated $PD_{50}$ [amine] $(nitrite]^2$ ($\mu mol^3$/kg bw/day) | Reference |
|---|---|---|---|---|---|---|
| Piperidine | | | | | | |
| 0.5 | 0.5 | $1.9 \times 10^{10}$ | 76 days | 0 | $5.8 \times 10^{11}$ | Sander *et al.* (1972) |
| Methylbenzylamine | | | | | | |
| 0.25 | 0.32 | $2.7 \times 10^9$ | | 100 | | |
| 0.25 | 0.16 | $6.8 \times 10^8$ | | 100 | | |
| 0.25 | 0.06 | $1.7 \times 10^8$ | | 100 | | |
| 0.25 | 0.06 | $9.6 \times 10^7$ | 152 days | 0 | $4.8 \times 10^8$ | Sander (1971) |
| 0.25 | 0.04 | $4.4 \times 10^7$ | | 0 | | |
| 0.25 | 0.02 | $1.1 \times 10^7$ | | 0 | | |
| 0.25 | 0.01 | $2.7 \times 10^6$ | | 0 | | |
| 0.25 | 0 | 0 | | 0 | | |
| Morpholine | | | | | | |
| 1 | 0.3 | $1.3 \times 10^{10}$ | 2 years | 65 | $1.5 \times 10^8$ | Mirvish *et al.* (1976) |
| 0.1 | 0.1 | $1.5 \times 10^7$ | | 97 | | |
| 0.005 | 0.1 | $7.6 \times 10^6$ | | 59 | | |
| 0.0005 | 0.1 | $7.6 \times 10^5$ | | 28 | | |
| 0.1 | 0.005 | $3.8 \times 10^5$ | 19-124 weeks | 3 | $1.5 \times 10^8$ | Shank & Newberne (1976) |
| 0.005 | 0.005 | $1.9 \times 10^4$ | | 2 | | |
| 0.1 | 0.0005 | $3.8 \times 10^3$ | | 1 | | |
| Aminopyrine | | | | | | |
| 0.1 | 0.1 | $1.4 \times 10^7$ | 30 weeks | 97 | | Lijinsky *et al.* (1973) |
| 0.25 | 0.25 | $2.2 \times 10^5$ | 50 weeks | 93 | $2.2 \times 10^7$ | |

[a] Calculated by assuming that in the previous experiments the average body weight (bw) of rats is 250 g, with a daily intake of 25 g food and 25 ml water

[b] Incidence in major target organ

is known that the pH of human stomach contents can vary from 1 to 7; (2) humans are as susceptible as rats to cancer induction by nitrosamines [the basis for the estimation of risk in humans reported by the National Research Council (1981), which estimated daily doses of NDMA and NPYR that correspond to various levels of risk for humans after lifetime exposure; their calculations were based on the dose-response studies in rats reported by Terracini *et al.* (1967) and Preussmann *et al.* (1976)]; (3) the effect of catalysts and inhibitors, or other factors that may affect nitrosation *in vivo*, is not taken into consideration.

Equation 2 was used to estimate the precursor dose quantities, [amine]$[nitrite]^2$, necessary to form various amounts of NDMA or NPYR that represent various levels of risk for humans. Data reported on the basis of body weight were used, although the National Research Council (1981) described carcinogenic doses on the basis of body surface area. Table 2 gives the calculated precursor dose quantities of dimethylamine and aminopyrine plus nitrite ($\mu mol^3$ per kg body weight per day) necessary for the formation of NDMA in humans and of pyrrolidine plus nitrite for NPYR. Thus, it was possible to calculate the amount of nitrite (mg per 60 kg body weight per person per day) necessary to form nitrosamines at amounts that result in a $10^{-4}$ level of risk for humans. The daily intakes of amine (dimethylamine, aminopyrine and pyrrolidine), by a subject of 60 kg body weight, were assumed to be 1, 0.1 and 0.01 g, respectively. It was further assumed that 1% of the total nitrate ingested (from vegetables and water) can be converted to nitrite *in vivo* (Spiegelhalder *et al.*, 1976; Tannenbaum *et al.*, 1976), and the daily nitrate intake which would yield that amount of nitrite was calculated (Table 3).

**Table 2. Estimated precursor dose quantities necessary for formation *in vivo* of *N*-nitrosodimethylamine (NDMA) and *N*-nitrosopyrrolidine (NPYR) that correspond to various levels of risk for humans[a]**

| Level of risk | Model[a,b] | Dose of NDMA[a] [nmol/kg bw/day] | Precursor dose quantity[c,d] | | Dose of NPYR[a] (nmol/kg bw per day) | Precursor dose quantity[e,d] [PYR] [nitrite]$^2$ (μmol$^3$/kg bw/day) |
|---|---|---|---|---|---|---|
| | | | [DMA] [nitrite]$^2$ (μmol$^3$/kg bw/day) | [AP] [nitrite]$^2$ (μmol$^3$/kg bw/day) | | |
| $10^{-2}$ | A | 789 | $2.5 \times 10^{10}$ | $2.2 \times 10^{7}$ | $3.1 \times 10^{3}$ | $1.5 \times 10^{11}$ |
| | B | | | | $8.3 \times 10^{2}$ | $7.3 \times 10^{9}$ |
| $10^{-4}$ | A | 7.89 | $7.2 \times 10^{5}$ | $6.2 \times 10^{2}$ | 31 | $4.1 \times 10^{6}$ |
| | B | | | | 72.2 | $2.8 \times 10^{7}$ |
| $10^{-6}$ | A | 0.0789 | 20 | $1.7 \times 10^{-2}$ | 0.31 | $1.1 \times 10^{2}$ |
| | B | | | | 6.52 | $1.2 \times 10^{5}$ |
| $10^{-8}$ | A | 0.000789 | $5.6 \times 10^{-4}$ | $4.8 \times 10^{-7}$ | $3.1 \times 10^{-3}$ | $3.1 \times 10^{-2}$ |
| | B | | | | 0.59 | $4.9 \times 10^{2}$ |

[a] Based on data from National Research Council (1981)

[b] Model used to estimate daily doses of nitrosamine that correspond to specific levels for humans (National Research Council, 1981)
A, Multihit with linear extrapolation from the risk level of $10^{-2}$
B, Multihit with 95% lower confidence limit on the estimated constant, h, for the multihit model

[c] Calculated by Equation 2 in this text, based on daily doses estimated by National Research Council (1981)

[d] DMA, dimethylamine; AP, aminopyrine; PYR, pyrrolidine

**Table 3. Estimated doses of nitrite and nitrate required to pose a $10^{-4}$ level of risk for humans**

| Amine | Precursor dose (μmol$^3$/60 kg per day) | Amine dose (g/person per day) | Sodium nitrite (mg/person per day) | Sodium nitrate (mg/person per day) |
|---|---|---|---|---|
| Dimethylamine | $4.7 \times 10^{7a}$ | 1 0.1 0.01 | 3.0 | 300 |
| | | | 9.6 | 960 |
| | | | 30 | 3 000 |
| Aminopyrine | $3.7 \times 10^{4a}$ | 1 | 0.20 | 20 |
| | | 0.1 | 0.64 | 64 |
| | | 0.01 | 2.0 | 200 |
| Pyrrolidine | $2.5 \times 10^{8a}$ | 1 | 9.2 | 920 |
| | | 0.1 | 29 | 2 900 |
| | | 0.01 | 92 | 9 200 |
| | $1.7 \times 10^{9b}$ | 1 | 24 | 2 400 |
| | | 0.1 | 76 | 7 600 |
| | | 0.01 | 240 | 24 000 |

[a] On the basis of multihit with linear extrapolation from the risk level of $10^{-2}$

[b] On the basis of multihit with 95% lower confidence limit on the estimated constant, h, for the multihit

The estimation of daily precursor doses necessary for endogenous formation of NDMA and NPYR that correspond to various levels of risk for humans is highly speculative; it may be wrong by several orders of magnitude, as the assumptions on which these calculations are made are not proven. However, it is noteworthy that similar figures have been reported by the National Research Council (1981), either for the precursor doses calculated to be necessary to give a carcinogenic risk of $10^{-4}$ or for actual nitrate intake in several countries, ranging from 240 to 800 mg/person per day. In addition, a wide range of drugs, agricultural chemicals (insecticides and herbicides), industrial materials and food components are known to react with nitrite to form *N*-nitroso compounds (Mirvish, 1975; Lijinsky, 1980; National Research Council, 1981). Therefore, a risk to humans, that may not be negligible, may result from simultaneous ingestion of nitrate and of 1 g/person per day of a nitrosatable amine like those listed in Table 3, particularly aminopyrine, which is used in human medicine as an analgesic and antipyretic drug. Gombar *et al.* (1983) have recently reported that the administration of $^{14}$C-aminopyrine with an extremely low level of nitrite resulted in an excretion of 7-[$^{14}$C-methyl]-guanine in the urine of rats. Spiegelhalder and Preussman (1984) detected NDMA in the urine of subjects who had ingested aminopyrine, but only after simultaneous administration of ethanol in order to reduce the metabolic conversion rate of the nitrosamine.

These observations further confirm that aminopyrine readily undergoes nitrosative deamination *in vivo* to yield NDMA.

### *Effects of inhibitors and catalysts on nitrosation of proline*

*Studies on experimental animals*. Administration of a 10-fold molar excess of ascorbic acid *versus* that of the proline or nitrite inhibited NPRO formation in rats by about 98%; α-tocopherol (vitamin E) was about 50% effective (Ohshima *et al.*, 1982b). After co-administration to rats of polyphenolic compounds (present in large quantities in the human diet) with proline and nitrite *per os*, both catalysis (resorcinol and catechin) and inhibition (chlorogenic acid) of the nitrosation of proline were observed (Pignatelli *et al.*, 1982). Inhibitory effects of polyphenol mixtures in betel nuts and beer on nitrosation have been observed (Pignatelli *et al.*, 1983; Stich *et al.*, 1983).

Endogenous transnitrosation from *N*-nitrosodiphenylamine to proline in rats in the presence or absence of thiocyanate (a catalyst of nitrosation) was demonstrated (Ohshima *et al.*, 1982a). *N*-Nitrosodiphenylamine produced bladder cancer in rats, and transnitrosation may be involved.

*Studies in humans*. In one human subject, intake of 1 g of ascorbic acid simultaneously with the precursors (proline and nitrate) totally inhibited nitrosation of proline *in vivo*. α-Tocopherol (500 mg) inhibited nitrosation *in vivo* by only about 50%. When orange juice (a possible source of vitamin C) was taken at the end of each nitrate-rich meal, followed by proline, urinary excretion of NPRO was also markedly suppressed (Ohshima & Bartsch, 1981a,b).

In a recently completed study (Lu *et al.*, 1984), conducted in the People's Republic of China, it was found that in a high-incidence area for oesophageal cancer (Linxian), with exposure to endogenous NPRO, NTCA and nitrate at a higher level than in the low-incidence Fanxian county, ingestion of vitamin C (three times a day, 100 mg after each meal) reduced the amount of NPRO and NTCA formed *in vivo* (and, by inference, total *N*-nitroso compounds) to the levels seen in the Fanxian subjects.

We (Bartsch *et al.*, 1984; Ohshima *et al.*, 1984) and others (Brunnemann *et al.*, 1984; Ladd *et al.*, 1984) have shown that, in smoking subjects, excretion of nitrosamino acids (NTCA, NMTCA, NPRO) is elevated compared to that of nonsmokers. This effect could be due to several factors, i.e., to increased saliva production rate (with more $NO_2^-$ reaching the stomach), or to exposure to formaldehyde through cigarette smoke, which may react subsequently with cysteine to form thiazolidine 4-carboxylic acid, or to higher salivary thiocyanate levels in smokers.

*Identification of new* N-*nitroso compounds in human urine*

During analyses of human urine samples by gas chromatography-thermal energy analysis for the presence of NPRO and *N*-nitrososarcosine, both used as indicators of endogenous nitrosation, several unidentified substances have frequently been detected after derivatization with diazomethane. The peaks disappeared after treatment with either ultra-violet irradiation (364 nm) or HBr/acetic acid, confirming the presence of *N*-nitroso compounds. By several separative procedures and by comparison with authentic material, a major unknown *N*-nitroso compound was shown to be *N*-nitrosothiazolidine 4-carboxylic acid (NTCA) (Ohshima *et al.*, 1983a; Tsuda *et al.*, 1983).

Two other unknown *N*-nitroso compounds were identified in a manner similar to that for NTCA, as isomers of *N*-nitroso-2-methylthiazolidine 4-carboxylic acid (NMTCA) (Tsuda *et al.*, 1983, 1984; Ohshima *et al.*, 1984). The isomeric reaction products of acetaldehyde and cysteine, following nitrosation, had the same retention times by gas chromatography-thermal energy analysis (after derivatization with diazomethane). By $^{1}H$-Fourier transform nuclear magnetic resonance examination of the synthetic products, it was possible to assign *trans* and *cis* orientation of the methyl group relative to the 4-carboxylic acid function (Ohshima *et al.*, 1984).

*Possible origin of NTCA and NMTCA in urine*

The amount of NTCA detected in 24-h urine collected from one human volunteer over a 15-day period varied from 2.4 to 49 μg (mean 11.9 μg)/day; NMTCA varied from traces to 16.1 μg (mean 9.9 μg)/day. In comparison with the amount of NPRO in the same urine samples, NTCA and NMTCA occurred at about 3.1- and 3.2-fold concentrations, respectively. NTCA and NMTCA have been detected in many human urine samples so far collected in the People's Republic of China, Finland, France and Italy.

In order to study the origin of NTCA and NMTCA in human urine, animal experiments have been carried out. When preformed NTCA or NMTCA was given orally to rats, more than 90% of the dose was recovered unchanged in the urine within the first 24-h (Table 4). Administration of thiazolidine 4-carboxylic acid or its 2-methyl derivative together with nitrite resulted in a significant increase in the excretion of NTCA and NMTCA. In addition, NTCA and NMTCA were found to be easily formed *in vivo* in rats after a dose of L-cysteine or nitrite together with formaldehyde or acetaldehyde, respectively (Table 5).

**Table 4. Urinary and faecal excretion of NTCA and NMTCA administered orally to rats**[a]

| Compound | Dose (μg/rat) | No. of rats | % Excretion | | |
|---|---|---|---|---|---|
| | | | Urine | Faeces | Total |
| NTCA | 100 μg | 4 | 92.9 ± 1.2 | 1.6 ± 0.6 | 94.4 ± 1.5 |
| NMTCA | 100 μg | 4 | 91.3 ± 1.3 | 1.5 ± 0.4 | 92.8 ± 1.0 |

[a] NTCA, *N*-nitrosothiazolidine 4-carboxylic acid; NMTCA, *N*-nitroso-2-methylthiazolidine 4-carboxylic acid

**Table 5. Formation of NTCA and NMTCA *in vivo* in rats[a]**

| Reagent (20 μmol/rat) | Number of rats | NMTCA (nmol/rat) | NTCA (nmol/rat) |
|---|---|---|---|
| Control (None, FA, AA, Cys, FA + Cys, AA + Cys, TCA, MTCA, alone) | 3 - 4 | < 0.06 | < 0.27 |
| $NaNO_2$ | 3 | 0.72 ± 0.47 | 0.63 ± 0.63 |
| Cys + $NaNO_2$ | 3 | ND[b] | 2.67 ± 1.10 |
| FA + $NaNO_2$ | 3 | ND[b] | 4.96 ± 1.42 |
| AA + $NaNO_2$ | 3 | 0.22 ± 0.09 | 0.77 ± 0.38 |
| FA + Cys + $NaNO_2$ | 3 | ND[b] | 913 ± 193 |
| AA + Cys + $NaNO_2$ | 3 | 331 ± 125 | ND[b] |
| TCA + $NaNO_2$ | 4 | -[c] | 4451 ± 325 |
| MTCA + $NaNO_2$ | 4 | 2061 ± 640 | -[c] |

[a] NTCA, *N*-nitrosothiazolidine 4-carboxylic acid; NMTCA, *N*-nitroso-2-methylthiazolidine 4-carboxylic acid; FA, formaldehyde; AA, acetaldehyde; Cys, L-cysteine; TCA, thiazolidine 4-carboxylic acid; MTCA, 2-methylthiazolidine 4-carboxylic acid

[b] ND, not detected

[c] -, not determined

From these results, three possible sources of NTCA and NMTCA in human urine may be considered: (1) intake of preformed *N*-nitroso compounds from food and subsequent excretion in urine [NTCA was recently detected in smoked food products (Helgason *et al.*, 1984]; (2) intake of (methyl) thiazolidine 4-carboxylic acid and subsequent nitrosation *in vivo*; (3) endogenous two-step synthesis by reaction of L-cysteine, an aldehyde and a nitrosating agent. Although we have not yet analysed their content in the diet ingested during the urine collection, their presence in the urine may originate partly from preformed compounds. However, as verified in preliminary experiments, intake of a diet supplemented with ascorbic acid reduced urinary excretion of NTCA and NMTCA (Fig. 1). As ascorbic acid is a well-known inhibitor of nitrosation, this suggests that some of these compounds may be formed endogenously (Ohshima *et al.*, 1984).

Thiazolidine 4-carboxylic acid and its 2-methyl derivative, the parent amine precursors, are known to be formed by in-vitro reaction of cysteine with formaldehyde and acetaldehyde; they have also been detected in the urine of rats treated with either formaldehyde (Neely, 1964) or acetaldehyde (Hemminki, 1982). Recently, structurally related *N*-nitrosothiazolidine has been reported in fried bacon (Kimoto *et al.*, 1982; Gray *et al.*, 1984) and smoked food products (Helgason *et al.*, 1984). Presumably this compound was formed in a similar manner to NTCA, i.e., by reaction of formaldehyde with cysteamine (instead of cysteine), and subsequent nitrosation by nitrite or by decarboxylation of NTCA.

*Toxicological significance of NTCA and NMTCA in urine*

Toxic and other adverse biological effects of NTCA and NMTCA have not been reported. In view of the apparently wide exposure of the general population (our studies), and the possibility that NTCA and NMTCA may undergo decarboxylation *in vivo* to yield *N*-nitrosothiazolidine and its 2-methyl derivative, further studies on their biological significance are especially desirable. In addition, the easily nitrosatable amino precursors, thiazolidine 4-carboxylic acid and its 2-methyl derivative, appear to be readily formed by reaction of formaldehyde or acetaldehyde with cysteine *in vivo* and *in vivo* (Neely, 1964; Hemminki, 1982); thus, by measuring the amounts of NTCA and NMTCA excreted in urine, it may be possible to monitor human exposure to precursors like formaldehyde, acetaldehyde and $NO_3^-/NO_2^-$.

**ACKNOWLEDGEMENTS**

We wish to thank J.C. Béréziat, M.C. Bourgade and J. Michelon for technical assistance and Y. Granjard for secretarial work. The TEA detector was lent by the National Cancer Institute of the United States under contract NO1 CP-55715.

## REFERENCES

Bartsch, H., Ohshima, H., Muaoz, N., Crespi, M. & Lu, S.H. (1983) *Measurement of endogenous nitrosation in humans: potential applications of a new method and initial results.* In: Harris, C.C. & Autrup, H.N., eds, *Human Carcinogenesis*, New York, Academic Press, pp. 833-856

Bartsch, H., Ohshima, H., Muaoz, N., Crespi, M., Cassale, V., Ramazotti, V., Lambert, R., Minaire, Y., Forichon, J. & Walters, C.L. (1984) In vivo *nitrosation, precancerous lesions and cancers of the gastrointestinal tract: on-going studies and preliminary results.* In: O'Neill, I.K., von Borstel, R.C., Long, J.E., Miller, C.T. & Bartsch, H., eds, N-*Nitroso Compounds: Occurrence, Biological Effects and Relevance to Human Cancer* (*IARC Scientific Publications No. 57),* Lyon, International Agency for Research on Cancer, pp. 955-962

Bellander, B.T.D., Osterdahl, B.-G. & Hagmar, L. (1984) *Nitrosation of piperazine in man.* In: O'Neill, I.K., von Borstel, R.C., Long, J.E., Miller, C.T. & Bartsch, H., eds, N-*Nitroso Compounds: Occurrence, Biological Effects and Relevance to Human Cancer* (*IARC Scientific Publications No. 57),* Lyon, International Agency for Research on Cancer, pp. 171-178

Brunnemann, K.D., Scott, J.C., Haley, N.J. & Hoffmann, D. (1984) *Endogenous formation of* N-*nitrosoproline upon cigarette smoke inhalation.* In: O'Neill, I.K., von Borstel, R.C., Long, J.E., Miller, C.T. & Bartsch, H., eds, N-*Nitroso Compounds: Occurrence, Biological Effects and Relevance to Human Cancer* (*IARC Scientific Publications No. 57*), Lyon, International Agency for Research on Cancer, pp. 819-828

Chu, C. & Magee, P.N. (1981) Metabolic fate of nitrosoproline in the rat. *Cancer Res.*, *41*, 3653-3657

Dailey, R.E., Braunberg, R.C. & Blaschka, A.M. (1975) The absorption, distribution and excretion of ($^{14}$C)-nitrosoproline by rats. *Toxicology*, *3*, 23-28

Gray, J.I., Skrypec, D.J., Mandagere, A.M., Booren, A.M. & Pearson, A.M. (1984) *Further factors influencing* N-*nitrosamine formation in bacon*. In: O'Neill, I.K., von Borstel, R.C., Long, J.E., Miller, C.T. & Bartsch, H., eds, N-*Nitroso Compounds: Occurrence, Biological Effects and Relevance to Human Cancer* (*IARC Scientific Publications No. 57*), Lyon, International Agency for Research on Cancer, pp. 301-309

Gombar, C.T., Zubroff, J., Strahan, G.D. & Magee, P.N. (1983) Measurement of 7-alkylguanine as an estimate of the amount of dimethylnitrosamine formed following administration of aminopyrine and nitrite to rats. *Cancer Res.*, *43*, 5077-5080

Helgason, T., Ewen, S.W.B., Jaffray, B., Stowers, J.M., Outram, J.R. & Pollock, J.R.A. (1984) *Nitrosamines in smoked meats and their relation to diabetes*. In: O'Neill, I.K., von Borstel, R.C., Long, J.E., Miller, C.T. & Bartsch, H., eds, N-*Nitroso Compounds: Occurrence, Biological Effects and Relevance to Human Cancer* (*IARC Scientific Publications No. 57*), Lyon, International Agency for Research on Cancer, pp. 911-920

Hemminki, K. (1982) Urinary sulfur containing metabolites after administration of ethanol, acetaldehyde and formaldehyde to rats. *Toxicol. Lett.*, *11*, 1-6

Kimoto, W.I., Pensabene, J.W. & Fiddler, W. (1982) Isolation and identification of *N*-nitrosothiazolidine in fried bacon. *J. Agric. food Chem.*, *30*, 757-760

Lijinsky, W. (1980) Significance of in-vivo formation of *N*-nitroso compounds. *Oncology*, *37*, 223-226

Lijinsky, W., Taylor, H.W., Snyder, C. & Nettesheim, P. (1973) Malignant tumours of liver and lung in rats fed aminopyrine or heptamethyleneimine together with nitrite. *Nature*, *244*, 176-178

Ladd, K.L., Newmark, H.L. & Archer, M.C. (1984) *Increased endogenous nitrosation in smokers*. In: O'Neill, I.K., von Borstel, R.C., Long, J.E., Miller, C.T. & Bartsch, H., eds, N-*Nitroso Compounds: Occurrence, Biological Effects and Relevance to Human Cancer* (*IARC Scientific Publications No. 57*), Lyon, International Agency for Research on Cancer, pp. 811-817

Lu, S.H., Bartsch, H. & Ohshima, H. (1984) *Recent studies on nitrosamine and oesophageal cancer*. In: O'Neill, I.K., von Borstel, R.C., Long, J.E., Miller, C.T. & Bartsch, H., eds, N-*Nitroso Compounds: Occurrence, Biological Effects and Relevance to Human Cancer* (*IARC Scientific Publications No. 57*), Lyon, International Agency for Research on Cancer, pp. 947-953

Mirvish, S.S. (1975) Formation of *N*-nitroso compounds: chemistry, kinetics, and in-vivo occurrence. *Toxicol. appl. Pharmacol.*, *31*, 325-351

Mirvish, S.S., Pelfrene, A.F., Garcia, H. & Shubik, P. (1976) Effect of sodium ascorbate on tumor induction in rats treated with morpholine and sodium nitrite, and with nitrosomorpholine. *Cancer Lett.*, *2*, 101-108

National Research Council (1981) *The health effects of nitrate, nitrite and* N-*nitroso compounds. Part I of a 2-part study by the Committee on Nitrite and Alternative Curing Agents in Food*, Assembly of Life Sciences, Washington, DC, National Academy Press

Neely, W.B. (1964) The metabolic fate of formaldehyde-$^{14}C$ intraperitoneally administered to the rat. *Biochem. Pharmacol.*, *13*, 1137-1142

Ohshima, H. & Bartsch, H. (1981a) Quantitative estimation of endogenous nitrosation in humans by monitoring *N*-nitrosoproline excreted in the urine. *Cancer Res.*, *41*, 3658-3662

Ohshima, H. & Bartsch, H. (1981b) *The influence of vitamin C on the in-vivo formation of nitrosamines*. In: Counsell, J.N. & Hornig, D.H., eds, *Vitamin C (Ascorbic Acid)*, London, Applied Science Publishers, pp. 215-224

Ohshima, H. & Bartsch, H. (1982) *Quantitative estimation of endogenous nitrosation in humans by measuring excretion of* N-*nitrosoproline in the urine*. In: Sugimura, T. & Kondo, S., eds, *Environmental Mutagens and Carcinogens*, New York, Alan R. Liss, pp. 577-585

Ohshima, H., Béréziat, J.C. & Bartsch, H. (1982a) *Measurement of endogenous* N-*nitrosation in rats and humans by monitoring urinary and faecal excretion of* N-*nitrosamino acids*. In: Bartsch, H., O'Neill, I.K., Castegnaro, M. & Okada, M., eds, N-*Nitroso Compounds: Occurrence and Biological Effects* (*IARC Scientific Publications No. 41*), Lyon, International Agency for Research on Cancer, pp. 397-411

Ohshima, H., Béréziat, J.C. & Bartsch, H. (1982b) Monitoring *N*-nitrosamino acids excreted in the urine and faeces of rats as an index for endogenous nitrosation. *Carcinogenesis*, *3*, 115-120

Ohshima, H., Friesen, M., O'Neill, I.K. & Bartsch, H. (1983a) Presence in human urine of a new *N*-nitroso compound, *N*-nitrosothiazolidine 4-carboxylic acid. *Cancer Lett.*, *20*, 183-190

Ohshima, H., Mahon, G.A.T., Wahrendorf, J. & Bartsch, H. (1983b) A dose-response study of *N*-nitrosoproline formation in rats and a deduced kinetic model for predicting carcinogenic effects caused by endogenous nitrosation. *Cancer Res.*, *43*, 5072-5076

Ohshima, H., O'Neill, I.K., Friesen, M., Pignatelli, B. & Bartsch, H. (1984) *Presence in human urine of new sulfur-containing* N-*nitrosamino acid;* N-*nitrosothiazolidine 4-carboxylic acid and* N-*nitroso 2-methylthiazolidine 4-carboxylic acid*. In: O'Neill, I.K., von Borstel, R.C., Long, J.E., Miller, C.T. & Bartsch, H., eds, N-*Nitroso Compounds: Occurrence, Biological Effects and Relevance to Human Cancer* (*IARC Scientific Publications No. 57*), Lyon, International Agency for Research on Cancer, pp. 77-85

Pignatelli, B., Béréziat, J.C., Descotes, G. & Bartsch, H. (1982) Catalysis of nitrosation *in vitro* and *in vivo* in rats by catechin and resorcinol and inhibition by chlorogenic acid. *Carcinogenesis*, *3*, 1045-1049

Pignatelli, B., Scriban, R., Descotes, G. & Bartsch, H. (1983) Inhibition of endogenous nitrosation of proline in rats by lyophilized beer constituents. *Carcinogenesis*, *4*, 491-494

Preussmann, R., Schmähl, D., Eisenbrand, G. & Port, R. (1976) *Dose-response study with* N-*nitrosopyrrolidine and some comments on risk evaluation of environmental* N-*nitroso compounds*. In: Tinbergen, B.J. & Krol, B., eds, *Proceedings of the Second International Symposium on Nitrite in Meat Products*, Wageningen, Centre for Agricultural Publishing and Documentation, pp. 261-268

Sander, J. (1971) Weitere Versuche zur Tumorinduktion durch orale Applikation niederer Dosen von *N*-Methylbenzylamin und Natriumnitrit. *Z. Krebsforsch.*, *76*, 93-96

Sander, J., Buerkle, G. & Schweinsberg, F. (1972) *Induction of tumors by nitrite and secondary amines or amides.* In: Nakahara, W., Takayama, S., Sugimura, T. & Odashima, S., eds, *Topics in Chemical Carcinogenesis*, Tokyo, University of Tokyo Press, pp. 297-310

Shank, R.C. & Newberne, P.M. (1976) Dose-response study of the carcinogenicity of dietary sodium nitrite and morpholine in rats and hamsters. *Food Cosmet. Toxicol.*, *14*, 1-8

Spiegelhalder, B. & Preussmann, R. (1984) *In-vivo formation of NDMA in humans after amidopyrine intake.* In: O'Neill, I.K., von Borstel, R.C., Long, J.E., Miller, C.T. & Bartsch, H., eds, N-*Nitroso Compounds: Occurrence, Biological Effects and Relevance to Human Cancer* (*IARC Scientific Publications No. 57*), Lyon, International Agency for Research on Cancer, pp. 179-183

Spiegelhalder, B., Eisenbrand, G. & Preussmann, R. (1976) Influence of dietary nitrate on nitrite content of human saliva: possible relevance to in-vivo formation of *N*-nitroso compounds. *Food Cosmet. Toxicol.*, *14*, 545-548

Stich, H., Ohshima, H., Pignatelli, B., Michelon, J. & Bartsch, H. (1983) Inhibitory effect of betel nut extracts on endogenous nitrosation in humans. *J. natl Cancer Inst.*, *70*, 1047-1050

Tannenbaum, S.R., Weisman, M. & Fett, D. (1976) The effect of nitrate intake on nitrite formation in human saliva. *Food Cosmet. Toxicol.*, *14*, 549-552

Terracini, B., Magee, P.N. & Barnes, J.M. (1967) Hepatic pathology in rats on low dietary levels of dimethylnitrosamine. *Br. J. Cancer*, *21*, 559-565

Tsuda, M., Hirayama, T. & Sugimura, T. (1983) Presence of *N*-nitroso-*L*-thioproline and *N*-nitroso-*L*-methylthioprolines in human urine as major *N*-nitroso compounds. *Gann*, *74*, 331-333

Tsuda, M., Kakizoe, T., Hirayama, T. & Sugimura, T. (1984) *New type of* N-*nitrosamino acids,* N-*nitroso*-L-*thioproline and* N-*nitroso*-L-*methylthioprolines, found in human urine as major* N-*nitroso compounds.* In: O'Neill, I.K., von Borstel, R.C., Long, J.E., Miller, C.T. & Bartsch, H., eds, N-*Nitroso Compounds: Occurrence, Biological Effects and Relevance to Human Cancer* (*IARC Scientific Publications No. 57*), Lyon, International Agency for Research on Cancer, pp. 87-94

# BACTERIAL URINARY ASSAY IN MONITORING EXPOSURE TO MUTAGENS AND CARCINOGENS

**H. Vainio**
*International Agency for Research on Cancer, Lyon, France*

**M. Sorsa**
*Institute of Occupational Health, Helsinki, Finland*

**K. Falck**
*Labsystems Oy, Helsinki, Finland*

**SUMMARY**

The bacterial bioassay procedure provides a sensitive test for the presence of mutagenic activity in urine. Its sensitivity for detecting the presence of individual chemicals may not be as high as that of specific analytical methods, but it has the following advantages: (1) many substances and metabolites may be active in a single assay, making it possible to detect mutagenic activity from unanticipated sources; (2) biological activity is demonstrated, rather than simply the presence of substances; or (3) the assay may also reflect the 'integrated' effect of multiple substances, although this capability has not been well characterized and (4) the assay can be easily coupled with a chemical analysis. The chief disadvantages of the test system are lack of sensitivity for certain specific substances, as compared to chemical techniques, and possible interference from substances normally present in urine (such as amino acids). The urinary mutagenicity assay is most useful when exposure to carcinogens and mutagens is suspected but when the specific chemical is unknown, when chemical analytical techniques are not available or when exposure is to undefined complex mixtures.

## INTRODUCTION

Recently, increasing attention has been paid to the development of monitoring methods by which human exposure to mutagens and carcinogens can be detected (Vainio *et al.*, 1981; Bridges *et al.*, 1982; Sorsa *et al.*, 1982b). Traditional methods have involved determinations of ambient levels of toxic substances in the environment. Biological monitoring involves evaluation of the body's absorption of chemicals to which it is exposed. Routine biological monitoring comprises regular measurements of validated indicators of exposure or effect.

Many alkylating anticancer drugs, aromatic amines and polycyclic aromatic hydrocarbons can cause the urine of treated animals to become mutagenic (Legator *et al.*, 1982). In most cases, the mutagenic activity is detectable only after the urine samples have been concentrated and treated with deconjugating enzymes, such as β-glucuronidase and arylsulfatase. The resulting metabolites are excreted as conjugates that are, normally, biologically inactive, although there are a few well-known exceptions in which reactive metabolites are generated, e.g., the glucuronide and sulfate conjugates of *N*-hydroxy-*N*-arylacetamides (Irving, 1971) and *N*-hydroxyphenacetin (Mulder *et al.*, 1977).

Body fluids, particularly human urine, have been assayed for mutagenicity for about ten years (Siebert & Simon, 1973). The most frequently used biological indicators of mutagenicity have been the bacterial strains routinely applied in mutagenicity testing. In some recent attempts, human lymphocyte cultures have also been used to test urine (Stiller *et al.*, 1982, 1983). The observation by Yamasaki and Ames (1977) that smokers' urine is mutagenic in the *Salmonella*/liver homogenate assay prompted a number of studies of mutagenic activity in populations exposed to various mutagens and carcinogens.

This review lists some of the advantages and disadvantages of assaying mutagenicity in urine, describes the situations that appear most suitable for application of these methods and shows how they can be used as tools for preventive medicine.

## BIOLOGICAL MONITORING OF HUMANS FOR EXPOSURE TO MUTAGENS AND/OR CARCINOGENS

### *Cigarette smoke*

Smoking is perhaps the best-studied modern carcinogenic risk factor - one for which the most extensive epidemiological evidence has been accumulated over the last three decades. The results of many studies on the mutagenicity and carcinogenicity of cigarette smoke and cigarette smoke condensate have led to the realization that smoking is one of the most important sources of human exposure to mutagens and/or carcinogens (Bridges *et al.*, 1979). Accordingly, Yamasaki and Ames (1977), employing XAD-2 resin to concentrate the urine and the *Salmonella* system as indicator, were able to show that smokers have mutagenic urine. The mutagenicity of the smokers' urine has since been confirmed by several research groups (Table 1). Also, there seems to be a dose-response relationship between the level of exposure to smoke and the extent of mutagenic activity in the urine (van Doorn *et al.*, 1979; Kriebel *et al.*, 1983).

**Table 1. Mutagenicity in smokers' urine as detected with bacterial assays**

| Test | Volume of urine[a] (ml) | Concentration method | Bacterial strain used | Metabolic activation[b] | Reference |
|---|---|---|---|---|---|
| Ames | 12.5 | XAD-2 | TA98 | + | Yamasaki & Ames (1977 |
| Ames | 25 | XAD-2 | TA1538 | + | van Doorn *et al.* (1979) |
| Ames | 12.5 | XAD-2 | TA98 | + | Møller & Dybing (1980) |
| Ames | 12.5 | XAD-2 | TA100 | - | Gelbart & Sontag (1980) |
| Ames | 24 | XAD-2 | TA98 | + | Hannan *et al.* (1981) |
| Ames | 15 | XAD-2 | TA1538 | + | Dolara *et al.* (1981) |
| Ames | 50 | XAD-2 | TA98 | + | Aeschbacher & Chappuis (1981) |
| Fluctuation test | 0.2 | XAD-2 | TA98 | + | Falck (1982b) |
| Ames | 1 | XAD-2 | TA98 | + | Kado *et al.* (1983b) |

[a] The smallest volume of urine that was found to have mutagenic activity

[b] Metabolic activation system comprising 9000 × *g* supernatant fractions of rat liver

Putzrath *et al.* (1981) analysed concentrates of smokers' urine further and concluded that the mutagens are a complex mixture of relatively nonpolar compounds. Concentrates of urine from cigarette smokers are usually mutagenic only after metabolic activation, suggesting that enzymatic splitting of conjugates may be necessary before activity can be detected. It has been observed, also, that the mutagens in smokers' urine act mainly on bacterial strains that are sensitive to frameshift mutagenesis (Falck *et al.*, 1980; Falck, 1982b).

Although cigarette smoke has been fractionated chemically and the mutagenicity and/or carcinogenicity of many of its components ascertained, the actual urinary metabolites of cigarette smoke have not been identified. For one smoker, in whom exceptionally high mutagenicity was found in urine, the major mutagenic compound was identified as 2-naphthylamine (and its metabolite 2-amino-7-naphthol), which was excreted as an unconjugated metabolite of a cigarette smoke component (Connor *et al.*, 1983).

Kado *et al.* (1983a) studied the kinetics of mutagen excretion in the urine of cigarette smokers. Peak urinary mutagenic activity appeared 3-5 h after the subject smoked one cigarette, and activity decreased to pre-smoking levels in approximately 12 h. Peak mutagenic activity in the urine of a one-pack-a-day smoker, collected over a 24-h period during which 19 cigarettes were smoked, appeared approximately 4 h after the first morning cigarette. On the basis of the excretion patterns of occasional smokers after smoking a single cigarette, the mutagens, as detected by the *Salmonella* assay, appear to be absorbed rapidly (3-5 h) and eliminated from the body following first-order kinetics.

Recently, it was reported that under experimental conditions that simulated passive smoking, inhaled compounds, also, led to mutagenic urine (Bos *et al.*, 1983). Thus, not only active but also passive smoking must be considered when the urine of a population is monitored for the mutagenic effects of other agents.

*Therapeutic drugs*

*Anticancer drugs*. The first reports of mutagenicity in urine from patients receiving anticancer drugs, especially cyclophosphamide, appeared more than ten years ago (Siebert &

Simon, 1973). Since then, it has become clear that many drugs used for cancer treatment are mutagenic in bacterial assays (*cf.* Benedict *et al.*, 1977; Pak *et al.*, 1979). Surprisingly, it has been reported, also, that nurses handling cytostatic drugs showed more mutagenic activity in urine than controls (Falck *et al.*, 1979). The activity was reduced after improved safety equipment was installed and when special attention was paid to proper working practices at the oncology unit (Falck, 1982a).

*Nitroimidazoles*. Nitroimidazoles are used for the treatment of infections caused by *Trichomonas vaginalis*, amoebae, anaerobic bacteria and *Coccidia* (Voogd, 1981). In addition to their antibacterial activity, these compounds have been identified as bacterial mutagens, also, and some of them are carcinogenic in rodent bioassays (Rustia & Shubik, 1972; Cohen *et al.*, 1973). The urine of patients treated with therapeutic doses of metronidazole or niridazole has been shown to be mutagenic to *Salmonella typhimurium* (Legator *et al.*, 1975). However, the mutagenic activity appears to be dependent on a specific bacterial nitroreductase activity (Blumer *et al.*, 1980; Speck *et al.*, 1981); and in mammalian cells, nitroreductase activity is very low under aerobic conditions. Metronidazole, for instance, although carcinogenic when given at high doses to mice and rats, is not mutagenic to mammalian cells *in vitro* (Voogd, 1981). It is not certain whether the carcinogenicity of the nitroimidazoles is due to catabolites formed by intestinal bacteria or to mammalian metabolism of these drugs. *In vivo*, both anaerobic gut flora and epidermal flora are rich in metabolically active bacteria capable of performing nitroreduction (Rosenkranz & Mermelstein, 1983), which may be responsible for reduction of the nitrogroup formed.

The results obtained with nitroimidazole drugs give an example of the difficulties in extrapolating data on mutagenicity in microbes to humans. However, the high mutagenic activity that is obtained with nitroimidazoles and other nitrocompounds in urine makes it possible to use bacteria as indicator organisms for assessment of exposure to these compounds.

*Occupational exposure*

Compared to exposures experienced in the general environment, *via* ambient air, drinking-water, etc., those encountered in the occupational environment are usually much more extreme and, usually, more easily controlled. In some occupational situations, the exposure is well characterized and the hazard is focussed on a single chemical entity, e.g., on benzene in benzene manufacturing plants. Therefore, preventive and hygienic measures can be selected easily, and environmental and biological monitoring should be straightforward. However, most occupational environments show a wide spectrum of chemicals to which a population may be exposed, and although there may be epidemiological evidence of an increased frequency of disease, no firm association with any single causal factor can be established. Results of urine mutagenicity assays applied in occupational situations are given in Table 2.

*Rubber workers*. The rubber industry was one of the earliest for which an association was documented between occupational exposure and cancer (IARC, 1982). Even since use of 2-naphthylamine was abolished in the rubber industry, epidemiological studies have revealed an excessive occurrence of various types of cancer.

The panorama of exposure in the rubber industry is enormously complex and undergoing constant change; therefore, it is difficult to adopt preventive health measures on the basis of monitoring single substances. A better way to monitor exposure is by nonspecific detection of a group of hazardous agents. This was the principle used in applying urinary mutagenicity studies for detection of exposure in different job categories in the rubber industry.

**Table 2. Use of the bacterial urine mutagenicity assay in monitoring of occupational exposures**

| Exposure | Bacterial strain used | Metabolic activation[a] | Result of study | Reference |
|---|---|---|---|---|
| Anaesthetic gases | TA1535 | - | + | McCoy *et al.* (1978) |
| | TA98<br>TA100<br>TA1535 | +/- | - | Baden *et al.* (1980) |
| Chemicals (various) | TA1538 | + | + | Dolara *et al.* (1981) |
| Coke oven emissions | TA98 | + | - | Møller & Dybing (1980) |
| Cytostatic drugs (nurses) | TA100<br>WP2uvrA | + | + | Falck *et al.* (1979) |
| | TA98<br>TA100<br>TA1535 | + | + | Anderson *et al.* (1982) |
| | TA100 | + | + | Theiss (1982) |
| | TA100 | - | + | Bos *et al.* (1982) |
| Cytostatic drugs (pharmacists) | TA100 | - | - | Staiano *et al.* (1981) |
| | WP2uvrA | + | ± | Kolmodin-Hedman *et al.* (1983) |
| Epichlorohydrin | TA1535 | - | + | Kilian *et al.* (1978) |
| Foundry fumes | TA98 | + | + | Schimberg *et al.* (1981) |
| Rubber chemicals | TA98 | + | + | Falck *et al.* (1980) |
| | WP2uvrA | | | Sorsa *et al.* (1982a) |

[a] Metabolic activation system comprising 9000 × *g* supernatant fraction of rat liver

Rubber workers, especially those working as chemical mixers, cleaners in the mixing department, vulcanizers and tyre builders, have more mutagenic activity in their urine than do groups of control ('nonexposed') workers. The mutagenic activity was detectable especially when urine was tested with the base-pair substitution strain *Escherichia coli* WP2uvrA (Falck *et al.*, 1980; Sorsa *et al.*, 1982a; Vainio *et al.*, 1982; Falck, 1983). After the workers had been absent for four weeks on annual leave, the mutagenic activity in their urine was at a similar level to that of the controls, indicating that the mutagenic activity is due to occupational exposure. Therefore, measurement of urinary mutagenic activity can be used to identify hazardous complex exposures in the rubber industry and to select appropriate industrial hygiene measures.

*Workers exposed to trinitrotoluene*. Trinitrotoluene is the most widely used explosive throughout the world, and it has been produced on an industrial scale since the beginning of the nineteenth century. A preliminary epidemiological survey in a Swedish factory where explosives, including trinitrotoluene, are prepared, suggested increased cancer mortality among the workers (Andersson *et al.*, 1980). Schepers (1971) reported that inhalation of trinitrotoluene by experimental animals causes specific pulmonary lesions, including lung cancer. Trinitrotoluene is a frameshift mutagen in *Salmonella typhimurium* (Won *et al.*, 1976), and mutagenicity has been detected, also, in the waste-water of a plant where trinitrotoluenes are manufactured (Spangord *et al.*, 1982).

Urine samples from workers in a chemical production factory were studied, both after a workshift (exposed sample) and after a holiday (unexposed sample), and the only group that showed increased levels of urinary mutagenicity comprised workers handling trinitrotoluene

(Table 3; Ahlborg *et al.*, 1983). However, mutagenic activity could be detected only with the frameshift tester strain TA98, and it was statistically significantly higher ($p < 0.01$) than the activity in the nonexposed samples, only without metabolic activation. The role of nitroreductase in the mutagenic responses of *Salmonella typhimurium* strain TA98 should be studied further, even though that strain is known to have fairly low activity (Blumer *et al.*, 1980).

**Table 3. Mutagenic activity detected in TA98 (mean ± SE) in urine from workers exposed to trinitrotoluene (TNT) and other workers in a chemical factory[a]**

| Group | No. | TA98 + S9[b] | | No. | TA98 − S9 | |
|---|---|---|---|---|---|---|
| | | Nonexposed | Exposed | | Nonexposed | Exposed |
| *TNT workers* | | | | | | |
| Smokers | 7 | 626 ± 206 | 750 ± 166 | 7 | 418 ± 279 | 1412 ± 370*c |
| Nonsmokers | 6 | 283 ± 142 | 711 ± 267*c | 7 | 49 ± 21 | 2156 ± 578**c |
| *Other workers* | | | | | | |
| Smokers | 34 | 501 ± 90 | 437 ± 107 | 33 | 159 ± 36 | 246 ± 51 |
| Nonsmokers | 26 | 274 ± 68 | 200 ± 53 | 27 | 134 ± 54 | 130 ± 35 |

[a] From Ahlborg *et al.* (1983)

[b] S9, metabolic activation system comprising 9000 × *g* supernatant fraction of liver from polychlorinated biphenyl-pretreated rats

[c*, **] Difference statistically significant (Student's t-test) as compared with the nonexposed samples in the group (*$p < 0.05$; **$p < 0.01$)

This example, in which trinitrotoluene exposure was detected in the urinary mutagenicity assay only without the addition of drug-metabolizing enzymes, demonstrates one aspect of the complexity of the test system. In complex exposure conditions, both frameshift and base-substitution tester strains should be used, and the tests should be carried out both with and without an exogenous metabolizing system.

## USE OF THE URINARY MUTAGENICITY ASSAY AS AN INDICATOR OF EXPOSURE

### *In epidemiological studies*

As yet, no report has been published on the use of urinary mutagenicity as an exposure indicator within the framework of an analytical epidemiological study. In prospective epidemiological studies, mutagenic activity in urine could be viewed either as an indicator of exposure or as an independent (prognostic) variable. If mutagenic activity is used as an index of exposure, the workers who have been thus classified as 'exposed' and 'nonexposed' can be monitored in follow-up studies. Of course, the observation would be valid only at the group level, since there are individuals who display mutagenic activity in urine without, subsequently, experiencing the assumed biological response. In order to make use of data from urinary mutagenicity assays, several questions must be answered.

(1) Are the methods for classifying individuals as 'exposed' or 'nonexposed' sufficiently developed? Can any urine sample that contains increased mutagenic activity be used to classify the individual from whom it was obtained as 'exposed'?

In our experience, a few individuals who show mutagenic activity in urine tested on bacteria (especially on frameshift-sensitive strains) have had no clearly identified exposure. An obvious cause of a spuriously positive result would be the presence of enhancers of bacterial growth, such as amino acids, in the sample (Gibson *et al.*, 1983). The activity may result from metabolic deviations or from other factors, such as those related to diet, ambient air, water, alcohol or drugs. The role of smoking in the mutagenicity of urine has been clearly established and should always be taken into consideration when evaluating the possible role of any other specific agent. This is true, probably, even for passive smoking. Fortunately, in some cases, it seems possible to differentiate between mutagenic activity due to smoking and that due to other exposure such as occupational exposure to chemicals, e.g., in the rubber industry (Falck *et al.*, 1980; Vainio *et al.*, 1982). However, in cases such as exposure to trinitrotoluene, which causes the same type of mutagenicity as does smoking (frame-shift) (see Table 3), smoking and nonsmoking groups must be treated separately.

(2) How well does the monitoring procedure distinguish between 'exposed' and 'nonexposed' individuals?

The sensitivity and specificity of the test may be of critical importance in categorizing people, e.g., for epidemiological analytical studies. If the procedure has little power to separate true 'exposed' from true 'nonexposed' workers, the contrasts in the group that are used to estimate the probabilities of subsequent events would tend to be diluted; the poorer the resolution, the more difficult it is to pick up differences in risk experience.

*In occupational health programmes*

Bacterial mutagenicity assays of urine from workers appear, therefore, to be applicable for monitoring exposure to mutagens and/or carcinogens. Table 4 lists some of the advantages and disadvantages of the urinary mutagenicity assay in human monitoring. Sample collection is noninvasive, samples can be obtained readily and repeatedly, and, in suitable exposure situations, urine testing can be integrated routinely into health surveillance programmes. One further advantage is that urine can be easily stored frozen and can be processed (concentrated, treated with enzymes) before testing, in order to make the assay more sensitive and specific.

**Table 4. Advantages and disadvantages of the urine mutagenicity assay**

| Advantages | Disadvantages |
|---|---|
| Noninvasive | Resolving power yet to be determined |
| Easy sampling and storage of samples | Unknown biological significance |
| Can be coupled with chemical analysis | Nonspecific as to the causal agent |
| Multiple substances active in a single assay | Limited to certain classes of mutagens |
| Biological activity is demonstrated | Possible interference by substances normally present in urine (such as amino acids) |
| Provides dose-response data | Decomposition or loss of activity during sample storage and preparation |
| Reasonable costs | Cannot detect cumulative exposures |

When this assay is used to monitor occupational exposure, special attention must be paid to possible confounding exposures, such as those occurring through drugs or living habits, since the assay is nonspecific, i.e., it is specific only to the nature of the exposure (mutagenicity) but not to the individual chemicals. However, one other advantage of this assay is that the mutagenicity data can be related to qualitative and quantitative chemical analysis of the urine, so that the nature of the exposure can be elucidated further.

The urinary mutagenicity assay is, thus, readily applicable to monitoring human exposure in certain situations. It can be used to identify hazards in workplaces, such as chemical production, and can be used, also, to aid the selection of appropriate hygienic measures.

Since quantitative risk assessment can be conceived as being composed of two aspects - the hazard and the exposure - exposure assessment is of critical importance. A presumption of hazard can be made if it can be shown that mutagens are excreted in urine. However, as most chemicals are excreted in conjugated 'detoxified' form, the presence of mutagenic activity in concentrated urine does not imply that there will necessarily be a further biological effect. Thus, data from the urinary mutagenicity assay should be considered an exposure indicator and not an effect indicator.

Since the urine mutagenicity assay is sensitive and is carried out easily at reasonable cost, it can, in appropriate situations, contribute substantially to the prevention of occupational disease.

## REFERENCES

Aeschbacher, H.V. & Chappuis, C. (1981) Non-mutagenicity of urine from coffee drinkers compared with that from cigarette smokers. *Mutat. Res.*, *89*, 161-177

Ahlborg, G., Jr, Bergström, B., Hogstedt, C., Falck, K., Einistö, P., Sorsa, M. & Vainio, H. (1983) *Genotoxisk påverkan på kemikaliearbetare* (*Report to the Swedish Work-Environment Fund (ASF 81-0196)*) (in Swedish)

Anderson, R.W., Puckett, W.H., Dana, W.J., Nguyen, T.V., Theiss, J.C. & Matney, T.S. (1982) Risk of handling injectable antineoplastic agents. *Am. J. Hosp. Pharm.*, *39*, 1881-1887

Andersson, K., Ahlborg, G., Jr & Axelson, O. (1980) *Dödlighet i lung-, ventrikel och blodtumörer på två bruksorter med tung metallindustri och kemisk industri.* (*Case-control study*) (*Report to the Swedish Work Environment Fund*), (*ASF projekt 78/123*) (in Swedish)

Baden, J.M., Kelley, M., Cheung, A. & Mortelmans, K. (1980) Lack of mutagens in urines of operating room personnel. *Anesthesiology*, *53*, 195-198

Benedict, W.F., Baker, M.S., Haroun, L., Choi, E. & Ames, B.N. (1977) Mutagenicity of cancer chemotherapeutic agents in the *Salmonella*/microsome test. *Cancer Res.*, *37*, 2209-2213

Blumer, J.L., Friedman, A., Meyer, L.W., Fairchild, E., Webster, L.T., Jr & Speck, W.T. (1980) Relative importance of bacterial and mammalian nitroreductases for niridazole mutagenesis. *Cancer Res.*, *40*, 4599-4605

Bos, R.P., Leenaars, A.O., Theuws, J.L.G. & Henderson, P.T. (1982) Mutagenicity of urine from nurses handling cytostatic drugs, influence of smoking. *Int. Arch. occup. environ. Health*, *50*, 359-369

Bos, R.P., Theuws, J.L.G. & Henderson, P.T. (1983) Excretion of mutagens in human urine after passive smoking. *Cancer Lett.*, *19*, 85-90

Bridges, B.A., Clemmesen, J. & Sugimura, T. (1979) Cigarette smoking - does it carry a genetic risk? *Mutat. Res.*, *65*, 71-81

Bridges, B.A., Butterworth, B.E. & Weinstein, I.B. (1982) *Indicators of Genotoxic Exposure* (*Banbury Report 13)*, Cold Spring Harbor, NY, Cold Spring Harbor Laboratory

Cohen, S.M., Ertürk, E., von Esch, A.M., Crovetti, A.J. & Bryan, G.T. (1973) Carcinogenicity of 5-nitrofurans, 5-nitroimidazoles, 4-nitrobenzene and related compounds. *J. natl Cancer Inst., 51,* 403-417

Connor, T.H., Ramanujam, V.M.S., Ward, J.B., Jr & Legator, M.S. (1983) The identification and characterization of a urinary mutagen resulting from cigarette smoke. *Mutat. Res.*, *113*, 161-172

Dolara, P., Mazzoli, S., Rosi, D., Buiatti, E., Baccetti, S., Turchi, A. & Vannucci, V. (1981) Exposure to carcinogenic chemicals and smoking increases urinary excretion of mutagens in humans. *J. Toxicol. environ. Health*, *8*, 95-103

van Doorn, R., Bos, R.P., Leijdekkers, C.M., Wagenaars-Zegers, M.A.P., Theuws, J.L.G. & Henderson, P.T. (1979) Thioether concentration and mutagenicity of urine from cigarette smokers. *Int. Arch. occup. environ. Health*, *43*, 159-166

Falck, K. (1982a) *Application of the Bacterial Urinary Mutagenicity Assay in Detection of Exposure to Genotoxic Chemicals* (*Academic dissertation*), Helsinki, Novoprint OY

Falck, K. (1982b) *Urinary mutagenicity caused by smoking.* In: Sorsa, M. & Vainio, H., eds, *Mutagens in Our Environment*, New York, Alan R. Liss, pp. 387-400

Falck, K. (1983) Biological monitoring of occupational exposure to mutagenic chemicals in the rubber industry. Use of the bacterial urinary mutagenicity assay. *Scand. J. Work environ. Health*, *9* (*Suppl. 2*), 39-42

Falck, K., Gröhn, P., Sorsa, M., Vainio, H., Heinonen, E. & Holsti, L.R. (1979) Mutagenicity in urine of nurses handling cytostatic drugs. *Lancet*, *i*, 1250-1251

Falck, K., Sorsa, M., Vainio, H. & Kilpikari, I. (1980) Mutagenicity in urine of workers in rubber industry. *Mutat. Res.*, *79*, 45-52

Gelbart, S.M. & Sontag, S.J. (1980) Mutagenic urine in cirrhosis. *Lancet*, *i*, 894-896

Gibson, J.F., Baxter, P.J., Hedmore-Whitty, R.B. & Gombertz, D. (1983) Urinary mutagenicity assay: a problem arising from the presence of histidine associated growth factors in XAD-2 prepared urine concentrates, with particular relevance to assays carried out using the bacterial fluctuation test. *Carcinogenesis*, *4*, 1471-1476

Hannan, M.A., Recio, L., Deluca, P.P. & Enoch, H. (1981) Co-mutagenic effects of 2-aminoanthracene and cigarette smoke condensate on smokers' urine in the Ames *Salmonella* assay system. *Cancer Lett.*, *13*, 203-212

IARC (1982) *IARC Monographs on the Evaluation of the Carcinogenic Risk of Chemicals to Humans*, Vol. 28, *The Rubber Industry*, Lyon, International Agency for Research on Cancer

Irving, C.C. (1971) Metabolic activation of *N*-hydroxy compounds by conjugation. *Xenobiotica*, *1*, 387-398

Kado, N.Y., Eisenstadt, E. & Hsieh, D.P.H. (1983a) The kinetics of mutagen excretion in the urine of cigarette smokers (Abstract). *Toxicol. Lett.*, *18* (*Suppl. 1*), 7

Kado, N.Y., Langley, D. & Eisenstadt, E. (1983b) A simple modification of the *Salmonella* liquid-incubation assay. Increased sensitivity for detecting mutagens in human urine. *Mutat. Res., 121,* 25-32

Kilian, D.J., Pullin, T.G., Connor, T.H., Legator, M.S. & Edwards, H.N. (1978) Mutagenicity of epichlorohydrin in the bacterial assay system: evaluation by direct in-vitro activity of urine from exposed humans and mice. *Mutat. Res.*, *53*, 72

Kolmodin-Hedman, B., Hartvig, P., Sorsa, M. & Falck, K. (1983) Occupational handling of cytostatic drugs. *Arch. Toxicol.*, *54*, 25-33

Kriebel, D., Commoner, B., Bolinger, D., Bronsdon, A., Gold, J. & Henry, J. (1983) Detection of occupational exposure to genotoxic agents with a urinary mutagen assay. *Mutat. Res.*, *108*, 67-79

Legator, M.S., Connor, T.H. & Stoeckel, M. (1975) Detection of mutagenic activity of metronidazole and niridazole in body fluids of humans and mice. *Science*, *188*, 1118-1119

Legator, M.S., Buedig, E., Batzinger, R., Connor, T.H., Eisenstadt, E., Farrow, M.G., Ficsor, G., Hsie, A., Seed, J. & Stafford, R.S. (1982) An evaluation of the host-mediated assay and body fluid analysis. *Mutat. Res.*, *98*, 319-374

McCoy, E.C., Hankel, R., Robins, K., Rosenkrantz, H.S., Ginfrida, J.G. & Bizzari, D.V. (1978) Presence of mutagenic substances in the urines of anesthesiologists. (Abstract). *Mutat. Res.*, *53*, 71

Mφller, M. & Dybing, E. (1980) Mutagenicity studies with urine concentrates from coke plant workers. *Scand. J. Work environ. Health*, *6*, 216-220

Mulder, G.J., Hinson, J.A. & Gillette, J.R. (1977) Generation of reactive metabolites of *N*-hydroxy-phenacetin by glucuronidation and sulfation. *Biochem. Pharmacol.*, *26*, 189-196

Pak, K., Iwasaki, T., Niyakawa, M. & Yoshida, O. (1979) The mutagenic activity of anticancer drugs and the urine of rats given these drugs. *Urol. Res.*, *7*, 119-127

Putzrath, R.M., Langley, D. & Eisenstadt, E. (1981) Analysis of mutagenic activity in cigarette smokers' urine by high performance liquid chromatography. *Mutat. Res.*, *85*, 97-108

Rosenkranz, H.S. & Mermelstein, R. (1983) Mutagenicity and genotoxicity of nitroarenes. All nitro-containing chemicals were not treated equal. *Mutat. Res.*, *114*, 217-267

Rustia, M. & Shubik, P. (1972) Induction of lung tumors and malignant lymphomas in mice by metronidazole. *J. natl Cancer Inst.*, *48*, 721-729

Schepers, G.W.H. (1971) Lung tumors of primates and rodents: Part II. *Ind. Med. Surg.*, *40 (2)*, 23-31

Schimberg, R., Skyttä, E. & Falck, K. (1981) Belastung von Eisengiessereiarbeitern durch mutagene polycyclische aromatische Kohlenwasserstoffe. *Staub-Reinhalt Luft*, *41*, 421-424

Siebert, D. & Simon, U. (1973) Genetic activity of metabolites in the ascites fluid and in the urine of a human patient treated with cyclophosphamide: induction of mitotic gene conversion in *Saccharomyces cerevisiae*. *Mutat. Res.*, *21*, 257-262

Sorsa, M., Falck, K. & Vainio, H. (1982a) *Detection of worker exposure to mutagens in the rubber industry by use of the urinary mutagenicity assay*. In: Sugimura, T. & Kondo, S., eds, *Environmental Mutagens and Carcinogens*, New York, Alan R. Liss, pp. 323-330

Sorsa, M., Hemminki, K. & Vainio, H. (1982b) Biological monitoring of exposure to chemical mutagens in the occupational environment. *Teratog. Carcinog. Mutag.*, *2*, 137-150

Spangord, R., Mortelmans, K., Griffin, A. & Simmon, V. (1982) Mutagenicity in *Salmonella typhimurium* and structure-activity relationships of wastewater components emanating from the manufacture of trinitrotoluene. *Environ. Mutag.*, *4*, 163-179

Speck, W.T., Blumer, J.L., Rosenkranz, E.J. & Rosenkranz, H.S. (1981) Effect of genotype on mutagenicity of niridazole in nitroreductose deficient bacteria. *Cancer Res.*, *41*, 2305-2307

Staiano, N., Gallelli, J.F., Adamson, R.H. & Thorgeirsson, S.S. (1981) Lack of mutagenic activity in urine from hospital pharmacists admixing antitumour drugs. *Lancet*, *i*, 615-616

Stiller, A., Obe, G., Riedel, L., Riehm, H. & Kappes, C. (1982) Mutagens in human urine: test with human peripheral lymphocytes. *Mutat. Res.*, *97*, 437-447

Stiller, A., Obe, G., Boll, I. & Pribilla, W. (1983) No elevation of the frequencies of chromosomal alterations as a consequence of handling cytostatic drugs. Analyses with peripheral blood and urine of hospital personnel. *Mutat. Res.*, *121*, 253-259

Theiss, J. (1982) Hospital personnel who handle anticancer drugs may face risks. *J. Am. med. Assoc.*, *247*, 11-12

Vainio, H., Sorsa, M., Rantanen, J., Hemminki, K. & Aitio, A. (1981) Biological monitoring in the identification of the cancer risk of individuals exposed to chemical carcinogens. *Scand. J. Work. environ. Health*, *7*, 241-251

Vainio, H., Falck, K., Mäki-Paakkanen, J. & Sorsa, M. (1982) *Possibilities for identifying genotoxic risks in the rubber industry: use of the urinary mutagenicity assay and sister chromatid exchange*. In: Bartsch, H. & Armstrong, B., eds, *Host Factors in Human Carcinogenesis* (*IARC Scientific Publications No. 39*), Lyon, International Agency for Research on Cancer, pp. 571-577

Voogd, C.E. (1981) On the mutagenicity of nitroimidazoles. *Mutat. Res.*, *86*, 243-277

Won, W.D., DiSalvo, L.H. & Ng, J. (1976) Toxicity and mutagenicity of 2,4,6-trinitrotoluene and its microbial metabolites. *Appl. environ. Microbiol.*, *31*, 576-580

Yamasaki, E. & Ames, B.N. (1977) Concentration of mutagens from urine by adsorption with the nonpolar resin XAD-2: cigarette smokers have mutagenic urine. *Proc. natl Acad. Sci. USA*, *74*, 3555-3559

# COMPARISON OF VARIOUS METHODOLOGIES WITH RESPECT TO SPECIFICITY AND SENSITIVITY IN BIOMONITORING OCCUPATIONAL EXPOSURE TO MUTAGENS AND CARCINOGENS

**P.H.M. Lohman, J.D. Jansen & R.A. Baan**
*Medical Biological Laboratory - TNO 2280 AA Rijswijk, The Netherlands*

## INTRODUCTION

The ultimate aim of biomonitoring should be to provide data that can be used to correlate quantitatively exposure to toxic agents with the risk of adverse effects on health and well-being (Fig. 1).

**Fig. 1.**

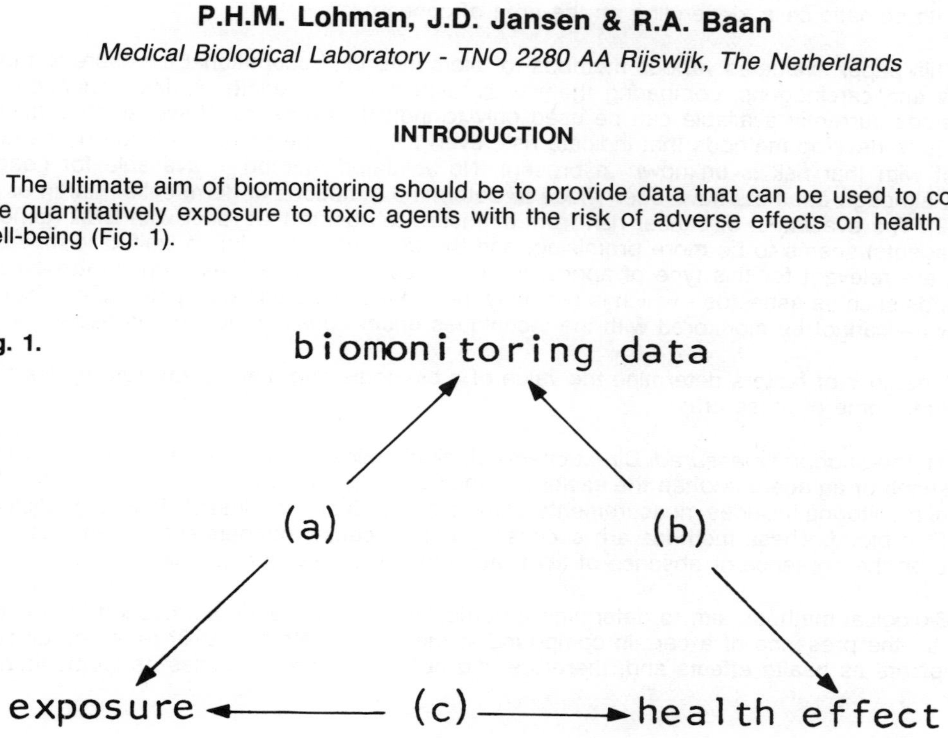

In classical biomonitoring, exposure to a certain chemical can be proven through detection of the compound - or its metabolite - in urine, blood, exhaled air, etc. After calibration of the method, i.e., when relation (a) in Figure 1 is known quantitatively, the level of exposure can be determined also. Even if the precise relationship to exposure is unknown, but the compound under study is suspected or known to induce unwanted health effects, biomonitoring may be useful in showing whether exposure is reduced in a given work situation. Results may indicate a possible health risk, if the relation between exposure and the effect [relation (c) in Fig. 1] has been established (usually on the basis of separate, independent evidence). If biomonitoring criteria are based on biological phenomena (so-called endpoints), that are consequences of the exposure and are causally related to the ultimate health effect, a more direct connection exists between biomonitoring data and risk [relation (b) in Fig. 1]. In this case, a direct risk assessment might be possible, without any need to determine the level of exposure first. This type of approach is very attractive for exposure to an unknown - but suspected - agent or mixture of agents. However, data obtained in biomonitoring studies have no predictive value with respect to the risk of adverse effects on health and well-being, unless there is a known, precise, quantitative relation between the endpoint and the health effect [i.e., through relations (a) and (c) or, directly, through relation (b) in Fig. 1].

For biomonitoring of populations exposed to mutagens and carcinogens, only a limited number of methods is available and most of these are still under development. The two health effects of particular concern are tumour formation and the occurrence of heritable disease. Tumour formation is the result of multiple changes in somatic cells, of which mutation induction is assumed to be one of the important steps; heritable disease arises from mutational events in germ cells. Therefore, when the direct approach is intended, biomonitoring after exposure to mutagens or carcinogens should comprise studies with germ cells or with somatic cells, depending on the area of concern.

This paper evaluates various methods for biomonitoring occupational exposure to mutagens and carcinogens, comparing them with respect to the validity in risk evaluation. All methods currently available can be used only to indicate exposure. However, the ultimate aim is to develop methods that indicate risk, even although the precise relation of the endpoint with that risk is unknown at present. No validated method is available for genetic risk, although attempts have been made to determine mutations in germ cells (Martin *et al.*, 1982). The evaluation of cancer risk from exposure to agents that react with DNA (genotoxic agents) seems to be more promising, and the present discussion is limited to methods that are relevant for this type of approach. This means, however, that nongenotoxic compounds such as asbestos - which is probably the most widespread occupational carcinogen known - cannot be monitored with the techniques under consideration in this paper.

A number of factors determine the value of a biomonitoring method for (future) risk estimation. Some of these are:

(1) *The endpoint measured*. Direct chemical, biochemical or physical determination of the presence of an agent is often the easiest method of determining exposure. Successful chemical monitoring includes measurements of metals (e.g., lead) and insecticides (e.g., dieldrin, DDT) in blood. These methods are successful only because independent evidence is available on the presence or absence of toxification at certain levels of the chemical.

Biological methods aim to determine a particular biological endpoint caused by, or related to, the presence of a certain compound in the body. Zielhuis (1978) regarded all such endpoints as health effects and, therefore, did not accept these studies as biomonitoring, which is performed not to detect but to prevent health effects. In his opinion, monitoring of these endpoints is to be compared with screening for cancer (e.g., of the cervix or bladder). However, many of the endpoints that are presently available for biomonitoring the effects of genotoxic agents are far from being established as adverse health effects, and comparison with tumour cells that are already manifest does not appear justified. Unfortunately, none of the biomonitoring methods presently available is calibrated in terms of human disease (i.e., risk). Consequently, all present biological methods should be considered as indicative of exposure only. Further independent data are required before the results of any of these methods can be related to a risk.

(2) *The sensitivity of the method.* In general the sensitivity of a method for a given compound is defined as the smallest quantity of the compound that can be detected above the background. Biological methods are indirect because they aim to measure a biological effect caused by the agent under study, instead of measuring chemically the concentration of a compound. As the amount of agent necessary for a certain effect may vary greatly with the agent, the sensitivity of a particular biological method depends strongly on the kind of compound to be monitored. Furthermore, a compound that induces this particular biological effect weakly may induce another endpoint strongly, and the opposite may be true for other agents. Therefore, it is difficult to generalize that one biological method is more sensitive than another.

An important factor, also, is that the signal-to-noise ratio in detecting a biological endpoint is not a fixed parameter, but depends strongly on background fluctuations - which are often high, indicating that the effect observed may be affected by other factors. In such a situation, method sensitivity will depend on the amount of data necessary to calibrate the background. Thus, when small differences are found between exposed subjects and controls, it will be difficult to decide whether these are caused by other occupational or nonoccupational factors. Theoretically, the sensitivity of a method can be improved by a proper correction for background effects, but often this is possible only at the expense of so much additional work that practical limitations prevent routine use of the method.

Lastly, the endpoints of biological methods are not persistent in many cases. This makes sensitivity dependent on the time of sampling, which may limit its applicability.

(3) *The specificity of the method.* Some methods may be highly specific for one compound, whereas others may be nonspecific for compounds but highly specific for genotoxicity in general (e.g., an increase in mutations). With exposure to a mixture of agents, the results of the latter type of method should be cumulative for all active chemicals present. Measurement of chromosomal aberrations in blood cells, for instance, is thought to reflect the total clastogenic effect of all agents to which the blood has been exposed. Measurement of mutagenicity in urine should not be expected to give information about cumulative effects of chemicals.

If biomonitoring pursues risk estimation - which it should, eventually, - the endpoints measured must be relevant to the adverse health effect of particular concern [relation (b) in Fig. 1]. In this case, it is not necessary to know the precise relationship to the actual exposure level [relation (a) in Fig. 1]. In order to assess health risk after exposure to a given compound or group of compounds, the specificity of the endpoint of a direct or indirect method must be established as valid for the agent under study. Secondly, there should be evidence that the endpoint indicates target cell exposure quantitatively and, lastly, the relationship between the target cell exposure and the magnitude of the adverse health effect (i.e., health risk) should be known quantitatively.

In this paper, currently available methods for biomonitoring occupational exposure to mutagens and carcinogens are evaluated with respect to these criteria.

## CURRENT BIOLOGICAL METHODS FOR BIOMONITORING

### *Chromosomal aberrations*

Undoubtedly the most well-developed method for biomonitoring of genotoxic exposure is the measurement of chromosomal aberrations in peripheral blood lymphocytes. This method was first applied, some decades ago, for the biomonitoring of populations exposed to ionizing radiation. For this purpose, the method has proved specific and sensitive enough for use as a 'biological dosimeter' in cases of accidental exposure to radiation (Lloyd *et al.*, 1975). This development was possible because no metabolism is involved and the results of in-vivo studies could be directly compared to those of in-vitro tests (Evans, 1982) which permitted experimental study of modifying factors.

The situation is different for chemicals. Because most chemicals undergo metabolic conversion *in vivo* into the ultimate reactant, the in-vivo and in-vitro results of chromosomal aberration tests are not directly comparable (Evans, 1982). Additional differences are: (1) that radiation-induced chromosomal aberrations are largely persistent, whereas those induced by most chemicals are reversible (Natarajan & Obe, 1980), with the possible exception of those resulting from exposure to benzene, and (2) that most chemicals seem to have a threshold for the induction of chromosomal aberrations (Jansen, 1983). Funes-Cravioto *et al.* (1975), for instance, reported an increase in chromosomal aberrations in workers exposed to high concentrations of vinyl chloride monomer; after exposure to low concentrations of the compound, no increased number of aberrations was found (Natarajan *et al.*, 1978). Additional evidence for a threshold with vinyl chloride monomer was provided by other investigators (Hansteen *et al.*, 1978; Anderson *et al.*, 1980). Furthermore, present epidemiological data suggest that angiosarcomas may be found in exposed workers only after high exposure to vinyl chloride monomer. Therefore, in this case, the presence or absence of chromosomal aberrations might indicate a health risk. The existence of a threshold and/or reversibility of the lesions can sometimes explain the seemingly conflicting data presented in the literature (Hook, 1981).

Although chromosomal aberrations are not specific for the agent, they are considered to be specific for the total clastogenic effect of a mixture of chemicals, which, in turn, may be related to a carcinogenic risk. In this sense, the method may be superior to other nonspecific methods. However, this view may be too optimistic; exposure to lead (Forni & Secchi, 1973), styrene (Meretoja *et al.*, 1977) and the tumour-promoter, phorbol myristate acetate (Emerit *et al.*, 1983), all seem to cause an in-vivo increase in chromosomal aberrations, whereas none of these compounds is known to be carcinogenic.

The major disadvantage of the method for biomonitoring purposes is that scoring of chromosomal aberrations is laborious and subjective. Therefore, for this test to be performed reliably, the laboratory must be thoroughly experienced and so organized as to obtain unbiased results. Under such conditions, one technician can analyse only about 200 blood samples per year, roughly half of these being controls. This consideration severely limits application of the method to large-scale population research.

An additional disadvantage is that the sensitivity of the method is decreased because the level of chromosomal aberrations is influenced by age, excessive alcohol consumption (alcoholics), cigarette smoking and viral infections. These factors should, therefore, be taken into account in choosing the control population for biomonitoring of occupationally exposed human populations (Evans *et al.*, 1979; Natarajan & Obe, 1980; Evans, 1982); the effect of these factors can be reduced by performing longitudinal studies.

*Sister chromatid exchanges*

In comparison with the chromosomal aberration method, this test is sometimes more sensitive, is much easier to perform and is equally nonspecific for the chemical in question. There are, however, several inherent difficulties, such as: (1) the relationship to DNA damage is not well known; (2) the relationship to clastogenicity or genotoxicity, in general, is not established; (3) reversion of the lesions is far quicker than that observed with chromosomal aberrations. The third factor may lead to variable results in studies of occupationally exposed populations, and renders the method less successful for biomonitoring than was originally expected (Lambert *et al.*, 1982). However, the sister chromatid exchange method can be of great help in solving questions that are formulated more specifically, as illustrated by the results with nurses handling cytostatics (Sorsa *et al.*, 1982).

*Micronucleus test in somatic cells*

This method can be applied successfully to diverse human tissues and is not unduly difficult or time-consuming (Heddle *et al.*, 1982; Stich *et al.*, 1983), but it has not been validated for biomonitoring purposes. Some preliminary results, reported by Stich *et al.* (1983), show a relationship between betel chewing and the induction of micronuclei in buccal mucosa. Micronuclei are thought to originate from aberrant cell division and unequal chromosome distribution, although Heddle *et al.* (1982) reported that (damaged) chromosomes in micronuclei of colon cells may separate without cell division. Therefore, in this case also, the biological endpoint needs further characterization.

The main advantages of this method are that it may be specific for clastogenic effects (Heddle *et al.*, 1982) and that it measures effects directly in the target tissue of exposed persons. In occupational biomonitoring where the target tissue for the chemical under study is frequently either unknown or unavailable for analysis, the latter advantage is lost.

*DNA repair studies in somatic cells*

Induction of DNA repair in somatic cells has been used in studies of rats exposed to carcinogens (Furihata & Matsushima, 1982) and humans with high cancer risk (Stich *et al.*, 1983). In principle, the method can be applied to cells from various human target organs and/or tissues (Stich *et al.*, 1983). Its significance is questionable, however, as only one step in the many, complicated, DNA-repair processes in somatic cells is presently used as a marker, *viz.*, DNA repair synthesis (unscheduled DNA synthesis). The biological endpoint, as such, is undefined and can reflect, for instance, either 'long-patch' or 'short-patch' excision repair, depending on the chemical (Hanawalt *et al.*, 1979; Lehmann & Karran, 1981). This affects both the specificity and sensitivity of the method of DNA repair synthesis. An example of this effect is given in Figures 2 and 3, which show the effects on survival, the induction of mutations (at the hypoxanthine-guanine phosphoribosyl transferase locus) and the occurrence of unscheduled DNA synthesis in a Chinese hamster cell line by various doses of three known mutagens. Although the relation between the extent of mutation induction and the percentage of surviving cells after exposure is similar for each of the three agents, vast differences are found between the compounds with regard to sensitivity in detection of unscheduled DNA synthesis, after exposures which are comparable for cytotoxicity.

**Fig. 2. DNA repair and mutation induction (at the hypoxanthine-guanine phosphoribosyl transferase locus) as a function of survival in Chinese hamster ovary cells**

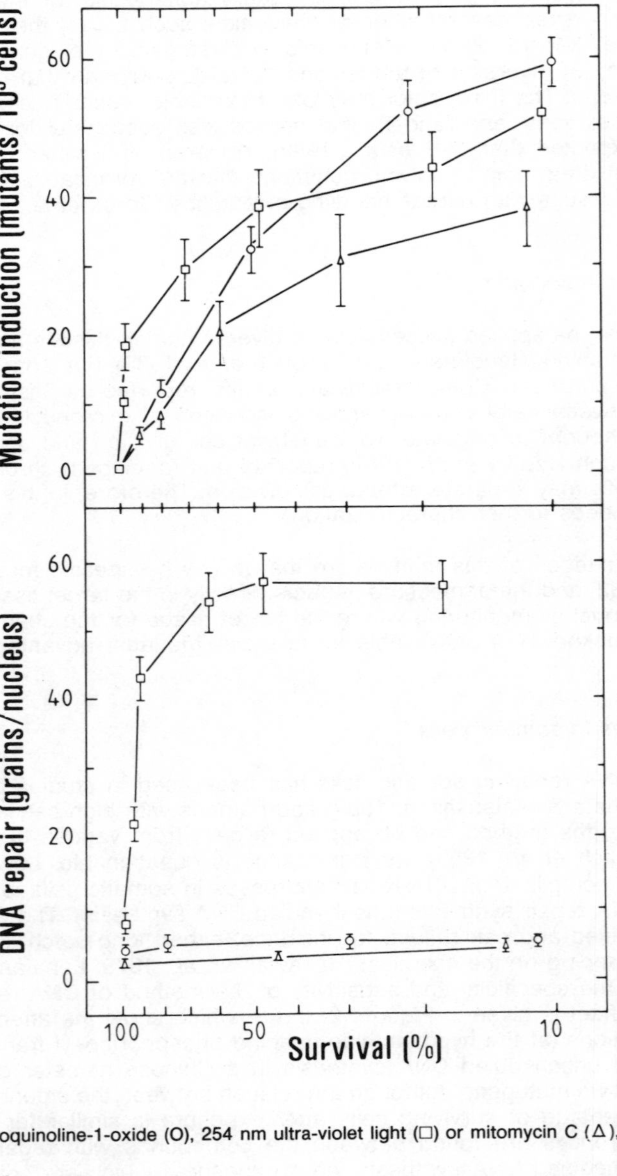


Cells treated with 4-nitroquinoline-1-oxide (O), 254 nm ultra-violet light (□) or mitomycin C (Δ), respectively.

**Fig. 3. The induction of unscheduled DNA synthesis in Chinese hamster ovary cells treated with mutagenic agents, as a function of dose**

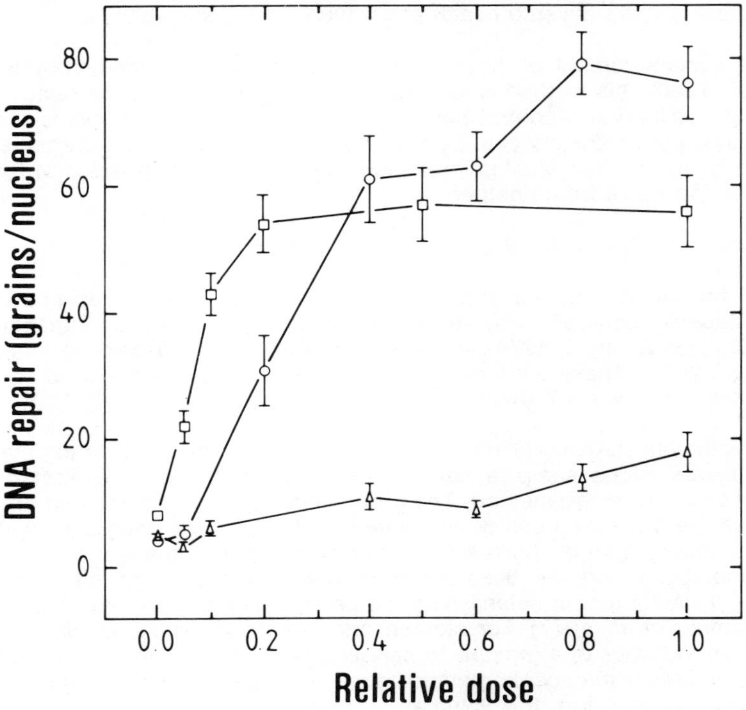


Dose is expressed in relative units, i.e., relative to the (highest) dose giving the maximum amount of unscheduled DNA synthesis. O, cells treated with 4-nitroquinoline-1-oxide for 1 h at 37°C. The maximal dose was 0.01 mmol/l; □, cells irradiated with 254 nm ultra-violet light. The maximal fluence was 12 J/m²; Δ, cells treated with mitomycin C for 1 h at 37°C. The maximal dose was 0.3 mmol/l.

Because of this lack of sensitivity and specificity, the application of this test to somatic cells of humans exposed to mutagens and/or carcinogens needs further validation (Butterworth *et al.*, 1982); its use will be probably restricted to monitoring only a limited number of known genotoxic agents.

*Determination of thioethers in urine*

Many carcinogens and mutagens are activated by metabolic conversion to the ultimate electrophilic carcinogen or mutagen. The electrophilic metabolites may react with DNA, but part of them will be conjugated and excreted harmlessly as thioethers in the urine. It is clear that increased excretion of thioethers indicates exposure to electrophilic compounds (van Doorn *et al.*, 1981; Henderson *et al.*, this volume). In an exposure situation that is precisely known (e.g., when a cancer patient is treated with a single cytostatic agent), the method can be specific and it can then be used to biomonitor exposure.

In an unknown situation, however, exposure to compounds with different genotoxic activities, and/or to nongenotoxic agents, may give a similar excretion of thioethers. In such a situation, the method is neither specific for a compound, nor indicative of the level of exposure nor of the expected adverse health effect (van Sittert, this volume).

As with the measurement of chromosomal aberrations or sister chromatid exchanges (Carrano *et al.*, 1980), this method is regarded as insensitive, due to the relatively wide fluctuation in data obtained in a control population. Again, the cause of this lack of sensitivity may be that the results are affected by the many endogenous and environmental factors, such as smoking habits, individual differences in metabolism, composition of food, etc., that distinguish one individual from another.

*Analysis of urine and faeces for the presence of mutagenic agents*

Urine and faeces of exposed groups can be analysed for the presence of mutagenic agents or mutagenic metabolites by various short-term tests, such as mutagenicity assays in bacteria (Durston & Ames, 1974), in *Drosophila* (Browning, 1973) or in mammalian cells (Amacher *et al.*, 1981). These tests can be performed in the presence or absence of bioactivation systems, e.g., liver microsomes.

In studies with urine, β-glucuronidase or arylsulfatase may be included to detect otherwise inactive conjugates. Urine samples can be either concentrated or extracted in order to increase the sensitivity of the method. The great advantages of this method are the ease of application and the fact that it can detect gene mutations. It has been used with much success to detect mutagenic metabolites produced by drugs in humans (Legator *et al.*, 1978). When these metabolites are identified and their mutagenicity in urine is known, the method appears to be a useful tool in establishing the precise metabolism and pharmacodynamics in humans (Conner *et al.*, 1977). For biomonitoring purposes, mutagenicity in urine can be used only as an indicator of exposure to certain agents, viz., to those chemicals that give rise to renally excreted mutagenic metabolites. It is unlikely that the method can be used for risk estimation (van Sittert, this volume).

Faecal mutagenic activity has been widely investigated to study the relation between diet and colon cancer. Although the effect of dietary variance on the mutagenic activity of faeces is well established, no relation has yet been found between the extent of mutagenic activity and the occurrence of colon cancer (Venitt, 1982). Because of the mutagenic differences that result from diet variance and because few chemicals - when introduced by inhalation or by percutaneous absorption - end up in the faeces, the method is not expected to be useful in biomonitoring populations occupationally exposed to genotoxic agents.

## NEW BIOCHEMICAL AND BIOLOGICAL METHODS FOR BIOMONITORING

Four promising methods, presently under development, may open new possibilities for biomonitoring mutagens and genotoxic carcinogens. Two of these methods aim at direct monitoring of human mutation rates and the other two measure the binding of specific chemicals to cellular macromolecules.

All are likely to be relevant for damage in cells of target organs and tissue. Since the endpoints can be measured directly in human material, without interference from the species differences, metabolism and other factors that distinguish a human from the test organism, these methods might be useful in measuring both exposure and risk.

None of the new tests has been validated so far. The two methods in which human mutation rates are measured are nonspecific for the chemical under test. The methods that measure chemical damage in cellular macromolecules (adduct formation) are specific for the genotoxic agent applied. However, risk assessment by these methods is limited because not all adduct formation is specific for the adverse health effect concerned. Each genotoxic compound induces a spectrum of lesions in cellular macromolecules (DNA, RNA and protein), but only a few of the adducts formed in the DNA may be relevant for mutation induction. Therefore, by use of molecular dosimetry on the sum of all adducts in each type of macromolecule, the method is expected to indicate exposure only. In general, risk estimations may be possible only if specific DNA lesions are monitored.

*The hypoxanthine-guanine phosphoribosyltransferase method*

Albertini *et al.* (1982) have developed a method that is sensitive enough to detect single-point mutations in white blood cells of exposed populations. The test is based on the detection of a mutation in a structural gene, located on the X-chromosome, that codes for the enzyme hypoxanthine-guanine phosphoribosyltransferase (HPRT). A selection method has been developed for white blood cells mutated at this locus, based on the fact that the mutant cells become resistant to either one of the toxic base analogues of normal DNA precursors, 8-azaguanine or 6-thioguanine. In mutant cells, these base analogues cannot be incorporated into the DNA due to the absence of the HPRT enzyme and, consequently, the mutants are not affected by base analogue toxicity.

The original protocol of the method was found to include an important artefact: proliferating cells were detected as thioguanine-resistant phenocopies of the real mutants, and the apparent mutation rate was increased by between one and two orders of magnitude. Recently, this problem has been overcome, and unequivocal results have been obtained with cell-cloning assays (Albertini, 1982; Strauss, 1982). However, the method may prove too cumbersome for application to large populations although there may be short-cuts. It is nonspecific for the chemical under study.

*Detection of mutations in blood-cell proteins*

*Haemoglobin-variants method*. The second method available to measure human mutation rates was developed by Mendelsohn *et al.* (1980). It is based on the detection of mutations in the haemoglobin (Hb) gene in erythropoietic stem cells *via* measurement of Hb variants in human red blood cells. The altered Hb molecules are detected with the help of monospecific antibodies that can recognize changes in certain amino acid sequences in the Hb molecule. The variant cells can be visualized with a fluorochrome, directly or indirectly attached to the mutant-specific antibody. The frequency of the mutant cells in blood is so low (about $10^{-7}$) that screening by hand is virtually impossible. Therefore, flow cytometric procedures are under development which will allow automatic, and relatively fast, detection of a small number of mutants in a suspension of red blood cells. Current procedures are able to analyse more than $10^6$ cells/sec.

For each identified Hb variant, a specific antibody must be raised. Because isolation of good-quality monoclonal antibodies is a time-consuming and laborious procedure, development of the Hb-variants method may be slow. However, when fully developed, the method should be eminently suitable for large-scale population research.

The great advantage of this method, as for the HPRT mutation test, is that it measures human mutation induction directly.

The method is nonspecific for the agent, but it has the potential to generate data that may be directly related to an observed health effect. In these respects, the Hb-variants method and the HPRT mutation test are comparable.

*The glycophorin A system.* A method similar to the Hb-variants technique is under development by Bigbee *et al.* (1983). This method studies the expression loss of the glycophorin A gene, through examination of a number of markers on the membrane-bound blood-cell protein, glycophorin A. Because the mutant phenotype can arise from a variety of mutational lesions, the frequency of variant cells should probably be 100-1000 times higher than the frequency of a single amino acid substitution in, for instance, Hb. Hence, such variants should be more easily detectable and they may better reflect the overall genetic damage in exposed persons. When developed for routine use, this method may become even more suitable for large-scale biomonitoring than the Hb-variants method.

*Measurement of haemoglobin alkylation in blood samples*

Ehrenberg *et al.* (1983) proposed the study of alkylation of Hb in order to measure exposure to alkylating agents. These investigators have developed sensitive chemical methods to detect this alkylation. The method has been applied to workers exposed to ethylene oxide and has proven sensitive enough to determine the specific exposure of some of the workers to the compound (Calleman *et al.*, 1978). For the specific case of exposure to ethylene oxide, it was shown that alkylation of Hb is a reliable indicator for overall DNA damage in all somatic cells of the body; Ehrenberg *et al.* (1983) have suggested that, on this basis, a risk estimation can be made for the carcinogenic effect of ethylene oxide. According to other authors (ICPEMC, 1983; Wright, 1983), the method needs further validation, for both heritable effects and cancer risk after exposure to ethylene oxide. Also, the findings obtained with ethylene oxide cannot be extrapolated to other alkylating agents, because other alkylating agents induce a different spectrum of DNA damage in somatic cells, so that the relation between DNA alkylation and protein alkylation must be determined experimentally for each compound. In each case, thorough validation studies are, therefore, required before the method can be applied to risk estimation. Of course, this difficulty is inherent to all chemical-specific methodology.

The advantage of the Hb-alkylation method is its extreme specificity to alkylating agents, which makes it a valuable tool for biomonitoring exposure to those agents. The method is also one of the most sensitive available to measure in-vivo exposure to a specific genotoxic chemical.

A disadvantage of the method is that during the studies on ethylene oxide, an unexpectedly high background fluctuation of Hb-alkylation was found in control human populations. Part of this background can, possibly, be attributed to previously indetectable endogenous production of ethylene oxide (ethylene oxide has been found in the exhaled air of unexposed rats - Filser & Bolt, 1983). Although the method is quite sensitive, this background fluctuation will lower the method's sensitivity for establishing low external exposures to alkylating agents.

*Immunochemical analysis of DNA adducts*

The most direct way to measure the primary effect of a compound is to detect and measure its reaction with the intracellular target. For mutagens, the target has been identified as DNA, and for genotoxic carcinogens, DNA is strongly suspected to be the target. Measuring at the level of DNA is called molecular dosimetry. In past years, several attempts have been made to design methods that can be used for molecular dosimetry *in vivo*, at subtoxic levels of exposure. The most promising methods are based on immunochemical detection of specific DNA adducts in blood cells. These methods may be sensitive enough for biomonitoring exposed human populations, and it seems possible to detect those DNA adducts that are indicative not only of exposure but also, possibly, of the adverse health effect concerned (Adamkiewicz *et al.*, this volume).

Immunochemical analysis of DNA adducts appears to be among the most specific of all techniques for biomonitoring because the antibodies recognize only one type of DNA adduct, originating from a known chemical, and, in most cases, the 'natural' background is expected to be very low. At the same time, the inherent drawback of this approach is that, for each type of adduct, a specific antibody must be raised. Even when a group of (immuno-)chemically related adducts can be studied with one antibody, assessment of the cross-reactivity of the antibody with the various adducts requires chemical characterization and purification of each individual DNA adduct.

For biomonitoring purposes, immunochemical methods must be able to detect a single DNA adduct in about $10^7$ - $10^8$ unmodified nucleosides (Baan *et al.*, 1984). In order to find out whether such sensitivity can be reached *in vivo*, a model experiment was made on rats treated with the well-known liver carcinogen, 2-acetylaminofluorene (Miller & Miller, 1975). DNA adducts were measured in hepatocytes of exposed animals and the results of the initial experiments are shown in Table 1. From these data, it can be concluded that, indeed, it will be possible to detect the 2-acetylaminofluorene DNA adduct at the required level of sensitivity.

**Table 1. Acetylaminofluorene adducts in liver DNA isolated from 2-acetylaminofluorene-treated rats[a]**

| Dose (mg/kg) | Time after start of treatment (h) | | |
|---|---|---|---|
| | 2 | 6 | 24 |
| 0.0 | 0 | 0 | 0 |
| 0.1 | ND[b] | ND[b] | ND[b] |
| 1.0 | 4 | 14 | 10 |
| 10.0 | 120 | 220 | 50 |
| 200.0 | 100 | 690 | 280 |

[a] Rats were treated (orally) with 2-acetylaminofluorene, dissolved in dimethyl sulfoxide (5 ml/kg). After 2, 6 and 24 h, the animals were killed. DNA was isolated from the livers and purified by means of caesium chloride equilibrium density gradient centrifugation. The amount of acetylaminofluorene adducts (given as the number of adducts/$10^8$ nucleotides) was determined immunochemically in a competitive enzyme-linked immunosorbent assay (Baan *et al.*, 1982; Mohn *et al.*, 1984).

[b] ND, not detectable

For large-scale screening, and for cases where only limited amounts of tissue or cells are available (human monitoring), the development of methods to analyse DNA adducts in a single cell is of interest. In this connection, immunofluorescence microscopy techniques are currently under investigation. Advanced computer technology is being applied to quantify the specific fluorescence signal in the nuclei of exposed cells in the microscopic preparation (Adamkiewicz *et al.*, this volume). Some preliminary data from a single-cell analysis are given in Figure 4. In this experiment, cultured human cells were exposed to *N*-acetoxy-2-acetylaminofluorene, and the genotoxic metabolite of 2-acetylaminofluorene and the resulting DNA adducts were measured using quantitative immunofluorescence microscopy. A dose-dependent increase of the fluorescence was found.

**Fig. 4. Quantitative immunofluorescence microscopy of DNA adducts formed in cultured human fibroblasts treated with *N*-acetoxy-2-acetylaminofluorene (AAAF)**

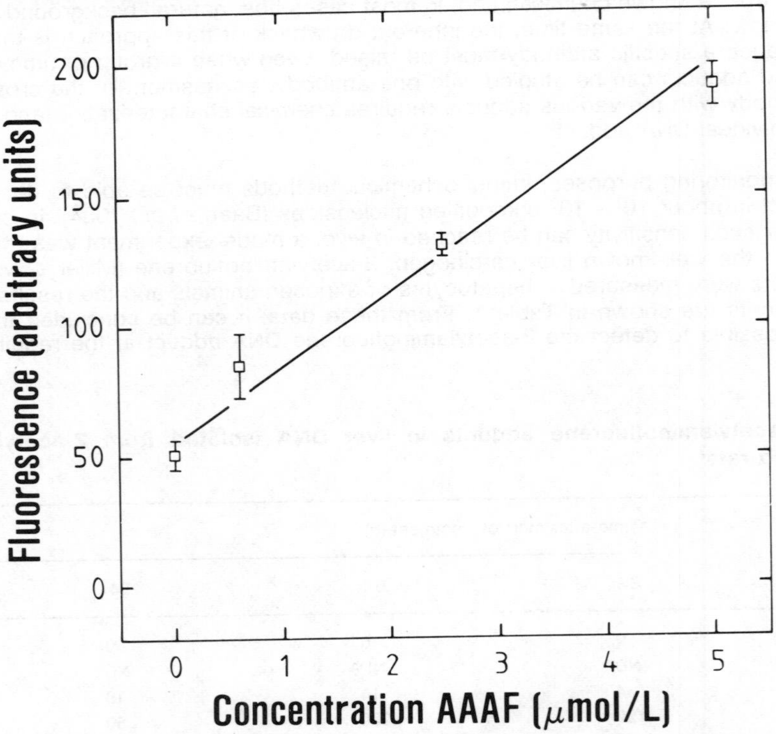


The cells were grown on coverslips, treated with the agent for 1 h at 37°C, washed and fixed with methanol/acetone. The preparations were treated with RNAse, proteinase K and alkali (0.07 N sodium hydroxide for 2 min). The slides were then incubated with anti-guanosine-acetylaminofluorene antiserum (Baan *et al.*, 1982) and fluorescein isothiocyanate-labelled second antibody. The fluorescence was measured at the single cell level (each point is the average of 35-40 cells).

## CONCLUDING REMARKS

Although many methods are currently successful in detecting exposure to mutagens and carcinogens, *no method* is available yet to quantify an increased carcinogenic or mutagenic risk in exposed populations. Some of the methods that measure a biological endpoint are thought to give qualitative information about possible induction of somatic mutation and/or chromosomal aberration, that may be relevant for cancer; those methods are listed as indicative of health effects in Table 2. None of the biological methods available today can indicate a possible heritable effect. Note that, because of the testis barrier, results in somatic cells cannot be extrapolated to germ cells (ICPEMC, 1983).

**Table 2. Methods for biomonitoring of occupationally exposed human populations**

| Method | Endpoint[b] | | Specificity[c] (chemical) | | Indication[d] (health effect) | |
|---|---|---|---|---|---|---|
| | r | n-r | s | n-s | i | n-i |
| Physico-chemical detection | | | | | | |
| - urine, faeces | | + | + | | | + |
| - blood | + | | + | | | + |
| Chromosomal aberration in blood cells | + | | | + | c | |
| Sister chromatid exchange in blood cells | + | | | + | | + |
| Micronuclei in blood cells | + | | | + | c | |
| DNA repair in somatic cells | + | | | + | | + |
| Thioethers in urine | | + | | + | | + |
| Mutagenic activity of: | | | | | | |
| - urine | | + | | + | | + |
| - faeces | | + | | + | | + |
| HPRT[e] mutants in blood cells | + | | | + | c | |
| Mutations in blood-cell proteins | + | | | + | c | |
| Alkylation of blood proteins | + | | + | | c | |
| DNA adducts in blood cells | + | | + | | c | |

[a] +, evidence available; c, future calibration may be possible for identified chemicals

[b] Indicates whether the endpoint of the method is relevant (r) or nonrelevant (n-r) for target cell exposure

[c] Indicates whether the method is specific (s) or nonspecific (n-s) to identify exposure to a known chemical. (*All* nonspecific methods can be made specific if applied under rigorously controlled exposure to one chemical only)

[d] Indicates whether the method is indicative (i) or nonindicative (n-i) of in-vivo mutagenicity or carcinogenicity

[e] Hypoxanthine-guanine phosphoribosyltransferase

As indicated in Table 2, all methods in which somatic cells are used have an endpoint that is relevant for the extent of exposure in target organs and tissues; the methods in which excreta are studied have endpoints that, as such, are irrelevant for target cell exposure. It is clear, also, that few methods are specific for the chemical under test. Methods that combine an irrelevant biological endpoint with nonspecificity for the chemical under test can be applied to future risk estimation only in a rigorously controlled situation, where exposure to known genotoxic chemicals is followed under precisely known conditions. For all methods, sensitivity depends on the chemical tested. If, however, the endpoint of a biological test is also influenced by 'natural' factors (e.g., high background fluctuation), the sensitivity of that method will be lowered, sometimes considerably, when required to establish additional occupational exposure to a known chemical.

The ultimate aim of biomonitoring occupationally exposed humans is, of course, estimation of risk - eventually, perhaps, even personal risk. Unfortunately, as stated, current methodology does not allow this. However, most techniques for biomonitoring genotoxic chemicals are still under development: the whole field is in its infancy. It is important, therefore, to speculate about future possibilities, on the basis of some of the preliminary results obtained with new methods described in this paper. One example is the detection of DNA adducts in patients treated with the cytostatic drug, cisplatin, studied in our laboratory (Fichtinger-Schepman *et al.*, 1982). Immunochemical techniques, in combination with high-performance liquid chromatography separation procedures, can be used for sensitive detection of individual adducts. Recent experiments indicate that several cisplatin-DNA adducts can be detected in nucleated blood cells obtained from patients treated with cisplatin (unpublished data). Since many other data on cisplatin-treated patients are known precisely [smoking habits, food intake, (other) medication] it will be interesting to carry out follow-up studies on the increasing number of people who have been cured by treatment with this or other cytostatics (IARC, 1981). It is hoped that the appearance of secondary cancers in some of these people, or of genetic disease in their offspring, may be correlated accurately with the medical history of each individual patient. This group, therefore, offers an opportunity for study of the etiology of these types of cancer, more precisely than hitherto possible. For the first time, an in-vivo calibration of a biomonitoring method may be possible.

Of course, only in rare cases is direct calibration of a method possible. In most other cases, calibration must be carried out by animal studies, viz., by comparing biomonitoring data and tumour formation (or heritable effects) in different species with biomonitoring data in humans (parallelogram approach - Sobels, 1982). In this connection, the use of cisplatin has the additional advantage that effects can be studied simultaneously in animals, allowing calibration of the parallelogram approach, also.

This outlook for the future of occupational biomonitoring of mutagens and carcinogens, resulting in quantitative risk estimation, may be too optimistic. At this moment, there is no proof that any of the endpoints measured, not even, for instance, that of the new methodology of measuring specific DNA adducts, is relevant for an increase in the number of mutations in germ cells, or for the appearance of a clinically verifiable tumour. Recent findings suggest that, at least for the induction of some tumours, an alteration at a specific site of DNA coding for a particular gene (oncogene) is related to the adverse health effect (Reddy *et al.*, 1982; Tabin *et al.*, 1982). Additional research is required to establish a relation between specific DNA adducts and an alteration at a specific DNA site (Neumann, this volume).

To summarize, current methodology can be successfully used in practice only to reduce exposure, in cases where a clearly positive effect is found with any of the described methods. At this moment, mutagenic or carcinogenic risk cannot be estimated on the basis of a positive result. Unfortunately, a negative response found with these methods can be used with even less confidence to define a no-risk situation after suspected exposure to a known genotoxic chemical; in such a case, the methods must be validated even more rigorously than with a positive response. However, newer methodology, now under development and discussed in this paper, may allow future evaluation of risks for health, perhaps even on a personal basis.

There is one group that cannot wait for the development of quantitative risk estimation, i.e., occupationally exposed workers who are studied with current biomonitoring techniques. These groups should be told clearly, *before* the investigation, that differences in response between individuals and between exposed persons and controls cannot be interpreted in terms of a quantifiable health risk, for instance, because of 'background fluctuations'. This information is far easier to accept before an investigation than afterwards, when individual data are known.

## ACKNOWLEDGEMENTS

This work was partly sponsored by the Koningin Wilhelmina Fonds, The Netherlands (project KWF-MBL 81-1) and the European Economic Community (EEC 533-N-(B)). The authors would like to thank Dr F. Berends for his help in preparing the manuscript, Dr M.A. Schoen for supplying data on the quantitative detection of DNA adducts at the single-cell level, and Mrs M.J. Lansbergen and Mrs A. Schuite for technical assistance.

## REFERENCES

Albertini, R.J. (1982) *Studies with T-lymphocytes: an approach to human mutagenicity monitoring*. In: Bridges, B.A., Butterworth, B.E. & Weinstein, I.B., eds, *Indicators of Genotoxic Exposure* (*Banbury Report 13*), Cold Spring Harbor, NY, Cold Spring Harbor Laboratory, pp. 393-412

Albertini, R.J., Sylvester, D.L. & Allen, E.F. (1982) *The 6-thioguanine-resistant peripheral blood lymphocyte assay for direct mutagenicity testing in humans*. In: Heddle, J.A., ed., *Mutagenicity: New Horizons in Genetic Toxicology*, New York, Academic Press, pp. 305-336

Amacher, D.E., Turner, G.N. & Ellis, J.H. Jr (1981) Detection of mammalian cell mutagenesis in urine from carcinogen-dosed mice. *Mutat. Res.*, *90*, 79-90

Anderson, D., Richardson, C.R., Weight, T.M., Purchase, I.F.H. & Adams, W.G.F. (1980) Chromosomal analysis in vinylchloride exposed workers. Results from analysis 18 and 42 months after an initial sampling. *Mutat. Res.*, *79*, 151-162

Baan, R.A., Schoen, M.A., Zaalberg, O.B. & Lohman, P.H.M. (1982) *The detection of DNA damages by immunological techniques*. In: Sorsa, M. & Vainio, H., eds, *Mutagens in Our Environment*, New York, Alan R. Liss, pp. 111-124

Baan, R.A., Lohman, P.H.M., Zaalberg, O.B., Schoen, M.A., Fichtinger-Schepman, A.M.J., Schutte, H.H. & van der Schans, G.P. (1984) *Future tools in biomonitoring*. In: Stich, H.F., ed., *Carcinogens and Mutagens in the Environment*, Vol. IV, *The Workplace*, Boca Raton, FL, CRC Press (in press)

Bigbee, W.L., Vanderlaan, M., Fong, S.S.N. & Jensen, R.H. (1983) Monoclonal antibodies specific for the *M*- and *N*-forms of human glycophorin A. *Mol. Immunol.*, *20*, 1353-1362

Browning, L.S. (1973) Mutagenicity of various chemicals and their metabolites in Drosophila. *Genetics*, *74*, 533

Butterworth, B.E., Doolittle, D.J., Working, P., Strom, S.C., Jirtle, R.L. & Michalopoulos, G. (1982) *Chemically-induced DNA repair in rodent and human cells*. In: Bridges, B.A., Butterworth, B.E. & Weinstein, I.B., eds, *Indicators of Genotoxic Exposure* (*Banbury Report 13*), Cold Spring Harbor, NY, Cold Spring Harbor Laboratory, pp. 101-114

Calleman, C.J., Ehrenberg, L., Jansson, B., Osterman-Golkar, S., Segerbäck, D., Svensson, K. & Wachtmeister, C.A. (1978) Monitoring and risk assessment by means of alkyl groups in hemoglobin in persons occupationally exposed to ethylene oxide. *J. environ. Pathol. Toxicol.*, *2*, 427-442

Carrano, A.V., Minkler, J.L., Stetka, D.G. & Moore, D.H. (1980) Variation in the baseline Sister Chromatid Exchange frequency in human lymphocytes. *Environ. Mutag.*, *2*, 325-337

Conner, T.H., Stoeckel, M., Errard, J. & Legator, M.S. (1977) The contribution of metronidazole and two metabolites to the mutagenic activity detected in urine of treated humans and mice. *Cancer Res.*, *37*, 629-633

van Doorn, R., Leijdekkers, C.M., Bos, R.P., Brouns, R.M.E. & Henderson, P.T. (1981) Detection of human exposure to electrophilic compounds by assay of thioether detoxication products in urine. *Ann. occup. Hyg.*, *24*, 77-92

Durston, W.E. & Ames, B.N. (1974) A simple method for the detection of mutagens in urine: studies with the carcinogen 2-acetylaminofluorene. *Proc. natl Acad. Sci. USA*, *71*, 737-741

Ehrenberg, L., Moustacchi, E. & Osterman-Golkar, S. (1983) Dosimetry of genotoxic agents and dose-response relationships of their effects, ICPEMC Working Paper 4/4. *Mutat. Res.*, *123*, 121-182

Emerit, I., Levy, A. & Cerutti, P. (1983) Suppression of tumour promoter phorbolmyristate acetate-induced chromosome breakage by antioxidants and inhibitors of arachidonic acid metabolism. *Mutat. Res.*, *110*, 327-335

Evans, H.J. (1982) *Cytogenetic studies on industrial populations exposed to mutagens*. In: Bridges, B.A., Butterworth, B.E. & Weinstein, I.B., eds, *Indicators of Genotoxic Exposure* (*Banbury Report 13*), Cold Spring Harbor, NY, Cold Spring Harbor Laboratory, pp. 325-340

Evans, H.J., Buckton, K.E., Hamilton, G.E. & Carothers, A. (1979) Radiation-induced chromosome aberrations in nuclear dockyard workers. *Nature*, *277*, 531-534

Fichtinger-Schepman, A.M.J., Lohman, P.H.M. & Reedijk, J. (1982) Detection and quantification of adducts formed upon interaction of diamminedichloroplatinum(II) with DNA, by anion-exchange chromatography after enzymatic degradation. *Nucleic Acids Res.*, *10*, 5345-5356

Filser, J.G. & Bolt, H.M. (1983) Exhalation of ethylene oxide by rats on exposure to ethylene. *Mutat. Res.*, *120*, 57-60

Forni, A. & Secchi, C.C. (1973) *Chromosome changes in preclinical and clinical lead poisoning and correlation with biochemical findings.*. In: *Proceedings of the International Symposium on Environmental Aspects of Lead, Amsterdam, 2-6 October 1972*, Luxembourg, Commission of the European Communities, pp. 473-483

Funes-Cravioto, F., Lambert, B., Lindsten, J., Ehrenberg, L., Natarajan, A.T. & Osterman-Golkar, S. (1975) Chromosome aberrations in workers exposed to vinylchloride. *Lancet*, *i*, 459

Furihata, C. & Matsushima, T. (1982) *Unscheduled DNA synthesis in rat stomach - short term assay of potential stomach carcinogens.* In: Bridges, B.A., Butterworth, B.E. & Weinstein, I.B., eds, *Indicators of Genotoxic Exposure* (*Banbury Report 13*), Cold Spring Harbor, NY, Cold Spring Harbor Laboratory, pp. 123-135

Hanawalt, P.C., Cooper, P.K., Ganesan, A.K. & Smith, C.A. (1979) DNA repair in bacteria and mammalian cells. *Ann. Rev. Biochem., 48,* 783-836

Hansteen, I., Hillestad, L., Thiis-Evensen, E. & Heldaas, S.S. (1978) Effects of vinylchloride in man - a cytogenetic follow-up study. *Mutat. Res., 51*, 271-278

Heddle, J.A., Blakey, D.H., Duncan, A.M.V., Goldberg, M.T., Newmark, H., Wargovich, M.J. & Bruce, W.R. (1982) *Micronuclei and related nuclear anomalies as a short-term assay for colon carcinogens.* In: Bridges, B.A., Butterworth, B.E. & Weinstein, I.B., eds, *Indicators of Genotoxic Exposure* (*Banbury Report 13*), Cold Spring Harbor, NY, Cold Spring Harbor Laboratory, pp. 367-377

Hook, E.B. (1981) Human teratogenic and mutagenic markers in monitoring about point sources of pollution. *Environ. Res., 25*, 178-203

IARC (1981) *IARC Monographs on the Evaluation of the Carcinogenic Risk of Chemicals to Humans*, Vol. 26, *Some Antineoplastic and Immunosuppressive agents*, Lyon, International Agency for Research on Cancer

ICPEMC (1983) Committee 4 final report: Estimation of genetic risks and increased incidence of genetic disease due to environmental mutagens. *Mutat. Res., 115*, 255-291

Jansen, J.D. (1983) *Measurement of chromosome-aberrations in man: dose response relationship.* In: Castellani, A., ed., *Course on the Use of Human Cells for the Assessment of Risk from Physical and Chemical Agents*, New York, Plenum Press, pp. 63-76

Lambert, B., Lindblad, A., Holmberg, K. & Francesconi, D. (1982) *The use of sister-chromatid exchange to monitor human populations for exposure to toxicologically harmful agents.* In: Wolff, S., ed., *Sister Chromatid Exchange*, New York, John Wiley & Sons, pp. 149-182

Legator, M.S., Truong, L. & Connor, T.H. (1978) *Analysis of body fluids including alkylation of macromolecules for detection of mutagenic agents.* In: Hollaender, A. & de Serres, F.J., eds, *Chemical Mutagens, Principles and Methods for their Detection*, New York, Plenum, 5, pp. 1-23

Lehman, A.R. & Karran, P. (1981) DNA repair. *Int. Rev. Cytol., 72*, 101-146

Lloyd, D.C., Purrat, R.J., Dolphin, G.W., Bolton, D. & Edwards, A.A. (1975) The relationship between chromosome aberrations and low LET radiation dose to human lymphocytes. *Int. J. Radiat. Biol., 28*, 75-90

Martin, R.H., Balkan, W., Burns, K. & Lin, C.C. (1982) Direct chromosomal analysis of human spermatozoa: results from 18 normal men. *Am. J. human Genet., 34*, 459-468

Miller, E.C. & Miller, J.A. (1975) *The metabolic activation and reactivity of carcinogenic aromatic amines and amides*. In: Bucalossi, P., Veronesi, U. & Cascinelli, N., eds, *Proc. 11th International Cancer Congress*, Vol. 2, Amsterdam, Excerpta Medica, pp. 3-8

Mendelsohn, M.L., Bigbee, W.L., Branscomb, E.W. & Stamatoyannopoulos, G. (1980) The detection and sorting of rare sickle-hemoglobin containing cells in normal human blood. *Flow Cytometry, IV*, 311-313

Meretoja, F., Vainio, H., Sorsa, M. & Härkönen, H. (1977) Occupational styrene exposure and chromosomal aberrations. *Mutat. Res., 56*, 193-197

Mohn, G.R., Kerklaan, P., van Zeeland, A.A., Ellenberger, J., Baan, R.A., Lohman, P.H.M. & Pons, F.-W. (1984) Methodologies for the determination of various genetic effects in permeable strains of *E. coli* K-12 differing in DNA repair capacity. Quantification of DNA-adduct formation, experiments with organ homogenates and hepatocytes, and animal-mediated assays. *Mutat. Res., 125*, 153-184

Natarajan, A.T., van Buul, P.P.W. & Raposa, T. (1978) *An evaluation of the use of the peripheral blood lymphocyte systems for assessing cytological effects induced* in vivo *by chemical mutagens*. In: Evans, H.J. & Lloyd, D.C., eds, *Mutagen-induced Chromosome Damage in Man*, Edinburgh, Edinburgh University Press, pp. 268-274

Natarajan, A.T. & Obe, G. (1980) Screening of human populations for mutations induced by environmental pollutants: use of human lymphocyte system. *Ecotoxicol. environ. Safety, 4*, 468-481

Reddy, E.P., Reynolds, R.K., Santos, E. & Barbacid, M. (1982) A point mutation is responsible for the acquisition of transforming properties by the T24 human bladder carcinoma oncogene. *Nature, 300*, 149-152

Sobels, F.H. (1982) *The parallelogram: an indirect approach for the assessment of genetic risks from chemical mutagens*. In: Bora, K.C., Douglas, G.R. & Nestmann, E.R., eds, *Chemical Mutagenesis, Human Population Monitoring and Genetic Risk Assessment*, Amsterdam, Elsevier Biomedical Press, pp. 323-327

Sorsa, M., Norppa, H. & Vainio, H. (1982) *Induction of Sister Chromatid Exchanges among nurses handling cytostatic drugs*. In: Bridges, B.A., Butterworth, B.E. & Weinstein, I.B., eds, *Indicators of Genotoxic Exposure* (*Banbury Report 13*), Cold Spring Harbor, NY, Cold Spring Harbor Laboratory, pp. 341-354

Stich, H.F., San, R.H.C. & Rosin, M.P. (1983) Adaptation of the DNA repair and micronucleus tests to human cell suspensions and exfoliated cells. *Ann. N.Y. Acad. Sci., 407*, 93-105

Strauss, G.H.S. (1982) *Direct mutagenicity testing: the development of a clonal assay to detect and quantitate mutant lymphocytes arising* in vivo. In: Bridges, B.A., Butterworth, B.E. & Weinstein, I.B., eds, *Indicators of Genotoxic Exposure* (*Banbury Report 13*), Cold Spring Harbor, NY, Cold Spring Harbor Laboratory, pp. 423-441

Tabin, C.J., Bradley, S.M., Bargmann, C.I., Weinberg, R.A., Papageorge, A.G., Scolnick, E.M., Dhar, R., Lowy, D.R. & Chang, E.H. (1982) Mechanism of activation of a human oncogene. *Nature, 300*, 143-149

Venitt, S. (1982) Mutagens in human faeces: are they relevant to cancer of the large bowel? *Mutat. Res.*, *98*, 265-286

Wright, A.S. (1983) *Molecular dosimetry techniques in human risk assessment: an industrial perspective.* In: Hayes, A.W., Schmell, R.C. & Miya, T.S., eds, *Developments in the Science and Practice of Toxicology, ICT*, Amsterdam, Elsevier Biomedical Press, pp. 311-318

Zielhuis, R.L. (1978) Biological monitoring. *Scand. J. Work Environ. Health*, *4*, 1-18

# EXPOSURE TO MUTAGENIC AROMATIC HYDROCARBONS OF WORKERS CREOSOTING WOOD

**R.P. Bos, F.J. Jongeneelen, J.L.G. Theuws & P.T. Henderson**

*Institute of Pharmacology, Toxicology Unit,*
*University of Nijmegen,*
*Nijmegen, The Netherlands*

## SUMMARY

Creosote P1 is mutagenic in the *Salmonella* microsome assay towards strains TA1537, TA1538, TA98 and TA100 in the presence of S9 mix. The mutagenic polycyclic aromatic hydrocarbons benzo[*a*]pyrene and benz[*a*]anthracene in this mixture are detected in concentrations of 0.18 and 1.1%, respectively.

Spot samples taken from contaminated surfaces in several areas of a wood-preserving industry were tested for mutagenicity. The positive results suggest that a wipe test can give a first indication of occupational exposure to mutagenic substances, particularly when greater exposure occurs *via* skin contact than *via* inhalation.

In urine of rats, mutagens appeared after treatment with creosote. However, no increase in mutagenicity could be detected in urine of creosote workers in relation to their work.

## INTRODUCTION

Creosote P1 is a mixture of oils that are separated during the distillation of coal-tar and consists principally of liquid and solid hydrocarbons. It is rich in materials boiling below 355°C but contains residual materials also.

Workers in the wood-preserving industry receive some level of exposure to components of creosote by inhalation of vapours or through skin contact. Apart from symptoms such as contact dermatitis and photosensitization, it is known that creosote causes skin cancer in humans and may cause cancer of the lung (Sax, 1981). Considering its source and the effects described in humans, it was striking that creosote P1 was denoted nonmutagenic in bacterial assays (US Department of Agriculture, 1981). For this reason we reinvestigated the mutagenic properties of this product in the *Salmonella*/microsome assay.

After mutagenic properties were found for creosote P1, it was suspected that the substance means a potential genetic hazard to persons who are occupationally involved in the preservative treatment of wood. In the present study, possibilities of detecting exposure to creosote were explored. At several places in the work area, contaminated surfaces were examined for the presence of mutagens. In addition, since it was found that exposure of Wistar rats to creosote, under experimental conditions, led to excretion of mutagenic products in urine, an attempt was made to monitor creosote workers by measurements of urinary mutagenicity.

## MATERIALS AND METHODS

### *Mutagenicity testing*

The mutagenicity test was performed according to Maron and Ames (1983) with *Salmonella typhimurium* strains TA1535, TA1537, TA1538, TA98 and TA100. Samples of 0.1 ml of dilutions of creosote or benzo[*a*]pyrene, or samples of 0.1 ml of the solution of the thin-layer chromatography (TLC) residues in dimethyl sulfoxide (see below), were added per plate.

Rat-liver 9000 $\times$ *g* supernatant fractions (S9) were prepared from male Wistar rats, pretreated with Aroclor 1254. The mix contained 0.1 ml S9 per ml.

A portion of S9 was centrifuged at 105 000 $\times$ *g* for 1 h and the resulting pellet resuspended in 0.15 M potassium chloride. The microsome mix contained 0.1 ml of this suspension per ml. The epoxide hydratase inhibitor, 1,2-epoxy-3,3,3-trichloropropane, was added at 2.3 μmol/plate (Bentley *et al.*, 1977).

### *Thin-layer chromatography*

A sample of 200 μl of a solution of creosote in ethanol (10 mg/ml) was applied to a TLC plate, in a continuous streak, with a line applicator.

Glass plates (20 $\times$ 20 cm) coated with a layer 0.25-mm thick of silicagel 60 obtained from Merck (Darmstadt, FRG) were employed. Pentane/diethylether (95:5, v/v) was used as a solvent system. Benzo[*a*]pyrene and benz[*a*]anthracene were visualized on the TLC plate under ultra-violet light of 254 nm. Some fractions on the TLC plate were scraped off and suspended in 5 ml of acetone. Silicagel residues were removed by centrifugation. The acetone was evaporated under nitrogen at 60°C, and the residues were dissolved in 0.25 ml of methanol for analytical purposes, or 1 ml of dimethyl sulfoxide for mutagenicity testing.

### *Impregnation process*

Preservation by impregnation with creosote is accomplished by a pressure method (full-cell process). After a charge of wood is taken into a cylinder, an initial vacuum is applied for a period of at least 15 min. At the end of this period, the vessel is filled with creosote that has a temperature of at least 70°C (mostly 85-100°C) while the vacuum is maintained. Next, the vacuum is released and pressure at a maximum of 10 kgf/cm$^2$ (981 kPa) is applied to the system. This is maintained until the required gross absorption of creosote has been achieved (maximum, 3 h). Absorption time varies with the species being treated. At the end of the pressure cycle, the pressure is reduced to atmospheric level and the preservative returned to storage, while the treated wood is often subjected to a final vacuum to remove excess creosote oil from the surface of the stock. The vacuum is released, the door of the vessel opened, and the treated wood removed.

*Workers*

We have studied the possible exposure of three workers. One of them is the operator of the cylinder; the other two move the wood in and out of the cylinder. The operator, especially, would be briefly exposed to creosote vapours when the cylinder door is opened after treatment. The workers may also be exposed to residual surface creosote through skin contact. Most of the time they wear gloves.

*Monitoring of contaminated surfaces*

To detect contamination of work areas with mutagenic creosote, we applied a method based on that described by Simmon and Peirce (1980) A 5-ml portion of acetone or alcohol was placed on a surface and the solvent mopped up with a paper tissue. The tissue was extracted twice with 25 ml of acetone or alcohol, which was then evaporated to dryness using a rotary evaporator. The total residue was dissolved in 5 ml of dimethyl sulfoxide, then 0.1 ml of this solution was added per plate and assayed for mutagenicity with the *S. typhimurium* tester strain TA98 in the presence of S9 mix.

*Urinary mutagenicity testing*

*Rats*. Male Wistar rats, weighing about 200 g, were purchased from TNO (Rijwswijk, The Netherlands). The animals were housed individually in stainless-steel metabolism cages, designed for separate collection of urine and faeces. The rats had free access to water and food (Hope Farms, Woerden, The Netherlands). Creosote, dissolved in olive oil, was injected intraperitoneally or given on the close-shaven skin in the neck at a dose of 250 mg/kg body weight. Control rats received only olive oil.

Urine samples were collected for 24 h and stored at -20°C until assayed. Before they were assayed, the individual samples were completed to 15 ml and sterilized by filtration through 0.2-μm membrane filters.

Mutagenicity of the urine samples was determined using the *S. typhimurium* strains TA98 and TA100. A sample of 0.1 ml of a full-grown suspension of the bacteria ($\pm$ 2 $\times$ 10$^9$ bacteria/ml) was added to a top agar that contained 0.3 ml of diluted urine, and an activating enzyme system was supplied. This system consisted of either 0.5 ml S9 mix or 0.1 ml of a sterile β-glucuronidase solution (1500 unit/ml), or a combination of both.

*Workers*. Urine samples from the workers were gathered during ten consecutive days, including two free weekends. During this period, two portions of urine were collected daily, one sample overnight and one sample between 10:00 and 16:00. Mutagenicity was determined according to van Doorn *et al.* (1979).

## RESULTS

### *Mutagenicity of creosote*

The mutagenicity of coal-tar creosote P1 towards *S. typhimurium* strains TA1537, TA1538, TA98 and TA100 is shown in Figure 1. Strain TA1535 appeared to be insensitive; with the other strains, mutagenic activity became evident after the addition of S9 mix. The toxicity of creosote for *S. typhimurium* was studied with strains TA98 and TA100: at 50 μg/plate, the survival of TA98 and TA100 was 100%.

**Fig. 1. Number of revertant colonies per plate of five *Salmonella typhimurium* strains as a function of the dose of creosote**

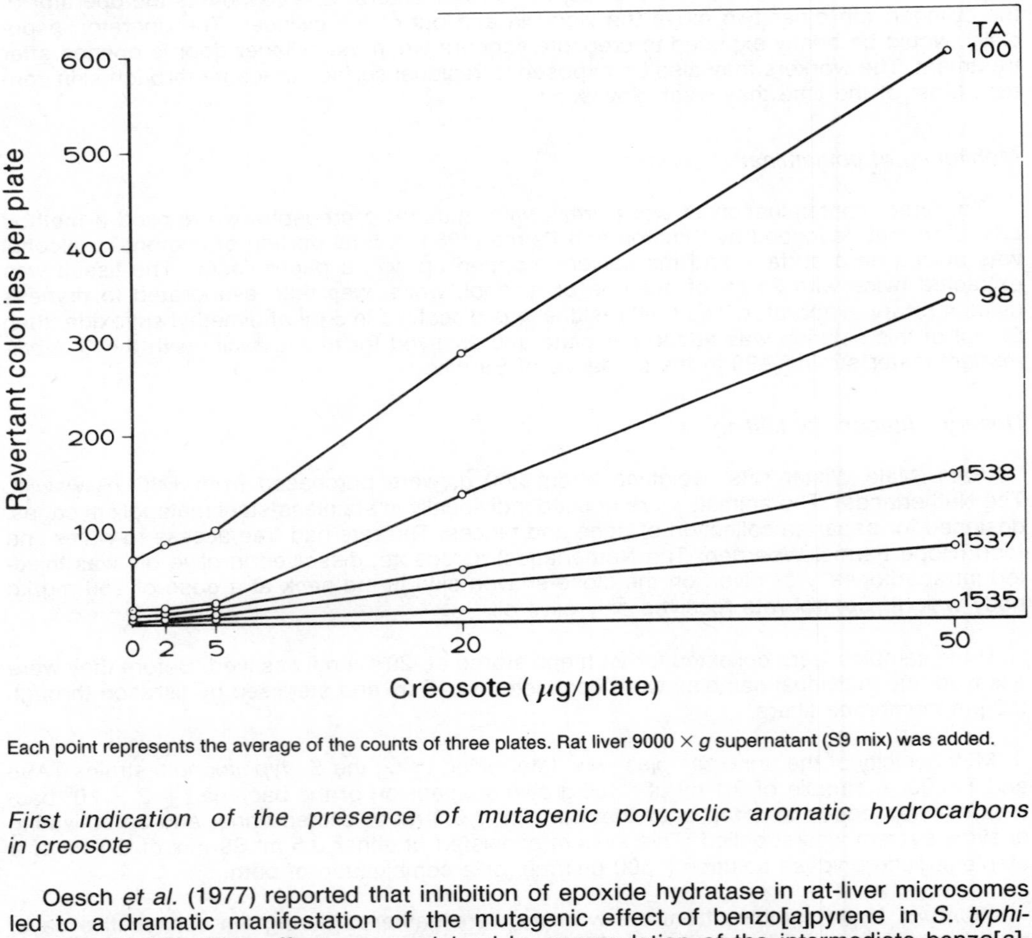


Each point represents the average of the counts of three plates. Rat liver 9000 × *g* supernatant (S9 mix) was added.

### *First indication of the presence of mutagenic polycyclic aromatic hydrocarbons in creosote*

Oesch *et al.* (1977) reported that inhibition of epoxide hydratase in rat-liver microsomes led to a dramatic manifestation of the mutagenic effect of benzo[*a*]pyrene in *S. typhimurium* TA1537. This effect was explained by accumulation of the intermediate benzo[*a*]-pyrene-4,5-oxide, which is mutagenic towards *S. typhimurium* TA1537 (Wood *et al.*, 1975).

Figure 2b shows that addition of the epoxide hydratase inhibitor, 1,1,1,-trichloropropane-2,3-oxide, to liver microsomes from Wistar rats pretreated with Aroclor 1254, substantially enhanced the mutagenic effect of benzo[*a*]pyrene. This effect was also found, to a lesser degree, when creosote was plated instead of benzo[*a*]pyrene (Fig. 2a). From these results, it may be concluded that polycyclic aromatic hydrocarbons are, at least in part, responsible for the mutagenicity of creosote P1.

**Fig. 2. The influence of the epoxide hydratase inhibitor 1,2-epoxy-3,3,3-trichloropropane (TCPO) on the number of *Salmonella typhimurium* TA1537 revertant colonies per plate as a function of the dose of creosote (a), or benzo[*a*]pyrene (b)**

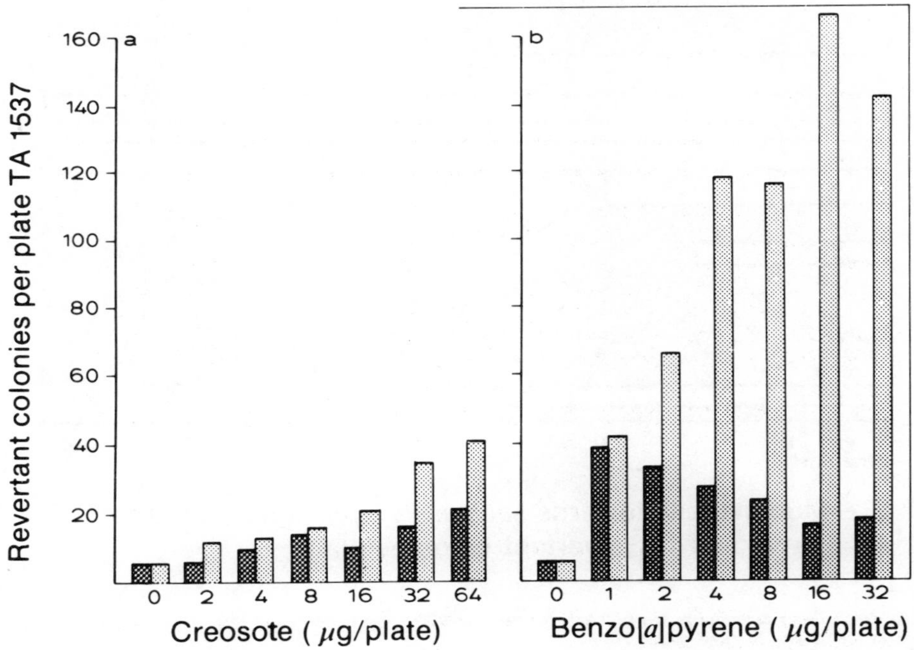


The dark hatched bars represent the number of revertant colonies in the absence of 1,2-epoxy-3,3,3-trichloropropane. Creosote and benzo[*a*]pyrene were plated in the presence of rat liver microsome mix. The values are averages of determinations in triplicate.

*Different mutagenic fractions of creosote*

After the application of creosote in a continuous streak to a TLC plate, several fractions recognized under ultra-violet light of 254 nm were scraped off and tested for mutagenicity towards *S. typhimurium* TA98 in the presence of S9 mix. The results are shown in Figure 3. In a preliminary experiment, no obvious mutagenicity was detected in the area between fraction 1 and fraction 3 and the area between fraction 7 and the front line. Fractions 1, 4 and 6 show most of the mutagenicity.

**Fig. 3. Number of revertant colonies per plate of *Salmonella typhimurium* TA98 as a function of the thin-layer chromatography (TLC) fraction**

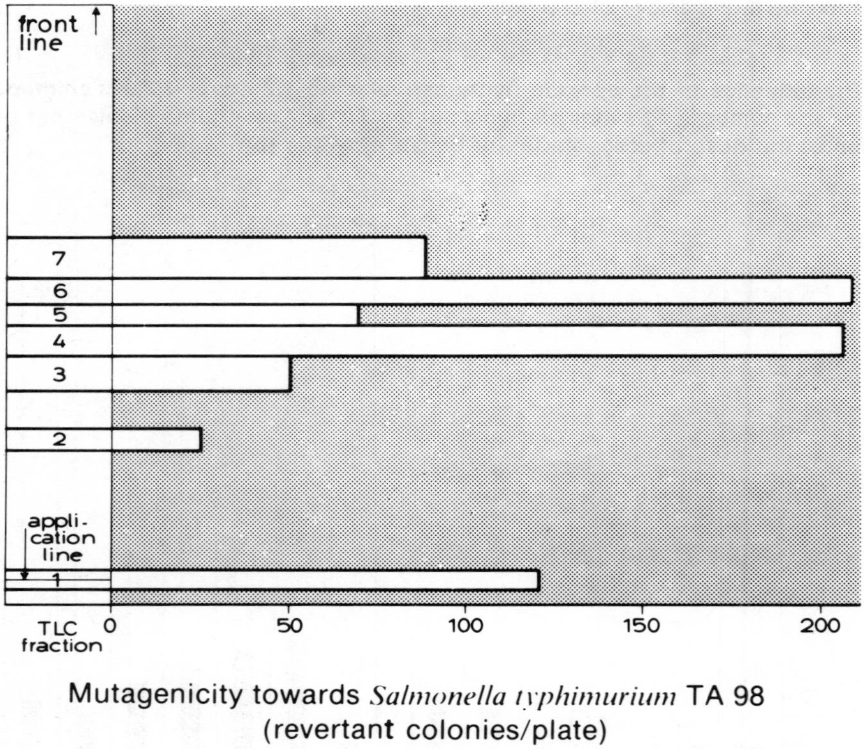


The mutagenicity was detected in the presence of rat liver 9000 × *g* supernatant (S9) mix. Mutagenicity values are averages of the counts of three plates.

### *Identification and quantification of benzo*[a]*pyrene and benz*[a]*anthracene*

The presence of benzo[*a*]pyrene and benz[*a*]anthracene in TLC fractions 4 and 6, respectively (Fig. 3), was confirmed by TLC, high-performance liquid chromatography and spectrophotofluorimetry (Bos, 1984).

Concentrations of benzo[*a*]pyrene and benz[*a*]anthracene in creosote type P1 were assayed by high-performance liquid chromatography equipped with fluorescence detection. It was calculated that creosote contains 1.8 g of benzo[*a*]pyrene and 11 g of benz[*a*]anthracene per kg.

*Detection of mutagenic substances in the work environment*

The work environment was examined for the presence of mutagenic substances. Spot samples from contaminated surfaces in several areas were tested for mutagenicity. These data are presented in Table 1. Areas 1,2,3 and 4 were on metal surfaces, with areas 1 and 2 on a grip of the door of the cylinder and on a handle very close to this door, respectively, area 3 on banisters about 3 m away from the cylinder and area 4 on a lorry used to transport wood in and out of the cylinder. Area 5 was on wood that was not treated with creosote, piled up about 15 m away from the cylinder. A negative control was made from the surface of a clean table in our laboratory while area on the surface of creosoted wood served as a positive control. The results from these samples show that the work environment is contaminated with mutagenic substances.

**Table 1. Contamination with mutagenic substances of some areas in the work environment[a]**

| Sample[c] | Number of *his*+ revertants/plate[b] | |
|---|---|---|
| | Solvent used for cleaning | |
| | Acetone | Alcohol |
| Area 1 | 510 ± 29 | 395 ± 8 |
| Area 2 | 558 ± 52 | 77 ± 2 |
| Area 3 | 435 ± 31 | 109 ± 5 |
| Area 4 | 231 ± 7 | 373 ± 20 |
| Area 5 | 181 ± 9 | 23 ± 1 |
| Negative control | 45 ± 3 | 31 ± 2 |
| Positive control | d | 324 ± 8 |
| Spontaneous | 19 ± 2 | |

[a] Experimental details given in Materials and Methods

[b] Mean values (± SE) of three determinations on the same sample

[c] Areas described in text

[d] No value available because of a toxic effect on the bacteria

It may be noted that extraction with acetone reveals higher mutagenic values than extraction with alcohol.

*Animal experiments*

Administration of creosote to rats resulted in the appearance of mutagens in urine, detectable with strains TA98 and TA100 (Table 2). The highest mutagenicity values were detected after the addition of S9 mix in the presence of β-glucuronidase to the urine.

**Table 2. Mutagenicity towards the *Salmonella typhimurium* strains TA98 and TA100 of urine from a creosote-treated rat after the addition of different enzyme preparations**

| Addition[a] | Number of *his*$^+$ revertants/plate[b] | | | |
|---|---|---|---|---|
| | Creosote[c] | | Control | |
| | TA98 | TA100 | TA98 | TA100 |
| None | 43 ± 2 | 188 ± 9 | 37 ± 1 | 175 ± 8 |
| S9 mix | 44 ± 5 | 202 ± 8 | 39 ± 4 | 151 ± 3 |
| S9 mix + β-glucuronidase | 123 ± 8 | 341 ± 10 | 43 ± 3 | 175 ± 12 |
| β-Glucuronidase | 76 ± 11 | 275 ± 18 | 47 ± 2 | 166 ± 2 |

[a] Experimental details given in Materials and Methods; 150 units of β-glucuronidase were used; S9, rat liver 9000 × *g* supernatant

[b] Mean values (±SE) of three measurements of the same urine

[c] Rats were injected intraperitoneally with a dose of 250 mg/kg body weight

Administration of creosote on close-shaven skin in the necks of rats resulted in urinary mutagenicity yielding about 70% of the number of revertants found after intraperitoneal injection.

*Absence of mutagenicity in urine of creosote workers*

We were interested in whether workers involved in the preservative treatment of wood with creosote had mutagenic urine. Using *S. typhimurium* TA98 in the presence of S9 mix and β-glucuronidase, no increase in urinary mutagenicity was detected that could be related to their work.

## DISCUSSION

The results of this study (Figure 1) are in contrast with earlier data on the mutagenic activity of coal-tar creosote P1 (US Department of Agriculture, 1981). The present demonstration of a mutagenic effect of creosote in *S. typhimurium* indicates a genetic hazard for humans exposed to this substance and correlates well with the carcinogenicity data of Sax (1981).

Mutagenicity of epoxides towards *Salmonella typhimurium* TA1537 is known only for some epoxides of polycyclic aromatic hydrocarbons (benzo[*a*]pyrene-4,5-oxide, dibenz[*a,h*]anthracene-5,6-oxide and chrysene-5,6-oxide) (McCann *et al.*, 1975; Wood *et al.*, 1975). In addition, a number of polycyclic aromatic hydrocarbons, such as benzo[*a*]pyrene, dibenz[*a,c*]anthracene, dibenz[*a,h*]anthracene, 3-methylcholanthrene and 7,12-dimethylbenz-[*a*]anthracene, are mutagenic towards *S. typhimurium* TA1537 in the presence of S9 mix (McCann *et al.*, 1975). From this, and from the results presented in Figure 2, it seems that mutagenic polycyclic aromatic hydrocarbons are present in creosote.

The method used may give a preliminary indication of the nature of the mutagenic principles present in various complex mixtures or environmental samples. The suitability of the procedure was confirmed by the demonstration of the presence of the mutagenic polycyclic aromatic hydrocarbons benzo[*a*]pyrene and benz[*a*]anthracene in creosote P1 by analytical procedures.

The results in Table 1 show clearly the presence of mutagenic substances in the work environment of a wood-preserving industry, most likely due to the presence of creosote.

This method of monitoring a work environment for the presence of mutagens is very easily carried out. It is based principally on the method of Simmon and Peirce (1980) who introduced it to detect the spill of carcinogens and mutagens in laboratories. The present results suggest that application of this wipe test in the work environment can give an indication of occupational exposure to mutagenic and carcinogenic substances that might be important, particularly in those cases where exposure may occur *via* skin contact rather than *via* inhalation.

The mutagenicity that was found in the urine of rats after intraperitoneal administration of creosote was most obvious in *S. typhimurium* TA98 when S9 mix was added to the urine in combination with β-glucuronidase. The appearance of mutagens in the urine of rats after administration of creosote *via* the skin demonstrated that mutagens present in creosote are able to pass through the skin of the rat. This observation, and the finding of Wheeler *et al.* (1981) that coal-tar treated psoriatic patients have mutagenic urine, make it reasonable to assume, in principle, that human exposure to creosote *via* the skin can result in mutagenicity in the urine.

Several studies have shown that the urinary mutagenicity assay is useful in detecting exposure to genotoxic substances. Falck (1982) detected mutagenicity in the urine of non-smoking nurses working in oncology departments. The mutagenicity was reduced when special safety measures were applied. Mutagens have also been detected in the urine of smokers (Yamasaki & Ames, 1977; Bos, 1984) and even in urine of passive smokers (Bos, 1984). However, in spite of substantial contamination of the work environment with mutagenic substances, no increase in urinary mutagenicity of the creosote workers was detected in relation to their work. This observation is in agreement with Møller and Dybing (1980). They did not detect urinary mutagenicity in coke-plant workers exposed to very high levels of polycyclic aromatic hydrocarbons. The appearance of mutagens in the urine of rats only after the administration of a relatively high dose of creosote (250 mg/kg) may indicate, also, that the urinary mutagenicity assay is less useful in detecting exposure to genotoxic polycyclic aromatic hydrocarbons. More selective methods of biological monitoring may give more information about the internal exposure.

**ACKNOWLEDGEMENTS**

Financial support was obtained from the General Directorate of Labour, Dutch Ministry of Social Affairs.

## REFERENCES


Bentley, P., Oesch, F. & Glatt, H. (1977) Dual role of epoxide hydratase in both activation and inactivation of benzo[*a*]pyrene. *Arch. Toxicol.*, *39*, 65-75

Bos, R.P. (1984) *Application of Bacterial Mutagenicity Assays in Genotoxicity Studies*, Thesis, University of Nijmegen

van Doorn, R., Bos, R.P., Leijdekkers, C.M., Wagenaars-Zegers, M.A.P., Theuws, J.L.G. & Henderson, P.T. (1979) Thioether concentration and mutagenicity of urine from cigarette smokers. *Int. Arch. occup. environ. Health*, *43*, 159-166

Falck, K. (1982) *Application of the Bacterial Urinary Mutagenicity Assay in Detection of Exposure to Genotoxic Chemicals* (*Academic dissertation*), Helsinki, Novoprint OY

Maron, D.M. & Ames, B.N. (1983) Revised methods for the *Salmonella* mutagenicity test. *Mutat. Res.*, *113*, 173-215

McCann, J., Choi, E., Yamasaki, E. & Ames, B.N. (1975) Detection of carcinogens as mutagens in the *Salmonella*/microsome test: assay of 300 chemicals. *Proc. natl Acad. Sci. USA*, *72*, 5135-5139

Møller, M. & Dybing, E. (1980) Mutagenicity studies with urine concentrates from coke plant workers. *Scand. J. Work Environ. Health*, *6*, 216-220

Oesch, F., Raphael, D., Schwind, H. & Glatt, H. (1977) Species differences in activating and inactivating enzymes related to the control of mutagenic metabolites. *Arch. Toxicol.*, *39*, 97-108

Sax, N.I. (1981) *Cancer Causing Chemicals*, New York, Van Nostrand/Rheinhold

Simmon, V.F. & Peirce, M.V. (1980) *Design, implementation and monitoring of laboratories for handling chemical carcinogens and mutagens*. In: Walter, D.B., ed., *Safe Handling of Chemical Carcinogens, Mutagens, Teratogens and Highly Toxic Substances*, Ann Arbor, MI, Ann Arbor Science Publishers, pp. 153-166

US Department of Agriculture (1981) *The Biologic and Economic Assessment of Pentachlorophenol, Inorganic Arsenicals, Creosote*, Vol. I, *Wood preservatives* (*Tech. Bull. No. 1658-I*), Washington DC

Wheeler, L.A., Saperstein, M.D. & Lowe, N.J. (1981) Mutagenicity of urine from psoriatic patients undergoing treatment with coal tar and ultraviolet light. *J. invest. Dermatol.*, *77*, 181-185

Wood, A.W., Goode, R.L., Chang, R.L., Levin, W., Conney, A.H., Yagi, H., Dansette, P.M. & Jerina, D.M. (1975) Mutagenic and cytotoxic activity of benzo[*a*]pyrene 4,5- 7,8- and 9,10-oxides and the six corresponding phenols. *Proc. natl Acad. Sci. USA*, *72*, 3176-3180

Yamasaki, E. & Ames, B.N. (1977) Concentration of mutagens from urine by adsorption with nonpolar resin XAD-2: cigarette smokers have mutagenic urine. *Proc. natl Acad. Sci. USA*, *74*, 3555-3559

# MUTAGENICITY STUDIES IN A TYRE PLANT: IN-VITRO ACTIVITY OF URINE CONCENTRATES AND RUBBER CHEMICALS

**R. Crebelli, E. Falcone, G. Aquilina & A. Carere[1]**

*Higher Institute of Health Rome, Italy*

**A. Paoletti & G. Fabri**

*Institute of Occupational Medicine, Catholic University Rome, Italy*

**SUMMARY**

A possible occupational contribution to urinary mutagenicity was studied in a tyre plant, by assaying concentrates of urine from 72 workmen and 23 controls for their activity in the Ames test and microtitre fluctuation test. The results show that smoking habits but not occupation are related to the appearance of a detectable urinary mutagenicity in strain TA98. A possible synergistic effect of occupation was, however, observed among tyre builders who were smokers.

Mutagenicity screening of 25 rubber chemicals, of major technological relevance and used in high volume in the workplace investigated, showed that three of them are weakly active in TA98 and TA100 (tetramethylthiuram disulfide) or TA98 alone (poly-*p*-dinitrosobenzene and mixed diaryl-*p*-phenylendiamines).

---

[1] To whom requests for reprints should be addressed

## INTRODUCTION

Epidemiological studies indicate that occupation in the rubber industry is significantly related to an increased risk of cancer in various organs (IARC, 1982). Biological monitoring techniques, i.e., urinary mutagenicity study and cytogenetic analysis, have been successfully applied previously to investigation of the individual exposure to genotoxins in such workplaces (Vainio *et al.*, 1982). This study describes the investigation of a possible relationship between genotoxic characteristics of raw materials in major use at the workplace and urinary mutagenicity in the employees. A concurrent determination by a sensitive bacterial test was made of the in-vitro activity of urine concentrates of workmen belonging to different job categories and of samples of raw materials handled by the subjects investigated.

## MATERIALS AND METHODS

### *Description of the plant*

All the plants are situated in a single wide shed. Materials are conveyed from one department to another on trolleys and conveyor belts. The mixing area consists of many weighing scales in series, provided with exhaust ventilation. Rubber chemicals are packed in plastic bags and put into Banbury mills together with natural or synthetic (styrene-butadiene polymer) rubber. Carbon blacks and oils are added automatically. Fume pollution in extruding, calendering and vulcanization areas is reduced by exhaust ventilation and by plastic sheets placed over the mills and along the press line.

### *Subjects investigated*

Urine samples from 72 workmen (44 smokers) and 23 control clerks (16 smokers) were collected on Thursday afternoon at the end of a shift. Workmen investigated belonged to the following job categories: chemical weighers (6 people), mostly exposed to rubber chemicals dust; Banbury mixers and calenderers (16), exposed to fumes of uncured rubber; adhesives (cement) compounding workers (3), exposed to organic solvents; tyre builders (22), with extensive skin contact with uncured rubber and solvents; vulcanizers (25), exposed to fumes of cured rubber.

After collection, urine was frozen quickly and stored in the dark at -20°C. Precautions were taken to avoid any chemical contamination during collection and storage.

### *Urinary mutagenicity study*

Urine concentrates were prepared by absorption on XAD-2 resin as described by Yamasaki and Ames (1977) after overnight incubation with 100 μg/ml *Escherichia coli* β-glucuronidase.

Concentrates equivalent to 10 ml of urine were assayed in both the Salmonella plate incorporation assay (Ames test) and the microtitre fluctuation test, carried out as described by Ames *et al.* (1975) and Gatehouse (1979), with strains TA1535, TA98 and TA100 in the presence of a 9000 $\times$ *g* supernatant of the livers from male Sprague-Dawley rats pretreated with Aroclor 1254 (S9).

Results were evaluated statistically by the methods of Gilbert (1980) and Mahon (1983) (microtitre and Ames test, respectively).

*Screening of rubber chemicals*

Samples of raw materials obtained from the chemical laboratory of the plant were tested for their in-vitro mutagenicity in the Ames test, employing TA1535, TA1537, TA98 and TA100 as tester strains and S9 for metabolic activation (Ames *et al.*, 1975).

Each of the chemicals was tested up to the highest nontoxic concentration. Positive controls as well as controls for S9 and sample sterility were routinely included.

Dimethyl sulfoxide was used as the solvent for all chemicals; mineral oils were tested as emulsions in Tween 80 (10% v/v).

## RESULTS

The results of the urinary mutagenicity study obtained from the microtitre fluctuation test and the plate incorporation assay are summarized in Figures 1 and 2.

**Fig. 1. Percent of urine concentrates that gave negative (nonsignificant, n.s.) and positive responses in the microtitre fluctuation test**

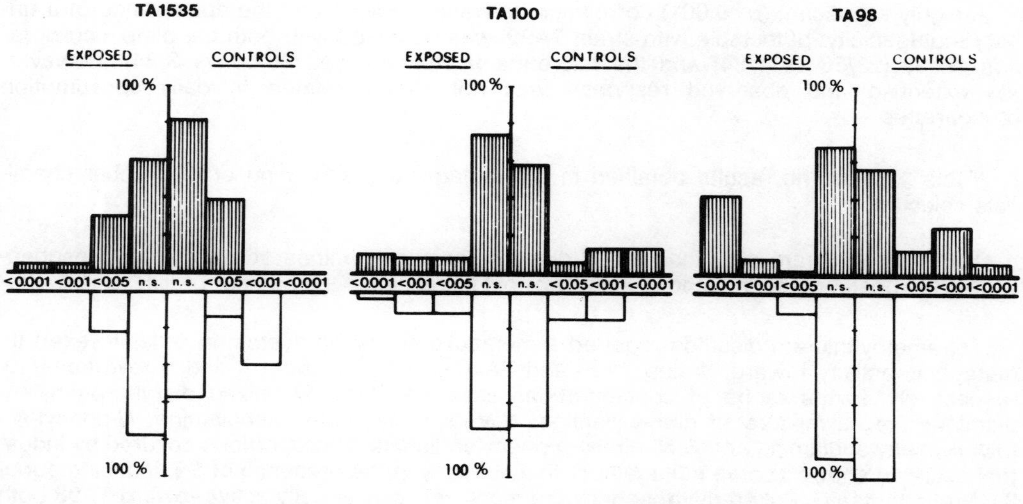


▨, smokers (n = 40 exposed; n = 6 controls); □, non-smokers (n = 27 exposed; n = 7 controls)

**Fig. 2. Percent of urine concentrates positive**

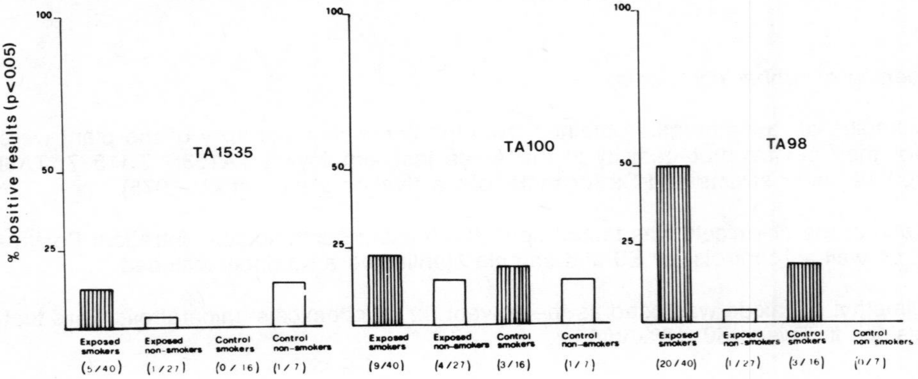


(p<0.05) in the Ames test

The possible correlation between urinary mutagenicity and occupation was investigated first by comparing the frequencies of mutagenic urine concentrates among workmen and control clerks. Other comparisons were set up, also, to take account of smoking habits, job category and the level of significance of the observed effects. No significant excess of positive samples among workmen was observed by any of these comparisons. One possible synergistic effect of occupation with smoking was seen when tyre builders who were smokers were compared with control smokers; the results in the Ames test showed an excess of samples mutagenic in TA98 that were statistically significant (8/11 *vs* 3/16, $p<0.01$). This indication, however, was not confirmed by the results of the microtitre fluctuation test (8/11 *vs* 7/16, nonsignificant).

A highly significant ($p<0.001$) correlation between smoking and the appearance of a urinary mutagenicity, detectable with strain TA98, was observed with both the plate incorporation assay (23/56 *vs* 1/34) and the microtitre fluctuation test (20/56 *vs* 3/34). However, the extent of the observed response was not clearly related to daily consumption of cigarettes.

Table 1 gives the results obtained in the mutagenicity screening of the rubber chemicals selected.

Tetramethylthiuram disulfide, mixed diaryl-*p*-phenylendiamines and poly-*p*-dinitrosobenzene were positive, i.e., induced at least two-fold increases in the number of *his*$^+$ revertants.

Tetramethylthiuram disulfide required a metabolic activating system in order to exert its mutagenic activity toward strains TA98 and TA100, inducing about 1 and 5 revertants/µg, respectively, in the range of concentrations assayed. Similarly, mixed diaryl-*p*-phenylendiamines (i.e., a mixture of diphenylamine, *N,N'*-diphenyl-*p*-phenylendiamine, *N*-phenyl-*N'*-tolyl-*p*-phenylendiamine and *N,N'*-ditolyl-*p*-phenylendiamine in proportions covered by industrial secrecy) showed some mutagenicity in TA98 only in the presence of S9, inducing about 0.07 revertants/µg. Poly-*p*-dinitrosobenzene, however, was equally active toward TA98 both in the presence and absence of S9, inducing about 0.5 revertants/µg in both cases.

**Table 1. Results of screening of rubber chemicals (in strains TA1535, TA1537, TA98 and TA100)**

| Chemical name | Purity | Use class | % in finished rubber goods | Range of conc. (μg/plate) | Activity |
|---|---|---|---|---|---|
| Dimethylpentyl-*p*-phenylendiamine | unknown | antioxidant | | 200-5000 | negative |
| Diaryl-*p*-phenylenediamines, mixed | 80% | antioxidant | 1-2 | 200-5000 | positive with TA98 |
| Isopropylphenyl-*p*--phenylendiamine | 95% | antioxidant | 3-4 | 20-500 | negative |
| Pentachlorothiophenol | 47% | peptizer | 0.1-0.2 | 40-1000 | negative |
| Diphenylguanidine | 96.5% | accelerator | 0.2-0.4 | 200-5000 | negative |
| Morpholinyl-benzothiazole-sulfenamide | 94% | accelerator | 1-1.5 | 40-200 | negative |
| Oxy-diethylene-benzothiazole-sulfenamide | unknown | accelerator | 1-2 | 10-1000 | negative |
| Polytrimethyldihydroquinoline | unknown | antioxidant | | 40-1000 | negative |
| Cyclohexylthiophthalimide | unknown | retardant | 0.1-0.2 | 2.5-40 | negative |
| Diisopropylbenzothiazolesulfenamide | 98.5% | accelerator | 1-2 | 10-200 | negative |
| Hexamethylenetetramine | unknown | accelerator | | 200-5000 | negative |
| Resorcinol | 99% | bonding | 3-4 | 200-5000 | negative |
| Poly-*p*-dinitrosobenzene | 25% | accelerator | 0.2-0.4 | 8-200 | positive with TA98 |
| Tetramethylthiuram disulfide | unknown | accelerator | 1-2 | 8-200 | positive with TA98 and TA100 |
| Phenolformaldehyde resin | unknown | softener | 3-5 | 200-5000 | negative |
| Stearic acid | unknown | softener | 2-3 | 40-1000 | negative |
| Naphthenic/paraffinic medium processing oil | | softener | | 10-50 μl | negative |
| Highly aromatic processing oil, denatured | | softener | | 10-50 μl | negative |
| Zinc oxide | 99% | activator | 3-5 | 1000-5000 | negative |
| Sulfur | | accelerator | 2-5 | 1000-3000 | negative |
| Dithiobisbenzothiazole | 95% | accelerator | 1-2 | 40-1000 | negative |
| Mercaptobenzothiazole | 94% | accelerator | 1-2 | 8-200 | negative |
| Cobalt naphthenate | unknown | bonding | 1-2 | 500-2500 | negative |
| *N*-Nitrosodiphenylamine[a] | unknown | retardant | | 40-1000 | negative |

[a] The use of *N*-nitrosodiphenylamine as a retardant has been suspended since 1980.

## DISCUSSION

The results of the urinary monitoring show that, under these experimental conditions, smoking habits but not occupation in the tyre plant investigated are related to detectable urinary mutagenicity. The biological significance of the observed synergistic effect of the occupation with smoking is questionable because statistical significance was reached only for a small group of employees, i.e., tyre builders.

Different conclusions were reached in previous investigations carried out in rubber industries in northern Europe; these showed that occupational exposure to xenobiotics in the rubber industry workplace is related positively to a remarkable urinary mutagenicity (Falck *et al.*, 1980; Sorsa *et al.*, 1982; Vainio *et al.*, 1982).

An explanation for this apparent discrepancy might rest on the different individual exposure to mutagenic xenobiotics in the workplaces investigated by us and by others. It is conceivable, in fact, that measures of industrial hygiene introduced recently in the plant considered here (see Description of the Plant) together with substitution of the most harmful chemicals (e.g., phenyl-2-naphthylamine and *N*-nitrosodiphenylamine) reduced levels of occupational exposure below those occurring in other workplaces.

Also, the low level of in-vitro mutagenic activities observed in the screening of rubber chemicals suggests that occupational exposure to mutagens in this workplace could be less relevant than that due, for example, to smoking habits and, therefore, remained undetected by the urinary mutagenicity assay.

The absence of detectable urinary mutagenicity, however, is by no means proof that there is no exposure to mutagenic xenobiotics; it simply indicates that previous exposure, if any, was below the lowest detectable. In order to obtain a more complete picture of the genotoxic risks in this workplace, other studies are to be undertaken, especially concerning possible exposure to newly formed genotoxins. Particular attention is to be paid to the volatile nitrosamines, which are scarcely detected in the urine assay (Ohshima *et al.*, 1982) but which are certainly highly significant toxicologically.

A comparison of the sensitivities of the test systems used in the urinary mutagenicity assays shows that the microtitre fluctuation test as used in the present study offered no clear advantage over the Ames plate incorporation assay, possibly because the microtitre test is more sensitive to cytotoxic and other confounding effects. In the microtitre fluctuation test, the small amount of cultural medium added for mutant expression and growth (one-tenth of the amount added in the high-performance macroscale fluctuation test) may be insufficient for adequate dilution of the urine concentrate, thus making the assay too sensitive to the effect of confounding factors present in the urinary concentrates.

## ACKNOWLEDGEMENTS

The technical assistance of U. Cervelli is gratefully acknowledged. This work was partially supported by the EEC [contract no. 530 EVN I(s)].

## REFERENCES

Ames, B.N., McCann, J. & Yamasaki, E. (1975) Methods for detecting carcinogens and mutagens with the *Salmonella*/mammalian-microsome mutagenicity test. *Mutat. Res.*, *31*, 347-364

Falck, K., Sorsa, M. & Vainio, H. (1980) Mutagenicity in urine of workers in rubber industry. *Mutat. Res.*, *79*, 45-52

Gatehouse, D.H. & Delow, G.F. (1979) The development of a 'microtitre'' fluctuation test for the detection of indirect mutagens, and its use in the evaluation of mixed enzyme induction of the liver. *Mutat. Res.*, *60*, 239-252

Gilbert, R.I. (1980) The analysis of fluctuation test. *Mutat. Res.*, *74*, 283-289

IARC (1982) *IARC Monographs on the Evaluation of the Carcinogenic Risk of Chemicals to Humans*, Vol. 28, *The Rubber Industry*, Lyon, International Agency for Research on Cancer

Mahon, G.A.T. (1983) *The statistical significance of the Ames test results*. In: *12th Meeting of the Contact Group Genetic Effects of Environmental Chemicals, London, January 6-7, 1983*, Luxembourg, Commission of the European Communities, pp. 109-115

Ohshima, H., Béréziat, J.C. & Bartsch, H. (1982) *Measurement of endogenous* N-*nitrosation in rats and humans by monitoring urinary and faecal excretion of of* N-*nitrosamino acids*. In: Bartsch, H., O'Neill, I.K., Castegnaro, M. & Okada, M., eds, N-*Nitroso Compounds: Occurrence and Biological Effects* (*IARC Scientific Publications No. 41*), Lyon, International Agency for Research on Cancer, pp. 397-411

Sorsa, M., Falck, K. & Vainio, H. (1982) *Detection of worker exposure to mutagens in the rubber industry by use of the urinary mutagenicity assay*. In: Sugimura, T. & Kondo, S., eds, *Environmental Mutagens and Carcinogens*, New York, Alan R. Liss, pp. 323-329

Vainio, H., Falck, K., Mäki-Paakkanen, J. & Sorsa, M. (1982) *Possibilities for identifying genotoxic risks in the rubber industry: use of the urinary mutagenicity assay and sister chromatid exchange*. In: Bartsch, H. & Armstrong, B., eds, *Host Factors in Human Carcinogenesis* (*IARC Scientific Publications No. 39*), Lyon, International Agency for Research on Cancer, pp. 571-577

Yamasaki, E. & Ames, B.N. (1977) Concentration of mutagens from urine by adsorption with the nonpolar resin XAD-2: cigarette smokers have mutagenic urine. *Proc. natl Acad. Sci. USA*, *74*, 3555-3559

# QUANTITATION OF CARCINOGEN-DNA ADDUCTS BY A STANDARDIZED HIGH-SENSITIVE ENZYME IMMUNOASSAY

**E. Kriek, M. Welling & C.J. van der Laken**

*Division of Chemical Carcinogenesis, Antoni van LeeuwenhoekHuis, The Netherlands Cancer Institute, Amsterdam, The Netherlands*

**SUMMARY**

A highly sensitive competitive enzyme immunoassay has been developed, allowing accurate determination of DNA containing the adducts *N*-(deoxyguanosin-8-yl)-*N*-acetyl-2-aminofluorene, *N*-(deoxyguanosin-8-yl)-2-aminofluorene, 1-[6-(2,5-diamino-4-oxopyrimidinyl-$N^6$-deoxyriboside)]-3-(2-fluorenyl)urea or *trans*-(7R)-$N^2$-[10-(7$\beta$,8$\alpha$,9$\alpha$-trihydroxy-7,8,9,10-tetrahydrobenzo[*a*]pyren)yl]deoxyguanosine. Standard amounts of the carcinogen-modified DNA preparations (25 fmol/500 ng) were coated on the wells of microtitre plates. Various amounts of the modified DNA preparations were used as inhibitors and added before binding of the antibodies (dilution 1:$10^6$). As second antibody, goat anti-rabbit IgG coupled to alkaline phosphatase was used. The amount of enzyme was determined with 4-methylumbelliferyl phosphate as substrate (Van der Laken *et al.*, 1982). The 50% inhibition values of the standard curves are in the range of 2-16 fmol for the enzyme immunoassays. It is possible to add up to 50 $\mu$g of unknown DNA to the wells, allowing for comparison with the same quantity of modified DNA. This extends the limit of detection to 1-6 adducts per $10^8$ nucleotides. The sensitivity of the assay seems sufficient to demonstrate exposure of humans to known chemical carcinogens.

## INTRODUCTION

During the last 15 years, considerable advances have been made in the identification of DNA bases that are structurally modified by chemical carcinogens. Modifications produced by a number of carcinogens, e.g., aflatoxin $B_1$, aromatic amines, benzo[*a*]pyrene and several *N*-nitroso compounds have been completely identified (Grover, 1979). Quantification of these products in tissues or cells has often been difficult because the binding levels to DNA, *in vivo* or in cultured cells, are extremely low. Until recently, this kind of analysis was possible only with the aid of radioactively labelled carcinogens that had been synthesized in the laboratory. Low specific radioactivity of certain carcinogens and the method of analysis of DNA, i.e., enzymatic hydrolysis followed by chromatography of the nucleosides, required relatively large amounts of DNA. Measurement of modified DNA bases (adducts) after exposure to low doses, or in small tissue and cell samples, was not possible in these cases. DNA from human tissues and cells exposed to nonradioactive carcinogens could not be measured at all. In recent years, highly sensitive immunological procedures have been developed, that are capable of detecting fmol amounts of carcinogen-modified DNA bases, by employing antibodies raised against protein-bound nucleoside adducts and carcinogen-modified DNA (Müller & Rajewsky, 1981; Poirier, 1981). These antibodies have been utilized in sensitive radioimmunoassays (Poirier *et al.*, 1979, 1980a; Müller & Rajewsky, 1980; Kriek *et al.*, 1982), enzyme-linked immunosorbent assays (Hsu *et al.*, 1980; Harris *et al.*, 1982; van der Laken *et al.*, 1982), as well as in immunohistochemical (Heyting *et al.*, 1983; Poirier *et al.*, 1983) and electron microscopic investigations (Daune *et al.*, 1980). Antibodies have been elicited, also, against *cis*-diamminedichloroplatinum(II)-modified DNA and are specific for *cis*-diamminedichloroplatinum(II)-DNA adducts formed *in vivo* and *in vitro* (Poirier *et al.*, 1982a). Antibodies raised against the bovine serum albumin conjugates of *N*-(guanosin-8-yl)-*N*-acetyl-2-aminofluorene (Guo-8-AAF) and *N*-(guanosin-8-yl)-2-aminofluorene (Guo-8-AF) have been employed to monitor the formation and removal of *N*-(deoxyguanosin-8-yl)-*N*-acetyl-2-aminofluorene (dGuo-8-AAF) and *N*-(deoxyguanosin-8-yl)-2-aminofluorene (dGuo-8-AF) in cultured cells and in intact animals (Poirier *et al.*, 1979, 1982b). Anti Guo-8-AAF antiserum cross-reacts with both guanine adducts (Kriek *et al.*, 1982) and has been used by Poirier *et al.* (1980b, 1982b) in radioimmunoassays, with [$^3$H]-Guo-8-AAF and [$^3$H]-Guo-8-AF, as tracers to quantitate both adducts in mixtures, by profile analysis and by comparison with appropriate standard curves. Because Guo-8-AF is unstable under certain conditions (Kriek & Westra, 1980) and we were unable to prepare the tracer [$^3$H]-Guo-8-AF in pure form, we decided to follow a different route for the determination of dGuo-8-AF in DNA. Antibodies were raised against the guanine imidazole ring-opened form of Guo-8-AF (roGuo-8-AF) and employed in a highly sensitive competitive enzyme immunoassay. DNA containing dGuo-8-AF (AF-DNA) was converted into the corresponding form with guanine imidazole ring-opened dGuo-8-AF moieties (roAF-DNA) for quantitation of this adduct.

A number of factors govern the sensitivity, accuracy and reproducibility of the competitive enzyme immunoassay, including coating of the microtitre plates with carcinogen-modified DNA, prevention of nonspecific binding of the antibodies, concentration of the antibodies, enzyme activity of the anti-rabbit IgG alkaline phosphatase conjugate and the reaction conditions during incubation with the substrate. The aim of this study was to establish optimal assay conditions for a highly sensitive and reproducible competitive enzyme immunoassay, in order to quantitate various carcinogen-DNA adducts, based on the procedure developed by Van der Laken *et al.* (1982).

## MATERIALS AND METHODS

### *Chemicals*

Calf thymus DNA (type I), nucleosides, bovine serum albumin and the F(ab$^1$)$_2$ fragment of goat anti-rabbit IgG (whole molecule) alkaline phosphatase conjugate were obtained from Sigma Chemical Co. (St Louis, MO, USA). The fetal calf serum was from Gibco Europe

(Glasgow, UK), and the 4-methylumbelliferyl phosphate di-lithium salt was from Boehringer Mannheim (Mannheim, FRG). Diethanolamine, ethylene dinitrilotetra-acetic acid disodium salt and gelatine were purchased from Merck (Darmstadt, FRG), and the Tween-20 was from Serva Feinbiochemica GMBH & Co. (Heidelberg, FRG).

*N*-Hydroxy-(ring-$^3$H)-*N*-acetyl-2-aminofluorene (specific radioactivity, 1.15 Ci/mmol) and *N*-hydroxy-(ring-$^3$H)-2-aminofluorene were synthesized from (ring-$^3$H)-2-nitrofluorene (Westra, 1981). *N*-Acetoxy-(ring-$^3$H)-*N*-acetyl-2-aminofluorene was prepared by acetylation in a mixture of acetic anhydride and dry pyridine (Irving & Veazey, 1969).

*Buffers*: Phosphate-buffered saline (PBS), 10 mM sodium phosphate monobasic, pH 7.4, 140 mM sodium chloride (Dulbecco A); PBS-Tween, 10 mM sodium phosphate monobasic, pH 7.4, 140 mM sodium chloride, 0.05% Tween-20; wash buffer, 10 mM sodium phosphate monobasic pH 7.4, 140 mM sodium chloride, 0.05% Tween-20 and 0.1% (w/v) gelatine; reaction buffer, 10 mM diethanolamine, pH 9.8, 1 mM magnesium chloride, 0.2 mM 4-methylumbelliferyl phosphate di-lithium salt; stop buffer, 10 mM ethylene dinitrilotetra-acetic acid disodium salt, 100 mM diethanolamine, pH 9.8. All buffers are autoclaved at 120°C for 20 min, except the reaction buffer. Magnesium chloride (0.1 M) was sterilized separately and then diluted with 10 mM ethanolamine, pH 9.8. 4-Methylumbelliferyl phosphate di-lithium salt was not sterilized, because it decomposes upon heating.

*Syntheses of modified nucleic acid components*

Modified nucleic acids and their components were synthesized as described in the references cited: *N*-(guanosin-8-yl)-*N*-acetyl-2-aminofluorene (Guo-8-AAF) and *N*-(deoxyguanosin-8-yl)-N-acetyl-2-aminofluorene (dGuo-8-AAF) (Kriek *et al.*, 1967); *N*-(guanosin-8-yl)-2-aminofluorene and its guanine imidazole ring-opened form (roGuo-8-AF) (Kriek & Westra, 1980). All compounds were purified by Sephadex LH-20 chromatography. Calf thymus DNA was modified *in vitro* with *N*-acetoxy-*N*-acetyl-2-aminofluorene at pH 7 and with *N*-hydroxy-2-aminofluorene at pH 5 (Kriek, 1979). AF-DNA was converted into roAF-DNA by heating in 0.1 N sodium hydroxide at 75°C for 2 h (Kriek & Westra, 1980). A series of standard DNA preparations modified with [$^3$H]-AAF and [$^3$H]-AF from 0.5 μmol/mol DNA-phosphate to 220 μmol/mol DNA-phosphate was prepared as described (Van der Laken *et al.*, 1982). A part of the standard [$^3$H]-AF-DNA series was converted into [$^3$H]-roAF-DNA, as described above. DNA modified with *anti*-(±) 7β,8α-dihydroxy-9α,10α-epoxy-7,8,9,10-tetrahydrobenzo[*a*]pyrene (BPDE I) (1.9% substitution of the bases) was a gift from Dr R.M. Santella (The Cancer Center, Institute of Cancer Research, Columbia University, New York, NY, USA).

*Immunological procedures*

The anti Guo-8-AAF antiserum raised in rabbits has been shown to be highly specific for dGuo-8-AAF and dGuo-8-AF in DNA, but does not cross-react with other AAF-adducts, rodGuo-8-AF, deoxyguanosine or DNA (Van der Laken *et al.*, 1982). Anti roGuo-8-AF was prepared in a similar fashion, by coupling roGuo-8-AF to bovine serum albumin, following oxidation with periodate, as described by Erlanger and Beiser (1964). Inhibition of the tracer-antibody binding by rodGuo-8-AF was determined in a radioimmunoassay, essentially as described by Müller and Rajewsky (1980). The affinity constant was calculated from the radioimmunoassay, as described by Müller (1980). Anti BP-DNA was a gift from Dr M.C. Poirier (Laboratory of Cellular Carcinogenesis and Tumor Promotion, Division of Cancer Cause and Prevention, National Cancer Institute, NIH, Bethesda, MD, USA).

*Enzyme-linked immunosorbent assay (ELISA).* The following procedure is the result of optimalization of the conditions and concentrations in the coating and in each of the following incubation steps adapted from Van der Laken *et al.* (1982):

(1) Polystyrene microtitre plates (3040 F Microtest II from Falcon, Oxnard, USA) are coated with 25 fmol adduct/500 ng DNA, by evaporation at 40°C overnight of 0.05 ml/well of a carcinogen-modified DNA solution in PBS.

(2) The wells are washed five times with wash buffer.

(3) To prevent nonspecific binding of the antibody, 0.2 ml/well of a solution of 0.05% IgG in PBS-Tween (0.05%) is added and the plates are incubated at 37°C for 1.5 h. During all incubations, the plates are covered with adhesive film (Falcon 3044).

(4) The wells are washed five times with wash buffer.

(5) Then 0.05 ml/well of the carcinogen-modified DNA (25 μg) in PBS is added, followed by 0.05 ml/well of the antiserum in 10% fetal calf serum in PBS (dilution $1{:}10^6$). The plates are incubated at 37°C for 1.5 h. It is important that the plates be slowly shaken during this incubation step.

(6) The plates are washed five times with wash buffer.

(7) To each well, 0.1 ml of goat anti-rabbit IgG alkaline phosphatase conjugate (dilution 1:1000 in wash buffer with 5% fetal calf serum) is added and the plates are incubated at 37°C for 1.5 h.

(8) The plates are washed five times with wash buffer and then twice with 0.1 M diethanolamine, pH 9.8.

(9) The reaction buffer is added (0.1 ml/well), and the plates are left at room temperature overnight in the dark.

(10) Aliquots (0.09 ml) from each well are diluted with 1.9 ml stop buffer, and the fluorescence is measured in an Aminco SPF-500 fluorescence spectrophotometer at 443 nm (excitation 366 nm), relative to a 100% converted substrate solution as standard. The degree of inhibition of antibody binding to the carcinogen-modified DNA on the solid phase is calculated by the formula:

$$\left[1-\frac{F_1 - F_2}{F_3 - F_2}\right] \times 100 = \text{inhibition (\%)},$$

where $F_1$ is the fluorescence in inhibitor-containing wells, $F_2$ is the fluorescence in wells with *unmodified* DNA coat and without inhibitor (100% inhibition), and $F_3$ is the fluorescence in wells without inhibitor (0% inhibition). Wells without inhibitor contain 25 μg of unmodified DNA. Samples are generally measured as duplicates and the inhibition values are within ± 10% of the mean.

## RESULTS

The rabbit antiserum raised against roGuo-8-AF showed cross-reactivity with dGuo-8-AF, and, to a lower degree, with dGuo-8-AAF and 2-aminofluorene. There was no cross-reaction with deoxyguanosine and DNA. The calculated antibody affinity constant for rodGuo-8-AF was $3.5 \times 10^7$ l/mol. This antiserum was employed in a radioimmunoassay to detect rodGuo-8-AF in DNA, and the detection limit was 450 fmol/0.1 ml, based on 50% inhibition of the tracer-antibody binding in the standard curve. Figure 1 shows typical standard curves of the rabbit antisera, specific for dGuo-8-AAF and rodGuo-8-AF, that are employed in the highly sensitive enzyme immunoassay; a standard curve for anti BP-DNA is presented in Figure 2. The detection limits for the enzyme immunoassay are given in Table 1, based on the 50% inhibition values of the standard curves. In this enzyme immunoassay, it is possible to add up to 50 μg of unknown DNA to each well, allowing for comparison with a similar quantity of standard modified DNA. Depending on the antibody employed, the detection limit of this enzyme immunoassay is now in the range of 1-10 adducts per $10^8$ nucleotides. The results show that carcinogen-DNA adducts can be quantitated at this level, accurately and with excellent reproducibility, provided that the assay is conducted under carefully controlled conditions. All determinations have to be performed relative to standard carcinogen-modified DNA preparations with known modification (calculated from radioactive labelling), preferably of a similar modification level to that of the DNA samples to be analysed.

**Fig. 1. Competitive enzyme immunoassay for the measurement of *N*-(deoxyguanosin-8-yl)-*N*-acetyl-2-aminofluorene (dGuo-8-AAF) (•), and DNA containing dGuo-8-AAF (AAF-DNA) (★) or 1-[6-(2,5-diamino-4-oxopyrimidinyl-*N*[6]-deoxyriboside)]-3-(2-fluorenyl)-urea (roAF-DNA) (○)**

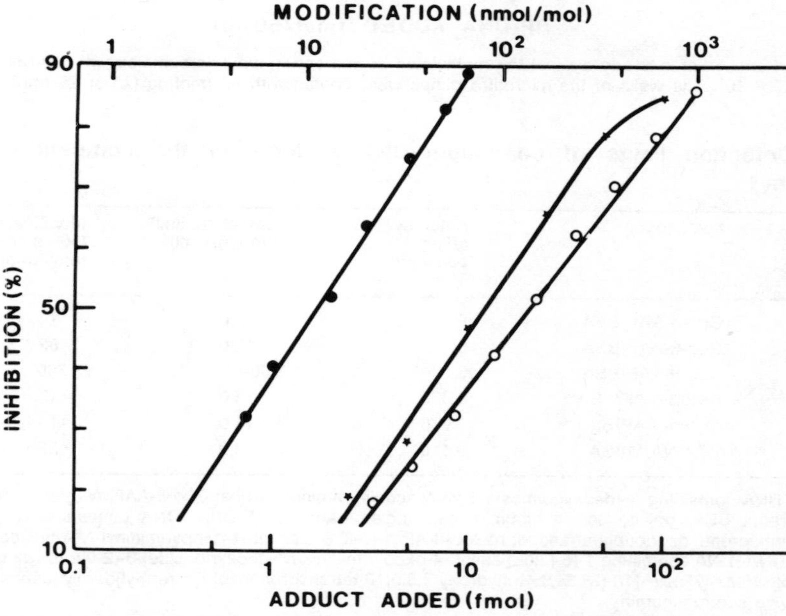


Rabbit-produced antisera are raised against the bovine serum albumin conjugates of *N*-(guanosin-8-yl)-*N*-acetyl-2-aminofluorene and 1-[6-(2,5-diamino-4-oxopyrimidinyl-*N*[6]-riboside)]-3-(2-fluorenyl)urea. Antiserum dilution was 1:$10^6$. The wells of the microtitre plates were coated with AAF-DNA or roAF-DNA (25 fmol/500 ng).

**Fig. 2. Competitive enzyme immunoassay for the measurement of DNA containing *trans*-(7R)-$N^2$-[10-(7β,8α,9α-trihydroxy-7,8,9,10-tetrahydrobenzo[*a*]pyren)-yl]-deoxyguanosine (BP-DNA)**

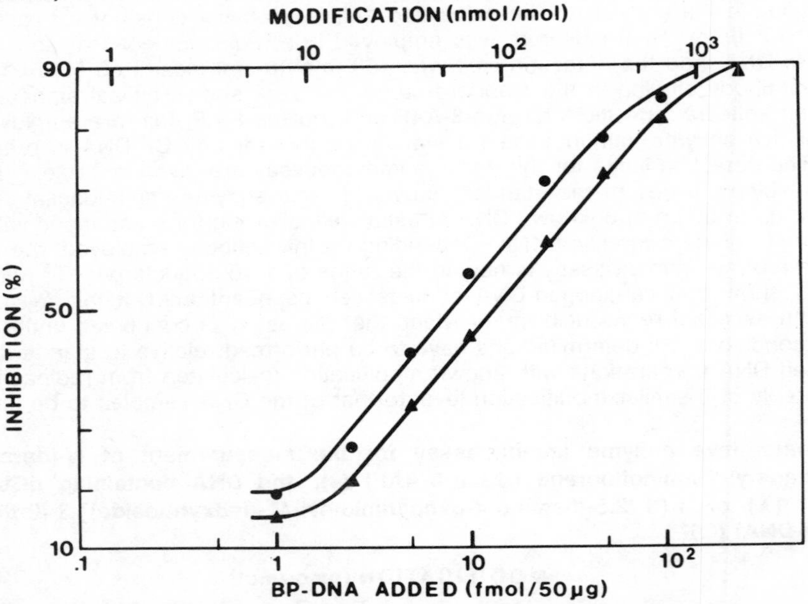


Rabbit-produced antiserum is raised against the methylated bovine serum albumin complex with BP-DNA. Antiserum dilution was 1:3 × $10^7$. The wells of the microtitre plates were coated with 50 fmol/ng (▲) or 25 fmol/0.5 ng (•) of BP-DNA.

**Table 1. Detection limits of carcinogen-DNA adducts in the competitive enzyme immunoassay[a]**

| Inhibitor | Immunogen | Antibody affinity constant (l/mol × $10^{-9}$) | Detection limit[b] (fmol/0.1 ml) | Modification (50 µg DNA) (nmol/mol) |
|---|---|---|---|---|
| AAF-DNA[c] | Guo-8-AAF/BSA | 6 | 2.0 | 14 |
| AAF-DNA(ss) | Guo-8-AAF/BSA | 6 | 12.0 | 82 |
| AF-DNA (ds) | Guo-8-AAF/BSA | 6 | 5500 | 37 700 |
| AF-DNA[d] | roGuo-8-AF/BSA | 0.035 | 16.0 | 110 |
| roAF-DNA | roGuo-8-AF/BSA | 0.035 | 16.0 | 110 |
| BP-DNA(ss) | BP-DNA/MBSA | 0.110 | 8.0 | 55 |

[a] AAF-DNA, DNA containing *N*-(deoxyguanosin-8-yl)-*N*-acetyl-2-aminofluorene; Guo-8-AAF, *N*-(guanosin-8-yl)-*N*-acetyl-2-aminofluorene; BSA, bovine serum albumin; ss, single stranded; AF-DNA, DNA containing *N*-(deoxyguanosin-8-yl)-2-aminofluorene; ds, double stranded; roGuo-8-AF, 1-[6-(2,5-diamino-4-oxopyrimidinyl-$N^6$-riboside]-3-(2-fluorenyl)urea; roAF-DNA, DNA containing 1-[6-(2,5-diamino-4-oxopyrimidinyl-$N^6$-deoxyriboside]-3-(2-fluorenyl)urea; BP-DNA, DNA containing *trans*-(7R)-$N^2$-[10-(7β,8α,9α-trihydroxy-7,8,9,10-tetrahydrobenzo[*a*]pyren)-yl]deoxyguanosine; MBSA, methylated bovine serum albumin

[b] At 50% inhibition of antibody binding to carcinogen-modified DNA on the solid phase

[c] Enzymatically hydrolysed preparation (Kriek *et al.*, 1982, procedure I)

[d] After conversion to roAF-DNA

Recently, instruments such as the 'Microfluor Reader' (Dynatech Laboratories Inc., Alexandria, VA, USA) and 'Fluoroscan' (Flow Laboratories Inc., McLean, VA, USA) have become available, which are capable of reading the fluorescence of 96 reactions in a short period of time. The enzyme immunoassay described in this study is directly applicable to instrumentation, but samples would have to be run in quadruplicate at least, in order to obtain the same accuracy as that obtained by manual reading of the fluorescence in the Aminco SPF-500 (Kriek *et al.*, unpublished data). The instruments facilitate large-scale immunoassays and processing of the data by computer.

The quality of the microtitre plates is an important factor in both the accuracy and the sensitivity of the assays. Polystyrene as well as polyvinyl chloride plates from different manufacturers were tested. The best results were obtained with the polystyrene plates from Falcon. Polyvinyl chloride plates appeared to be very inefficient in coating of the DNA samples. Pretreatment of the plates with *O*-(diethylaminoethyl) (DEAE)-dextran gave irreproducible results and a high background fluorescence (Van der Laken & Kriek, unpublished data). These observations are in opposition to those reported by Harris *et al.* (1982), but we have no explanation, as yet, for the observed differences.

The competitive enzyme immunoassay has been applied successfully for analysis of mixtures containing AAF-DNA, AF-DNA and roAF-DNA, employing anti roGuo-8-AF as single antibody (Kriek *et al.*, 1983). A detailed description of this analysis will be published in a separate article (Kriek & Welling, in preparation).

## ACKNOWLEDGEMENTS

The authors wish to thank Drs J. Nagel and A.M. Hagenaars, Unit for Immunochemistry, National Institute of Public Health, Bilthoven, The Netherlands, for the immunizations. Our thanks are due to Dr M.C. Poirier, Laboratory of Cellular Carcinogenesis and Tumor Promotion, National Cancer Institute, NIH, USA for a gift of anti BP-DNA, and to Dr R.M. Santella, The Cancer Center/Institute of Cancer Research, Columbia University, New York, USA, for a sample of BP-DNA.

## REFERENCES

Daune, M.P., de Murcia, G., Freund, A.M., Fuchs, R.P.P., Lang, M.C., Leng, M. & Sage, E. (1980) *Structural modifications and specific recognition by antibodies of carcinogenic aromatic amines*. In: Pullman, B., Ts'o, P.O.P. & Gelboin, H., eds, *Carcinogenesis: Fundamental Mechanisms and Environmental Effects*, Dordrecht, D. Reidel, pp. 193-205

Erlanger, B.F. & Beiser, S.M. (1964) Antibodies specific for ribonucleosides and ribonucleotides and their reaction with DNA. *Proc. natl Acad. Sci. USA*, *52*, 68-74

Grover, P.L. (1979) *Chemical Carcinogens and DNA*, Boca Raton, FL, CRC Press

Harris, C.C., Yolken, R.H. & Hsu, I.-C. (1982) Enzyme immunoassays: applications in cancer research. *Methods Cancer Res.*, *20*, 213-243

Heyting, C., van der Laken, C.J., van Raamsdonk, W. & Pool, C.W. (1983) Immunohistochemical detection of $O^6$-ethyldeoxyguanosine in the rat brain after in-vivo applications of *N*-ethyl-*N*-nitrosourea. *Cancer Res.*, *43*, 2935-2941

Hsu, I.-C., Poirier, M.C., Yuspa, S.H., Yolken, R.H. & Harris, C.C. (1980) Ultrasensitive enzymatic radioimmunoassay detects femtomoles of acetylaminofluorene-DNA adducts. *Carcinogenesis*, *1*, 455-458

Irving, C.C. & Veazey, R.A. (1969) Persistent binding of 2-acetylaminofluorene to rat liver DNA *in vivo* and consideration of the mechanism of binding of *N*-hydroxy-2-acetylaminofluorene to rat liver nucleic acids. *Cancer Res.*, *29*, 1799-1804

Kriek, E. (1979) Effect of pH on the ratio of substitution products in DNA after reaction with the carcinogen *N*-acetoxy-2-acetylaminofluorene. *Cancer Lett.*, *7*, 141-146

Kriek, E. & Westra, J.G. (1980) Structural identification of the pyrimidine derivatives formed from *N*-(deoxyguanosin-8-yl)-2-aminofluorene in aqueous solution at alkaline pH. *Carcinogenesis*, *1*, 459-468

Kriek, E., Juhl, U., Miller, J.A. & Miller, E.C. (1967) *8*-(*N*-2-Fluorenylacetamido)guanosine, an arylamidation reaction product of guanosine and the carcinogen *N*-acetoxy-*N*-2-fluorenylacetamide in neutral solution. *Biochemistry*, *6*, 177-182

Kriek, E., van der Laken, C.J., Welling, M. & Nagel, J. (1982) *Immunological detection and quantification of the reaction products of 2-acetylaminofluorene with guanine in DNA*. In: Bartsch, H. & Armstrong, B., eds, *Host Factors in Human Carcinogenesis* (*IARC Scientific Publications No. 39*), Lyon, International Agency for Research on Cancer, pp. 541-549

Kriek, E., Welling, M. & van der Laken, C.J. (1983) Quantitation of 2-fluorenylamine-modified DNA using antibodies specific to guanine imidazole ring-opened *N*-(guanosin-8-yl)-2-fluorenylamine. *Proc. Am. Assoc. Cancer Res.*, *24*, Abstract 247

van der Laken, C.J., Hagenaars, A.M., Hermsen, G., Kriek, E., Kuipers, A.J., Nagel, J., Scherer, E. & Welling, M. (1982) Measurement of $O^6$-ethyldeoxyguanosine and *N*-(deoxyguanosin-8-yl)-*N*-acetyl-2-aminofluorene in DNA by high-sensitive enzyme immunoassays. *Carcinogenesis*, *3*, 569-572

Müller, R. (1980) Calculation of average antibody affinity in anti-hapten sera from data obtained by competitive radioimmunoassay. *J. immunol. Methods*, *34*, 345-352

Müller, R. & Rajewsky, M.F. (1980) Immunological quantification by high-affinity antibodies of $O^6$-ethyldeoxyguanosine in DNA exposed to *N*-ethyl-*N*-nitrosourea. *Cancer Res.*, *40*, 887-896

Müller, R. & Rajewsky, M.F. (1981) Antibodies specific for DNA components structurally modified by chemical carcinogens. *J. Cancer Res. clin. Oncol.*, *102*, 99-113

Poirier, M.C. (1981) Antibodies to carcinogen-DNA adducts. *J. natl Cancer Inst.*, *67*, 515-519

Poirier, M.C., Dubin, M.A. & Yuspa, S.H. (1979) The formation and removal of specific acetylaminofluorene-DNA adducts in mouse and human cells measured by radioimmunoassay. *Cancer Res.*, *39*, 1377-1381

Poirier, M.C., Santella, R.M., Weinstein, I.B., Grunberger, D. & Yuspa, S.H. (1980a) Quantitation of benzo[*a*]pyrene-deoxyguanosine adducts by radioimmunoassay. *Cancer Res.*, *40*, 412-416

Poirier, M.C., Williams, G.M. & Yuspa, S.H. (1980b) Effect of culture conditions, cell type and species of origin on the distribution of acetylated and deacetylated deoxyguanosine C-8 adducts of *N*-acetoxy-2-acetylaminofluorene. *Mol. Pharmacol.*, *18*, 581-587

Poirier, M.C., Lippard, S.J., Zwelling, L.A., Ushay, H.M., Kerrigan, D., Thill, C.C., Santella, R.M., Grunberger, D. & Yuspa, S.H. (1982a) Antibodies elicited against *cis*-diammine-dichloroplatinum(II)-modified DNA are specific for *cis*-diammine-dichloroplatinum-DNA adducts formed *in vivo* and *in vitro*. *Proc. natl Acad. Sci. USA*, *79*, 6443-6447

Poirier, M.C., True, B. & Laishes, B.A. (1982b) The formation and removal of (guan-8-yl)-DNA-2-acetylaminofluorene adducts in liver and kidney of male rats given dietary 2-acetylaminofluorene. *Cancer Res.*, *42*, 1317-1321

Poirier, M.C., Hunt, J.M., True, B. & Laishes, B.A. (1983) *Kinetics of DNA adduct formation and removal in liver and kidney of rats fed 2-acetylaminofluorene*. In: Rydström, J., Montelius, J. & Bengtsson, M., eds, *Extrahepatic Drug Metabolism and Chemical Carcinogenesis*, Amsterdam, Elsevier, pp. 479-488

Westra, J.G. (1981) A rapid and simple synthesis of reactive metabolites of carcinogenic aromatic amines in high yield. *Carcinogenesis*, *2*, 355-357

# BIOLOGICAL MONITORING OF WORKERS EXPOSED TO POLYCHLORINATED BIPHENYL COMPOUNDS IN CAPACITOR ACCIDENTS

**M. Luotamo, J. Järvisalo & A. Aitio**
*Institute of Occupational Health, Helsinki, Finland*

**O. Elo & P. Vuojolahti**
*Oy Enso-Gutzeit Ab, Helsinki, Finland*

## SUMMARY

For biological monitoring of workers exposed to polychlorinated biphenyl (PCB) compounds, a gas-chromatographic method for analysis of PCBs in serum was developed. The quantitation is performed with 15 pure isomers, and extraction and concentration factors are corrected with the aid of asymmetric PCB compounds added to each specimen. By this method, humans with no known occupational exposure to PCB compounds showed PCB levels in serum that were always $<3$ µg/l. Workers exposed in capacitor accidents have shown maximal concentrations close to 50 µg/l.

## INTRODUCTION

There are few studies on biological monitoring of occupational exposure to polychlorinated biphenyl (PCB) compounds and for analytical reasons, it is difficult to compare those studies. The composition of the PCBs, to which workers have been exposed in capacitor and transformer manufacture and repair, have gradually changed during past decades towards mixtures with lesser chlorination (Maroni *et al.*, 1981; Smith *et al.*, 1982; Wolff *et al.*, 1982a,b).

A special type of exposure to PCB compounds may occur during accidents where transformers or capacitors containing PCB products are damaged, e.g., under the influence of fire or an electric arc. However, there is no literature on the uptake of PCBs by the workers exposed in these incidents.

After a capacitor accident in a paper mill in Finland, we developed methods for biological monitoring of workers in these accidents. We also developed a method of analysis of PCB compounds in serum, where the quantitation uses pure PCB congeners.

## SUBJECTS AND METHODS

Since August 1982, we have assessed occupational exposure to PCBs by an analysis of human specimens in 24 capacitor accidents. Typically, specimens have been collected from electricians, maintenance workers, persons cleaning the polluted area and firemen. The venous blood specimens have been collected from subjects fasted overnight.

To define the levels of the PCB compounds in the working population in general, serum specimens were obtained from 47 forestry and metal industry workers with no known exposure to PCB compounds at work (group I).

Group II comprised 15 workers exposed in a capacitor accident that happened in a paper mill. Twenty capacitors were damaged in a fire and about 100 l of PCB (Clophen A30) were liberated, burned or spread to the adjacent rooms. The capacitor room was heavily contaminated with PCBs, and polychlorinated dibenzofurans were also found. These workers were exposed during cleaning procedures on the accident night and successive specimens were collected from them.

Group III comprised an additional 150 workers from the factory who had been potentially exposed during the weeks following the accident, through contamination of the air conditioning. The workers in both groups II and III were given charcoal, orally, for two weeks after the accident.

## SPECIMEN PREPARATION AND ANALYSIS

Details of the analytical method will be published elsewhere (M. Luotamo *et al.*, in preparation).

Blood specimens were drawn after an overnight fast, and the serum was separated and frozen. After thawing, the serum was extracted with *n*-hexane-diethylether and the extract was then purified with sulfuric acid and on silica columns. The eluate was concentrated under the flow of nitrogen. Losses during extraction and purification were corrected with the aid of 2,4,6-trichloro- and 2,3,4,5,6-pentachlorobiphenyls, added to each serum sample as internal standards. An aliquot of the concentrate was injected onto a Hewlett-Packard 5880A gas chromatograph equipped with an OV-1701 silica capillary column and an $Ni^{63}$ electron capture detector.

PCB compounds coeluting with 2,2'-di-, 2,4'-di-, 4,4'-di-, 2,5,2'-tri-, 2,4,4'-tri-, 3,4,2'-tri-, 2,4,2'4'-tetra-, 2,3,2'5'-tetra-, 2,3,2'3'-tetra-, 2,4,3'4'-tetra-, 3,4,3'4'-tetra-, 2,4,6,3'5'-penta-, 2,4,5,2'5'-penta-, 2,4,5,2'3'-penta- and 2,3,4,2'5'-pentachlorobiphenyl (Analabs) were quantified separately, and a sum of these compounds was recorded. The recovery of the added chlorobiphenyls was around 80%, and the day-to-day variation coefficient around 20%.

The pattern of the chosen PCB congeners closely resembled that of Clophen A30.

## RESULTS

*Group I*

The PCB levels in serum of the 47 workers without known PCB exposure ranged from 0.3 to 2.9 μg/l and was log-normally distributed (Fig. 1). For practical purposes, an upper reference limit of 3 μg/l was adopted for monitoring of exposure to PCBs.

**Fig. 1. Frequency distribution of serum polychlorinated biphenyl (PCB) concentrations in nonexposed humans (Group I)**

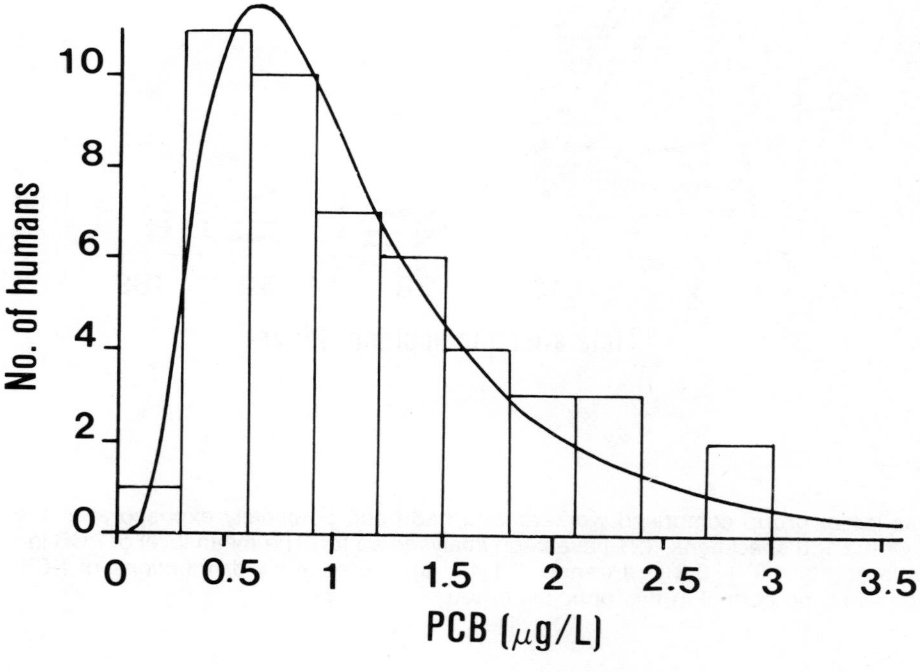


*Group II*

Figure 2 shows the concentrations of PCB compounds in various workers after the accident. Initially, high levels (range, 3.5 to 48.3 μg/l) of serum PCBs were detected, but each showed a rapid decrease from the maximal level to the reference level of 3 μg/l.

**Fig. 2. Serum polychlorinated biphenyl (PCB) concentrations in 15 humans exposed in a capacitor accident (Group II)**

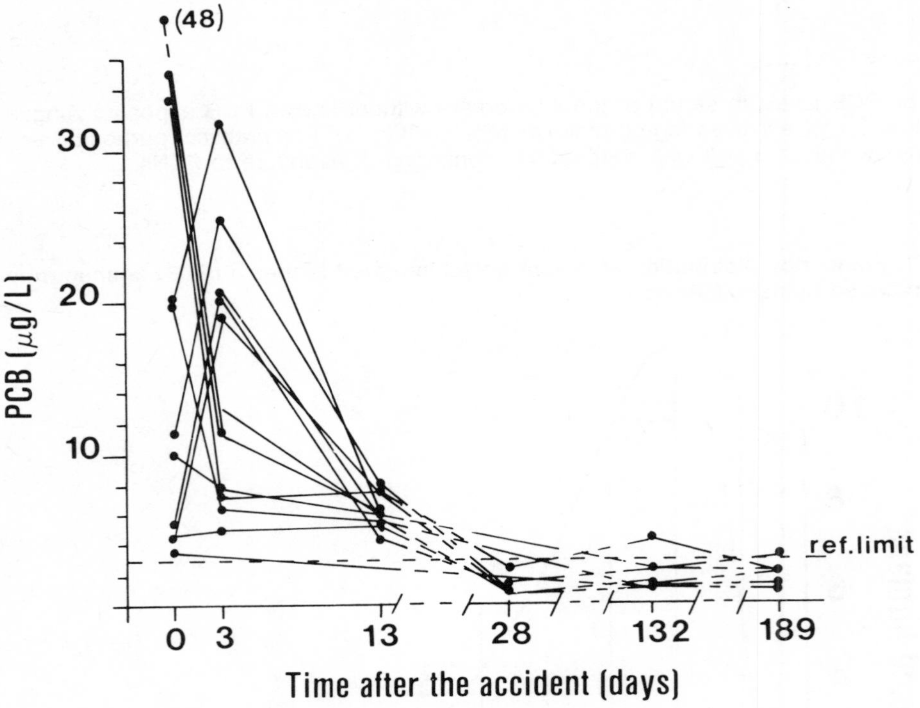


*Group III*

This larger group comprised workers who had been potentially exposed after the accident. Of the 150 specimens, 121 have been analysed so far. The mean level of PCB in serum in this group is 1.3 $\pm$ 0.8 μg/l (range, 0.1-4.1 μg/l). Frequency distributions of PCB compounds were log-normal in this population, too.

## DISCUSSION

Our experience suggests that, while exposure to PCB compounds in capacitor accidents may lead to uptake of these compounds in the human body, the sum concentration of the compounds decreases rapidly. However, various compounds probably show different kinetics and toxicity. Therefore, the analysis should be directed towards specific monitoring of the toxicologically important congeners.

Groups I and III show similar results; only three subjects in group III exceeded the adopted reference limit of 3 μg/l. However, it is clear that analytical aspects must be specially emphasized when applying PCB determinations for biological monitoring of exposure. Analyses of PCB compounds lack sufficient quality control; there are no commercial quality control materials. In addition, there are no quality control schemes to test laboratory performance or to show the level of comparability between various countries and various laboratories.

## ACKNOWLEDGEMENTS

The authors wish to thank the Finnish Work Environment Fund for financial support, and also the workers for the smooth cooperation.

## REFERENCES

Maroni, M., Colombi, A., Cantoni, S., Ferioli, E. & Foa, V. (1981) Occupational exposure to polychlorinated biphenyls in electrical workers. I. Environmental and blood polychlorinated biphenyls concentrations. *Br. J. ind. Med.*, *38*, 49-54

Smith, A.B., Schloemer, J., Lowry, L.K., Smallwood, A.W., Ligo, R.N., Tanaka, S., Stringer, W., Jones, M., Hervin, R. & Glueck, C.J. (1982) Metabolic and health consequences of occupational exposure to polychlorinated biphenyls. *Br. J. ind. Med.*, *39*, 361-369

Wolff, M.S., Fischbein, A., Thornton, J., Rice, C., Lilis, R. & Selikoff, I.J. (1982a) Body burden of polychlorinated biphenyls among persons employed in capacitor manufacturing. *Int. Arch. occup. environ. Health*, *49*, 199-208

Wolff, M.S., Thornton, J., Fischbein, A., Lilis, R. & Selikoff, I.J. (1982b) Disposition of polychlorinated biphenyl congeners in occupationally exposed persons. *Toxicol. appl. Pharmacol.*, *62*, 294-306

# AN ENZYME-LINKED IMMUNOSORBENT PROCEDURE FOR ASSAYING AFLATOXIN $B_1$


**C.N. Martin, R.C. Garner[1] & F. Tursi**

*Cancer Research Unit, University of York, Heslington, York Y01 5DD, UK*

**J.V. Garner**

*Microtest Research Ltd, University Road, Heslington, York Y01 5DU, UK*

**H.C. Whittle & R.W. Ryder**

*Medical Research Council Laboratories, Fajara, Nr Banjul, The Gambia, West Africa*

**P. Sizaret & R. Montesano**

*International Agency for Research on Cancer Lyon, France*



## SUMMARY

A polyclonal rabbit antibody preparation against aflatoxin $B_1$ ($AFB_1$), produced by immunization with a bovine serum albumin-$AFB_1$ conjugate, has been used to develop an enzyme-linked immunosorbent assay (ELISA). $AFB_1$-ovalbumin, obtained by reacting either $AFB_1$-8,9-dichloride or -8,9-dibromide with ovalbumin, was used to coat each well of a polyvinyl microtitre dish. Rabbit antibody, diluted 1:100 000, was added to each well and left to attach for 90 min, then the excess was washed off with phosphate-buffered saline/Tween.



[1] To whom correspondence should be addressed

Anti-rabbit IgG coupled to peroxidase was then added, left for 90 min and the excess washed away. Residual peroxidase activity was assayed using tetramethylbenzidine as substrate; the reaction was quenched at the end of the 15-min incubation period with 2 N sulfuric acid. Inhibitor studies to assay $AFB_1$ and related compounds in urine involved prior incubation of the diluted anti-$AFB_1$ antibody with inhibitor for 60 min at 37°C, prior to dispensing into the multi-well plates. Inhibitor studies with $AFB_1$ showed inhibition over a concentration range of $10^{-1}$ to $10^{-5}$ μg/ml. The minimum detectable concentration was approximately $10^{-5}$ μg/ml (0.032 pmol/ml). Anti-$AFB_1$ antibody was also inhibited by iro-$AFB_1$-DNA, one of the forms of $AFB_1$-DNA, as well as by $AFB_1$-guanine. Undiluted urine from normal subjects inhibited antibody binding to plates; this inhibition could be prevented by either dilution or extraction procedures. Analysis of urine samples from The Gambia indicated that people with primary hepatocellular carcinoma and their immediate contacts were as likely to have $AFB_1$ in their urine as controls in the same population; the mean $AFB_1$ concentration was 21.6 ng/ml (range, 0.0-145) for exposed persons and 77 ng/ml (range, 0-730) for controls.

## INTRODUCTION

Primary liver cancer is the most common cancer in Asia and Africa (Waterhouse *et al.*, 1982); a major etiological factor for this type of cancer has been identified as hepatitis B virus infection (Okuda & MacKay, 1982). The evidence of a causal association between liver cancer and aflatoxin intake, although strongly suggested by various epidemiological studies (Linsell, 1979), is limited by the fact that the link between cancer cases and exposure to aflatoxins is derived from environmental data, and is thus indirect. However, until recently, it was not possible to contemplate procedures, other than environmental monitoring, that might be used to assess levels of carcinogenic exposure to humans. Recently, a novel approach has been proposed, in which antibodies to various carcinogen-nucleic acid adducts have been used to determine the level of these adducts in DNA samples obtained from animal or human tissues by radio- or enzyme-immunoassays. The advantage of these procedures is that they are sensitive, highly specific, cheap, relatively rapid to perform and permit the monitoring of exposure to carcinogens on an individual basis and in a large number of human subjects (Müller *et al.*, 1982; IARC-IPCS Working Group, 1982; Adamkiewicz *et al.*, this volume). A complementary approach is to prepare antibodies against carcinogens which may cross-react not only with the parent compounds but also with metabolites or DNA adducts.

This paper is a report on these studies, using $AFB_1$ as a model, and includes, also, a preliminary report on assaying urine samples from persons in The Gambia thought to be exposed to $AFB_1$.

## MATERIALS AND METHODS

### *Chemicals and reagents*

$AFB_1$ was purchased from Makor Chemicals Ltd (Jerusalem, Israel), bovine serum albumin, ovalbumin and Tween-20 from Sigma Chemical Co. Ltd (Poole, Dorset, UK) and polyvinyl microtitre plates from Northumbria Biologicals Ltd (Cramlington, Northumberland, UK) (Costar 96-well assay plates). $AFB_1$-bovine serum albumin and $AFB_1$-ovalbumin were prepared by reacting the respective protein with $AFB_1$-8,9-dichloride or -dibromide. The rabbit anti-$AFB_1$ antibody preparation was the C-antiserum material obtained by Sizaret *et al.* (1982). 3,3',5,5'-Tetramethyl benzidine was purchased from the Aldrich Chemical Co. Ltd (Gillingham, Dorset, UK).

### *ELISA procedure*

50-μl aliquots of $AFB_1$-ovalbumin dissolved in phosphate-buffered saline (PBS, 200 ng/ml) were dispensed into each well of 96-well polyvinyl plates. The plates were dried overnight by incubating at 37°C without lids to evaporate the water, then stored at -20°C until use. Plates have been stored for as long as three months without any apparent adverse effect.

In preparation for an assay, the above plates were washed six times with PBS/Tween (8.0 g sodium chloride, 200 mg potassium chloride, 200 mg potassium phosphate, monobasic, 0.15 mg anhydrous sodium phosphate, dibasic, 0.5 ml Tween-20, distilled water to 1 l). 100-μl aliquots of 3% ovalbumin dissolved in PBS were added to each well and the plates were left for 60 min at room temperature. This procedure reduces nonspecific binding of antibody to plates. After 60 min, the protein solution was removed by flicking the plates and the plates washed a further four times in PBS/Tween. the plates were then ready for use.

To carry out a competitive inhibitor assay, the rabbit $AFB_1$ antisera was diluted 1:50 000 in 1% bovine serum albumin dissolved in 0.01 M phosphate buffer:0.15 M sodium chloride, pH 7.2 (buffer A). An equal volume of inhibitor in PBS was added to the diluted antisera and the whole was mixed and incubated at 37°C for 60 min. 50-μl aliquots of these mixtures were then dispensed into vertical rows of the prepared microtitre plates and left to shake slowly at room temperature for 90 min. The excess antibody was washed off with PBS/Tween and a 1:10 000 dilution of anti-rabbit IgG-peroxidase conjugate diluted in buffer A was added (Miles Laboratories Ltd, Slough, UK). This was left in contact with the wells for a further 90 min at room temperature, then the excess was washed off in PBS/Tween (six times). After the last wash, a single distilled-water wash was performed and the 3,3',5,5'-tetramethybenzidine substrate was added (5 mg tetramethylbenzidine in 0.5 ml dimethyl sulfoxide, 1.47 g trisodium citrate, 0.41 g sodium acetate, 10 μl 100 volume hydrogen peroxide, water to 50 ml - 50 μl/well). The colour was allowed to develop for 15 min and the reaction quenched by the addition of 50 μl 2 N sulfuric acid. The absorbance of each well was read at 450 nm using an automated plate-reader (Titertek Multiscan, Flow Laboratories Ltd, Rickmansworth, Herts, UK). Inhibition values were calculated as a percentage of the optical density when no inhibitor was present.

### *Urine collection and assay*

Freshly voided urine samples were collected from primary hepatocellular carcinoma patients in The Gambia. Samples were frozen as soon as possible after collection and transported to York still frozen. Here, they were stored at -90°C until assayed. For each liver cancer case, urine samples were also collected from a relative within the same household. Controls were matched with cases for age, sex and community, and samples were also collected from relatives of controls. For the purposes of this pilot study, the above groups were divided into those exposed, i.e., cancer cases, their relatives and the relevant controls. Samples were thawed prior to ELISA and tested neat or by serial dilution in PBS. Some samples were also concentrated by passing 20 ml through an activated Sep-Pak $C^{18}$ cartridge (Waters Associates, Macclesfield, Cheshire, UK). Material absorbed to the column was eluted with 80% v/v methanol, then the solvent was removed under reduced pressure and the residue re-dissolved in 0.5 ml PBS. This material was used in ELISA inhibitor studies as a neat solution and at various serial dilutions. A standard curve using $AFB_1$ was derived from the same ELISA plate as the particular urine sample being tested.

## RESULTS

The antibody used in this study was characterized previously by radioimmunoassay (Sizaret *et al.*, 1982). In order to use the antibody in ELISA, the hapten $AFB_1$ had to be coupled to a protein that was immunologically distinct from that used for antibody production. We used ovalbumin. Conjugation of this with $AFB_1$-8,9-dichloride or -8,9-dibromide yielded a conjugate with an absorption spectrum that was similar to $AFB_1$ at wavelengths over 300 nm. The position of the 360-nm absorption maximum was moved to 400 nm on addition of the alkali, indicating that the attachment of the $AFB_1$ was through a Schiff base at the $C^8$-position. A checkerboard titration experiment, in which the concentration of $AFB_1$-ovalbumin absorbed to the plate and dilutions of the antisera were varied, indicated that a reasonable colour reaction was obtained when 10 ng of antigen were absorbed per well and the antisera were diluted to 1:100 000. An inhibitor curve using these conditions is shown in Figure 1. The minimum detectable $AFB_1$ concentration is between $10^{-4}$ and $10^{-5}$ µg/ml. The

**Fig. 1. Enzyme-linked immunosorbent (ELISA) inhibition curve for aflatoxin $B_1$ ($AFB_1$) using rabbit $AFB_1$ antisera and $AFB_1$ as inhibitor**

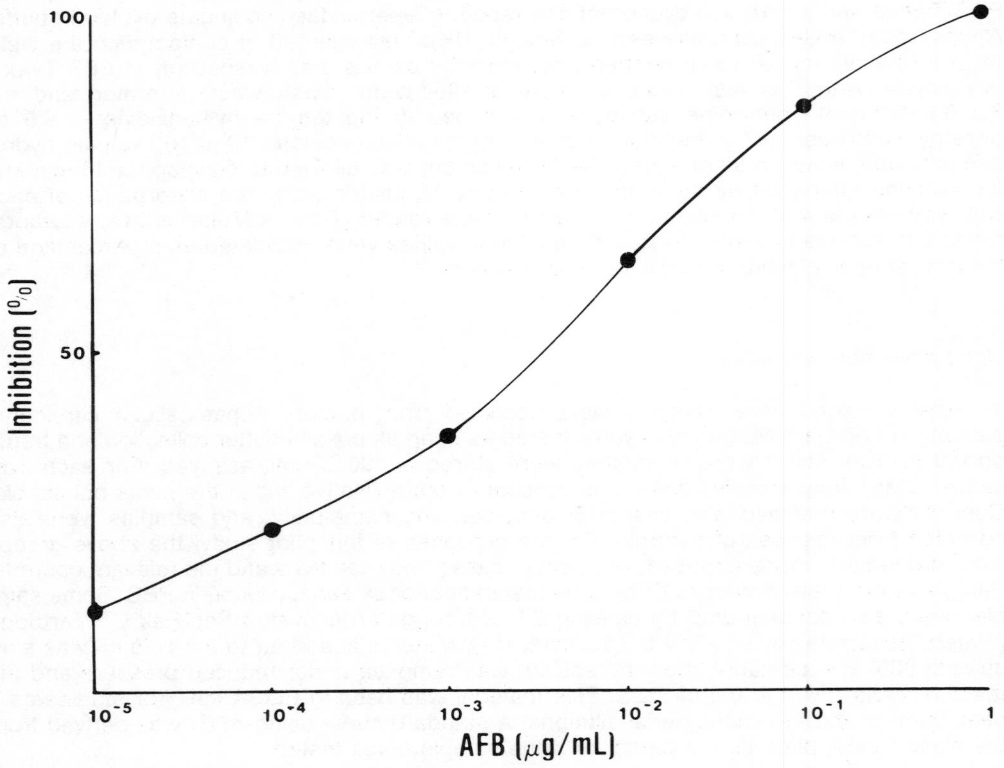

assay appears reproducible from day to day. Figure 2 shows an inhibition curve obtained when $AFB_1$-reacted DNA was used as an inhibitor; this DNA was obtained by reacting DNA, $AFB_1$ and 3-chloroperoxybenzoic acid together, as previously described (Garner *et al.*, 1979). All the guanine residues in this particular preparation were in the ring-opened form. For one

**Fig. 2. Enzyme-linked immunosorbent (ELISA) inhibition curve using rabbit aflatoxin $B_1$ ($AFB_1$) antisera and iro-$AFB_1$-DNA as inhibitor**

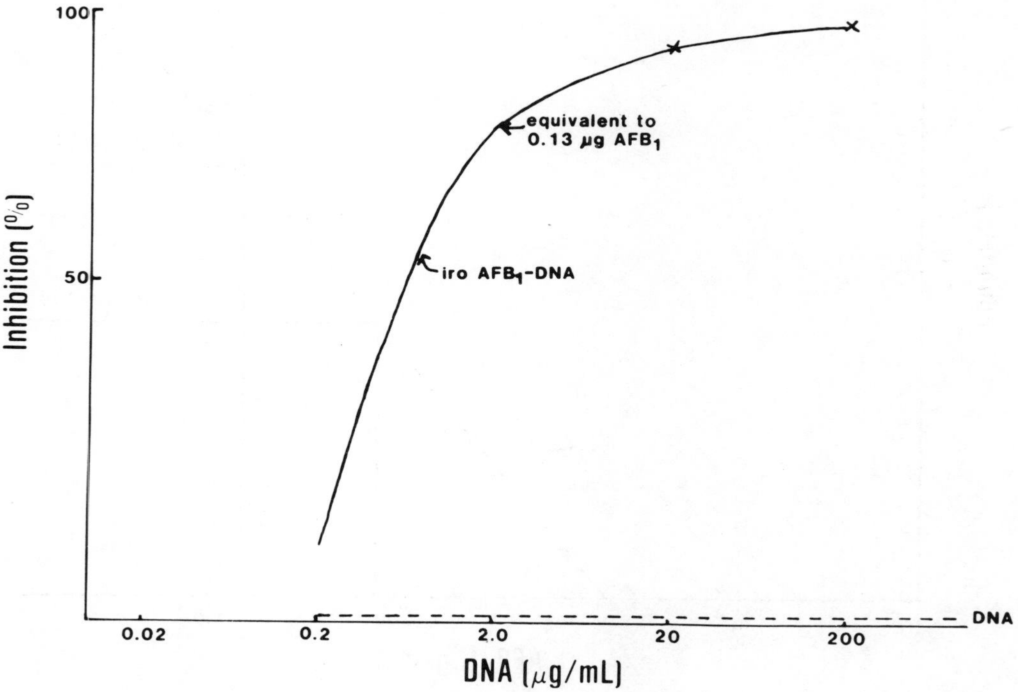

experiment, urine from an individual who had not knowingly been exposed to $AFB_1$ was spiked with $AFB_1$ and the sample extracted using a Sep-Pak $C^{18}$ cartridge. Figure 3 presents the results from an inhibitor study in which control and spiked urine were compared.

**Fig. 3. Enzyme-linked immunosorbent (ELISA) inhibition study of control urine spiked with aflatoxin $B_1$ ($AFB_1$)**

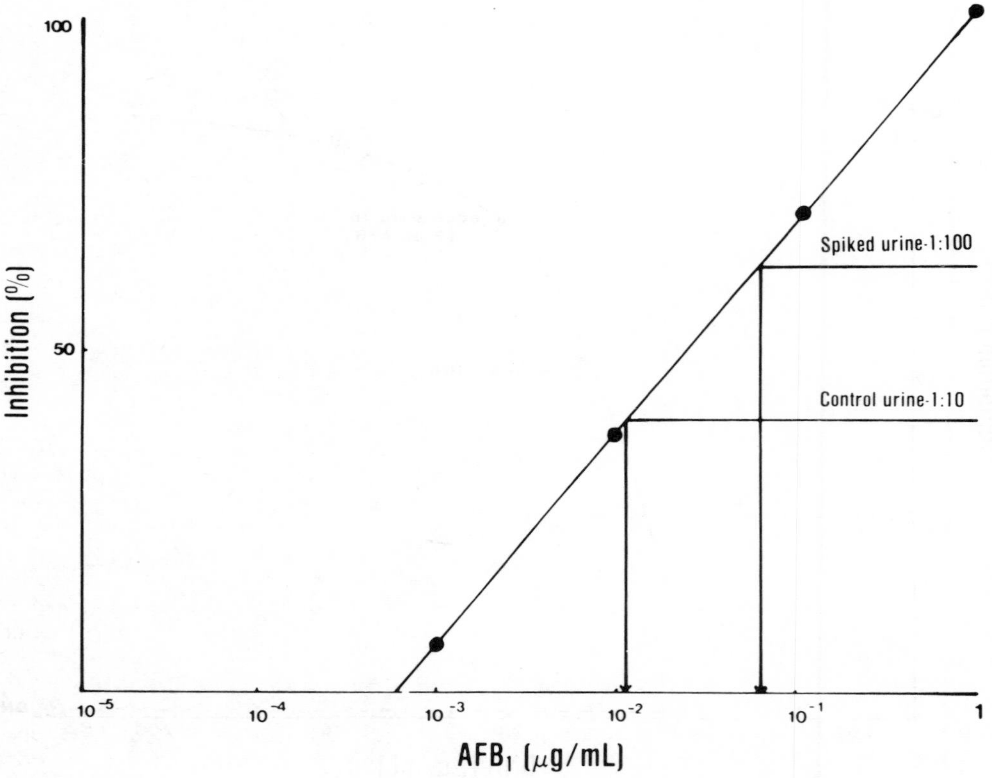


Using the ELISA procedure described above, we examined 24 human urine samples obtained from The Gambia (Table 1). Samples were assayed neat and at several dilutions. $AFB_1$ concentration equivalents in the urine were calculated from standard inhibitor curves derived from the same plates simultaneously with the testing of the unknowns.

**Table 1. Summary of enzyme-linked immunosorbent assay results for the Gambian urine samples tested by serial dilution**

| Sample population (exposed ?) | AFB₁ equivalents (ng/ml) |
|---|---|
| 30-1 | 0.00 |
| 30-7 | 0.00 |
| 32-1 | 16.5 |
| 32-3 | 0.00 |
| 40-1 | 6.25 |
| 40-15 | 13.0 |
| 42-1 | 16.5 |
| 44-1 | 4.8 |
| 44-14 | 35.0 |
| 45-1 | 0.00 |
| 45-2 | 145 |
| Mean, 21.6 (range, 0.0-145) | |
| *Controls* | |
| C30-1 | 0.0 |
| C30-2 | 165 |
| C37-1 | 0.00 |
| C37-2 | 0.00 |
| C39-2 | 0.00 |
| C39-7 | 0.00 |
| C40-1 | 46.0 |
| C42-1 | 730 |
| C42-2 | 41.0 |
| C43-1 | 0.00 |
| C43-2 | 18.5 |
| C44-1 | 0.00 |
| C44-2 | 0.00 |
| Mean, 77 (range, 0.00-730) | |

## DISCUSSION

This paper reports on an ELISA for $AFB_1$ which has been used to examine the urine of people thought to be exposed to $AFB_1$ in The Gambia as a result of ingesting contaminated groundnuts. The data are preliminary in that, while some 200 urine samples have been collected, the results for only 24 are presented here. It is hoped to test the remainder of the samples systematically over the coming months. The early results reported are encouraging to the extent that, using immunoassay measurements, the $AFB_1$ exposed persons have antibody-inhibitory activity in their urine. As far as we are aware, only one other systematic study of $AFB_1$-related materials in human urine has been published (Autrup *et al.*, 1983). In this study, physico-chemical procedures were used to monitor for $AFB_1$-guanine, the postulated excretion product that arises from excision repair or depurination of $AFB_1$-DNA. Our study, probably, estimates $AFB_1$, any of its metabolites or adducts and, possibly, other aflatoxins such as $AFG_1$ or $AFG_2$. Definitive assay of the true aflatoxin content in the samples tested here can come only from chromatographic analysis and assay of subfractions for antibody inhibitory activity. If peaks of activity coinciding with $AFB_1$, its metabolites or $AFB_1$-guanine were found, this would confirm these preliminary findings.

The use of immunoassay for human monitoring is an exciting prospect. Nevertheless, some caution is needed. Several reports have been published on the use of poly- or monoclonal antibodies to assay carcinogen adducts. In only a few cases have these antibodies been used for screening. We have found that the mouse monoclonal antibodies we ourselves have obtained (Hertzog *et al.*, 1983), while relatively specific, do not have sufficiently high association constants for them to be useful. The problem appears to exist with some monoclonals and may not be so prominent for polyclonals. However, the latter are less specific, which has both advantages and disadvantages for screening purposes.

Nonspecific inhibition of this particular antibody preparation was observed in some control urine samples, and investigations are underway at present to clarify this matter.

In conclusion, these preliminary studies have shown that it is possible to measure $AFB_1$ equivalents in human urine using immunoassay, and that urine from persons thought to be exposed to $AFB_1$ contains antibody inhibitory material, justifying the pursuit of this approach in measuring human exposure to $AFB_1$.

## ACKNOWLEDGEMENTS

The work is supported by the Yorkshire Cancer Research Campaign. Part of this work was funded by an Interagency Agreement between the International Agency for Research on Cancer, Lyon, France, and the Cancer Research Unit, University of York, and by the Ministère de la Santé, France.

## REFERENCES

Autrup, H., Bradley, K.A., Shamsuddin, A.K.M., Wakhisi, J. & Wasunna, A. (1983) Detection of putative adduct with fluorescence characteristics identical to 2,3-dihydro-2-(7'-guanyl)-3-hydroxy aflatoxin $B_1$ in human urine collected in Murang'a district, Kenya. *Carcinogenesis*, *4*, 1193-1195

Butler, W.H. (1974) *Aflatoxin*. In: Purchase, I.F.H., ed., *Mycotoxins*, Amsterdam, Elsevier, pp. 1-28

Garner, R.C., Martin, C.N., Lindsay-Smith, J.R., Coles, B.F. & Tolsen, M.R. (1979) Comparison of aflatoxin $B_1$ and aflatoxin $G_1$ binding to cellular macromolecules *in vitro*, *in vivo* and after peracid oxidation: characterisation of the major nucleic acids. *Chem. biol. Interact.*, *26*, 57-73

Haugen, A., Groopman, J.D., Hsu, I.-C., Goodrich, G.R., Wogan, G.N. & Harris, C.C. (1981) Monoclonal antibody to aflatoxin $B_1$-modified DNA detected by enzyme immunoassay. *Proc. natl Acad. Sci. USA*, *78*, 4124-4127

Hertzog, P.J., Lindsay-Smith, J.R. & Garner, R.C. (1982) Production of monoclonal antibodies to guanine imidazole ring-opened aflatoxin $B_1$-DNA, the persistent DNA adduct *in vivo*. *Carcinogenesis*, *3*, 825-828

Hertzog, P.J., Shaw, A., Lindsay-Smith, J.R. & Garner, R.C. (1983) Improved conditions for the production of monoclonal antibodies to carcinogen-modified DNA for use in enzyme-linked immunosorbent assays (ELISA). *J. immunol. Meth.*, *62*, 49-58

IARC-IPCS (1982) Development and possible use of immunological techniques to detect individual exposure to carcinogens: International Agency for Research on Cancer/International Programme on Chemical Safety Working Group Report. *Cancer Res.*, *42*, 5236-5239

Kew, M.C. (1978) *Hepatoma and the HBV.* In: Vyas, G.N., Cohen, S.N. & Schmid, O., eds, *Viral Hepatitis. Etiology, Epidemiology, Pathogenesis and Prevention*, Philadelphia, The Franklin Institute Press

Levine, L., Seaman, E., Hammerschlag, E. & van Vunakis, H. (1966) Antibodies to photoproducts of deoxyribonucleic acids irradiated with ultraviolet light. *Science*, *153*, 1666-1667

Linsell, C.A. (1979) Environmental chemical carcinogens and liver cancer. *J. Toxicol. environ. Health*, *5*, 183-191

Müller, R., Adamkiewicz, J. & Rajewsky, M.F. (1982) *Immunological detection and quantification of carcinogen-modified DNA components.* In: Bartsch, H. & Armstrong, B., eds, *Host Factors in Human Carcinogenesis*, (*IARC Scientific Publications No. 39*), Lyon, International Agency for Research on Cancer, pp. 463-479

Okuda, K. & Mackay, I., eds (1982) *Hepatocellular Carcinoma, Workshops on the Biology of Human Cancer*, *Report No. 17*, Geneva, UICC

Poirier, M.C. (1981) Antibodies to carcinogen-DNA adducts. *J. natl Cancer Inst.*, *67*, 515-519

Poirier, M.C., Weinstein, I.B. & Blobstein, S. (1977) Detection of carcinogen-DNA adducts by radioimmunoassay. *Nature*, *270*, 186-188

Poirier, M.C., Santella, R., Weinstein, I.B., Grunberger, D. & Yuspa, S.H. (1980) Quantitation of benzo[*a*]pyrene-deoxyguanosine adducts by radioimmunoassay. *Cancer Res.*, *40*, 412-416

Saffhill, R., Strickland, P.T. & Boyle, J.M. (1982) Sensitive radioimmunoassays for $O^6$-n-butyldeoxyguanosine, $O^2$-n-butylthymidine and $O^4$-n-butylthymidine. *Carcinogenesis*, *3*, 547-552

Sizaret, P., Malaveille, C., Montesano, R. & Frayssinet, C. (1982) Detection of aflatoxins and related metabolites by radio-immunoassay. *J. natl Cancer Inst.*, *69*, 1375-1381

Waterhouse, J., Muir, C., Shanmugaratnam, K. & Powell, J. (1982) *Cancer Incidence in Five Continents*, Vol. IV (*IARC Scientific Publications No. 42*), Lyon, International Agency for Research on Cancer

# SESSION III

Chairman: P. Westerholm
Rapporteur: M. Sorsa

# CHROMOSOMAL ABERRATIONS IN MONITORING EXPOSURE TO MUTAGENS-CARCINOGENS

**A. Forni**

*Institute of Occupational Medicine, Clinica del Lavoro 'L. Devoto', University of Milan, Milan, Italy*

## SUMMARY

Determination of chromosomal aberration rates in cultured lymphocytes is an established method of monitoring populations occupationally or environmentally exposed to known or suspected mutagenic-carcinogenic agents. Subjects and controls for chromosome studies should be properly selected. Methods of culturing lymphocytes should be standardized, to minimize technical factors which might affect the yield of aberrations. Types of aberrations, their significance, criteria for scoring, reporting and statistical analysis are discussed.

From the available experience, the following points must be taken into consideration: (1) Several factors may create confounding, due to nonspecificity of chromosomal aberrations. (2) Persistence of aberrations for years and decades in long-lived lymphocytes makes them an indicator of past damage, but limits the value of the test in monitoring present low-level exposures in subjects with past exposure(s). (3) No clear-cut dose-response relationships have so far been demonstrated, but there is evidence that the method might not be sensitive enough for monitoring very low-level chronic exposures. (4) Results should be evaluated mainly on a group basis. (5) The method is time-consuming, and therefore expensive.

Chromosomal damage indicates a biological effect on the genome, the implications of which for carcinogenesis and mutagenesis are still unknown. It is generally believed that increased rates of chromosomal aberration in a population may indicate an increased cancer risk for the group, but not for the single individual presenting excess damage.

## INTRODUCTION

Determination of induced chromosomal aberration rates is one of the methods presently available for monitoring populations exposed to mutagenic-carcinogenic agents. This method was made available in the early 1960s; it was first used in studies of the biological effects in humans of exposure to a known mutagen, such as ionizing radiations (Bender & Gooch, 1962; Buckton *et al.*, 1962). The experience gained in radiation cytogenetics was subsequently applied to assess possible genetic damage from exposure to chemicals known or suspected to be carcinogenic in man, and benzene was one of the first of these substances to be studied (review in Forni, 1979). Data accumulated over the years; at the beginning of the 1970s, at least 50 substances capable of inducing structural chromosomal aberrations in human cells *in vivo* or *in vitro* were listed by Shaw (1970). In the last few years, more and more studies of groups occupationally or environmentally exposed to various inorganic and organic chemicals, or to mixtures of chemicals (see, e.g., Funes Cravioto *et al.*, 1977; Sorsa *et al.*, 1983), have been reported, sometimes with conflicting results. The knowledge gained shows, increasingly, that evaluation of the cytogenetic effects of chemicals is more complex than that of ionizing radiations, due to the different mechanisms by which different chemicals induce chromosomal aberrations (Wolff, 1982).

The purpose of this paper is not to review completely the available data, an impossible task in the space allotted, but to summarize and critically discuss the methodological problems presented by the use of chromosomal aberrations to monitor subjects exposed to mutagens-carcinogens, and to give some practical indications of the significance and limits of this test.

## METHODOLOGICAL ISSUES

*Choice of subjects and controls*

The choice of the *subjects* for cytogenetic studies is a matter of interest more for the medical staff or the epidemiologists than for the cytogeneticists, but the criteria for choice should be discussed by the different experts before starting a study.

Criteria might differ according to different aims. In preliminary studies, in order to detect possible chromosomal damage from exposure to a certain chemical, subjects with heavy or long-term exposure, or with poisoning should be chosen. When the potential of a chemical to induce chromosomal aberrations in humans has been established, cytogenetic studies can be used to monitor groups of subjects with lower exposures.

The groups should be as homogeneous as possible, but splitting into very small groups should be avoided, so that the data obtained are more reliable and suitable for statistical analysis. The mathematical basis for sample-size determination is discussed by Archer *et al.* (1981).

Before a subject is admitted to a group, anamnesic data should be collected regarding sex, age, life habits (smoking, alcohol consumption, etc.), detailed occupational history, medical X-ray exposure (especially therapeutic), recent viral diseases and vaccinations, consumption of drugs, in order to discard those subjects who have had multiple exposures to known chromosome-damaging agents, unless the effect of multiple factors is to be studied. Relevant clinical and laboratory findings should be recorded. For occupationally exposed subjects, exposure data should be available.

In follow-up studies, the history and laboratory data should be updated at every check-up.

The choice of *controls* is also very important. Since age, and possibly sex, seem to influence the rates of 'spontaneous' chromosomal aberrations (Evans, 1982; Hedner *et al.*, 1982), it seems a good policy to match study subjects with healthy controls of the same sex and approximately the same age (within $\pm$ 5 yr), not exposed to the substance under study or to any known chromosome-damaging agent (Forni *et al.*, 1971a,b).

The form used for obtaining data on exposed subjects can be used for controls too, and the same criteria for rejection should be applied. Some authors use on-site controls, i.e., subjects working in the same factory, but in different departments (Tough *et al.*, 1970; O'Riordan & Evans, 1974). If the principle is correct, care must be taken with selection when the effect of volatile substances is under study, since workers not directly exposed could have minor or occasional exposure which might reduce possible differences between 'exposed' and 'controls'.

Ideally, when the cytogenetic effect of ubiquitous pollutants is under study and methods of biological monitoring are available, exposure should be measured in the controls also, at the time of blood collection for chromosome studies.

However, it becomes increasingly difficult to obtain proper controls as more knowledge is gained on the numerous factors that may create confounding.

Ideally, chromosome studies on exposed subjects and controls should be performed concurrently, in order to minimize differences due to technical factors (see next section). In addition, every laboratory can compare results with those overall values of aberrations accumulated over years of experience with control groups. This must be done when *individual* cytogenetic studies are requested.

In follow-up studies, the subjects may serve as their own controls, if pre-employment chromosome studies are performed and factors other than occupational exposure can be ruled out (Forni *et al.*, 1976; Evans *et al.*, 1979). In long-term follow-up, however, the effect of age should be taken into account (Evans *et al.*, 1979).

### *Material for chromosomal aberration studies*

In order to assess rates of acquired chromosomal aberrations, it is necessary that numerous dividing cells be available.

*Peripheral blood lymphocytes* can be cultured in the presence of phytohaemagglutinin, a substance able to transform the $G_o$ circulating T lymphocytes into cycling cells which undergo mitosis. They represent the cell system used most extensively for chromosomal aberration studies. Besides the ease of availability of peripheral blood, which *per se* is important in population monitoring, the rationale for using lymphocyte cultures is that the long-lived T lymphocytes undergo mitosis *in vivo* only when stimulated mainly by specific antigens and, therefore, chromosome damage that has occurred *in vivo* in the $G_o$ cell can be detected at the first-second mitosis *in vitro*, even months and years later. Therefore, the rates of chromosomal aberrations in cells scored at $M_1$-metaphase *in vitro* can be considered representative of the in-vivo situation. However, since the cells are cultured, even for a short time only, the possibility of culture-borne aberrations must be taken into account.

Cells which divide spontaneously can be obtained from *haematopoietic bone marrow* but, obviously, this material cannot be used to monitor populations for exposure to carcinogens, except in individual cases where blood disorders are present. Due to the rapid turnover of bone-marrow cells, severely damaged cells disappear rather quickly after damage, having undergone only one or a few mitoses. Therefore, bone marrow cannot be used to assess past damage, but only on-going toxicity or formation of abnormal clones.

Other cell systems, like skin fibroblast cultures or germinal cells, are studied still more rarely and are not, therefore, discussed.

*Lymphocyte cultures and chromosome preparations*

The original technique of Moorhead *et al.* (1960) required separation of leucocyte-rich plasma from red blood cells, but present techniques are mainly micromethods, using a few drops of whole heparinized blood (Buckton & Evans, 1973; Evans & O'Riordan, 1975).

Cultures should be initiated within a few hours of blood collection. If a longer delay is necessary for practical reasons, blood should be stored at 4-25°C, since an excess of either cold or heat can increase chromatid aberrations. A delay of more than 24 h in setting up cultures should be avoided, since it might decrease cell viability and, hence, chromosomal aberration yield (Buckton & Evans, 1973).

Various commercial culture media are available, with different abilities to support cell growth. It is imperative to use the same medium for the whole duration of the study at least, and it is advisable to set up two to three cultures from each individual, in order to obtain sufficient numbers of metaphases to be scored.

The medium should be supplemented with (fetal) calf serum or with AB human serum or autologous plasma. Cultures with autologous plasma appear to be most representative of the in-vivo situation of the subject under study. When heterologous or homologous serum is used, the same batch of serum should be used for the whole study, a difficult task in long-term studies.

Obviously, media and sera should be virus-free, since certain viruses can induce chromatid-type aberrations.

*Culture time* is a critical factor, since it is desirable to score metaphases at first mitosis *in vitro*. This is best obtained by harvesting cells after 44-52 h of culture, including 2-3 h of demecolcine or vinblastine treatment to obtain metaphase arrest. The number of dividing cells at 48 h, however, varies with different media and different subjects, and hence the rate of second mitosis at 72 h is also different.

Longer culture times may affect chromosomal aberration yields in two directions. Cells with heavy chromosomal damage can die out after 1-2 mitoses *in vitro*, thus decreasing the aberration yield. On the other hand, culture-borne chromatid aberrations can give rise to 'derived' chromosome-type aberrations. While some authors have observed a significant decrease in chromosome-type aberration frequencies with increasing culture time (Buckton & Pike, 1964), others find no significant differences in aberration rates when scoring metaphases at 48 and 72 h, and even pool the data obtained (Funes-Cravioto *et al.*, 1975; Purchase *et al.*, 1978). In cytogenetic studies of women occupationally exposed to excess lead, the present author found a greater increase in aberrations (mainly of chromatid type) at 72 h in exposed over control subjects (Forni *et al.*, 1980).

Since exposure to clastogenic chemicals can induce mitotic delay, it might be advisable to harvest cultures at different times, but it would be better to report the data separately.

To ensure that only $M_1$-metaphases are scored, it has been proposed, also, that cultures be treated with bromodeoxyuridine, followed by differential staining, as is done for determining sister chromatid exchanges, and to score only uniformly stained metaphases for chromosomal aberrations. Although it has been shown that this method gives more consistent dose-response curves for radiation-induced aberrations *in vitro* (Scott & Lyons, 1979), the technique does not seem to be practical in human population studies and introduces a new variable, that of bromodeoxyuridine treatment.

Chromosome preparations are usually stained by the standard Giemsa technique. Although banded preparations allow detection of more abnormal metaphases than do the conventionally stained ones (Ohtaki *et al.*, 1982), it does not seem advisable to use banded methods for routine studies of acquired chromosomal changes. This is due to the impossibility, in banded preparations, of recognizing some types of aberrations, like gaps, and due to the enormous increase in scoring time, and hence in cost.

Since technical differences may be responsible for interlaboratory differences in aberration yields, the major technical details should be reported. To avoid intralaboratory differences, whatever method and material is used, it must be applied for the whole study, both to the subjects to be monitored and to the controls.

### *Types and classification of structural chromosomal aberrations*

The general term 'chromosomal aberrations' applied to metaphase chromosomes includes both *chromosome*-type aberrations, equally involving both chromatids, and *chromatid*-type aberrations involving a single chromatid.

Chromosome-type aberrations generally represent damage in $G_o$-$G_1$ cells that was maintained in the resting cell and duplicated during the S-phase.

Chromatid-type aberrations, on the contrary, are produced during the S or $G_2$ phases of the cell cycle.

Ionizing radiations, and some radiomimetic substances, are able to induce aberrations in every phase of the cell cycle. The double-strand breaks and subsequent rearrangements that are induced *in vivo* in the resting lymphocytes become apparent at first mitosis *in vitro* as chromosome-type aberrations.

Other agents, however, including most chemicals, induce chromatid-type aberrations, independent of the phase of the cell cycle at which damage occurs. In fact, these so-called S-dependent agents induce damage which, if not repaired, gives rise to aberration only during the next S-phase, and thus induces a chromatid-type aberration.

Moreover, chromatid aberrations can be induced during culture, if a damaging agent is present in the culture medium, or under certain conditions (e.g., high temperature, changes in pH, etc.).

First-mitosis chromatid-type aberrations can originate 'derived' chromosome-type aberrations at second mitosis (Evans & O'Riordan, 1975; Archer *et al.*, 1981). Therefore, S-dependent chemicals that had acted on lymphocytes *in vivo* long before cell sampling may be responsible for increased chromosome-type aberrations at first mitosis *in vitro*, if cells had undergone and survived one or more mitoses *in vivo*.

Several classifications of structural chromosomal changes are described in detail in the literature (Buckton & Evans, 1973; Evans & O'Riordan, 1975; ISCN, 1978; Archer *et al.*, 1981). Examples of the main types of scorable aberrations are reported in Figure 1.

**Fig. 1. a-c, Chromatid-type aberrations; d-k, chromosome-type aberrations**

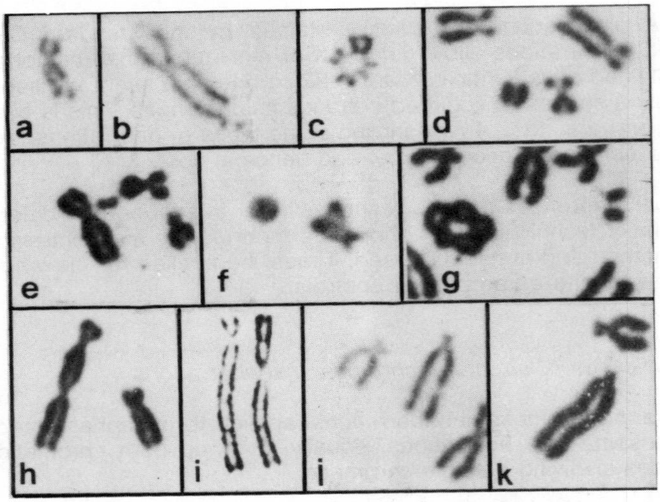


a, gap; b, break of long arm and gap in short arm; c, exchange; d, acentric fragment; e, double minute (interstitial deletion); f, small ring; g, multicentric ring on the left and acentric fragment on the right; h, dicentric chromosome; i, partial karyotype showing large abnormal chromosome resulting from pericentric inversion in chromosome No. 2; j and k, Dq+ chromosome due to translocation

While some types of aberration are easily scored (acentric fragments, dicentrics, rings) (Fig. 1, d-h), other types, like reciprocal translocations and pericentric inversions, are evident in conventional preparations only when they give rise to grossly abnormal chromosomes, with peculiar arm ratios (Fig. 1, i-k). Therefore, these aberrations, which are probably the most important from a biological point of view, are always underscored, unless banded preparations are studied.

*Scoring of aberrations and reporting*

All metaphases that appear scorable under low power should be analysed under oil immersion. Some authors analyse only cells with 46 chromosomes, others only cells with at least 45 centromeres, still others cells with 46 $\pm$ 2 centromeres. An analysis of only those cells with 46 centromeres seems too restrictive; it excludes hypo- and hyperdiploid cells which might contain abnormal chromosomes. Whatever criterion is used, however, must be maintained. Metaphases chosen under low power should be rejected only if they do not meet the prestated requirements, or if there is too much overlapping of chromosomes.

All types of aberrations should be recorded and reported separately. If, in addition, the percentages of abnormal metaphases are reported, the presence of more than one aberration per cell can be demonstrated.

For practical reasons, in reporting and for statistical analysis, aberrations can be grouped. For example, the Edinburgh cytogeneticists (Buckton *et al.*, 1962; Buckton & Pike, 1964) used to group (1) cells with chromatid-type aberrations (B cells); (2) cells with chromosome-type aberrations of unstable type ($C_u$), comprising acentrics, dicentrics and rings; (3) cells with chromosome-type aberrations of stable type ($C_s$), comprising abnormal monocentric chromosomes due to inter- or intrachanges (reciprocal translocations, pericentric inversions, etc.). This type of grouping is still used in radiation cytogenetics and in other population studies (Evans *et al.*, 1979; Anderson *et al.*, 1981). If grouping is used, the criteria for grouping should be indicated.

According to some authors, chromatid and isochromatid gaps should be recorded, but reported separately from other aberrations and excluded from the number of aberrations per cell (or % cells), because their significance is unknown (Archer *et al.*, 1981), and because they often represent technical artefacts (Evans & O'Riordan, 1975). However, a correlation was found between chromatid breaks and gaps in subjects exposed to heavy metals (Nordenson *et al.*, 1978).

The number of metaphases to be analysed may vary according to the size of the groups under study, and the rate of aberration in the controls. At least 100, preferably 200, cells from each individual in the group should be scored, in order to obtain data that might detect statistically significant differences from the controls. The problem of sample size, with its mathematical basis, has been stressed by Archer *et al.* (1981).

If more than one culture from each individual has been set up, an equal number of mitoses from each culture should be scored.

Scoring should be carried out blindly, in order to avoid bias. If more than one scorer is involved in a study, interscorer variability should be reduced by equal distribution of slides from the same individual between scorers, and by double-checking of aberrations.

### *Statistical analysis*

Chromosomal data obtained from exposed groups should be compared with those of controls.

The simplest and most frequently used method is the chi square test, which is applied to proportions of abnormal metaphases (and of the different types of aberrations) among the total cells studied in the exposed and the control groups, in $2 \times 2$ tables. If numbers are small, Fisher's exact test can be used.

Some studies compare numbers of individuals in groups with different proportions of aberrations (Brandom *et al.*, 1978), or examine proportions of individuals with aberration frequencies above a cut-off point.

On some occasions, other tests may be appropriate, e.g., the Wilcoxon matched-pairs test if matched controls are used (Forni *et al.*, 1971a,b).

More sophisticated statistical analyses should be used in other cases, e.g., analysis of variance when differences within and between groups must be analysed (Forni *et al.*, 1976) or multiple linear regression when several variables are simultaneously included in the analysis and their interactions evaluated (Evans *et al.*, 1979).

If reliable measures of exposure are available, the possibility of a relationship between dose and response should be investigated.

## VALUE AND LIMITS OF CHROMOSOMAL ABERRATION STUDIES

From the available data on chromosomal aberration studies in lymphocytes, there is no doubt, at present, that this test can be used as an indicator of genetic damage to human cells. Some observations, however, should be kept in mind when programming an investigation or evaluating the results in relation to a possible exposure. Some of these points have been stressed on several occasions (Forni, 1978; Archer *et al.*, 1981; Evans, 1982; Wolff, 1982).

(1) Induced chromosome aberrations are nonspecific, hence it is important to rule out exposure to other possible agents that might be responsible for increased rates of chromosomal changes. Some of these factors, however, may presently be unknown. A widespread confounding factor with clastogenic effects that have been demonstrated only recently is cigarette smoking (Vijayalaxmi & Evans, 1982). The background rates of 'spontaneous' aberration present in control cultures may represent the effect of various, mainly unknown, environmental factors, besides technical artefacts. Moreover, in cases of multiple exposure, little or nothing is known of the possible interaction of the various agents.

(2) Among both controls and exposed subjects, dispersion of the data is observed, with occasional subjects presenting high rates of aberrations, not justified by the anamnesic data. Whether this is due to a different susceptibility and/or exposure to unknown clastogenic agents remains to be established. The presence of single subjects with high proportions of aberration may strongly influence the overall rate of aberration, if the group is small.

(3) The long half-life of T lymphocytes is responsible for the persistence of chromosomal damage for years and even decades after cessation of exposure, shown both in irradiated subjects (Bender & Gooch, 1962; Buckton *et al.*, 1962; Goh, 1968) and in groups exposed to chromosome-damaging chemicals, such as benzene (Tough *et al.*, 1970; Forni *et al.*, 1971a,b) and vinyl chloride (Anderson *et al.*, 1981).

Chromosomal aberration cannot be used, therefore, to monitor present low-level exposure in groups of subjects with past heavier exposure to the same or other chromosome-damaging agents.

(4) For chronic exposure to clastogenic chemicals, generally no clear-cut dose-response relationships can be demonstrated (Forni *et al.*, 1971a; Forni & Secchi, 1973; Funes-Cravioto *et al.*, 1975), although some positive association has been reported (Nordenson *et al.*, 1978; Purchase *et al.*, 1978). In some studies on groups of subjects exposed to benzene, positive findings have been demonstrated for high-level exposure and negative findings for low-level exposure (Tough *et al.*, 1970; reviewed by Forni, 1979). Failure to demonstrate definite relationships may be due to several factors such as (a) difficulty in measuring the exposure(s) and/or different modality of exposure (e.g., low-chronic *versus* peak exposure); (b)

different individual susceptibility to chromosomal damage and capacity to repair damage or eliminate damaged cells; (c) the impossibility of evaluating the total damage from chronic exposure and the loss of damaged cells with time, a factor particularly important in the evaluation of past exposures; (d) sampling problems, due to relatively few cells scored per individual, which may influence, mainly, the rates of the less frequent aberrations. In some studies, levelling of aberration rates has been observed also, which suggests that some kind of equilibrium mechanism might intervene after a certain length of exposure (Forni *et al.*, 1976; Funes-Cravioto *et al.*, 1977). In radiation cytogenetics, Evans *et al.* (1979) were able to demonstrate a linear correlation between $C_u$ cells and cumulative doses in subjects with chronic low-level occupational exposure to ionizing radiations, by pooling a huge number of data obtained in a 10-year follow-up. However, they stressed, correctly, that the correlation can be interpreted only on a group basis.

(5) The method might not be sensitive enough for chronic low-level exposures, such as those that generally now occur in industrial or environmental settings, unless very large groups of subjects (and controls) are studied, and/or larger numbers of mitoses per individual are scored. Neither condition is practical for several reasons, such as increased heterogeneity of the groups (and hence of confounding) with increased numbers of subjects studied, and increases in volume of work and costs, beyond feasibility, both for the individual cytogenetic examination and for the overall study.

(6) Chromosome studies should be evaluated mainly on a group basis, except for particular situations (e.g., individuals with signs of poisoning due to accidents, or subjects with evidence of abnormal clones).

(7) As far as type of aberration is concerned, the easily scored 'unstable' aberrations evaluated at first mitosis *in vitro* are a better indicator of damage than the underestimated 'stable' aberrations, even though the latter are more important from a biological point of view, since they might give rise to abnormal clones. In studying the cytogenetic effects of chemicals, chromatid-type aberrations should be considered also, although the possible culture effects must be reduced to a minimum, as previously stressed.

(8) Well-performed cytogenetic studies are costly, since they are time-consuming and require the work of skilled personnel. They should, therefore, be limited to selected groups.

## MEDICAL IMPLICATIONS

The implications for carcinogenesis and mutagenesis of increased rates of chromosomal aberrations in lymphocytes in subjects exposed to genotoxic agents are still under discussion.

The suggestion that increased rates of chromosomal aberrations might, somehow, favour the occurrence of cancer or leukaemia is supported by the observation that among subjects with congenital diseases like Fanconi's anaemia and Bloom's syndrome, both characterized by the presence of abnormally high rates of chromosomal breakage and rearrangement, there is an increased incidence of leukaemia and neoplasm (German, 1972). It is possible that in people exposed to chromosome-damaging agents, like ionizing radiations or benzene, which are both leukaemogenic, chromosomal changes in bone-marrow cells might give rise to aberrant clones directly responsible for the occurrence of leukaemia or might provide an unbalanced and unstable chromosome complement more susceptible to other leukaemogenic factors; occurrence of leukaemia might also be favoured by concomitant impairment of the immune system. But these are only suggestions, and the presence of abnormal cell clones, in the absence of any sign of disease, has been described both in irradiated subjects (Goh, 1968) and in subjects over-exposed to benzene (Forni *et al.*, 1971a,b) in studies that were followed up for several years.

A correlation between the prevalence of chromosomal aberration in lymphocytes and cytological abnormalities in the specific target tissue (i.e., bronchial cells) has been demonstrated in uranium miners, a population at high risk for lung cancer (Brandom *et al.*, 1978).

However, it is generally believed that increased proportions of chromosomal aberrations indicate exposure correlated to an increased risk for the group, not for the single subjects who have excess aberrations, as it appears from long-term studies on the survivors of Hiroshima and Nagasaki (Archer *et al.*, 1981).

It is even more difficult to define the significance of increased rates of chromosomal aberration in somatic cells, like lymphocytes, for mutagenesis and teratogenesis which involve germinal and/or embryo cells. In fact, the study of germinal cells is very difficult; there is no knowledge of the amount of the toxic chemical (or, better, of its active metabolites) that reaches the germinal cells, whether these cells have the same or a different susceptibility to chromosomal damage compared to somatic cells, and whether the damaged cells are still viable and capable of giving rise to a viable embryo. No transmissible chromosomal aberration has been detected in the offspring of atomic bomb survivors (Awa *et al.*, 1968). Other methods, such as estimates of spontaneous abortions, should be applied.

Finally, if chromosomal aberrations indicate genetic damage, more subtle changes of the genetic material, not scorable even with the most sophisticated microscopic techniques, and also point mutations, should be searched for, since they might have more implications for both the individual and the offspring.

## REFERENCES

Anderson, D., Richardson, C.R., Purchase, I.F.H., Evans, H.J. & O'Riordan, M.L. (1981) Chromosomal analysis in vinyl chloride exposed workers: comparison of the standard technique with the sister-chromatid exchange technique. *Mutat. Res.*, *83*, 137-144

Archer, P.G., Bender, M., Bloom, A.D., Brewen, J.G., Carrano, A.V. & Preston, R.J. (1981) *Report of Panel 1: Guidelines for cytogenetic studies in mutagen-exposed human populations.* In: Bloom, A.D., ed., *Guidelines for Studies of Human Populations Exposed to Mutagenic and Reproductive Hazards*, New York, March of Dimes Birth Defects Foundation, pp. 1-35

Awa, A.A., Bloom, A.D., Yoshida, M.C., Neriishi, S. & Archer, P.G. (1968) Cytogenetic study of the offspring of atom bomb survivors. *Nature*, *218*, 367-368

Bender, M.A. & Gooch, P.C. (1962) Persistent chromosome aberrations in irradiated human subjects. *Radiat. Res.*, *16*, 44-53

Brandom, W.F., Saccomanno, G., Archer, V.E., Archer, P.G. & Bloom, A.D. (1978) Chromosome aberrations as a biological dose-response indicator of radiation exposure in uranium miners. *Radiat. Res.*, *76*, 159-171

Buckton, K.E. & Pike, M.C. (1964) Time in culture. An important variable in studying in-vivo radiation-induced chromosome damage in man. *Int. J. Radiat. Biol.*, *8*, 439-452

Buckton, K.E. & Evans, H.J., eds (1973) *Methods for the Analysis of Human Chromosome Aberrations*, Geneva, World Health Organization

Buckton, K.E., Jacobs, P.A., Court Brown, W.M. & Doll, R. (1962) A study of the chromosome damage persisting after X-ray therapy for ankylosing spondylitis. *Lancet*, *ii*, 676-682

Evans, H.J. (1982) Chromosomal mutations in human populations. *Cytogenet. Cell Genet.*, *33*, 48-56

Evans, H.J. & O'Riordan, M.L. (1975) Human peripheral blood lymphocytes for the analysis of chromosome aberrations in mutagen tests. *Mutat. Res.*, *31*, 135-148

Evans, H.J., Buckton, K.E., Hamilton, G.E. & Carothers, A. (1979) Radiation-induced chromosome aberrations in nuclear-dockyard workers. *Nature*, *277*, 531-534

Forni, A. (1978) Significance and limitations of the study of chromosome aberrations in occupational medicine (Italian). *Med. Lav.*, *69*, 331-340

Forni, A. (1979) Chromosome changes and benzene exposure. A review. *Rev. environ. Health*, *3*, 5-17

Forni, A. & Secchi, G.C. (1973) *Chromosome changes in preclinical and clinical lead poisoning and correlation with biochemical findings.* In: *Proceedings of the International Symposium on Environmental Health Aspects of Lead, Amsterdam, October 1972*, Luxembourg, Council of the European Communities, pp. 473-482

Forni, A., Pacifico, E. & Limonta, A. (1971a) Chromosome studies in workers exposed to benzene or toluene or both. *Arch. environ. Health*, *22*, 373-378

Forni, A.M., Cappellini, A., Pacifico, E. & Vigliani, E.C. (1971b) Chromosome changes and their evolution in subjects with past exposure to benzene. *Arch. environ. Health*, *23*, 385-391

Forni, A., Cambiaghi, G. & Secchi, G.C. (1976) Initial occupational exposure to lead. Chromosome and biochemical findings. *Arch. environ. Health*, *31*, 73-78

Forni, A., Sciame', A., Bertazzi, P.A. & Alessio, L. (1980) Chromosome and biochemical studies in women occupationally exposed to lead. *Arch. environ. Health*, *35*, 139-146

Funes-Cravioto, F., Lambert, B., Lindsten, J., Ehrenberg, L., Natarajan, A.T. & Osterman-Golkar, S. (1975) Chromosome aberrations in workers exposed to vinyl chloride. *Lancet*, *i*, 459

Funes-Cravioto, F., Zapata-Gayon, C., Kolmodin-Hedman, D., Lambert, B., Lindsten, J., Norberg, E., Nordenskjöld, M., Olin, R. & Swensson, Å. (1977) Chromosome aberrations and sister-chromatid exchange in workers in chemical laboratories and a rotoprinting factory and in children of women laboratory workers. *Lancet*, *ii*, 322-325

German, J. (1972) *Genes which increase chromosomal instability in somatic cells and predispose to cancer*. In: Steinberg, A.G., Bearn, A.G., Motulsky, A.G. & Childs, B., eds, *Progress in Medical Genetics*, Vol. 8, New York, Grune & Stratton, pp. 61-102

Hedner, K., Högstedt, B., Kolnig, A.-M., Mark-Vendel, E., Strömbeck, B. & Mitelman, F. (1982) Sister chromatid exchanges and structural chromosome aberrations in relation to age and sex. *Human Genet.*, *62*, 305-309

Goh, K.-O. (1968) Total-body irradiation and human chromosomes: cytogenetic studies of the peripheral blood and bone marrow leukocytes seven years after total-body irradiation. *Radiat. Res.*, *35*, 155-170

ISCN (1978) An International System for Human Cytogenetic Nomenclature. *Cytogenet. cell Genet.*, *21*, 309-404

Moorhead, P.S., Nowell, P.C., Mellman, W.J., Battips, D.M. & Hungerford, D.A. (1960) Chromosome preparations of leukocytes cultured from human peripheral blood. *Exp. Cell Res.*, *20*, 613-616

Nordenson, I., Beckman, G., Beckman, L. & Nordström, S. (1978) Occupational and environmental risks in and around a smelter in northern Sweden. IV. Chromosomal aberrations in workers exposed to lead. *Hereditas*, *88*, 263-267

Ohtaki, K., Shimba, H., Awa, A.A. & Sofuni, T. (1982) Comparison of type and frequency of chromosome aberrations by conventional and G-staining methods in Hiroshima atomic bomb survivors. *J. Radiat. Res.*, *23*, 441-449

O'Riordan, M.L. & Evans, H.J. (1974) Absence of significant chromosome damage in males occupationally exposed to lead. *Nature*, *247*, 50-53

Purchase, I.F.H., Richardson, C.F., Anderson, D., Paddle, G.M. & Adams, W.G.F. (1978) Chromosomal analyses in vinyl-chloride exposed workers. *Mutat. Res.*, *57*, 325-334

Scott, D. & Lyons, C.Y. (1979) Homogenous sensitivity of human peripheral blood lymphocytes to radiation-induced chromosome damage. *Nature*, *278*, 756-758

Shaw, M.W. (1970) Human chromosome damage by chemical agents. *Ann. Rev. Med.*, *21*, 409-432

Sorsa, M., Mäki-Paakkaanen, J. & Vainio, H. (1983) A chromosome study among worker groups in the rubber industry. *Scand. J. Work environ. Health*, *9* (Suppl. 2), 43-47

Tough, I.M., Smith, P.G., Court Brown, W.M. & Harnden, D.G. (1970) Chromosome studies on workers exposed to atmospheric benzene. The possible influence of age. *Eur. J. Cancer*, *6*, 49-55

Vijayalaxmi & Evans, H.J. (1982) In-vivo and in-vitro effects of cigarette smoke on chromosomal damage and sister chromatid exchange in human peripheral blood lymphocytes. *Mutat. Res.*, *92*, 321-332

Wolff, S. (1982) Difficulties in assessing the human health effects of mutagenic carcinogens by cytogenetic analyses. *Cytogenet. cell Genet.*, *33*, 7-13

# MONITORING OF SISTER CHROMATID EXCHANGE AND MICRONUCLEI AS BIOLOGICAL ENDPOINTS

**M. Sorsa**

*Institute of Occupational Health, Helsinki, Finland*

## SUMMARY

Cytologically visible damage in human chromosomes can be detected as structural chromosomal aberrations, numerical changes in genome, sister chromatid exchanges (SCE) or as micronucleated cells. The importance of in-vivo cytogenetic damage that is induced in human cells is that it indicates that similar alterations may have occurred in other tissues, either in somatic or in germinal cells.

SCEs represent symmetrical exchanges between sister chromatids; generally, they do not result in alteration of the chromosome morphology or the genetic information. Although the detection method is highly sensitive as an in-vitro screening test, in monitoring studies, it seems to be restricted to cases where the exposing agents are strong alkylating compounds (e.g., ethylene oxide, cytostatic drugs) or to some multi-exposure conditions (e.g., cigarette smoking, laboratory work, rubber industries).

Micronuclei arise from acentric chromosome fragments or lagging whole chromosomes. They have been detected in a variety of dividing human cells, including exfoliated epithelial cells, bone-marrow cells and lymphocytes. Positive responses of induction of cells with micronuclei have been obtained in studies with cells of the buccal mucosa (chewers of betel or tobacco) or cultured lymphocytes of some groups occupationally exposed to agents like styrene or ethylene oxide.

## INTRODUCTION

In preventive medicine, biological monitoring methods are used for early detection of harmful exposure. In the conventional sense of the term, 'biological monitoring' comprises regular measurements among a group of exposed people, using validated measurements to estimate the internal dose. Exposure to mutagenic agents may be monitored for the biological effects, also, of the exposure (Vainio *et al.*, 1981). In practice, cytogenetic methods, as yet, have rarely fulfilled the requirement for consecutive regular measurements (Sorsa *et al.*, 1982a). However, with increasing use of cytogenetic surveillance to monitor the effects of genotoxic exposures, cytogenetic methods may, gradually, be developed for routine use. Development implies both standardization and methodological improvements, such as automation, to replace the presently tedious and relatively insensitive cytogenetic techniques.

## CYTOGENETIC SURVEILLANCE TECHNIQUES

Cytogenetic damage can be visualized only in proliferating cell populations. In principle, three types of cytogenetic change can be distinguished, viz., structural alterations, intrachromosomal exchanges and alterations in the chromosome number.

Somatic chromosomal aberrations have been used for four decades already, as an indicator of exposure to chromosome-breaking agents. Most of the early work on induced clastogenicity dealt with irradiation (Lea, 1946). For ionizing radiation, the dose-response relationship has been characterized well enough for the frequency of dicentric and fragmented chromosomes to be used as a biological dosimeter of the radiation dose received. The use of structural chromosomal aberrations to detect chemical exposure is discussed in this volume by Dr A. Forni.

### *Sister chromatid exchanges*

Induction of sister chromatid exchanges (SCEs) has been used only in the last decade, to detect the exposure of people to hazardous mutagenic chemicals. The first detection of reciprocal exchanges of segments between sister chromatids of a chromosome this year (1983) coincides with the quarter-of-a-century jubilee honouring the first findings of J.H. Taylor (1958). It is now possible to visualize an SCE by treatment of the replicating chromosome, with either physical (e.g., tritiated thymidine) or chemical (e.g., bromodeoxyuridine) agents so that the sister chromatids and their exchanged segments become distinguishable from one another. Most mutagenic agents induce SCEs (Wolff, 1982a), except those that break DNA directly (e.g., ionizing radiation, bleomycin).

As an in-vitro assay, the SCE method has been shown to be exquisitely sensitive, to the extent that, for many mutagenic chemicals, only 1% of the lowest concentration needed to produce effective chromosomal aberration is needed to induce a significant response in SCEs (Perry & Evans, 1975; Stetka & Wolff, 1976). However, the DNA lesions that induce SCE are not necessarily the same as those that induce point mutations or chromosome breakage (Carrano *et al.*, 1978).

For human in-vivo monitoring, the SCE techniques have been used, so far, only to a limited extent (see Table 1). The human exposures most extensively studied for SCE responses in peripheral lymphocytes are those of cancer patients under chemotherapy (Düker, 1981; Gebhart *et al.*, 1980; Lambert *et al.*, 1982). These studies of controlled exposure to mutagenic chemicals suggest that the damage that is observed as increased SCE frequency in the cultured lymphocytes of the exposed persons is acutely induced, but it disappears a few months after therapeutic exposure has ceased. This fact must be considered when SCE techniques are applied to occupational monitoring (see below).

*Micronuclei*

Micronuclei are either acentric chromosome fragments or whole chromosomes lagging behind in the normal karyokinesis. These fragments or chromosomes are excluded from the postmitotic daughter cells and form additional small nuclei or micronuclei. Thus, the induction of micronucleated cells represents either chromosomal breakage (clastogenicity) or a failure in the spindle fibre mechanism (turbagenicity). The conventional micronucleus test in rodents is designed to detect damage in the polychromatic erythrocytes of bone marrow. Extensive chemical evidence obtained from this test shows that micronucleus formation is qualitatively correlated closely with chromosomal breakage or loss (Heddle *et al.*, 1983). In humans, micronucleated cells have been visualized, likewise, in dividing tissues, e.g., in bone marrow *in vivo* (Krogh Jensen & Nyfors, 1979) and in cultured lymphocytes *in vitro* (Heddle *et al.*, 1978). Recent animal experiments on micronucleated spermatids of rodents exposed to mutagens (Lähdetie, 1983; Tates *et al.*, 1983) suggest that, in principle, testicular germinal cells, also, can be used to detect micronuclei.

The original findings of Countryman and Heddle (1976), for micronucleated human lymphocytes induced by in-vitro exposure to ionizing radiation, have been modified recently by Högstedt *et al.* (1983a), who were the first to show that human in-vivo exposure to cigarette smoke can be detected as an increased frequency of micronuclei in the cultured lymphocytes of smokers. The method has been further improved by using lymphocyte concentrates and analysing the cells without hypotonic treatment, so that the cytoplasm stays intact and the preparations are easier to score (Högstedt, 1984). An increased frequency of micronucleated lymphocytes has been found among workers exposed to rather low (~ 30 ppm) levels of styrene (Högstedt *et al.*, 1983b).

A most interesting recent application of the micronucleus assay has been reported by Stich *et al.* (1982a,b) on exfoliated human cells of the oral cavity. Micronuclei were analysed from the buccal mucosa of Indian betel-quid eaters (Stich *et al.*, 1982b) and users of powdered (*Khaini*) tobacco (Stich *et al.*, 1982a). In both cases, the betel or tobacco chewers exhibited higher frequencies of micronucleated epithelial cells than the control group of the same ethnic origin. Epidemiological studies, in both groups that chewed betel or tobacco, suggest a causal relationship between oral cancer and the habit of tobacco chewing (Stich *et al.*, 1982a). Thus, the micronucleus test on the buccal smear indicates chromosomal damage in the same target tissue.

The dose-response relationship and the relevance of the findings of micronucleated mucosal cells for the individual risk of oral cancer in tobacco and betel chewers are, as yet, unclear. However, these studies are an unusual example of estimation of induced cytogenetic damage in the tissue from which the neoplastic transformation and carcinomas develop.

## APPLICATION OF SCE AND MICRONUCLEI TECHNIQUES IN OCCUPATIONAL MONITORING

Compared to studies on structural chromosomal aberrations, very little information is yet available on the possible association of occupational carcinogenic or mutagenic agents with SCEs or micronuclei. Published results, most of which originate from recent years, are compiled in Table 1.

**Table 1. Published results on occupational exposure and determinations of sister chromatid exchanges (SCE) and micronuclei (MN)**

| Major exposing chemical | Result of study[a] | | Reference |
|---|---|---|---|
| | SCE | MN | |
| Anaesthetic gases | - | ·· | Husum & Wulf (1980) |
| Asbestos | + | ·· | Rom *et al.* (1983) |
| Benzene | + | ·· | Watanabe *et al.* (1980) |
| Bichromate production | - | ·· | Imreh & Radulescu (1982) |
| Carbon disulfide | ± | ·· | Nordenson *et al.* (1980)<br>Sorsa *et al.* (1982a) |
| Chromic acid | + | ·· | Sarto *et al.* (1982) |
| Cytostatic drugs | + | ·· | Norppa *et al.* (1980)<br>Sorsa *et al.* (1982b) |
| Epoxy resins | - | ·· | Mitelman *et al.* (1980) |
| Ethylene oxide | + | + | Högstedt *et al.* (1983b)<br>Yager *et al.* (1983) |
| Lead smelting | (+) | ·· | Mäki-Paakkanen *et al.* (1981) |
| Organic solvents | + | ·· | Funes-Cravioto *et al.* (1977) |
| Pentachlorophenol | - | ·· | Bauchinger *et al.* (1982a) |
| Petroleum products | ·· | + | Högstedt *et al.* (1981) |
| Phenoxy acid herbicides | - | ·· | Linnainmaa (1983) |
| Rubber chemicals | + | ·· | Sorsa *et al.* (1982c) |
| Styrene | + | + | Anderson *et al.* (1980)<br>Högstedt *et al.* (1979, 1983b) |
| 2,4,5-T (accident) | - | ·· | Blank *et al.* (1983) |
| Tetrachloroethylene | - | ·· | Ikeda *et al.* (1980) |
| Toluene | ± | ·· | Bauchinger *et al.* (1982b)<br>Mäki-Paakkanen *et al.* (1980) |
| Vinyl chloride | + | ·· | Anderson *et al.* (1980, 1981) |
| Welding (manual metal arc) fumes | - | ·· | Husgafvel-Pursiainen *et al.* (1982) |
| Xylene | - | ·· | Haglund *et al.* (1980) |

[a] +, positive results; (+), suggestive positive results; -, negative results; ··, no data

Of some seventy studies on occupational cytogenetics, very few have been confirmed by several independent findings. The Nordic working group on occupational chromosomal damage (Nordic Council of Ministers, 1984, Project no. 171.12-2.4) discussed the available studies recently, and concluded that, at present, only seven occupational agents are confirmed as causal for chromosomal damage in humans (at exposures above lowest effective concentrations) (Table 2). Knowledge of human cytogenetic surveillance is thus very limited. Further studies, using different cytogenetic parameters for the same exposure group, are urgently needed.

When the chromosome-damaging effects of a chemical have been established and occupational exposure to that chemical remains inevitable, every effort should be made to minimize the exposure. Occupational health personnel should make use of the available possibilities for cytogenetic monitoring.

A major cause of the reluctance to apply monitoring procedures to occupationally exposed populations is the expense and the resources required. Thus, automation of cytogenetic analyses, as well as further progress in methods to detect induced DNA damage, are both urgently needed for preventive medicine in the field of occupational health.

**Table 2. Human chromosome-damaging agents confirmed by several occupational studies[a]**

| Agent | Cytogenetic information[b] | | |
|---|---|---|---|
| | CA | SCE | MN |
| Alkylating cytostatics | + | + | + |
| Benzene | + | + | + |
| Epichlorohydrin | + | .. | .. |
| Ethylene oxide | + | + | + |
| Ionizing radiation | + | - | + |
| Styrene | + | (+) | + |
| Vinyl chloride | + | + | .. |

[a] From Report of Working Group, Nordic Council of Ministers, 1984

[b] CA, chromosomal aberrations; SCE, sister chromatid exchanges; MN, micronuclei

## CONFOUNDING FACTORS AND VARIABLES TO BE CONTROLLED IN CYTOGENETIC STUDIES

Over seventy occupational cytogenetic studies, mostly on structural chromosomal aberrations, have been published, with negative, contradictory and positive findings (Working Group of Project No. 171.12.-2.4, Nordic Council of Ministers, 1984). It is unfortunate that many of these studies do not fulfil the criteria of a reliable cytogenetic study.

The major problem in designing well-conducted studies of this type concerns the control subjects. Obviously, the manifestation of cytologically detectable chromosomal damage induced by a clastogenic chemical is, often, the result of a variety of metabolic activations and inactivations, of processes of repair and replication, all of which are genetically controlled enzymatic functions. The fact that there is a large interindividual and interorgan variability in enzymatic capacities and toxicological responses, has been realized only recently, most distinctly among homo- and heterozygotes of some genetic disorders (Setlow, 1978; Berg, 1979). It is quite obvious that both inherited predisposing factors and uncontrolled 'lifestyle' factors affect the level of 'spontaneous' chromosomal aberration and SCEs in all individuals (Crossen, 1982).

The need for valid controls is essential in all cytogenetic studies of exposed persons. More effort should be made, also, to design studies for prospective cytogenetic monitoring, when every individual would serve as a control (prior to exposure) for himself.

Some of the confounding factors in cytogenetic studies are listed in Table 3. The technical problems, including variations induced by culture medium, antibiotics, mitogen, spindle inhibitor, preparative methods and the skill of the analyst can be overcome by practice (Das & Sherma, 1983). Special problems in SCE studies include the concentration of bromodeoxyuridine, as well as differences in the capabilities of various chemicals to cause cell-cycle delay and in the induction of repair or the persistence of lesions, all of which may selectively affect the level of SCEs being scored (Wolff, 1982b).

**Table 3. Confounding and controllable factors in occupational cytogenetic studies**

EXPOSURE CONDITIONS

- identification of true exposure
- estimation of dose exposed

INDIVIDUAL VARIABILITY

- genetic factors
- lifestyle factors (e.g., smoking, use of drugs)
- health (viral infections, diagnostic X-rays)

CONDITIONS DURING CULTURE

- culture time
- culture medium and chemicals used
- time between blood sampling and culture
- persistence of the mutagen in the blood sample

SCORING OF RESULTS

- variation between scorers
- interpretation of damage scored

CONTROL GROUP

- methodological control
- group-matched controls (age, sex, smoking)

The preculturing confounding effects (Table 3) can be controlled, partly at least, by careful interviewing of the subjects for other possible clastogenic exposures (drugs, X-rays, vaccinations, viral infections). Smoking has been shown to increase SCE frequency in several studies (Lambert *et al.*, 1982), and heavy smokers, also, show increased chromosomal aberration rates (Madle *et al.*, 1981). Thus, smoking is a 'lifestyle' factor that must be recorded carefully and taken into account when the control subjects are selected.

## SIGNIFICANCE OF SOMATIC CHROMOSOMAL DAMAGE

The most relevant targets for assessing human risk caused by exogenous genotoxic exposure are human cells. Thus, cytogenetic methods that assess human cells for in-vivo exposure are of utmost importance in the identification of hazardous exposure, even though damage as such in the cell targets most frequently used, peripheral blood lymphocytes, has, probably, no health significance for the individual. The major relevance of the finding of increased chromosomal damage in lymphocytes of an exposed group lies in its indicative value - similar damage may have occurred in the DNA of other, potentially more vulnerable, cells.

Much indirect evidence suggests the importance of genetic damage in the process of malignant transformation (Sorsa, 1980). Recent data concerning the localization of mammalian and human oncogenes and the role of chromosomal rearrangements in their activation (Logan & Cairn, 1982; Yunis, 1983) has given further support to the association of chromosomal damage with malignancies.

The significance of the SCEs observed in somatic cells depends on their value as indicators of DNA damage and on the quality of empirical correlation of clastogenicity with the SCE-induction ability of chemicals. Of the carcinogenic chemicals studied so far, about 80% are known to induce SCEs *in vitro* or *in vivo* (Latt *et al.*, 1981). No health effect is known to be associated with SCEs as such.

The hitherto meagre knowledge of micronuclei in somatic cells does not allow any specific clinical consequences to be linked to this phenomenon. As for SCEs, the significance of findings about micronucleated cells depends on their value as indicators for damage induced in DNA. According to results from experimental systems, there is an extremely close relationship between micronucleated cells and chromosomal breakage (Heddle *et al.*, 1983).

Induced chromosomal damage (structural aberrations, SCEs or numerical alterations) that has a properly documented association with specific exposure should always be considered an adverse sign of exposure to mutagenic and potentially carcinogenic environmental agents, for the population studied. Even though the health consequences for the individual cannot be estimated on the basis of present knowledge, the findings, at the group level, must be interpreted as risk for potential manifestation of genotoxic endpoints.

## REFERENCES

Anderson, D., Richardson, C.R., Weight, T.M., Purchase, I.F.H. & Adams, W.G.F. (1980) Chromosomal analyses in vinyl chloride exposed workers. Results from analysis 18 and 42 months after an initial sampling. *Mutat. Res.*, *79*, 151-162

Anderson, D., Richardson, C.R., Purchase, I.F.H., Evans, H.J. & O'Riordan, M.L. (1981) Chromosomal analyses in vinyl chloride exposed workers: comparison of the standard technique with the sister-chromatid exchange technique. *Mutat. Res.*, *83*, 137-144

Andersson, H.C., Tranberg, E.A., Uggla, A.H. & Zetterberg, G. (1980) Chromosomal aberrations and sister-chromatid exchanges in lymphocytes of men occupationally exposed to styrene in a plastic-boat factory. *Mutat. Res.*, *73*, 387-401

Bauchinger, M., Dresp, J., Schmid, E. & Hauf, R. (1982a) Chromosome changes in lymphocytes after occupational exposure to pentachlorphenol (PCP). *Mutat. Res.*, *102*, 83-88

Bauchinger, M., Schmid, E., Dresp, J., Kolin-Gerresheim, J., Hauf, R. & Suhr, E. (1982b) Chromosome changes in lymphocytes after occupational exposure to toluene. *Mutat. Res.*, *102*, 439-445

Berg, K. (1979) *Inherited variation in susceptibility and resistance to environmental agents.* In: Berg, K., ed., *Genetic Damage in Man Caused by Environmental Agents*, New York, Academic Press, pp. 1-25

Blank, C.E., Cooke, P. & Potter, A.M. (1983) Investigations for genotoxic effects after exposure to crude 2,4,5-trichlorophenol. *Br. J. ind. Med.*, *40*, 87-91

Carrano, A.V., Thompson, L.H., Lindl, P.A. & Minkler, J.L. (1978) Sister chromatid exchange as an indicator of mutagenesis. *Nature*, *271*, 551-553

Crossen, P.E. (1982) Variation in the sensitivity of human lymphocytes to DNA-damaging agents measured by sister chromatid exchange frequency. *Human Genet.*, *60*, 19-23

Countryman, P.J. & Heddle, J.A. (1976) The production of micronuclei from chromosome aberration in irradiated cultures of human lymphocytes. *Mutat. Res.*, *41*, 321-332

Das, B.C. & Sharma, T. (1983) Reduced frequency of baseline sister chromatid exchanges in lymphocytes grown in antibiotics and serum-excluded culture medium. *Human Genet.*, *64*, 249-253

Düker, D. (1981) Investigations into sister chromatid exchange in patients under cytostatic therapy. *Human Genet.*, *58*, 198-203

Funes-Cravioto, F., Zapata-Gayon, C., Kolmodin-Hedman, D., Lambert, B., Lindsten, J., Nordberg, E., Nordenskjöld, M., Olin, R. & Swensson, A. (1977) Chromosome aberrations and sister-chromatid exchange in workers in chemical laboratories and a rotoprinting factory and in children of women laboratory workers. *Lancet*, *ii*, 322-325

Gebhart, E., Windolph, B. & Wopfner, F. (1980) Chromosome studies on lymphocytes of patients under cytostatic therapy. II. Studies using the BUDR-labelling technique in cytostatic interval therapy. *Human Genet.*, *56*, 157-167

Haglund, U., Lundberg, I. & Zech, L. (1980) Chromosome aberrations and sister chromatid exchanges in Swedish paint industry workers. *Scand. J. Work environ. Health*, *6*, 291-298

Heddle, J.A., Lue, C.B., Saunders, E.F., Benz, R.D. (1978) Sensitivity to five mutagens in Fanconi's anemia as measured by the micronucleus method. *Cancer Res.*, *38*, 2983-2988

Heddle, J.A., Hite, M., Kirkhart, B., Mavournin, K., MacGregor, J.T., Newell, G.W. & Salamone, M.F. (1983) The induction of micronuclei as a measure of genotoxicity. *Mutat. Res.*, *123*, 61-118

Högstedt, B. (1984) Micronuclei in lymphocytes with preserved cytoplasm - a method for assessment of cytogenetic damage in man. *Mutat. Res.*, *130*, 63-72

Högstedt, B., Hedner, K., Mark-Vendel, E., Mitelman, F., Schutz, A. & Skerfving, S. (1979) Increased frequency of chromosome aberrations in workers exposed to styrene? *Scand. J. Work environ. Health*, *5*, 333-335

Högstedt, B., Gullberg, B., Mark-Vendel, E., Mitelman, F. & Skerfving, S. (1981) Micronuclei and chromosome aberrations in bone marrow cells and lymphocytes of humans exposed mainly to petroleum vapors. *Hereditas*, *94*, 179-187

Högstedt, B., Gullberg, B., Hedner, K., Kolnig, A.M., Mitelman, F., Skerfving, S. & Widegren, B. (1983a) Chromosome aberrations and micronuclei in bone marrow cells and peripheral blood lymphocytes in humans exposed to ethylene oxide. *Hereditas*, *98*, 105-113

Högstedt, B., Åkesson, B., Axell, K., Gullberg, B., Mitelman, F., Pero, R.W., Skerfving, S. & Welinder, H. (1983b) Increased frequency of lymphocyte micronuclei in workers producing reinforced polyester resin with low exposure to styrene. *Scand. J. Work environ. Health*, *49*, 271-276

Husgafvel-Pursiainen, K., Kalliomäki, P.-L. & Sorsa, M. (1982) A chromosome study among stainless steel welders. *J. occup. Med.*, *24*, 762-766

Husum, B. & Wulf, H.C. (1980) Sister chromatid exchanges in lymphocytes in operating-room personnel. *Acta anaesth. scand.*, *24*, 22-24

Ikeda, M., Koizumi, A., Watanabe, T., Endo, A. & Sato, K. (1980) Cytogenetic and cytokinetic investigation on lymphocytes from workers occupationally exposed to tetrachloroethylene. *Toxicol. Lett.*, *5*, 251-256

Imreh, S. & Radulescu, D. (1982) Cytogenetic effects of chromium *in vivo* and *in vitro*. *Mutat. Res.*, *97*, 192-193

Krogh Jensen, M. & Nyfors, A. (1979) Cytogenetic effect of methotrexate on human cells *in vivo*. *Mutat. Res.*, *64*, 339-343

Lähdetie, J. (1983) Meiotic micronuclei induced by adrianmycin in male rats. *Mutat. Res.*, *119*, 79-82

Lambert, B., Bredberg, A., McKenzie, W. & Sten, M. (1982) Sister chromatid exchange in human populations: the effect of smoking, drug treatment and occupational exposure. *Cytog. Cell Genet.*, *33*, 62-67

Latt, S.A., Allen, J., Bloom, S.E., Carrano, A., Falke, E., Kram, D., Schneider, E., Schreck, R., Tice, R., Whitfield, B. & Wolff, S. (1981) Sister chromatid exchanges: a report of the Gene-Tox Program. *Mutat. Res.*, *87*, 17-62

Lea, D.E. (1946) *Actions of Radiations on Living Cells*, Cambridge, Cambridge University Press

Linnainmaa, K. (1983) Sister chromatid exchanges among workers occupationally exposed to phenoxy acid herbicides 2,4-D and MCPA. *Teratog. Carcinog. Mutag.*, *3*, 269-279

Logan, J. & Cairns, J. (1982) The secrets of cancer. *Nature*, *300*, 104-105

Madle, S., Korte, A. & Obe, G. (1981) Cytogenetic effects of cigarette smoke condensates *in vitro* and *in vivo*. *Human Genet.*, *59*, 349-352

Mäki-Paakkanen, J., Husgafvel-Pursiainen, K., Kalliomäki, P.-L., Tuominen, J. & Sorsa, M. (1980) Toluene-exposed workers and chromosome aberrations. *J. Toxicol. environ. Health*, *6*, 775-781

Mäki-Paakkanen, J., Sorsa, M. & Vainio, H. (1981) Chromosome aberrations and sister chromatid exchanges in lead-exposed workers. *Hereditas*, *94*, 269-275

Mitelman, F., Fregert, S., Hedner, K. & Hillbertz-Nilsson, K. (1980) Occupational exposure to epoxy resins has no cytogenetic effect. *Mutat. Res.*, *77*, 345-348

Nordenson, I., Beckman, G., Beckman, L., Rosenhall, L. & Stjernberg, N. (1980) Is exposure to sulphur dioxide clastogenic? *Hereditas*, *93*, 161-164

Nordic Council of Ministers (1984) *Report of Working Group for Project 171.12.-2.4*

Norppa, H., Sorsa, M., Vainio, H., Gröhn, P., Heinonen, E., Holsti, L. & Nordman, E. (1980) Increased sister chromatid exchange frequencies in lymphocytes of nurses handling cytostatic drugs. *Scand. J. Work environ. Health*, *6*, 299-301

Perry, P. & Evans, H.J. (1975) Cytological detection of mutagen-carcinogen exposure by sister chromatid exchanges. *Nature*, *258*, 121-125

Rom, W.N., Livingston, G.K., Casey, K.R., Wood, S.D., Egger, M.J., Chiu, G.L. & Jerominski, L. (1983) Sister chromatid exchange frequency in asbestos workers. *J. natl Cancer Inst.*, *70*, 45-48

Sarto, F., Comianto, I., Bianchi, V. & Levis, A.G. (1982) Increased incidence of chromosomal aberrations and sister chromatid exchanges in workers exposed to chromic acid ($CrO_3$) in electroplating factories. *Carcinogenesis*, *3*, 1011-1016

Setlow, R.B. (1978) Repair-deficient human disorders and cancer. *Nature*, *271*, 713-717

Sorsa, M. (1980) Somatic mutation theory. *J. Toxicol. environ. Health*, *6*, 977-982

Sorsa, M., Hemminki, K. & Vainio, H. (1982a) Biological monitoring of exposure to chemical mutagens in the occupational environment. *Teratog. Carcinog. Mutag.*, *2*, 137-150

Sorsa, M., Kolmodin-Hedman, B. & Järventaus, H. (1982b) No effect of sulphur dioxide exposure, in aluminium industry, on chromosomal aberrations or sister chromatid exchanges. *Hereditas*, *97*, 159-161

Sorsa, M., Mäki-Paakkanen, J. & Vainio, H. (1982c) Identification of mutagen exposures in the rubber industry by the sister chromatid exchange method. *Cytogenet. Cell Genet.*, *33*, 68-73

Stetka, D.G. & Wolff, S. (1976) Sister chromatid exchange as an assay for genetic damage induced by mutagen-carcinogens. II. In-vitro test for compounds requiring metabolic activation. *Mutat. Res.*, *41*, 343-350

Stich, H.F., Curtis, J.R., Parida, B.B. (1982a) Application of the micronucleus test to exfoliated cells of high cancer risk groups: tobacco chewers. *Int. J. Cancer*, *30*, 553-559

Stich, H.F., Stich, W. & Parida, B.B. (1982b) Elevated frequency of micronucleated cells in the buccal mucosa of individuals at high risk for oral cancer: betel quid chewers. *Cancer Lett.*, *17*, 125-134

Tates, A.D., Dietrich, A.J.J., de Vogel, N., Neuteboom, I., Bos, A. (1983) A micronucleus method for detection of meiotic micronuclei in male germ cells of mammals. *Mutat. Res.*, *121*, 131-138

Taylor, J.H. (1958) Sister chromatid exchanges in tritium-labeled chromosomes. *Genetics*, *43*, 515-529

Vainio, H., Sorsa, M., Rantanen, J., Hemminki, K. & Aitio, A. (1981) Biological monitoring in the identification of the cancer risk of individuals exposed to chemical carcinogens. *Scand. J. Work environ. Health*, *7*, 241-251

Watanabe, T., Endo, A., Kato, Y., Shima, S., Watanabe, T. & Ikeda, M. (1980) Cytogenetics and cytokinetics of cultured lymphocytes from benzene-exposed workers. *Int. Arch. occup. environ. Health*, *46*, 31-41

Wolff, S., ed. (1982a) *Sister Chromatid Exchange*, New York, John Wiley & Sons

Wolff, S. (1982b) Difficulties in assessing the human health effects of mutagenic carcinogens by cytogenetic analyses. *Cytogenet. Cell Genet.*, *33*, 7-13

Yager, J.W., Hines, C.J. & Spear, R.C. (1983) Exposure to ethylene oxide at work increases sister chromatid exchanges in human peripheral lymphocytes. *Science*, *219*, 1221-1223

Yunis, J.J. (1983) The chromosomal basis of human neoplasia. *Science*, *221*, 227-236

# CHROMOSOMAL CHANGES IN CANCER IN RELATION TO EXPOSURE TO CARCINOGENIC AGENTS

**F. Mitelman**

*Department of Clinical Genetics, University Hospital, Lund, Sweden*

## INTRODUCTION

Increasing evidence indicates that most neoplastic conditions arise from interactions between external factors and a predisposing genotypic/phenotypic susceptibility, that may be attributed to, e.g., increased susceptibility of target cells to transformation, inability to repair certain kinds of DNA damage or reduced capacity to destroy malignant cells. Much progress has been made in our understanding of the relationship to cancer development of such predisposing host factors and environmental carcinogens. However, one basic mechanism that remains to be elucidated is the role played by chromosomal aberrations in the causation and/or evolution of neoplasia, especially the influence of external factors on the karyotypic pattern of tumour cells.

Chromosomal aberrations in eukaryotic cells, that are induced by either ionizing radiation, viruses or chemical agents, are generally accepted as indicators of a mutagenic and, therefore, a potentially malignant event. However, the relationship between induced chromosomal damage and the carcinogenic process is far from clarified. In fact, although it is generally accepted that the chromosomal aberrations found in most experimental and human neoplasms are crucial steps in tumour evolution, visible chromosomal abnormality is not apparently, a prerequisite of neoplastic change, and there is, so far, no answer to the fundamental question of whether the chromosomal aberrations are primary causal factors or secondary consequences of tumour development. Also, there is very little information about the possible influence of external or internal host factors on the karyotypic pattern of malignant cells. The aims of this paper are to review the information that has been obtained from recent cytogenetic studies of malignant disorders and to discuss presently available evidence for a relationship between acquired chromosomal changes and external, possibly causal, factors. Before the significance of chromosomal aberration is discussed, a brief review is made of the types of chromosomal change that may be found in tumour cells.

## CHROMOSOME PATTERNS IN NEOPLASIA

Chromosomal aberrations have been described in more than 5000 cases of various malignant disorders, that have been studied over the last ten years with chromosome banding techniques (Mitelman, 1983a). The most important interrelated facts that have emerged from recent cytogenetic studies are:

(1) The aberrations are strictly nonrandom within each group of tumours. This has been demonstrated for all types of cancer and leukaemia, in experimental animals and in humans, where the number of subjects studied has been sufficient to draw conclusions (Mitelman & Levan, 1981). Consistent specific aberrations have been discovered in an increasing number of tumour types; the most characteristic aberrations are summarized in Table 1. Cancer-associated, nonrandom, specific aberrations are convincing evidence for the fundamental role of chromosomal change in the initiation of the malignant process.

**Table 1. Specific chromosome aberrations in human neoplasia**

| Tumour type | Chromosome aberration | References |
|---|---|---|
| Myeloproliferative disorders | | |
| Chronic myeloid leukaemia | t(9;22)(q34;q11) | Nowell & Hungerford (1960), Rowley (1973a) |
| Acute myeloid leukaemia | t(8;21)(q22;q22) | Rowley (1973b) |
| Acute promyelocytic leukaemia | t(15;17)(q22;q21) | Rowley *et al.* (1977a) |
| Acute monocytic leukaemia | t(9;11)(p22;q23) | Berger *et al.* (1982), Hagemeijer *et al.* (1982) |
| Refractory anaemia | del(5)(q12q31) | Van den Berghe *et al.* (1974), Sokal *et al.* (1975) |
| Polycythemia vera | del(20)(q11) | Reeves *et al.* (1972) |
| Lymphoproliferative disorders | | |
| Burkitt's lymphoma | t(8;14)(q24;q32) | Manolov & Manolova (1972), Zech *et al.* (1976) |
| Acute lymphocytic leukaemia | t(9;22)(q34;q11)<br>t(4;11)(q21;q23)<br>t(8;14)(q24;q32) | Third International Workshop on Chromosomes in Leukemia (1981) |
| Chronic lymphocytic leukaemia | +12 | Gahrton *et al.* (1980), Morita *et al.* (1981) |
| Solid tumours | | |
| Meningioma | -22 | Mark *et al.* (1972), Zankl & Zang (1972) |
| Cervical carcinoma | ?1q | Atkin & Baker (1979) |
| Ovarian carcinoma | t(6;14)(q21;q24) | Wake *et al.* (1980) |
| Neuroblastoma | del(1)(p32) | Brodeur *et al.* (1981), Gilbert *et al.* (1982) |
| Pleomorphic adenoma | t(3;8)(p21;q12) | Mark *et al.* (1982) |
| Small-cell lung carcinoma | del(3)(p14-23) | Whang-Peng *et al.* (1982) |
| Seminoma | i(12p) | Atkin & Baker (1982) |
| Ewing's sarcoma | t(11;22)(q24;q12) | Aurias *et al.* (1983), Turc-Carel *et al.* (1983) |
| Malignant melanoma | der(6)(q21q23) | Trent *et al.* (1983), Becher *et al.* (1983) |
| Nasopharyngeal carcinoma | der(3)(q25q27) | Mitelman *et al.* (1983) |

(2) A survey of the distribution of cancer-associated aberrations shows that only a few of the specific marker chromosomes are pathognomonic for a particular malignant disorder (Mitelman, 1981, 1983b). Two characteristic aberrations deserve special mention: the 14q+ marker chromosome produced by the t(8;14)(q24;q32) and the reciprocal translocation t(15;17)(q22;q21), each being limited to just one cell type, the B-cell lymphocyte and the promyelocyte, respectively. Although all the other types of specific marker chromosomes are, usually, much more frequent in one, or even two, disorders, they may, characteristically, be found in several other tumour types, also. Thus, there appears to be little or no target-cell specificity with the specific chromosomal aberrations in neoplastic disorders.

(3) The chromosomes that are involved, preferentially, in aberrations in different tumour types tend to cluster to certain chromosome types. Thus, some chromosomes, e.g., Nos 1, 8 and 14, seem to be particularly prone to aberration during malignant development, whereas others, such as chromosomes 2, 4, 10, 18, 19, X and Y, show no selective involvement (Mitelman & Levan, 1981; Mitelman, 1983b). One interpretation of this is that the chromosomes most often affected carry genetic material which is important to the regulation of cell proliferation and/or differentiation, and which needs to be manipulated in the process of malignant transformation. It is, certainly, striking that the locations of all oncogenes that have been mapped to specific chromosomal regions so far, agree with the breakpoints of characteristic cancer-associated chromosomal aberrations (Rowley, 1983).

(4) There are many known cases of experimental and human neoplasms where specific chromosome segments tend to be involved in duplication or deletion. Often, the same type of change affects many cases of the same malignant disease, even though the exact size of the added or deleted segment may be variable. Thus, common duplicated segments have been observed in chromosome 2 of rat sarcomas induced by polycyclic hydrocarbons (Levan *et al.*, 1974) and in chromosome 1 of patients with various myeloproliferative disorders (Rowley, 1977). A similar finding was made for deletions of chromosomes 5, 6 and 7, in samples taken from different human malignancies (Mitelman & Levan, 1981). In this material, although the breakpoints showed little obvious specificity, common deleted segments were found in all three chromosomes. In fact, among the 125 cases scrutinized, only two were outside the common deleted segments. Again, it may be speculated that genes of importance in the transformation of a normal cell to a malignant cell are concentrated in these regions.

(5) Results from experimental studies suggest, that at least in certain instances, the inducing factor may produce specific chromosome patterns (Mitelman, 1980, 1981). It is extremely difficult to find avenues of approach to this question as far as human neoplasias are concerned. Tumours associated with known environmental carcinogens are uncommon, and none has been subjected to chromosome analysis, so far. However, some observations suggest that such correlations may, in fact, exist in humans. This aspect is discussed in more detail below.

## EXPOSURE TO ENVIRONMENTAL FACTORS

The possibility that external factors influence the karyotypic pattern of malignant human cells was first suggested by a retrospective study (Mitelman *et al.*, 1978) of 56 adult patients with acute nonlymphocytic leukaemia (ANLL). Clonal chromosomal aberrations were significantly more common in patients with previous occupational exposure to potentially mutagenic/carcinogenic agents, than in patients with no such history of occupational exposure. More recent evaluations of 93 patients from Lund, Sweden, 69 patients from Rome, Italy, and 74 patients from Chicago, USA (Mitelman *et al.*, 1981; Golomb *et al.*, 1982), confirmed the initial observation (Table 2). The studies indicated, also, that certain specific karyotypic aberrations, in particular monosomy 5 (-5), deletions of the long arm of chromosome 5 (5q-), monosomy 7 (-7) and deletions of the long arm of chromosome 7 (7q-), were more frequent in patients with previous occupational exposure to chemical solvents, insecticides and/or petroleum products or their combustion residues.

**Table 2. Incidence of chromosomal abnormality in relation to occupational exposure in acute nonlymphocytic leukaemia**

| Centre | Abnormal karyotype (%) | No. of patients |
|---|---|---|
| Lund | | |
| Exposed | 68.6 | 35 |
| Nonexposed | 25.8 | 58 |
| Rome | | |
| Exposed | 88.3 | 17 |
| Nonexposed | 38.5 | 52 |
| Chicago | | |
| Exposed | 75.0 | 16 |
| Nonexposed | 43.1 | 58 |
| Total | | |
| Exposed | 75.0 | 68 |
| Nonexposed | 35.7 | 168 |

In the three centres, case records were reviewed of all patients from whom successful chromosome preparations had been obtained by banding techniques, and the patients were divided into two groups: exposed and nonexposed. Only occupational exposure was considered; social and personal habits, including hobbies, were not evaluated. Patients previously treated for other malignancies were excluded, as were patients who had shown a myeloproliferative condition prior to the acute leukaemic phase. Of the total number of patients analysed during the peirod of study, approximately 30% were excluded because chromosome examination failed or was ambiguous, or because no data were available regarding professional exposure.

The relationship between occupational exposure and chromosomal abnormality in ANLL was examined recently by a large, prospective, collaborative study of patients collected at 16 centres; the study was reviewed at the Fourth International Workshop on Chromosomes in Leukemia (1984). The collected data cover all patients with ANLL who were studied at the participating centres during the period January 1, 1980, to June 30, 1982. On admission to hospital, each patient was interviewed regarding previous and present occupations, according to a standardized questionnaire agreed upon by the participants. Before the workshop, the occupations were reviewed and each patient was classified by the group in Lund according to criteria previously used (Mitelman *et al.*, 1978, 1981). Of 361 patients available for evaluation, 239 were considered nonexposed and 122 were considered exposed to potentially mutagenic/carcinogenic agents.

In general, the results of the prospective collaborative study confirmed the findings of previous retrospective studies. Thus, clonal chromosomal aberrations were significantly more common in leukaemic cells of occupationally exposed patients than in nonexposed patients. In addition, the workshop material demonstrated that this relationship was evident only in patients 30 years of age or older, which may indicate that a significant number of ANLL cases in this age group is, in fact, related to occupational hazards. Furthermore, the material revealed that the incidence of certain characteristic aberrations may be different in different occupational groups. Thus, among the patients 30 years of age or older who were exposed to chemicals, a remarkable over-representation of the karyotypic aberrations -5/-7 was observed, and a notably high incidence of the specific translocation t(8;21) was found in a group of patients who had worked with minerals. A relatively large proportion of patients with this abnormality was observed, also, among those who were occupationally exposed to petrol products or their combustion residues. In both these exposure groups - patients exposed to petrol products and those who had worked with minerals - clonal abnormalities other than those specifically recorded were particularly common. Most of those exposed to minerals had been working in mines, i.e., in an environment that was, probably, contaminated with combustion residues of petrol products. It is, therefore, possible that such exposure may be associated also with other specific, but as yet unrecognized, cytogenetic abnormalities.

## EXPOSURE TO CYTOTOXIC FACTORS

During the past decade, late development of acute leukaemia, almost always of the non-lymphocytic types, has been reported with increasing frequency after previous cytotoxic and/or radiation therapy for a primary malignancy, usually malignant lymphoma or multiple myeloma. The first case of secondary ANLL studied with the banding technique was reported by Lundh *et al.* (1975), and Rowley *et al.* (1977b) reported on a banding study of a series of 10 patients who had ANLL following treatment for malignant lymphoma. Recently, the University of Chicago group updated its findings (Rowley *et al.*, 1981), in a report of 26 patients with ANLL or a dysmyelopoietic syndrome as a second malignancy. Of the 26 previously treated patients, 16 had Hodgkin's disease, 5 had poorly differentiated lymphocytic lymphoma, 3 had multiple myeloma, 1 had cervical carcinoma and 1 had squamous-cell carcinoma of the lung. A total of 15 of these patients had previously received both radiotherapy and chemotherapy, 7 had only chemotherapy and 4 had only radiotherapy. All but one of the 26 patients had an abnormal karyotype and one or both of two consistent chromosome changes were noted in 23 of these 25 patients: monosomy 5 was present in 11 patients, and a deletion of the long arm of one chromosome 5 in 3 patients; monosomy 7 was found in 20 patients; 11 patients had monosomy 7 as well as either -5 or 5q-. Thus, the incidence of chromosomal abnormality, as well as the incidence of the particular changes -5/5q- and -7, were decidedly more common in secondary leukaemia than in ANLL *de novo*. Similar findings have been reported by other investigators (Whang-Peng *et al.*, 1979; Pedersen-Bjergaard *et al.*, 1981) and were confirmed recently by the combined material reviewed at the Fourth International Workshop on Chromosomes in Leukemia (1984). This pattern of abnormality is very similar to that in patients with ANLL *de novo*, who have had documented exposure to environmental mutagens and/or carcinogens. Obviously these results support the concept that nonrandom chromosomal changes may be influenced by external factors. More information is needed on this important aspect, in particular, comparative studies of primary malignancies and secondary leukaemias following different treatments. In this context, it may be relevant to note that intensive chemotherapy during the chronic phase of chronic myeloid leukaemia has been found to produce clones with preferential engagement of chromosome 1 (Alimena *et al.*, 1979).

## INFLUENCE OF GENETIC AND/OR GEOGRAPHIC FACTORS

An indirect approach, which may give some indication of the influence of external factors on human tumour chromosomes, is the study of chromosomal aberrations in relation to the geographic area from which the tumours are derived. An uneven geographic distribution of specific chromosomal aberrations may be compatible with either genetic heterogeneity and/or an uneven geographic distribution of environmental carcinogens, that may be endemic in specific regions. In either case, it should be possible to test the relative importance of genetic *versus* environmental factors.

Since the introduction of chromosome-banding techniques in 1970, we have tried, systematically, to collect and survey data on banded human tumour chromosomes, including data on the geographic provenances of all cases recorded. The material has been collected from three main sources: published cases, unpublished cases kindly communicated by numerous colleagues from all over the world and unpublished cases from our own laboratory. The accumulated data comprise, at present, almost 5000 cases in which neoplastic cells have been submitted to complete karyotypic analysis by chromosome banding and in each of which at least one chromosomal aberration has been revealed and identified. By 1981, the complexity of the material had forced us to adopt computer methods for assembling, indexing and revising (Mitelman, 1983a).

Detailed surveys of the geographic distribution of various chromosomal abnormalities in different tumour types have been reported by Mitelman and Levan (1978, 1981) and Mitelman (1981, 1983b). These surveys have revealed notable differences in the incidence of specific aberrations for all tumour types where the number of subjects studied was sufficient to permit conclusions. A few examples are briefly considered here.

In ANLL, the incidence of monosomy 5 or deletion of the long arm of chromosome 5 is about 20% in the USA, which is about twice the incidence in Europe and Australia, whereas not one case has been reported from Japan. Similarly, monosomy 7 or a deletion of the long arm of chromosome 7 appears to be extremely rare in Japan, whereas the frequency in both Europe and the USA is about 20%. The incidence of the particular 8;21 translocation that is characteristic of well-differentiated acute myeloid leukaemia is less than 10% in Australia, Europe and the USA, but has been reported in more than 30% of the cases from Japan. In another subtype of ANLL, acute promyelocytic leukaemia, the incidence of the t(15;17)(q22;q21), i.e., reciprocal translocation involving the long arms of chromosomes 15 and 17, varies from 0 to 100%, both between and within different continents.

In meningiomas, about one-third of the tumours with aberration display a loss of one chromosome 8, in addition to the characteristic and consistent aberration of this disorder: monosomy 22. Two unselected series of patients studied in the Federal Republic of Germany and Sweden are available for comparison. Monosomy 22 is found in over 95% of the tumours with aberration from both regions. However, monosomy 8 has been found in 65% of the Swedish cases, but in none of the German ones.

These data admittedly suffer from certain weaknesses. The results are, in some instances, based on only a few cases and may be biased, also, since many published cases may represent selections from larger numbers analysed. Also, technical differences which, theoretically, could influence the outcome of chromosome analysis, cannot be excluded. Nevertheless, some of the correlations are suggestive indeed and indicate that real geographic differences in chromosomal aberration may exist. As mentioned, such geographic heterogeneity may be taken to indicate heterogeneity in the distribution of etiological factors. Significant results will, no doubt, be obtained as more data are accumulated.

## CONCLUDING REMARKS

The study of chromosomal aberration and its significance in the development and progression of tumours has become a rapidly expanding branch of cancer research. The use of chromosome analysis by banding techniques has already brought a clearer understanding of such dynamic processes as cell competition, selection and adaptation, all of which operate within a malignant cell population and may help our understanding of basic tumour biology. The information available is, however, in many respects still fragmentary, and only certain generalizations can be advanced concerning the role of chromosomal change in carcinogenesis. Thus, among human neoplasms, still very little information is available on solid tumours. About 70% of all cases so far analysed with banding techniques belong to one group of haematological disorders - the myeloproliferative diseases. Lymphoproliferative disorders comprise about 20% and solid tumours only about 10% of the cases studied so far. Also, most solid tumours analysed have been effusions, i.e., highly progressed tumour-cell populations containing karyotypic changes that are, usually, extremely complex, which makes interpretation of the karyotypic findings difficult unless a very large amount of material is studied. Thus, the data presented in this paper are clearly incomplete, both for chromosomal aberration in relation to exposure to environmental and toxic agents and for geographic heterogeneity of the chromosomal changes. Large, collaborative, cytogenetic studies are urgently needed to fill the gaps in our knowledge. This laborious and time-consuming work will certainly be rewarding and is the only way to elucidate the exciting question of whether the pattern of chromosome variation in malignant cells is determined or influenced by external inducing factors.

## ACKNOWLEDGEMENTS

Original work presented in this review was supported by the Swedish Cancer Society and the John and Augusta Persson Foundation for Medical Research.

## REFERENCES

Alimena, G., Brandt, L., Dallapiccola, B., Mitelman, F. & Nilsson, P.G. (1979) Secondary chromosome changes in chronic myeloid leukemia: Relation to treatment. *Cancer Genet. Cytogenet.*, *1*, 79-85

Atkin, N.B. & Baker, M.C. (1979) Chromosome 1 in 26 carcinomas of the cervix uteri. *Cancer*, *44*, 604-613

Atkin, N.B. & Baker, M.C. (1982) Specific chromosome change, i(12p), in testicular tumours? *Lancet*, *ii*, 1349

Aurias, A., Rimbaut, C., Buffe, D., Dubousset, J. & Mazabraud, A. (1983) Chromosomal translocations in Ewing's sarcoma. *New Engl. J. Med.*, *309*, 496-497

Becher, R., Gibas, Z. & Sandberg, A.A. (1983) Chromosome 6 in malignant melanoma. *Cancer Genet. Cytogenet.*, *9*, 173-175

Berger, R., Bernheim, A., Saigaux, F., Daniel, M.-T., Valensi, F. & Flandrin, G. (1982) Acute monocytic leukemia chromosome studies. *Leukemia Res.*, *6*, 17-26

Brodeur, G.M., Green, A.A., Hayes, F.A., Williams, K.J., Williams, D.L. & Tsiatis, A.A. (1981) Cytogenetic features of human neuroblastomas and cell lines. *Cancer Res.*, *41*, 4678-4686

Fourth International Workshop on Chromosomes in Leukemia (1984) Karyotype and occupational exposure to potential mutagenic/carcinogenic agents in acute nonlymphocytic leukemia. *Cancer Genet. Cytogenet.*, *11*, 249-360

Gahrton, G., Robert, K.-H., Friberg, K., Zech, L. & Bird, A.G. (1980) Extra chromosome 12 in chronic lymphocytic leukaemia. *Lancet*, *i*, 146-147

Gilbert, F., Balaban, G., Moorhead, P., Bianchi, D. & Schlesinger, H. (1982) Abnormalities of chromosome 1p in human neuroblastoma tumors and cell lines. *Cancer Genet. Cytogenet.*, *7*, 33-42

Golomb, H.M., Alimena, G., Rowley, J.D., Vardiman, J.W., Testa, J.R. & Sovik, C. (1982) Correlation of occupation and karyotype in adults with acute nonlymphocytic leukemia. *Blood*, *60*, 404-411

Hagemeijer, A., Hählen, K., Sizoo, W. & Abels, J. (1982) Translocation (9;11)(p21;q23) in three cases of acute monoblastic leukemia. *Cancer Genet. Cytogenet.*, *5*, 95-105

Levan, G., Ahlström, U & Mitelman, F. (1974) The specificity of chromosome A2 involvement in DMBA-induced rat sarcomas. *Hereditas*, *77*, 263-280

Lundh, B., Mitelman, F., Nilsson, P.G., Stenstam, M. & Söderström, N. (1975) Chromosome abnormalities identified by banding technique in a patient with acute myeloid leukaemia complicating Hodgkin's disease. *Scand. J. Haematol.*, *14*, 303-307

Manolov, G. & Manolova, Y. (1972) Marker band in one chromosome 14 from Burkitt lymphomas. *Nature*, *237*, 33-34

Mark, J., Levan, G. & Mitelman, F. (1972) Identification by fluorescence of the G chromosome lost in human meningiomas. *Hereditas*, *71*, 163-168

Mark, J., Dahlenfors, R., Ekedahl, C. & Stenman, G. (1982) Chromosomal patterns in a benign human neoplasm, the mixed salivary gland tumour. *Hereditas*, *96*, 141-148

Mitelman, F. (1980) Cytogenetics of experimental neoplasms and non-random chromosome correlations in man. *Clin. Haematol.*, *9*, 195-219

Mitelman, F. (1981) *Tumor etiology and chromosome pattern - evidence from human and experimental neoplasms*. In: Arrighi, F.E., Rao, P.N. & Stubblefield, E., eds, *Genes, Chromosomes and Neoplasia*, New York, Raven Press, pp. 335-350

Mitelman, F. (1983a) Catalogue of chromosome aberrations in cancer. *Cytogenet. Cell Genet.*, *36*, 1-515

Mitelman, F. (1983b) *Chromosome pattern in human cancer and leukemia*. In: Rowley, J.D. & Ultmann, J.E., eds, *Chromosomes and Cancer: From Molecules to Man*, New York, Academic Press, pp. 61-84

Mitelman, F. & Levan, G. (1978) Clustering of aberrations to specific chromosomes in human neoplasms. III. Incidence and geographic distribution of chromosome aberrations in 856 cases. *Hereditas*, *89*, 207-232

Mitelman, F. & Levan, G. (1981) Clustering of aberrations to specific chromosomes in human neoplasms. IV. A survey of 1871 cases. *Hereditas*, *95*, 79-139

Mitelman, F., Brandt, L. & Nilsson, P.G. (1978) Relation among occupational exposure to potential mutagenic/carcinogenic agents, clinical findings, and bone marrow chromosomes in acute nonlymphocytic leukemia. *Blood*, *52*, 1229-1237

Mitelman, F., Nilsson, P.G., Brandt, L., Alimena, G., Gastaldi, R. & Dallapiccola, B. (1981) Chromosome pattern, occupation, and clinical features in patients with acute nonlymphocytic leukemia. *Cancer Genet. Cytogenet.*, *4*, 197-214

Mitelman, F., Mark-Vendel, E., Mineur, A., Giovanella, B. & Klein, G. (1983) A 3q+ marker chromosome in EBV-carrying nasopharyngeal carcinomas. *Int. J. Cancer*, *32*, 651-655

Morita, M., Minowada, J. & Sandberg, A.A. (1981) Chromosomes and causation of human cancer and leukemia. XLV. Chromosome patterns in stimulated lymphocytes of chronic lymphocytic leukemia. *Cancer Genet. Cytogenet.*, *3*, 293-306

Nowell, P.C. & Hungerford, D.A. (1960) A minute chromosome in human granulocytic leukemia. *Science*, *132*, 1497

Pedersen-Bjergaard, J., Philip, P., Thing Mortensen, B., Ersböll, J., Jensen, G., Panduro, J. & Thomsen, M. (1981) Acute nonlymphocytic leukemia, preleukemia, and acute myeloproliferative syndrome secondary to treatment of other malignant diseases. Clinical and cytogenetic characteristics and results of in-vitro culture of bone marrow and HLA typing. *Blood*, *57*, 712-723

Reeves, B.R., Lobb, D.S. & Lawler, S.D. (1972) Identity of the abnormal F-group chromosome associated with polycythemia vera. *Humangenetik*, *14*, 159-161

Rowley, J.D. (1973a) A new consistent chromosomal abnormality in chronic myelogenous leukemia identified by quinacrine fluorescence and Giemsa staining. *Nature*, *243*, 290-293

Rowley, J.D. (1973b) Identification of a translocation with quinacrine fluorescence in a patient with acute leukemia. *Ann. Genet.*, *16*, 109-112

Rowley, J.D. (1977) Mapping of human chromosomal regions related to neoplasia: evidence from chromosomes 1 and 17. *Proc. natl Acad. Sci. USA*, *74*, 5729-5733

Rowley, J.D. (1983) Human oncogene locations and chromosome aberrations. *Nature*, *301*, 290-291

Rowley, J.D., Golomb, H.M. & Dougherty, C. (1977a) 15/17 translocation, a consistent chromosomal change in acute promyelocytic leukaemia. *Lancet*, *i*, 549-550

Rowley, J.D., Golomb, H.M. & Vardiman, J.W. (1977b) Nonrandom chromosomal abnormalities in acute nonlymphocytic leukemia in patients treated for Hodgkin disease and non-Hodgkin lymphomas. *Blood*, *50*, 759-770

Rowley, J.D., Golomb, H.M. & Vardiman, J.W. (1981) Nonrandom chromosome abnormalities in acute leukemia and dysmyelopoietic syndromes in patients with previously treated malignant disease. *Blood*, *58*, 759-767

Sokal, G., Michaux, J.L., van den Berghe, H., Cordier, A., Rodhain, J., Ferrant, A., Moriau, M., de Bruyere, M. & Sonnet, J. (1975) A new hematologic syndrome with a distinct karyotype: the 5q- chromosome. *Blood*, *46*, 519-533

Third International Workshop on Chromosomes in Leukemia (1981) Chromosomal abnormalities in acute lymphoblastic leukemia. *Cancer Genet. Cytogenet.*, *4*, 101-110

Trent, J.M., Rosenfeld, S.B. & Meyskens, F.L. (1983) Chromosome 6q involvement in human malignant melanoma. *Cancer Genet. Cytogenet.*, *9*, 177-180

Turc-Carel, C., Philip, I., Berger, M.-P., Philip, T. & Lenoir, G.M. (1983) Chromosomal translocation in Ewing's sarcoma. *New Engl. J. Med.*, *309*, 497-498

Van den Berghe, H., Cassiman, J.J., David, G., Fryns, J.P., Michaux, J.L. & Sokal, G. (1974) Distinct haematological disorder with deletion of long arm of No. 5 chromosome. *Nature*, *251*, 437-438

Wake, N., Hreschchyshyn, H.M., Piver, S.M., Matsui, S.-I. & Sandberg, A.A. (1980) Specific cytogenetic changes in ovarian cancer involving chromosomes 6 and 14. *Cancer Res.*, *40*, 4512-4518

Whang-Peng, J., Knutsen, T., O'Donnell, J.F. & Brereton, H.D. (1979) Acute nonlymphocytic leukemia and acute myeloproliferative syndrome following radiation therapy for non-Hodgkin's lymphoma and chronic lymphocytic leukemia. Cytogenetic studies. *Cancer*, *44*, 1592-1600

Whang-Peng, J., Bunn, P.A., Jr, Kao-Shan, C.S., Lee, E.C., Carney, D.N., Gazdar, A. & Minna, J.D. (1982) A nonrandom chromosomal abnormality, del 3p(14-23), in human small cell lung cancer (SCLC). *Cancer Genet. Cytogenet.*, *6*, 119-134

Zankl, H. & Zang, K.D. (1972) Cytological and cytogenetical studies on brain tumors. IV. Identification of the missing G chromosome in human meningiomas as No. 22 by fluorescence technique. *Humangenetik*, *14*, 167-169

Zech, L., Haglund, U., Nilsson, K. & Klein, G. (1976) Characteristic chromosomal abnormalities in biopsies and lymphoid-cell lines from patients with Burkitt and non-Burkitt lymphomas. *Int. J. Cancer*, *17*, 47-56

# DETECTION OF DOMINANT ENZYME MUTANTS IN MICE: MODEL STUDIES FOR MUTATIONS IN MAN

**W. Pretsch & D.J. Charles**

*Gesellschaft für Strahlen- und Umweltforschung,*
*Institut für Genetik, Abteilung für Säugetiergenetik,*
*D-8042 Neuherberg, Federal Republic of Germany*

## SUMMARY

After intraperitoneal injection with the mutagens procarbazine hydrochloride (PHCl) or *N*-ethyl-*N*-nitrosourea (ENU), male (101/EL×C3H/EL)$F_1$ mice were mated with untreated test-stock females. The offspring were screened for induced mutations that cause alterations of two different enzyme properties. Charge modifications were analysed by separation of liver enzymes by isoelectric focussing on polyacrylamide gels. Banding patterns of six enzyme systems were checked by using the agar contact replica technique and specific activity stainings. No mutant was found in 5278 offspring of the control group. After paternal treatment of spermatogonia, the mutant frequency was one mutant in 5630 offspring (600 mg PHCl per kg body weight), one mutant in 1892 offspring (160 mg/kg ENU), two mutants and two presumed mutants, which died before genetic confirmation, in 4136 offspring (250 mg/kg ENU). No mutant was detected with PHCl and the two ENU doses in the postspermatogonial treatment groups of 469, 1088, and 2020 offspring, respectively. Specific activities of 10 erythrocyte enzymes were measured in the blood with an automatic enzyme analyser. To date, in the spermatogonial treatment group, findings have been: no mutant in 3610 controls, seven mutants in 3509 offspring (80 mg/kg ENU), five mutants in 800 offspring (160 mg/kg ENU) and four mutants in 759 offspring (250 mg/kg ENU). Of these 16 independent mutants, 10 caused reduced enzyme activity and six increased it.

## INTRODUCTION

The aims of our research are to detect gene mutations in mice that lead to alterations in enzymes, and to determine the corresponding mutation frequencies. For these purposes, the present investigation used two different techniques: structural changes (charge modifications) in the protein were detected in liver proteins by the polyacrylamide gel isoelectric focussing method (Pretsch *et al.*, 1982), and functional changes (activity alterations) in the protein were ascertained by measuring specific enzyme activities in erythrocytes with an automatic enzyme analyser (Charles & Pretsch, 1982). The animals used in these experiments were screened for specific-locus and dominant cataract mutations (Ehling, 1980). Therefore, this was a unique opportunity to compare results obtained in mice with several test systems, examining different genetic endpoints (Charles & Pretsch, 1982).

The study used two mutagens with well-known potency in mice, procarbazine hydrochloride (PHCl) and *N*-ethyl-*N*-nitrosourea (ENU). Both have been shown to induce morphological specific-locus mutations in mice (Ehling & Neuhäuser, 1979; Russell *et al.*, 1979), and ENU has been shown to induce dominant cataract mutations, also (Ehling *et al.*, 1982; Favor, 1983). Therefore, both mutagens seem well suited to the production of new gene mutations.

Since there are homologies between mouse and man for many of the examined loci, experiments with mice can supply important information for humans. For instance, in man, heritable enzyme alterations are often connected with clinical symptoms. In other mammals, animals with genetic disorders related to enzyme alterations have been observed, e.g., the histidinaemic mutant in the mouse described by Kacser *et al.* (1973). Thus, the mouse mutant can serve as a model for the corresponding disease in man.

This paper presents the results of the application of the two approaches (proof of structural and functional protein changes) to the detection of chemically induced germ-cell mutations in mice.

## MATERIALS AND METHODS

### *Treatment and mating of mice*

(101/EL×C3H/EL)$F_1$ hybrid male mice, 12 weeks old, were injected intraperitoneally with either 600 mg PHCl (supplied by Hoffman-La Roche, Grenzach-Whylen, FRG) per kg body weight or with 80 mg/kg, 160 mg/kg or 250 mg/kg ENU (Serva, Heidelberg, FRG). Shortly before the injections, PHCl was dissolved in water and ENU in 66 mM phosphate buffer (pH 6.0), concentrated so that the males received 0.5 ml injected volume. The control males were injected with an equal volume of the solvent. Each male was then caged with an untreated T stock female (Ehling, 1978); out of a total of 350 males, 50 were mated when 12 months old with HT stock females (Lyon & Morris, 1966). Offspring from T stock matings were kept under specific pathogen-free conditions until three weeks of age, and thereafter maintained conventionally. HT stock matings and offspring were maintained under conventional conditions exclusively. Presumed mutants were outcrossed to strain C3H/EL, and at least 25 offspring were examined for genetic confirmation of the altered phenotype.

### *Investigation for enzyme-charge modifications*

A piece of liver was removed by biopsy, under ether anaesthesia, from each $F_1$ offspring at eight weeks of age. The tissues were homogenized in distilled water and centrifuged at 100 000 $\times$ *g* and 4°C for 45 min. Polyacrylamide gel isoelectric focussing was performed on commercially available gels (pH 3.5-9.5, gel concentration T = 5%, degree of cross-linkage C = 3%), which were prefocussed for 30 min at 4°C and 20 W. Replicas of the polyacrylamide gel were obtained by direct contact with 1.1% agar-coated plates (agar contact replica technique) (Pretsch *et al.*, 1982). Banding patterns of haemoglobin and of six enzyme systems [lactate dehydrogenase, sorbitol dehydrogenase, glycerol-3-phosphate dehydrogenase, phosphoglucomutase, diaphorase, nonspecific esterases] were visualized by specific activity stainings. Banding alterations can be seen readily.

### *Investigation for enzyme-activity alterations*

For each $F_1$ offspring, at four to six weeks of age, 0.1 ml blood was taken from the orbital sinus. Erythrocytes were lysed by dilution. In an automatic enzyme analyser (ACP 5040, Eppendorf, Hamburg, FRG), haemoglobin content and activity for ten enzymes were determined: lactate dehydrogenase, triose phosphate isomerase, malate dehydrogenase, glucose phosphate isomerase, phosphoglycerate kinase, phosphoglyceromutase, glyceraldehyde phosphate dehydrogenase, glucose-6-phosphate dehydrogenase, pyruvate kinase and glutathione reductase.

## RESULTS

The frequencies of observed enzyme mutants were enhanced after chemical mutagen treatment of spermatogonia (Table 1). Except for the experiments with 600 mg/kg PHCl and 160 mg/kg ENU (electrophoretical mutants), the increases are significantly different from that of the controls. In offspring from postspermatogonia (tested only for electrophoretical alterations), no mutant could be found.

**Table 1. Frequency of chemically induced dominant enzyme mutations in mice**

| Paternal treatment[a] | Germ cell stage[b] | No. of offspring[c] | Electrophoretical mutants | *p*-value | No. of offspring | Activity mutants[d] | *p*-value |
|---|---|---|---|---|---|---|---|
| Control | pg + g | 5812 | 0 | - | 3610 | 0 | - |
| 600 mg/kg PHCl | pg | 469 | 0 | | | | |
| | g | 7506 | 1 | <0.4 | | | |
| 80 mg/kg ENU | pg | - | - | | - | - | |
| | g | - | - | | 3509 | 7 | < 0.01 |
| 160 mg/kg ENU | pg | 1088 | 0 | | - | - | |
| | g | 1892 | 1 | < 0.08 | 800 | 5 | < 0.001 |
| 250 mg/kg ENU | pg | 2020 | 0 | | - | - | |
| | g | 4254 + 865[e] | 5 + 1[e] | < 0.01 | 759 | 5[f] | < 0.001 |

a PHCl, procarbazine hydrochloride; ENU, *N*-ethyl-*N*-nitrosourea
b pg, postspermatogonia; g, spermatogonia
c Offspring from T stock females
d Charles & Pretsch (1983)
e Offspring from HT stock females
f Includes one cluster of two mutations

The eight observed electrophoretical mutants concern charge alterations of lactate dehydrogenase (LDH 1049, LDH 1592, LDH 6980, LDH 10866), phosphoglucomutase (GPDH 4009, GPDH 4953), phosphoglucomutase (PGM 1068) and haemoglobin (HAEM 259) (Table 2A). Mutants LDH 1592 and GPDH 4009 have been characterized genetically and biochemically in previous studies and are described briefly below, as examples of electrophoretic mutants.

**Table 2. Description of enzyme mutants recovered in offspring derived from chemically treated male mice**

A. Electrophoretical mutants

| Paternal treatment[a] | Mutant[b] | Comments |
|---|---|---|
| 600 mg/kg PHCl | LDH 1592 | for detailed description see Pretsch & Charles (1980), Charles & Pretsch (1981) |
| 160 mg/kg ENU | LDH 10866 | additional bands at the basic pole of the PAGIF[c] gel; no alteration of LDH-specific activity |
| 250 mg/kg ENU | LDH 1049 | several weak banding deficiencies after PAGIF; reduced LDH-specific activity (about 50% of wild type) |
| | LDH 6980 | died after liver biopsy - presumed mutation |
| | GPDH 4009 | for detailed description see Pretsch & Charles (1984) |
| | GPDH 4953 | identical PAGIF pattern to GPDH 4009; died after liver biopsy - presumed mutation |
| | PGM 1068 | additional band at the basic pole of the PAGIF gels in heterozygotes |
| | HAEM 259 | additional bands at the basic pole of the PAGIF gel; heterozygotes and homozygotes indistinguishable with PAGIF |

B. Activity mutants

| Paternal treatment[a] | Mutant[b] | Specific activity[d] +/− | Specific activity[d] −/− | Comments |
|---|---|---|---|---|
| 80 mg/kg ENU | TPI 3502 | 45 | | |
| | PGAM 12601 | 400 | 850 | |
| | MDH 14202 | 125 | | |
| | PGAM 26405 | 190 | 270 | |
| | GPI 27201 | 65 | 25 | |
| | LDH 29804 | 65 | 20 | no charge modification |
| | PGAM 32101 | 250 | | |
| 160 mg/kg ENU | MDH 3859 | 60 | 15 | no. of -/- reduced |
| | PK 3878 | 135 | | |
| | MDH 4099 | 63 | | -/- lethals |
| | MDH 4272 | 70 | 40 | no. of -/- reduced |
| | GPI 4314 | 65 | 15 | fertility of female -/- reduced |
| 250 mg/kg ENU | LDH 2014 | 55 | 1 | no charge modification locus for LDH-A mutated |
| | LDH 9546 | 55 | | no. of -/- sharply reduced |
| | G6PD 10168 | 60 | | |
| | G6PD 10168 | 60 | 15 (Hemizygotes: 20) | mutation is X-linked no charge modification heat stability normal |
| | PK 10681 | 165 | 240 | no charge modification heat stability normal |
| | PK 10683 | 170 | 230 | as for PK 10681 |

[a] PHCl, procarbazine hydrochloride; ENU, *N*-ethyl-*N*-nitrosourea

[b] LDH, lactate dehydrogenase; GPDH, glycerol-3-phosphate dehydrogenase; PGM, phosphoglucomutase; HAEM, haemoglobin, TPI, triose phosphate isomerase; PGAM, phosphoglyceromutase; MDH, malate dehydrogenase; GPI, glucose phosphate isomerase; PK, pyruvate kinase; G6PD, glucose-6-phosphate dehydrogenase

[c] Polyacrylamide gel isoelectric focussing

[d] Wild type activity, 100%; +/-, heterozygous mutant; -/-, homozygous mutant

In the mutant LDH 1592, LDH bands in the wild-type pattern near the acidic pole shift to the basic pole in the mutant pattern (Fig. 1A). By intercrossing heterozygous mutants, fully viable and fertile homozygotes were obtained; their polyacrylamide-gel-isoelectric-focussing banding patterns are characterized by a much lower LDH staining intensity (Fig. 1A). The mutation was codominantly expressed with 100% penetrance. The banding alteration was not organ specific, but was detectable by gel electrophoresis in different tissues and concerned the bands corresponding to subunit LDH-A. In liver, muscle and spleen, specific LDH activity decreased to about 50% of the wild type in the heterozygous mutant and to about 9% in the homozygous mutant. In kidney, heart, brain and lung, LDH activity was less diminished. In thermolability experiments, the mutant liver LDH was much more heat-sensitive than that of the wild type. From these biochemical properties it was concluded that the induced mutation affects the LDH locus which codes for subunit A (LDH of muscle type), i.e., Ldh-1 (Pretsch & Charles, 1980; Charles & Pretsch, 1981).

The protein of mutant GPDH 4009 was charged more positively than that of the wild type (Fig. 1B). The mutation was expressed codominantly with 100% penetrance. Heterozygous and homozygous mutants had distinguishable banding patterns after isoelectric focussing. The mutated allele gave the GPDH protein greater heat stability; specific GPDH activity was not altered by the mutation, and the $K_m$ was not changed (Pretsch & Charles, 1984).

For the ENU-induced enzyme-activity mutants, the dose-response curve seems to be nonlinear, i.e., a plateau is observed after the 160 mg/kg dose (Table 1; Charles & Pretsch, 1983).

Of 16 independent enzyme-activity mutants that were detected (Charles & Pretsch, 1983), ten reduced and six increased enzyme activity (Table 2B). The mutations were dominantly expressed. For each mutant, heterozygotes were crossed *inter se*; in some cases, the number of homozygotes was reduced (malate dehydrogenases MDH 3859 and MDH 4272 and LDH 9546) or, in another case homozygotes were not viable (malate dehydrogenase MDH 4099). The pyruvate kinase mutants, PK 10681 and PK 10683, are probably expressions of an identical mutation. In four mutants, the banding pattern after isoelectric focussing was studied and no charge modification was detected.

## DISCUSSION

A gene mutation causing an amino acid substitution in a protein may have two consequences: if the charge of the protein is altered, the result will be a new banding pattern after electrophoresis. this type of variation is common for a large number of different gene products. It seems that, if enough individuals are tested, a charge variant will eventually be discovered for any given protein. When these variant proteins have been purified and sequenced, such as the 'A' variant of glucose-6-phosphate dehydrogenase, single amino acid substitutions have been found (Yoshida, 1967), which are likely to be the product of a single base change (point mutation).

If there is an alteration, for example of the amino acid sequence in the active site of the enzyme, the activity of that enzyme will be affected. However, the determination of enzyme activity after electrophoresis is relatively inaccurate and insensitive. In routine screening for loci without parental differences, the 50% loss of activity, that may be expected when an autosomal mutation results in a completely inactive product, may be easily overlooked. The diminution of LDH activity to about 50% in the heterozygotes of the mutant LDH 1592 cannot be seen after isoelectric focussing (Fig. 1A).

**Fig. 1. Isoelectric focussing of mouse liver lactate dehydrogenase (A) and glycerol-3-phosphate dehydrogenase (B)**

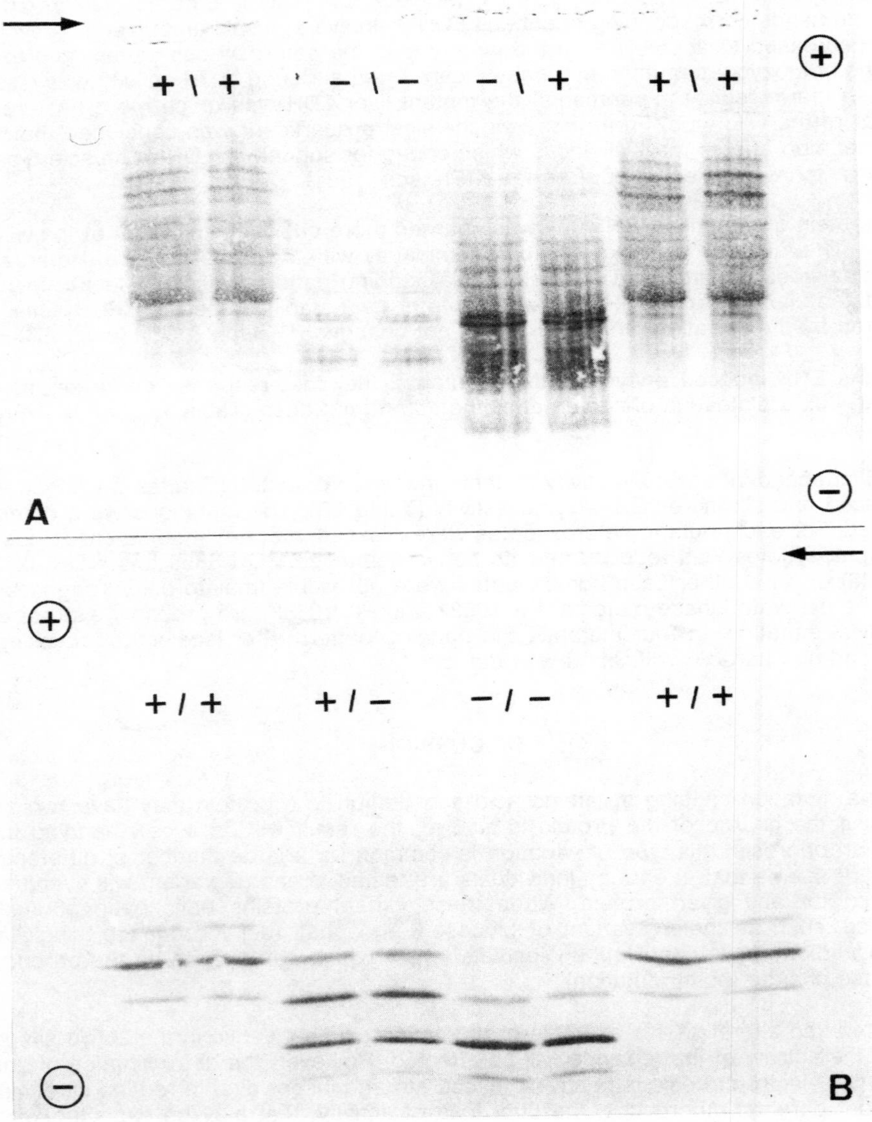


+/+, wild types; +/-, heterozygous mutants; -/-, homozygous mutants. Samples were applied in duplicate near the anodal end of the polyacrylamide gel (arrow).

Enzyme charge modifications and activity alterations may occur together or individually in one mutant. For example, mutant LDH 1592 (Pretsch & Charles, 1980) showed simultaneous appearance of a charge modification and a decreased enzyme activity (Fig. 1A). In contrast, mutant GPDH 4009 (Fig. 1B) showed altered banding patterns with normal enzyme activity (Pretsch & Charles, 1984). In the present study, to date, none of the tested mutants that show altered enzyme activity shows a charge modification also (Table 2A).

Theoretically, about 25% of all possible base substitutions in DNA lead to the transposition of one amino acid by another of different charge. Base substitutions that gave rise to equivalent codons cannot be detected by electrophoresis. Those that lead to the replacement of one amino acid by another of similar charge are, probably, not easily detectable. Therefore, it may be expected that only 10-15% of point mutations can be observed with the electrophoretic approach (Neel *et al.*, 1980a). There are two possible explanations for the detection of 5- to 13-fold fewer electrophoretical mutants than activity mutants in our study (Table 1; 160 and 250 mg/kg ENU groups): the number of loci screened for enzyme-activity mutants may be higher than that for enzyme-charge mutants (due to detection of regulatory mutations in addition to structural mutations) or there may be a higher probability that point mutations are expressed as activity alterations than as charge alterations.

Mohrenweiser (1981), in screening offspring from normal human populations, found a less pronounced difference between frequencies of variants determined by activity measurements or electrophoresis than we did: the average frequency of enzyme-activity variants was two to three times the frequency observed for electrophoretic variants of erythrocyte enzymes in the same population.

An advantage of the methods described in this paper is that they may be applied not only to experimental animals but also to humans. The electrophoresis technique was used in a search for spontaneous mutations affecting protein structure in a combination of human populations (Harris *et al.*, 1974; Neel *et al.*, 1980a,b). The indicators were haemoglobin, a set of serum proteins and erythrocyte enzymes. No mutation was found in a total of 522 119 locus tests. This observation excludes, at the 95% level of probability, a mutation rate greater than $0.6 \times 10^{-5}$ per locus per generation (Neel *et al.*, 1980b). Only one probable mutation was found in a total of 289 868 locus tests by Neel *et al.* (1980a) in the offspring of proximally exposed parents after the atomic bombings of Hiroshima and Nagasaki.

Enzyme activity determinations were employed recently to detect spontaneous deficiency variants in erythrocytes of newborn infants (Mohrenweiser, 1981). This type of genetic abnormality is found in the homozygous condition often associated with significant health consequences. The frequency of enzyme-deficiency variants was 2.4 per 1000, some two to three times above the average frequency of rare electrophoretic variants for erythrocyte enzymes (Neel *et al.*, 1980b).

As there are homologies between the mouse and humans for many of the loci examined in this study, experiments with PHCl- and ENU-treated mice can supply information for humans, since a large number of different mutants in offspring are induced by these compounds. For the above-mentioned LDH 1592 mutant, in addition to an altered LDH banding pattern and activity, nonsphaerocytic haemolytic anaemia was observed in the homozygotes. These homozygotes may, therefore, be model animals for this hereditary disease in humans.

As the animals used in these experiments were also screened for recessive specific-locus and dominant cataract mutations, results obtained with these four different test systems were compared. For the 250 mg/kg ENU dose, the specific-locus test (screening for recessive mutations at seven loci) yielded 54 independent mutations in 9352 offspring (Ehling *et al.*, 1982). The dominant cataract mutation test screens an unspecified number of loci, assumed to be at least 20 on the basis of human data (Ehling *et al.*, 1982); 17 dominant cataract mutations were observed in 9352 offspring (Favor, 1983). The mutations causing alterations of enzyme charge and activity in this study, are coded by at least 22 and 15 loci, respectively. On a per-locus basis (assuming the minimal number of loci for the test for dominant cataracts and enzyme mutants and considering a cluster of mutations as one independent mutational event), the rates for recessive specific-locus mutations, dominant cataract mutations, electrophoretical mutations and enzyme-activity mutations are $82.5 \times 10^{-5}$, $9.1 \times 10^{-5}$, $5.3 \times 10^{-5}$ and $35.1 \times 10^{-5}$, respectively. Electrophoretical mutants have the lowest observed mutation rate, whereas that of the studied enzyme-activity mutants lies between those of the specific loci and of the dominant cataracts.

## ACKNOWLEDGEMENTS

We are grateful for review and comments by Dr U.H. Ehling during the writing of this manuscript. The skilful technical assistance of Gunthild Borgwedel and Andreas Werle is appreciated.

## REFERENCES

Charles, D.J. & Pretsch, W. (1981) A mutation affecting the lactate dehydrogenase locus Ldh-1 in the mouse. I. Genetical and electrophoretical characterization. *Biochem. Genet.*, *19*, 301-309

Charles, D.J. & Pretsch, W. (1982) Activity measurements of erythrocyte enzymes in mice. Detection of a new class of gene mutations. *Mutat. Res.*, *97*, 177-178

Charles, D.J. & Pretsch, W. (1984) Dose response relationship of ENU-induced enzyme activity mutants. *Mutat. Res.* (in press)

Ehling, U.H. (1978) *Specific-locus mutations in mice*. In: Hollaender, A. & de Serres, F.J., eds, *Chemical Mutagens*, Vol. 5, New York, Plenum Press, pp. 233-256

Ehling, U.H. (1980) Induction of gene mutations in germ cells of the mouse. *Arch. Toxicol.*, *46*, 123-138

Ehling, U.H. & Neuhäuser, A. (1979) Procarbazine-induced specific-locus mutations in male mice. *Mutat. Res.*, *59*, 245-256

Ehling, U.H., Favor, J., Kratochvilova, J. & Neuhäuser-Klaus, A. (1982) Dominant cataract mutations and specific-locus mutations in mice induced by radiation or ethylnitrosourea. *Mutat. Res.*, *92*, 181-192

Favor, J. (1983) A comparison of the dominant cataract and recessive specific-locus mutation rates induced by treatment of male mice with ethylnitorosurea. *Mutat. Res.*, *110*, 367-382

Harris, H., Hopkinson, D.A. & Robson, E.B. (1984) The incidence of rare alleles determining electrophoretic variants: data on 43 enzyme loci in man. *Ann. human Genet.*, *37*, 237-253

Kacser, H., Bulfield, G. & Wallace, M.E. (1973) Histidinaemic mutant in the mouse. *Nature*, *244*, 77-79

Lyon, M.F. & Morris, T. (1966) Mutation rates at a new set of specific loci in the mouse. *Genet. Res.*, *7*, 12-17

Mohrenweiser, H.W. (1981) Frequency of enzyme deficiency variants in erythrocytes of newborn infants. *Proc. natl Acad. Sci. USA*, *78*, 5046-5050

Neel, J.V., Satoh, C., Hamilton, H.B., Otake, M., Goriki, K., Kageoka, T., Fujita, M., Neriishi, S. & Asakawa, J. (1980a) Search for mutations affecting protein structure in children of atomic bomb survivors: preliminary report. *Proc. natl Acad. Sci. USA*, *77*, 4221-4225

Neel, J.V., Mohrenweiser, H.W. & Meisler, M.H. (1980b) Rate of spontaneous mutation at human loci encoding protein structure. *Proc. natl Acad. Sci. USA*, *77*, 6037-6041

Pretsch, W. & Charles, D. (1980) *Genetical and biochemical characterization of a dominant mutation of mouse lactate dehydrogenase*. In: Radola, B.J., ed., *Electrophoresis '79*, Berlin, Walter de Gruyter, pp. 817-824

Pretsch, W. & Charles, D.J. (1984) An inherited variant of mouse *sn*-glycerol-3-phosphate dehydrogenase detected by isoelectric focusing: genetical and biochemical analyses. *Biochem. Genet.*, *22*, 419-428

Pretsch, W., Charles, D.J. & Narayanan, K.R. (1982) The agar contact replica technique after isoelectric focusing as a screening method for the detection of enzyme variants. *Electrophoresis*, *3*, 142-145

Russell, W.L., Kelly, E.M., Hunsicker, P.R., Bangham, J.W., Maddux, S.C. & Phipps, E.L. (1979) Specific-locus test shows ethylnitrosourea to be the most potent mutagen in the mouse. *Proc. natl Acad. Sci. USA*, *76*, 5818-5819

Yoshida, A. (1967) A single amino acid substitution (asparagine to aspartic acid) between normal (B+) and the common negro variant (A+) of human glucose-6-phosphate dehydrogenase. *Proc. natl Acad. Sci. USA*, *57*, 835-840

# DNA REPAIR IN RELATION TO BIOLOGICAL MONITORING OF EXPOSURE TO MUTAGENS AND CARCINOGENS

**F. Laval & J. Huet**

*Groupe 'Radiochimie de l'ADN', Institut Gustave Roussy, 94805 Villejuif, France*

## INTRODUCTION

Since numerous difficulties arise in measuring the carcinogenic potential of chemicals in animals, and in measuring chemical mutagenicity in cells *in vitro*, different tests have been proposed to determine these effects.

Most chemical carcinogens become electrophilic through an enzymatic activation process, or are electrophilic reagents which react with DNA, RNA and proteins (Miller & Miller, 1974). They cause DNA damage and induce the various mechanisms of repair. Therefore, determination of DNA repair has been suggested to predict carcinogenicity.

Combes (1983) studied the efficacy of DNA repair assays in predicting genotoxicity in bacteria; a review of the literature showed good correlation between DNA repair and mutagenicity in bacterial systems.

In order to evaluate DNA repair in predicting carcinogenic potential of chemicals in mammalian cells, the following are discussed here: (1) The different repair pathways known in mammalian cells. (2) The methods commonly used to detect DNA repair. (3) The role of unrepaired lesions in mutagenesis.

## DNA REPAIR PATHWAYS

DNA repair pathways, described in bacterial systems, are summarized in Figure 1. They can be divided into two classes: (1) Repair pathways where the damage is removed without rupture of the phosphodiester chain. (2) Repair pathways where excision of the lesion is followed by incision of the phosphodiester chain and synthesis of a new patch of DNA.

**Fig. 1. DNA repair pathways**

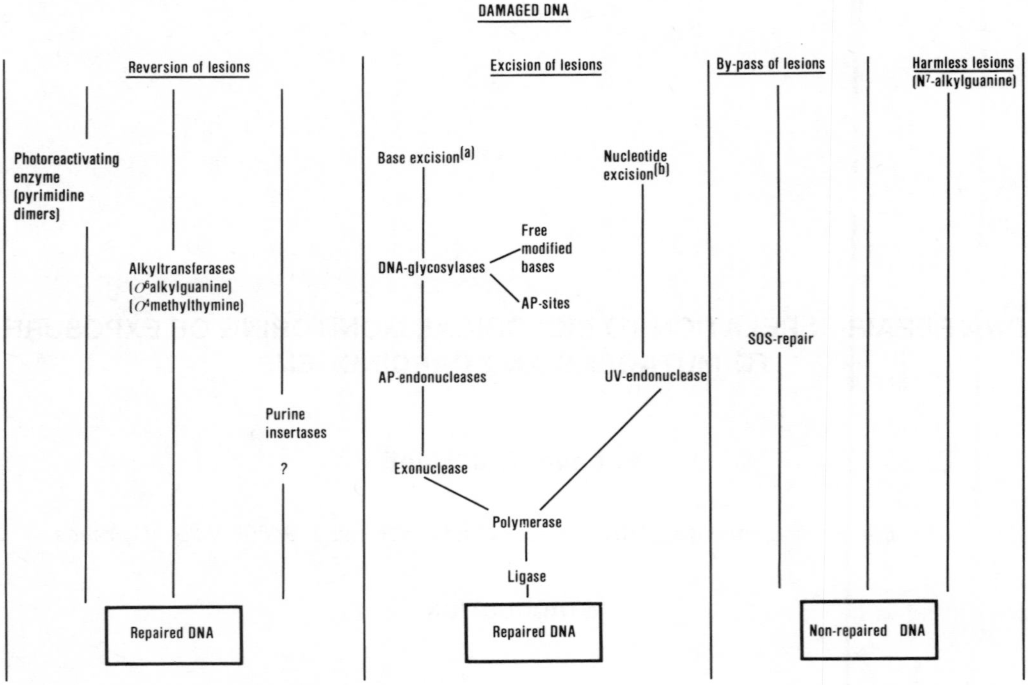


(a) occurs for small lesions: 3-methyladenine, 7-methylguanine, 3-methylguanine, rom$^7$-guanine, hypoxanthine, etc.
(b) occurs for bulky lesions: ultra-violet-induced lesions, psoralen adducts, aflatoxin adducts, etc.

*Reversion of the lesions*

Three different types of protein are involved in this repair pathway.

*Photoreactivating enzymes.* Pyrimidine dimers generated in DNA by ultra-violet light can be eliminated during subsequent irradiation with light at a wave-length above 300 nm. The photoreactivating enzyme requires a cofactor of low molecular weight for activity. This repair process has been described in human cells (Sutherland *et al.*, 1975).

*Insertases.* It has been reported that an enzyme can reinsert purines directly, at apurinic sites in DNA, in *Escherichia coli* (Livneh *et al.*, 1979) and in mammalian cells (Deutsch & Linn, 1979). However, there is controversy about the activity of a purine insertase in *E. coli* (Kataoka & Sekiguchi, 1982).

*Alkyl transferases.* The mutagenic effect of simple methylating and ethylating agents on mammalian cells appears to be due largely to formation of miscoding $O^6$-alkylguanine residues in the cellular DNA (Newbold *et al.*, 1980). Extracts of rat organs can demethylate the $O^6$-methylguanine formed when the cellular DNA is alkylated by *N*-methyl-*N*-nitrosourea (Pegg, 1978). The protein, which is not an enzyme, catalyses the transfer of the methyl group from $O^6$-methylguanine in DNA to a protein-bound cysteine residue. This protein has been purified from mammalian tissues and exists in human placenta (Mehta *et al.*, 1981; Pegg *et al.*, 1982; Harris *et al.*, 1983). Normal human cells are able to repair $O^6$-methylguanine residues, but some human tumours and tumour-derived cell lines, and many cell lines established from cells transformed with DNA tumour viruses, are deficient in this form of repair. Such cells, which are anomalously sensitive to alkylating agents, have been termed $Mer^-$ (Yarosh *et al.*, 1983) or $Mex^-$ (Sklar & Strauss, 1981).

This repair activity has been described also in human lymphoid cell lines, which contain 10 000 to 25 000 molecules of the methyltransferase per cell, whereas repair-deficient cell lines do not contain detectable amounts of this protein (Harris *et al.*, 1983).

Pretreatment of *E. coli* with low doses of *N*-methyl-*N'*-nitro-*N*-nitrosoguanidine induces increased survival and resistance to mutation in cells that are challenged with high concentrations of the mutagen. This phenomenon, which is transient, is called the 'adaptive response' (Samson & Cairns, 1977). It is correlated with an increased disappearance of $O^6$-methylguanine from adapted cells, whereas $N^3$-methyladenine is eliminated at a similar rate in adapted and nonadapted bacteria (Schendel & Robins, 1978). There is evidence of an adaptive process in animals (Montesano, 1981).

Adaptation of mammalian cells with low levels of alkylating agents induces cell resistance against the lethal effect of these agents (Samson & Schwartz, 1980; Kaina, 1982). This pretreatment renders the cells resistant, also, to the mutagenic effect of *N*-methyl-*N*-nitro-*N*-nitrosoguanidine and increases the rate of $O^6$-methylguanine removal from cellular DNA (Laval & Laval, 1984). In cross-reactivity experiments with different mutagens, an adaptive response was obtained for cytotoxicity but not for mutagenicity (Laval & Laval, 1984).

*Excision repair*

Excision repair has been described in a variety of mammalian cells, although it occurs to different extents (Hart & Setlow, 1974). Two types of excision repair (Fig. 1) have been described: nucleotide- and base-excision repair.

In base-excision repair, the damaged bases are recognized and excized by a specific DNA-glycosylase, yielding an apurinic or an apyrimidinic site (AP site). This site is then recognized by an AP endonuclease which nicks the DNA; an exonuclease excises the AP site, yielding a gap which is filled by a DNA polymerase, and the continuity of the phosphodiester chain is restored by a polynucleotide ligase. The sequential action of a DNA glycosylase and an AP endonuclease on alkylated DNA was first demonstrated in *Micrococcus luteus* (Laval, 1977).

Enzymology of DNA repair has been reviewed by Laval and Laval (1980) and Lindahl (1982). DNA glycosylases specific for uracil (Leblanc *et al.*, 1982), 3-methyladenine (Laval, 1977) and 7-methylguanine (Laval *et al.*, 1981) have been purified from *M. luteus*. Several glycosylases have been described in mammalian cells: two uracil-DNA glycosylases, a cytoplasmic and a nuclear one (Leblanc & Laval, 1982), and a glycosylase specific for alkylated bases (Cathcart & Goldthwait, 1981; Margison & Pegg, 1981; Singer & Brent, 1981) that is present in the cell nuclei and removes 3-methyladenine and 7-methylguanine. A 3-methyladenine glycosylase has been isolated from chicken erythrocytes (Price *et al.*, 1983). A glycosylase specific to formamidopyrimidine residues has been described in rodent-cell extracts (Margison & Pegg, 1981). A hypoxanthine DNA glycosylase, that excises hypoxanthine arising from spontaneous deamination of adenine, has been isolated from calf thymus (Karran & Lindahl, 1980). A urea DNA glycosylase, that excises urea arising from fragmentation of thymine, has been described in calf thymus (Breimer, 1983).

The action of DNA glycosylases generates AP sites that can also occur by spontaneous hydrolysis. AP endonucleases cleave DNA specifically at AP sites, and have been purified to homogeneity from HeLa cells (Kane & Linn, 1981), rat liver (Verly *et al.*, 1981), placenta (Shaper & Grossman, 1980), and lymphoblasts (Brent, 1983).

In nucleotide-excision repair, the lesion is removed from DNA as an oligonucleotide. Ultra-violet-light-induced lesions and bulky lesions (e.g., those caused by aflatoxin B1 adducts) are repaired by this pathway. In bacteria, two mechanisms have been documented. In *M. luteus* and *E. coli* infected with phage T4, a pyrimidine-dimer-specific glycosylase incises the glycosidic bond, generating an AP site which is, in turn, nicked by a specific endonuclease (reviewed by Lindahl, 1982). In *E. coli*, a specific multi-subunit endonuclease, coded by the *uvrA*, *uvrB* and *uvrC* genes, recognizes ultra-violet-induced and bulky lesions and excises a patch, 12-13 nucleotides long, containing the lesion (Sancar & Rupp, 1983). In mammalian cells, neither of these two mechanisms has been described, although it is clear from studies performed on cells from patients with xeroderma pigmentosum, that there is a mechanism that recognizes and nicks ultra-violet-damaged DNA (Cleaver, 1968).

## HOW TO DETECT DNA REPAIR

Table 1 summarizes the methods that are used most commonly to detect DNA repair. The different steps in the excision-repair pathway described above can be followed, either by measuring the number of DNA single-strand breaks occurring during and after removal of the lesion, or by measuring newly-synthesized patches of DNA. This latter process is known either as repair replication or as unscheduled DNA synthesis (UDS).

UDS has been demonstrated *in vivo* after ultra-violet irradiation or drug treatment. Autoradiography has shown UDS in human germ cells after irradiation with high doses of ultra-violet light (Chandley & Kofman-Alfaro, 1971). *In-vivo*, DNA repair has been shown in early spermatid stages in the mouse, with four mutagens that are chemical homologues: methyl- ethyl- propyl- and isopropyl-methane sulfonate (Sega *et al.*, 1976).

Excision repair was measured by UDS in synchronous populations of WI-38, human, diploid fibroblasts after exposure to sodium bisulfite or to methyl methane sulfonate; UDS was enhanced 4- to 30-fold during cell proliferation, and the maximum enhancement occurred prior to the induction of DNA replication (Gupta & Sirover, 1980). These results show that human cells actively modulate this repair pathway.

Reproducibility of UDS measurements has been shown in human leucocytes treated with *N*-acetoxy-2-acetylaminofluorene (Pero & Lund-Pero, 1983). Using rat hepatocytes, Lonati-Galligani *et al.* (1983) showed that false negative responses are sometimes obtained with weak UDS inducers when cells are not functioning well. The suitability of the UDS determination to screen potential carcinogens has been reviewed recently (Althaus *et al.*, 1982).

Replicative DNA synthesis can interfere with UDS measurement. It can be suppressed by hydroxyurea (Evans & Norman, 1968), but not completely (Smith & Hanawalt, 1976), and hydroxyurea may affect repair synthesis (Williams, 1977). For accurate determinations of UDS in rats treated with *N*-nitrosodimethylamine, a method that separates replicating and nonreplicating nuclei has been used (Craddock *et al.*, 1976).

**Table 1. How to detect DNA repair**

*At the cytological level:*

- measurement of unscheduled DNA synthesis (UDS)

*At the molecular level:*

- measurement of DNA repair replication:
  - isopycnic centrifugation
  - 5-bromodeoxyuridine photolysis
- analysis of modified DNA:
  - determination of DNA breaks
    - sedimentation at alkaline pH
    - sedimentation at neutral pH
    - alkaline elution
  - determination of DNA cross-links
    - isopycnic centrifugation
    - alkaline elution
    - hydroxy apatite
  - analysis of the tertiary structure
    - sedimentation of nucleoids
    - viscoelasticity
- determination of sites sensitive to enzymes:
  - enzyme specific to single-stranded regions in DNA (S1 nuclease)
  - enzyme specific to DNA damage (DNA glycosylases, AP endonucleases, ultra-violet endonucleases)
- disappearance of modified bases:
  - high-performance liquid chromatography analysis
  - immunological detection

Incorporation of the thymidine analogue, 5-bromodeoxyuridine, and centrifugation of the cellular DNA in caesium chloride density gradients separates repair replication from semi-conservative DNA synthesis (Pettijohn & Hanawalt, 1964). This method has been used after a variety of mammalian-cell treatments (Roberts *et al.*, 1971; Laval, 1980), and can be used, also, to measure the extent of repair by photolysis of 5-bromodeoxyuridine incorporated during repair (Regan *et al.*, 1971).

Measurement of single-stranded DNA is another technique for detection of DNA damage and repair. DNA breaks occur either directly (i.e., after X-irradiation) or as a result of the action of enzymes involved in excision repair. The size of the single-strand DNA can be measured by centrifugation in alkaline sucrose gradients (McGrath & Williams, 1966) or by alkaline elution (Kohn, 1979).

Sucrose-gradient centrifugation has been used to measure the damage caused by a variety of agents, both *in vitro* (Nikaido & Fox, 1976; Abbondandolo *et al.*, 1982; Hurt *et al.*, 1983) and *in vivo* (Su *et al.*, 1983). This technique was used, also, to monitor the disappearance of DNA damage in repair-deficient cell lines, i.e., ataxia telangiectasia cells treated with 4-nitroquinoline-1-oxide (Smith & Paterson, 1980), Fanconi's anaemia cells treated with mitomycin C (Fujiwara & Tatsumi, 1975), Bloom's syndrome skin-fibroblasts treated with near ultra-violet and $\gamma$-rays (Hirschi *et al.*, 1981).

The alkaline elution technique (Kohn, 1979) measures the elution rate of DNA through a membrane filter; the rate is a function of the molecular weight of DNA. This technique has been used to determine DNA breaks in cultured cells (Iqbal *et al.*, 1976; Swenberg *et al.*, 1976; Shiloh *et al.*, 1983) and in animals (Parodi *et al.*, 1978; Petzold & Swenberg, 1978; Russo *et al.*, 1982; Cesarone *et al.*, 1983). Alkaline elution allows, also, determination of DNA cross-linking (Erickson *et al.*, 1980; Sognier & Hittelman, 1983), and measurement of enzyme-sensitive lesions in damaged DNA (Fornace, 1982). Birnboim (1982) used the technique to show that a tumour promoter, phorbol myristate acetate, induced DNA strand breakage in human leucocytes.

Repair of DNA single-strand breaks can be assessed by measuring the sedimentation rate of nucleoids. When mammalian cells are treated with high salt concentration and non-ionic detergent, chromosomal DNA is released within nucleoids that are depleted of proteins. Nucleoid sedimentation in neutral sucrose gradients corresponds to the superhelical properties of DNA, which can be assessed by sedimentation in the presence of increasing concentrations of ethidium bromide. This method allows detection of the changes in DNA supercoiling that occur during DNA repair, and is extremely sensitive, since DNA damage induced by X-ray doses below 100 rad can be detected (Cook & Brazell, 1976).

Using nucleoid sedimentation, Michel and Laval (1982) showed that repair of $\gamma$-ray-induced single-strand breaks was limited in resting human lymphocytes. The nucleoid sedimentation rate is modified, also, after treatment with alkylating agents (Mattern *et al.*, 1981).

In mammalian cells, DNA ligase activity increases after treatment with DNA-damaging agents (Mezzina *et al.*, 1982; Saucier & Laval, 1983) and could be used to detect DNA lesions. Furthermore, hypersensitivity of human cells (46 BR) to a variety of DNA-damaging agents has been correlated with a defect in the ligation step of excision repair (Teo *et al.*, 1983).

Other techniques that measure DNA damage and repair have been described, viz., a new viscosimetric determination of DNA damage (Parodi *et al.*, 1982); the determination in damaged DNA of sites sensitive to S1-nuclease (Leonard & Merz, 1982) or to specific repair enzymes, i.e., *M. luteus* ultra-violet/endonuclease (Hull & Kantor, 1983); the production of herpes virus in ultra-violet-irradiated human fibroblasts (Coppey *et al.*, 1979).

DNA repair can be assessed by measuring the disappearance of lesions that can be determined as radioactive adducts or by radioimmunoassays. Beranek *et al.* (1980) described a quantitative method using high-performance liquid chromatography analysis of methylated and ethylated products in DNA. For instance, DNA modifications introduced by haloethyl-nitrosoureas were studied by high-performance liquid chromatography (Tong *et al.*, 1982). By this method, a close parallel between alkylphosphotriester formation and micronucleus formation was shown by den Engelse *et al.* (1983) in livers of rats treated with *N*-nitrosodiethylamine, *N*-nitrosodimethylamine and ethylmethanesulfonate; the mutagenicity of *N*-methyl-*N*-nitrosourea and *N*-methyl-*N*'-nitro-*N*-nitrosoguanidine in human diploid fibroblasts was related to the efficiency with which cells remove $O^6$-methylguanine from their DNA (Medcalf & Wade, 1983).

Using antibodies specific for DNA damage, repair of ultra-violet-induced lesions has been studied in mouse and human skin *in vivo* (Eggset *et al.*, 1983). Müller and Rajewsky (1983) correlated elimination of $O^6$-ethylguanine, detected with specific antibodies, with carcinogenicity of *N*-ethyl-*N*-nitrosourea in different rat tissues. Antiserum was prepared against benzo[*a*]pyrene adducts and was used to quantitate the extent of cellular DNA modification in epidermal cells exposed to this agent (Hsu *et al.*, 1981). The suitability of immunological techniques for detection of exposure to carcinogens has been reviewed by Montesano *et al.* (1982).

## ROLE OF UNREPAIRED DAMAGE

It is believed that excision repair is an error-free process, and experimental data suggest that mutagenesis could be due to replication of a template containing unrepaired damage. This is supported by the finding that cells from inherited disorders are unable to repair certain types of DNA lesions and present a high incidence of mutations (Setlow, 1978).

Therefore, the role of unrepaired damage seems preponderant in mutagenesis. This role has been studied *in vitro*, by using different, modified, synthetic polynucleotides as the template for DNA polymerase and analysing the reaction products. The results (Table 2) show that unrepaired damage can be a block to DNA replication, thus inducing cell death. However, when DNA synthesis is not arrested by the lesion, some lesions, at least in the case of *E. coli* polymerase I, induce the incorporation of an incorrect base. These results suggest strongly that some remaining lesions are promutagenic.

**Table 2. Effects of chemical methylation of polydeoxyribonucleotides on DNA synthesis by *E. coli* DNA polymerase I**

| Methylated bases | Inhibition of synthesis | Base-pairing properties | References |
|---|---|---|---|
| *N*-7 Guanine | no | unmodified | Abbot & Saffhill (1979)<br>Boiteux & Laval (1982) |
| *N*-3 Adenine | yes | noninstructing | Miyaki *et al.* (1983)<br>Abbot & Saffhill (1977) |
| *O*-6 Guanine | no | pairs with T | Abbot & Saffhill (1977)<br>Miyaki *et al.* (1983) |
| *O*-4 Thymine | no | pairs with G | Abbot & Saffhill (1977)<br>Singer *et al.* (1983) |
| rom$^7$ Guanine | yes | noninstructing | Boiteux & Laval (1983) |
| Apurinic/apyrimidinic sites | yes | ambiguous pairs with AG T, C | Boiteux & Laval (1982)<br>Kunkel *et al.* (1983)<br>Sagher & Strauss (1983) |

## CONCLUSION

In mammalian cells, indication and quantification of an insult by a chemical can be monitored by one of the sensitive methods used to detect DNA repair. However, it is questionable whether these determinations are, in all cases, accurate indicators of the mutagenic potential of a chemical compound.

Some complications may arise when DNA repair is determined in order to predict mutagenesis. Variations of the amount of repair among cell lines have been described (Fox *et al.*, 1983). When the cells are treated with a chemical compound, mechanisms other than direct action with DNA can produce similar results, e.g., interactions with enzymes that reduce the fidelity of DNA replication, as described for chromium compounds (Bianchi *et al.*, 1983). Some chemicals (e.g., 3-aminobenzamide, cytosine arabinoside) are repair inhibitors (reviewed by Downes *et al.*, 1983) and, therefore, can modify the cell response. Other chemicals (uncouplers of oxidative phosphorylation) increase the mutagenicity of X-rays without modifying DNA repair (Laval, 1980).

Other difficulties arise through the organization of eukaryotic DNA. If it is accepted that repair processes in mammalian cells are similar to those in prokaryotes, efficiency of repair can differ in the different tissues of the body, and some regions of DNA seem more accessible to damage or more accessible to repair (Price *et al.*, 1983).

The most serious question about the prediction of mutagenicity by methods that measure DNA repair and do not directly quantify the lesion, is that these determinations correspond to the damage that is subject to excision repair. But, when cells face a lesion, it can be either excised or reversed. Among the lesions induced by alkylating agents, at least six are mutagenic (Singer & Kusmierek, 1982), and among these, $O^6$-methylguanine is repaired by an alkyltransferase through the reversion pathway. Although this type of damage is one of the most mutagenic, it is not detected by measurements of DNA repair.

This example, and the role of unrepaired lesions in mutagenesis, show how important it is to detect, in addition to DNA repair, the presence of DNA damage, using, for instance, high-performance liquid chromatography analysis or immunological techniques.

## ACKNOWLEDGEMENTS

The authors thank Dr S. Boiteux for helpful discussions. This work was supported by grants from the Centre National de la Recherche Scientifique and the Institut National de la Santé et de la Recherche Médicale, and a Research Contract from Institut Gustave Roussy (82 D 9).

## REFERENCES

Abbondandolo, A., Dogliotti, E., Lohman, P.H.M. & Berends, F. (1982) Molecular dosimetry of DNA damage caused by alkylation. *Mutat. Res.*, *92*, 361-377

Abbot, P.J. & Saffhill, R. (1977) DNA synthesis with methylated poly(dAT) template: possible role of $O^4$-methylthymidine as promutagenic base. *Nucleic Acids Res.*, *4*, 761-769

Abbot, P.J. & Saffhill, R. (1979) DNA synthesis with methylated poly(dC.dG) templates, evidence for a competitive nature to miscoding by *O*,6-methylguanine. *Biochem. biophys. Acta*, 562, 51-61

Althaus, F.R., Lawrence, S.D., Sattler, G.L., Longfellow, D.G. & Pitot, H.C. (1982) Chemical quantification of unscheduled DNA synthesis in cultured hepatocytes as an assay for the rapid screening of potential chemical carcinogens. *Cancer Res.*, *42*, 3010-3015

Beranek, D.T., Weis, C.C. & Swenson, D.H. (1980) A comprehensive quantitative analysis of methylated and ethylated DNA using high pressure liquid chromatography. *Carcinogenesis*, *1*, 595-606

Bianchi, V., Celotti, L., Lanfranchi, G., Majone, F., Marin, G., Montaldi, A., Sponza, G., Tamino, G., Venier, P., Zantedeschi, A. & Levis, A.G. (1983) Genetic effects of chromium compounds. *Mutat. Res.*, *117*, 279-300

Birnboim, H.C. (1982) DNA strand breakage in human leukocytes exposed to a tumor promoter, phorbol myristate acetate. *Science*, *215*, 1247-1249

Boiteux, S. & Laval, J. (1982) Coding properties of poly(deoxycytidilic acid) templates containing uracil or apyrimidinic sites: in-vitro modulation of mutagenesis by deoxyribonucleic acid repair enzymes. *Biochemistry*, *21*, 6746-6751

Boiteux, S. & Laval, J. (1983) Imidazole open ring 7-methylguanine: an inhibitor of DNA synthesis. *Biochem. biophys. Res. Commun.*, *110*, 552-558

Breimer, L.H. (1983) Urea-DNA glycosylase in mammalian cells. *Biochemistry*, *22*, 4192-4197

Brent, T.P. (1983) Properties of a human lymphoblast AP-endonuclease associated with activity for DNA damaged by ultra-violet light, X-rays or osmium tetroxide. *Biochemistry*, *22*, 4507-4512

Cathcart, R. & Goldthwait, D.A. (1981) Enzymic excision of 3-methyladenine and 7-methylguanine by a rat liver nuclear fraction. *Biochemistry*, *20*, 273-280

Cesarone, C.F., Bolognesi, C. & Santi, L. (1983) DNA damage induced *in vivo* in various tissues by nitrochlorobenzene derivatives. *Mutat. Res.*, *116*, 239-246

Chandley, A.C. & Kofman-Alfaro, S. (1971) 'Unscheduled' DNA synthesis in human germ cells following UV irradiation. *Exp. Cell Res.*, *69*, 45-48

Cleaver, J.E. (1968) Defective repair replication of DNA in xeroderma pigmentosum. *Nature*, *218*, 652-656

Combes, R.D. (1983) An analysis of the efficacy of bacterial DNA-repair assays for predicting genotoxicity. *Mutat. Res.*, *108*, 81-92

Cook, P.R. & Brazell, I.A. (1976) Conformational constraints in nuclear DNA. *J. Cell Sci.*, *22*, 287-302

Coppey, J., Nocentini, S., Menezes, S. & Moreno, G. (1979) Herpes virus production as a marker of repair in UV irradiated human skin cells of different origin. *Int. J. Radiat. Biol.*, *36*, 1-10

Craddock, V.M., Henderson, A.R. & Ansley, C.M. (1976) Repair replication of DNA in the intact animal following treatment with dimethylnitrosamine and with methyl methanesulphonate, studied by fractionation of nuclei in a zonal centrifuge. *Biochem. biophys. Acta*, *447*, 53-64

Deutsch, W.A. & Linn, S. (1979) DNA binding activity from cultured human fibroblasts that is specific for partially depurinated DNA and that inserts purines into apurinic sites. *Proc. natl Acad. Sci. USA*, *76*, 141-144

Downes, C.S., Collins, A.R.S., Johnson, R.T. (1983) International workshop on inhibition of DNA repair. *Mutat. Res.*, *112*, 75-83

Eggset, G., Volden, G. & Krokan, H. (1983) UV-induced DNA damage and its repair in human skin *in vivo* studied by sensitive immunohistochemical methods. *Carcinogenesis*, *4*, 745-750

den Engelse, L., Floot, B.G.J., De Brij, R.-J. & Tates, A.D. (1983) The induction of chromosomal damage in rat hepatocytes and lymphocytes. II. Alkylation damage and repair of rat-liver DNA after diethylnitrosamine, dimethylnitrosamine and ethylmethanesulphonate in relation to clastogenic effects. *Mutat. Res.*, *107*, 153-166

Erickson, L.C., Bradley, M.O., Ducore, J.M., Ewig, R.A.G. & Kohn, K.W. (1980) DNA cross-linking and cytotoxicity in normal and transformed human cells treated with antitumor nitrosoureas. *Proc. natl Acad. Sci. USA*, *77*, 467-471

Evans, R.G. & Norman, A. (1968) Radiation-stimulated incorporation of thymidine into the DNA of human lymphocytes. *Nature*, *217*, 455-456

Fornace, A.J. Jr. (1982) Measurement of *M. luteus* endonuclease-sensitive lesions by alkaline elution. *Mutat. Res.*, *94*, 263-276

Fox, M., Bloomfield, M.E., Hopkins, J. & Boyle, J.M. (1983) Differential responses of nascent DNA synthesis and chain elongation in V79 and V79/79 cells exposed to UV light and chemical carcinogens. *Carcinogenesis*, *4*, 261-268

Fujiwara, Y. & Tatsumi, M. (1975) Repair of mitomycin C damage to DNA in mammalian cells and its impairment in Fanconi's anemia cells. *Biochem. biophys. Res. Comm.*, *66*, 592-598

Gupta, P.K. & Sirover, M.A. (1980) Sequential stimulation of DNA repair and DNA replication in normal human cells. *Mutat. Res.*, *72*, 273-284

Harris, A.L., Karran, P. & Lindahl, T. (1983) $O^6$-Methylguanine-DNA methyltransferase of human lymphoid cells: structural and kinetic properties and absence in repair-deficient cells. *Cancer Res.*, *43*, 3247-3252

Hart, R.W. & Setlow, R.B. (1974) Correlation between deoxyribonucleic acid excision repair and life-span in a number of mammalian species. *Proc. natl Acad. Sci. USA*, *71*, 2169-2173

Hirschi, M., Netrawali, M.S., Remsen, J.F. & Cerutti, P.A. (1981) Formation of DNA single-strand breaks by near-ultraviolet and $\gamma$-rays in normal and Bloom's syndrome skin fibroblasts. *Cancer Res.*, *41*, 2003-2007

Hsu, I.-C., Poirier, M.C., Yuspa, S.H., Grunberger, D., Weinstein, I.B., Yolken, R.H. & Harris, C.C. (1981) Measurement of benzo[*a*]pyrene-DNA adducts by enzyme immunoassays and radioimmunoassay. *Cancer Res.*, *41*, 1091-1095

Hull, D.R. & Kantor, G.J. (1983) Evidence that DNA excision-repair in xeroderma pigmentosum group A is limited but biologically significant. *Mutat. Res.*, *112*, 169-179

Hurt, M.M., Beaudet, A.L. & Moses, R.E. (1983) Repair response of human fibroblasts to bleomycin damage. *Mutat. Res.*, *112*, 181-189

Iqbal, Z.M., Kohn, K.W., Ewig, R.A.G. & Fornace, A.J., Jr (1976) Single-strand scission and repair of DNA in mammalian cells by bleomycin. *Cancer Res.*, *36*, 3834-3838

Kaina, B. (1982) Enhanced survival and reduced mutation and aberration frequencies induced in V79 chinese hamster cells pre-exposed to low levels of methylating agents. *Mutat. Res.*, *93*, 195-211

Kane, C.M. & Linn, S. (1981) Purification and characterization of an Apurinic/Apyrimidinic endonuclease from Hela cells. *J. biol. Chem.*, *256*, 3405-3414

Karran, P. & Lindahl, T. (1980) Hypoxanthine in deoxyribonucleic acid: generation by heat-induced hydrolysis of adenine residues and release in free form by a deoxyribonucleic acid glycosylase from calf thymus. *Biochemistry*, *19*, 6005-6011

Kataoka, H. & Sekiguchi, M. (1982) Are purine bases enzymatically inserted into depurinated DNA in *E. coli*? *J. Biochem.*, *92*, 971-973

Kohn, K.W. (1979) DNA as a target in cancer chemotherapy: measurement of macromolecular DNA damage produced in mammalian cells by anticancer agents and carcinogens. *Meth. Cancer Res.*, *16*, 291-345

Kunkel, T.A., Schaaper, R.M. & Loeb, L.A. (1983) Depurination - induced infidelity of deoxyribonucleic acid synthesis with purified deoxyribonucleic acid replication proteins *in vitro*. *Biochemistry*, *22*, 2378-2384

Laval, F. (1980) Effect of uncouplers on radiosensitivity and mutagenicity in X-irradiated mammalian cells. *Proc. natl Acad. Sci. USA*, *77*, 2702-2705

Laval, F. & Laval, J. (1984) Adaptive response in mammalian cells: cross reactivity of different pretreatments on cytotoxicity as contrasted to mutagenicity. *Proc. natl Acad. Sci. USA*, *81*, 1062-1066

Laval, J. (1977) Two enzymes are required for strand incision in repair of alkylated DNA. *Nature*, *269*, 829-832

Laval, J. & Laval, F. (1980) *Enzymology of DNA repair*. In: Montesano, R., Bartsch, H. & Tomatis, L., eds, *Molecular and Cellular Aspects of Carcinogen Screening Tests* (*IARC Scientific Publications No. 27*), Lyon, International Agency for Research on Cancer, pp. 55-73

Laval, J., Pierre, J. & Laval, F. (1981) Release of 7-methylguanine residues from alkylated DNA by extracts of *M. luteus* and *E. coli*. *Proc. natl Acad. Sci. USA*, *78*, 852-855

Leblanc, J.P. & Laval, J. (1982) Comparison at the molecular level of uracil-DNA glycosylases from different origins. *Biochimie*, *64*, 735-738

Leblanc, J.P., Martin, B., Cadet, J. & Laval, J. (1982) Uracil-DNA glycosylase. Purification and properties of uracil DNA-glycosylase from *Micrococcus luteus*. *J. biol. Chem.*, *257*, 3477-3483

Leonard, J.C. & Merz, T. (1982) Repair of single-strand breaks in normal and trisomic lymphocytes. *Mutat. Res.*, *105*, 417-422

Lindahl, T. (1982) DNA repair enzymes. *Ann. Rev. Biochem.*, *51*, 61-87

Livneh, Z., Elad, D. & Sperling, J. (1979) Enzymatic insertion of purine bases into depurinated DNA *in vitro*. *Proc. natl Acad. Sci. USA*, *76*, 1089-1093

Lonati-Galligani, M., Lohman, P.H.M. & Berends, F. (1983) the validity of the autoradiographic method for detecting DNA repair synthesis in rat hepatocytes in primary culture. *Mutat. Res.*, *113*, 145-160

Margison, G.P. & Pegg, A.E. (1981) Enzymatic release of 7-methylguanine from methylated DNA by rodent liver extracts. *Proc. natl Acad. Sci. USA*, *78*, 861-865

Mattern, M.R., Paone, R.F. & Day, R.S. (1981) Human tumor strains defective in the repair of alkylated DNA fail to regenerate rapidly-sedimenting nucleoids after *N*-methyl-*N'*-nitro-*N*-nitrosoguanidine treatment. *Carcinogenesis, 2,* 1215-1218

McGrath, R.A. & Williams, R.W. (1966) Reconstruction *in vivo* of irradiated *Escherichia coli* deoxyribonucleic acid: the rejoining of broken pieces. *Nature*, *212*, 534-535

Medcalf, A.S.C. & Wade, M.H. (1983) Comparison of mutagenicity of *N*-methyl-*N*-nitrosourea and *N*-ethyl-*N*-nitrosourea in human diploid fibroblasts. *Carcinogenesis*, *4*, 115-118

Mehta, J.R., Ludlum, D.B., Renard, A. & Verly, W. (1981) Repair of $O^6$-ethylguanine in DNA by a chromatin fraction from rat liver: transfer of the ethyl group to an acceptor protein. *Proc. natl Acad. Sci. USA*, *78*, 6766-6770

Mezzina, M., Nocentini, S. & Sarasin, A. (1982) DNA ligase activity in carcinogen-treated human fibroblasts. *Biochimie*, *64*, 743-748

Michel, S. & Laval, F. (1982) Repair of $\gamma$-ray induced single-strand breaks in human lymphocytes. *Biochimie*, *64*, 749-751

Miller, E.C. & Miller, J.A. (1974) *Biochemical mechanisms of chemical carcinogenesis*. In: Busch, H., ed., *The Molecular Biology of Cancer*, New York, Academic Press, pp. 377-402

Miyaki, M., Suzuki, K., Aihara, M. & Ono, T. (1983) Misincorporation in DNA synthesis after modification of template or polymerase by MNNG, MMS and UV radiation. *Mutat. Res.*, *107*, 203-218

Montesano, R. (1981) Alkylation of DNA and tissue specificity in nitrosamine carcinogenesis. *J. supramol. Struct. cell. Biochem.*, *17*, 259-273

Montesano, R., Rajewsky, M.F., Pegg, A.E. & Miller, E. (1982) Development and possible use of immunological techniques to detect individual exposure to carcinogens: International Agency for Research on Cancer/International Programme on Chemical Safety Working Group Report. *Cancer Res.*, *42*, 5236-5239

Müller, R. & Rajewsky, M.F. (1983) Elimination of *O*6-ethylguanine from the DNA of brain, liver, and other rat tissues exposed to ethylnitrosourea at different stages of prenatal development. *Cancer Res.*, *43*, 2897-2904

Newbold, R.F., Warren, W., Medcalf, A.S. & Amos, J. (1980) Mutagenicity of carcinogenic methylating agents is associated with a specific DNA modification. *Nature*, *283*, 596-599

Nikaido, O. & Fox, B.W. (1976) The time course of DNA repair following methyl nitro-nitrosoguanidine (MNNG) treatment of P388F lymphoma cells in culture. *Chem.-biol. Interact.*, *14*, 37-45

Parodi, S., Taningher, M., Santi, L., Cavanna, M., Sciaba, L., Maura, A. & Brambilla, G. (1978) A practical procedure for testing DNA damage *in vivo*, proposed for a pre-screening of chemical carcinogens. *Mutat. Res.*, *54*, 39-46

Parodi, S., Balbi, C., Taningher, M., Pala, M., Russo, P., Abelmoschi, M.L. & Santi, L. (1982) Decreased viscosity of rat-liver DNA treated by 3'-methyl-4-dimethylaminoazobenzene, detected with a new viscometric approach. *Mutat. Res.*, *106*, 91-99

Pegg, A.E. (1978) Enzymatic removal of *O*6-methylguanine from DNA by mammalian cell extracts. *Biochem. biophys. Res. Commun.*, *84*, 166-173

Pegg, A.E., Roberfroid, M., von Bahr, C., Foote, R.S., Mitra, S., Bresil, H., Likhachev, A. & Montesano, R. (1982) Removal of *O*6-methylguanine from DNA by human liver fractions. *Proc. natl Acad. Sci. USA*, *79*, 5162-5165

Pero, R.W. & Lund-Pero, M. (1983) The productibility of individual estimations in the covalent binding of *N*-acetoxy-2-acetylaminofluorene to DNA and in the induction of unscheduled DNA synthesis. *Mutat. Res.*, *120*, 219-224

Pettijohn, D. & Hanawalt, P. (1964) Evidence for repair-replication of ultraviolet damaged DNA in bacteria. *J. mol. Biol.*, *9*, 395-410

Petzold, G.L. & Swenberg, J.A. (1978) Detection of DNA damage induced *in vivo* following exposure of rats to carcinogens. *Cancer Res.*, *38*, 1589-1598

Price, J.A., Heller, E. & Goldthwait, D.A. (1983) The release of 3-methyladenine from nucleosomal DNA by a 3-methyladenine DNA glycosylase. *Carcinogenesis*, *4*, 145-152

Regan, J.D., Setlow, R.B. & Ley, R.D. (1971) Normal and defective repair of damaged DNA in human cells: a sensitive assay utilizing the photolysis of bromodeoxyuridine. *Proc. natl Acad. Sci. USA*, *68*, 708-712

Roberts, J.J., Pascoe, J.M., Smith, B.A. & Crathorn, A.R. (1971) Quantitative aspects of the repair of alkylated DNA in cultured mammalian cells. *Chem.-biol. Interact.*, *3*, 49-68

Russo, P., Pala, M., Nicolo, G., Santi, L. & Parodi, S. (1982) DNA damage in liver of rats treated with nitrofurantoin. *Mutat. Res.*, *105*, 377-382

Sagher, D. & Strauss, B. (1983) Insertion of nucleotides opposite apurinic/apyrimidinic sites in deoxyribonucleic acid during in-vitro synthesis: uniqueness of adenine nucleotides. *Biochemistry*, *22*, 4518-4526

Samson, L. & Cairns, J. (1977) A new pathway for DNA repair in *Escherichia coli*. *Nature*, *267*, 281-283

Samson, L. & Schwartz, J.L. (1980) Evidence for an adaptive DNA repair pathway in CHO and human skin fibroblasts. *Nature*, *287*, 861-863

Sancar, A. & Rupp, W.D. (1983) A novel repair enzyme: uvrABC excision nuclease of *Escherichia coli* cuts a DNA strand on both sides of damaged region. *Cell*, *33*, 249-260

Saucier, J.M. & Laval, F. (1983) DNA ligase activity of fibroblasts and lymphocytes. *Biochem. biophys. Res. Commun.*, *116*, 657-662

Schendel, P.F. & Robins, P.E. (1978) Repair of $O^6$-methylguanine in adapted *Escherichia coli*. *Proc. natl Acad. Sci. USA*, *75*, 6017-6020

Sega, G.A., Owens, J.G. & Cumming, R.B. (1976) Studies on DNA repair in early spermatid stages of male mice after in-vivo treatment with methyl-, ethyl-, propyl-, and isopropyl methanesulfonate. *Mutat. Res.*, *36*, 193-212

Setlow, R.B. (1978) Repair-deficient human disorders and cancer. *Nature*, *271*, 713-717

Shaper, N.L. & Grossman, L. (1980) Purification and properties of the human placental apurinic/apyrimidinic endonuclease. *Meth. Enzymol.*, *65*, 216-223

Shiloh, Y., van der Schans, P., Lohman, P.H. & Becker, Y. (1983) Induction and repair of DNA damage in normal and ataxia telangiectasia skin fibroblasts treated with neocarzinostatin. *Carcinogenesis*, *4*, 917-921

Singer, B. & Brent, T.P. (1981) Human lymphoblasts contain DNA glycosylase activity excising *N*-3 and *N*-7 methyl and ethylpurines but not $O^6$-alkylguanines or 1-alkyladenines. *Proc. natl Acad. Sci. USA*, *78*, 856-860

Singer, B. & Kusmierek, J.T. (1982) Chemical mutagenesis. *Ann. Rev. Biochem.*, *52*, 655-693

Singer, B., Sagi, J. & Kusmierek, J.T. (1983) *Escherichia coli* DNA polymerase I can substitute $O^2$-methyldeoxythymidine or $O^4$-methyldeoxythymidine for deoxythymidine in primed poly(dA-dT).poly(dA-dT) synthesis. *Proc. natl Acad. Sci. USA*, *80*, 4884-4888

Sklar, R. & Strauss, B. (1981) Removal of $O^6$-methylguanine from DNA of normal and xeroderma pigmentosum-derived lymphoblastoïd cell lines. *Nature*, *289*, 417-420

Smith, C.A. & Hanawalt, P.C. (1976) Repair replication in human cells: simplified determination utilizing hydroxyurea. *Biochem. biophys. Acta*, *432*, 336-347

Smith, P.J. & Paterson, M.C. (1980) Defective DNA repair and increased lethality in ataxia telangiectasia cells exposed to 4-nitroquinoline-1-oxide. *Nature*, *287*, 747-749

Sognier, M.A. & Hittelman, W.N. (1983) Loss of repairability of DNA interstrand crosslinks in Fanconi's anemia cells with culture age. *Mutat. Res., 108*, 383-393

Su, C.M., Brash, D.E., Chang, M.J.W., Hart, R.W. & D'Ambrosio, S.M. (1983) Induction of single-strand breaks plus alkali-labile bonds by *N*-nitrosoureas in rat tissues *in vivo*: ethylnitrosourea versus benzylnitrosourea. *Mutat. Res., 108*, 1-12

Sutherland, B.M., Rice, M. & Wagner, E.K. (1975) Xeroderma pigmentosum cells contain low levels of photoreactivating enzyme. *Proc. natl Acad. Sci. USA, 72*, 103-107

Swenberg, J.A., Petzold, G.L. & Harback, P.R. (1976) In-vitro DNA damage/alkaline elution assay for predicting carcinogenic potential. *Biochem. biophys. Res. Commun., 72*, 732-738

Teo, I.A., Broughton, B.C., Day, R.S., James, M.R., Karran, P., Mayne, L.V. & Lehmann, A.R. (1983) A biochemical defect in the repair of alkylated DNA in cells from an immunodeficient patient (46BR). *Carcinogenesis, 4*, 559-565

Tong, W.P., Kohn, K.W., Ludlum, D.B. (1982) Modifications of DNA by different haloethylnitrosoureas. *Cancer Res., 42*, 4460-4464

Verly, W.G., Colson, P., Zocchi, G., Goffin, C., Liuzzi, M., Buchsenschmidt, G. & Muller, M. (1981) Localisation of the phosphodiester bond hydrolysed by the major apurinic/apyrimidinic endodeoxyribonuclease from rat-liver chromatin. *Eur. J. Biochem., 118*, 195-201

Williams, G.M. (1977) The detection of chemical carcinogens by unscheduled DNA synthesis in rat liver primary cell cultures. *Cancer Res., 37*, 1845-1851

Yarosh, D.B., Foote, R.S., Mitra, S., Day, R.S. (1983) Repair of $O^6$-guanine in DNA by demethylation is lacking in $Mer^-$ human tumor cell strains. *Carcinogenesis, 4*, 199-205

# IDENTIFYING AGENTS THAT DAMAGE HUMAN SPERMATOGENESIS: ABNORMALITIES IN SPERM CONCENTRATION AND MORPHOLOGY

**A.J. Wyrobek**

*Lawrence Livermore National Laboratory, Biomedical Sciences Division, University of California, CA 94550, USA*

## INTRODUCTION

Exposure of the human male to radiation or to certain chemicals can result in abnormalities in sperm production, fertility or reproductive outcome. About 100 different chemical agents and mixtures have been studied for these effects in exposed men. However, for about 90% of these exposures, data are available only for effects on sperm production (sperm concentration, motility and morphology; reviewed by Wyrobek *et al.*, 1983a). Effects of in-vivo chemical exposure on male fertility and abnormal reproductive outcome have been assessed for very few agents (reviewed by Edmonds *et al.*, 1981) and usually not for the same exposure regimens as those studied for effects on sperm production. As yet, there has been no systematic attempt to evaluate, in exposed men, the relationship between induced abnormalities in sperm production and subsequent fertility changes or increased risk of abnormal reproductive outcome.

Sperm tests for reduced sperm concentration, reduced motility and abnormal morphology have a long history of clinical use as procedures for the assessment of infertility in domesticated animals and man. On the basis of these applications, it is generally agreed that abnormal sperm test results reflect damage to spermatogenesis, have some predictive value for fertility, but have an as yet ambiguous relationship to induced heritable genetic damage.

Human sperm tests are receiving growing attention by those interested in monitoring reproductive consequences associated with human exposure to mutagens and carcinogens. Some of the advantages of studying human sperm are: (1) sperm cells are easy to obtain in large numbers; (2) sperm can be studied also in model animals, where the mechanisms and the quantitative relationship to reproductive outcome can be studied in detail; and (3) sperm tests generally require small cohort sizes in order to detect induced effects compared to questionnaire-based studies.

This paper gives a brief overview of the human tests for changes in sperm concentration and morphology, particularly as they apply to the study of potential occupational and environmental hazards.

**Table 1. Results of agents tested for their effects on human sperm concentration[a,b]**

| Agents showing decreases in sperm counts | | Agents with inconclusive evidence of decreases[c] | Agents showing no response[d] |
|---|---|---|---|
| Acridinyl anisidide (C) | Methotrexate (C) | Centchroman (C) | Anaesthetic gases (2,A) |
| Alcoholic beverages (2,A) | MOPP (mechlorethamine, vincristine, procarbazine and prednisone) (2,C) | Cimetidine (A) | Bromocriptine (B) |
| Aspartic acid (C) | MVPP (mechlorethamine, vinblastine, procarbazine and prednisone) (3,C) | Colchicine (4,A-C) | Carbaryl (A) |
| Carbon disulfide (2,A)[e] | Norethandrolone (C) | Diethylstilboestrol (6,A-E) | Diphenyl hydantoin (A) |
| Chlorambucil (4,C) | Norethindrone (C) | Hydroxymethyl nitrofurantoin (C) | Doxycycline (A) |
| Chlorambucil and mechlorethamine (C) | Norethindrone, norethandrolone and testosterone (C) | Kepone (2,E) | Epichlorohydrin (2,A) |
| Chlorambucil, mechlorethamine and azathioprine (C) | 1-Norgestrel and testosterone enanthate (2,A) | Nitrofurantoin (C) | Fluoxymesterone (B) |
| Cyclophosphamide (8,C-D) | Norgestrienone and testosterone (2,C) | Norethandrolone and testosterone (C) | Glycerine production compounds (A) |
| Cyclophosphamide and colchicine (C) | Prednisolone (2,C) | Tobacco smoke (3,A-B) | Lysine (C) |
| Cyclophosphamide and prednisone (3,C) | Progesterone (C) | Trimeprimine (C) | Methyl testosterone (2,C) |
| Cyclophosphamide, prednisone and azathioprine (C) | Propafenone (C) | | Niridazole (C) |
| CVP (cyclophosphamide, vincristine and prednisone) (D) | 13-Ethyl-17α-ethynyl-17-hydroxy-gona-4,9,11-trien-3-one (R2323) and testosterone (C) | | Nerethindrone and testosterone (C) |
| CVVP (cyclophosphamide, vincristine, prednisone and procarbazine) (C) | Sulfasalazine (6,A-D) | | 1-Orinthine (C) |
| Cyproterone acetate (5,A-C) | Testosterone (3,B-D)[e] | | Pentoxyfilline (A) |
| Danazol and methyl testosterone (2,C) | Testosterone cyclopentylpropionate (C) | | Polybrominated biphenyls (2,B) |
| Danazol and testosterone enanthate (2,C) | Testosterone enanthate (6,A-C)[e] | | Tryptophan (C) |
| Dibromochloropropane (9,A-C) | Testosterone propionate (3,C) | | WIN 59941[f] (E) |
| Dibromochloropropane and ethylene dibromide (A) | Toluene diamine and dinitrotoluene (A) | | |
| | VACAM (vincristine, adriamycin, cyclophosphamide, actinomycin D and medroxyprogesterone acetate) (C) | | |

| | |
|---|---|
| Enovid (C) | |
| Gossypol (D) | *N,N'*-Bis(dichloroacetyl)-*N,N'*-diethyl-1,4-xylylenediamine (WIN 13099) (C) |
| Lead (A) | WIN 13099 and diethylstiboestrol (C) |
| Luteinizing hormone releasing hormone analogue (A) | *N,N'*-Bis(dichloroacetyl)-*N,N'*-diethyl-1,6-hexanediamine (WIN 17416) (C) |
| Marijuana (5,A-E) | *N,N'*-Bis(dichloroacetyl)-1,8-octanediamine (WIN 18446) (2,C) |
| Medroxyprogesterone acetate (C) | |
| Medroxyprogesterone acetate and testosterone enanthate (6,C) | |
| Medroxyprogesterone acetate and testosterone propionate (C) | |
| Megestrol acetate and testosterone (2,C-D) | |
| Metanedienone (A) | |

[a] Adapted from Wyrobek *et al.* (1983a), in which details and references are given

[b] The number in parentheses is the number of studies evaluated; letters in the parentheses represent the level of statistical confidence that can be placed on the result. In cases where the classification depended on only one study, only the letter representing statistical confidence is given. Statistical criteria: A, statistical criteria and analyses were presented in the original paper; B, statistics appear to have been done (as indicated by *p* values, etc.), but no indication is given of the method used; C, statistical analyses were not done in the original paper, but sufficient data are presented for analysis; this includes case studies in which sufficient data for counts and motility were presented for a conservative comparison with historical control values. Descriptive statements of men becoming azoospermic were regarded as quantitative; D, statistical analyses were not done in the original paper and the data presented were insufficient for analysis; E, descriptive statement only, no data presented.

[c] Agents evaluated showed consistently small or questionable effects on sperm count, gave conflicting results with no apparent explanation or were presented with inadequate statistical support.

[d] Agents that showed no change in sperm count at the exposure levels and sampling times tested

[e] At least one of the studies reported a negative response, apparently due to lower exposure doses

[f] Placebo in study of WIN compounds

## APPLICATIONS OF SPERM TESTS

As recently reviewed (Wyrobek *et al.*, 1983a), human sperm tests have been used to assess the spermatogenesis effects of about 90 individual chemical agents or mixtures of agents.

### *Sperm concentration*

Table 1 lists 78 chemical exposures that were studied for their effects on human sperm concentration; 51 showed significant decreases, 10 showed inconclusive evidence of decreased sperm concentration and 17 showed no apparent change in sperm concentration. This list does not include nine drugs that gave some indication of increased sperm concentration after treatment.

### *Sperm motility*

Although sperm motility may be one of the best performance evaluations of spermatogenic function in relation to fertility, it is subjective and very sensitive to time and to temperature after collection. Thus, sperm motility is a very difficult parameter to measure in a human field-study, especially when samples are collected at home. Results based on its application are not summarized here (detailed by Wyrobek *et al.*, 1983a).

### *Sperm morphology*

Table 2 lists 44 chemical exposures that were studied for their effects on the proportion of morphologically abnormal sperm; 17 agents showed significant increases, 9 showed inconclusive evidence of an increase and 17 showed no change. Only one agent, vitamin B12, appeared to be able to decrease the incidence of abnormally shaped sperm.

### *Occupational and environmental chemicals*

Inspection of the types of agents listed in Tables 1 and 2 shows that approximately 85% are experimental or therapeutic drugs, 5% are recreational drugs (alcohol, tobacco, marijuana) and only 10% are occupational or environmental agents. Table 3 lists the occupational and environmental agents. Of 12 agents and mixtures studied, five gave evidence of a detrimental effect on sperm production, two gave inconclusive results and five showed no effect at the doses tested.

## SPERM MORPHOLOGY TESTING

### *Variability and sample-size calculations*

Subjectivity and variations in scoring have been major concerns with human sperm morphology testing.

**Table 2. Agents tested for their effects on human sperm morphology[a,b]**

| Agents showing increase in the proportion of abnormally shaped sperm | Agents with inconclusive evidence of increase[c] | Agent suggestive of decrease | Agents showing no response[d] |
|---|---|---|---|
| Acridinyl anisidide (A) | Carbaryl (A) | Vitamin B-12 (2,D) | Anaesthetic gases (2,A) |
| Alcoholic beverages (A) | Centchroman (E) | | Bromocriptine (B) |
| Carbon disulfide (2,A)[f] | Clomiphene citrate (5,A-E)[e] | | Chlorambucil (2,D-E) |
| Cyproterone acetate (3,A-C) | Cyclophosphamide and prednisone (C) | | Dibromochloropropane (4,A-E) |
| Danazol and methyl testosterone (C) | Diethylstilboestrol (5,A-C) | | Dibromochloropropane and ethylene dibromide (A) |
| Danazol and testosterone enanthate (C) | Diphenyl hydantoin (A) | | Fluoxymesterone (B) |
| Gossypol (D) | Kepone (2,E) | | Glycerine production compounds (A) |
| Lead (A) | Metronidazole (C) | | Hydroxymethyl nitrofurantoin (C) |
| Marijuana (3,B-C) | Tobacco smoke (3,A-B) | | Kallikrein (7,A-E) |
| Medroxyprogesterone acetate (C) | | | Luteinizing hormone releasing factor (E) |
| Metanedienone (A) | | | Methyl testosterone (C) |
| Methotrexate (C) | | | Niridazole (D) |
| 1-Norgestrel and testosterone enanthate (E) | | | Norethandrolone and testosterone (C) |
| Sulfasalazine (4,A-D) | | | Polybrominated biphenyls (2,B) |
| Testosterone enanthate (3,A-C)[f] | | | Tamoxifen (E) |
| *N-N'*-Bis(dichloroacetyl)-*N,N'*-diethyl-1,6-hexanediamine (WIN 17416) (E) | | | Testosterone (E) |
| *N-N'*-Bis(dichloroacetyl)-1,8-octane-diamine (WIN 18446) (2,C-E) | | | Toluene diamine and dinitrotoluene (A) |

[a] Adapted from Wyrobek *et al.* (1983a), in which details and references are given

[b] The number in parentheses is the number of studies evaluated; letters in the parentheses represent the level of statistical confidence that can be placed on the result. In cases where the classification depended on only one study, only the letter representing statistical confidence is given. Statistical criteria: A, statistical criteria and analyses were presented in the original paper; B, statistics appear to have been done (as indicated by *p* values, etc.), but no indication is given of the method used; C, statistical analyses were not done in the original paper, but sufficient data are presented for analysis; this includes case studies in which sufficient data for counts and motility were presented for a conservative comparison with historical control values. Descriptive statements of men becoming azoospermic were regarded as quantitative; D, statistical analyses were not done in the original paper and the data presented were insufficient for analysis; E, descriptive statement only, no data presented.

[c] Agents for which papers evaluated showed a small or questionable increase in abnormally shaped sperm, gave conflicting results with no apparent explanation or were presented with inadequate statistical support.

[d] Agents that showed no change in sperm morphology at the exposure levels and sampling times tested

[e] At the highest doses tested, clomiphene citrate decreases the percentage of morphologically normal sperm. At lower doses several studies found no effects while one suggests some improvement

[f] At least one of the studies reported a negative response, apparently due to lower exposure doses

**Table 3. Occupational and environmental exposures studied with human sperm tests[a]**

| Positive effects | Inconclusive effects | No effect observed |
|---|---|---|
| Alcoholic beverages (chronic alcoholism) | Carbaryl | Anaesthetic gases |
| Carbon disulfide | Kepone | Epichlorohydrin |
| Dibromochloropropane | | Glycerine production compounds |
| Dibromochloropropane and ethylene dibromide | | Polybrominated biphenyls |
| Lead | | Wastewater treatment at oil refinery[b] |
| Marijuana | | |
| Toluene diamine and dinitrotoluene | | |

[a] Includes studies on sperm counts and morphology; adapted from Wyrobek *et al.* (1983a), in which details and references are given

[b] See Rosenberg *et al.* (1984) for study of waste-water treatment workers

Figure 1 indicates the decision process we use to classify fixed and stained human sperm into shape categories (Wyrobek *et al.*, 1982). By comparing readings from a coded set of sperm smears scored over a four-year period, we have demonstrated that this method can be highly reproducible (see coefficient of variation in Table 4), and that it can be used to compare results of studies separated by periods of years.

**Table 4. Human sperm morphology: sources of variation[a]**

| Source | No. of groups | No. of men per group | No. of sperm samples per man | No. of readings per sample | Coefficient of variation (%) |
|---|---|---|---|---|---|
| Within slide | 1 | 12 | 1 | 6 to 29 | 6 |
| Within person | 1 | 6 | 6 to 10 | 1 | 9 |
| Person-to-person | 5 | 14-35 | 1 | 1 | 30 |

[a] Based on scoring 500 sperm per sample, using the criteria of Wyrobek *et al.* (1982)

In studies of presumably unexposed men, we consistently observed interperson variations in sperm concentration that were considerably larger than variations observed in the proportion of morphologically abnormal sperm (Wyrobek *et al.*, 1982). These quantitative differences were used to estimate sample-size requirements for both sperm concentration and morphology tests in a cross-sectional study design. We found that, compared to sperm concentration, sperm morphology required from five to ten times fewer men, in order to detect particular percentage differences in mean value between an exposed group and an unexposed group. Furthermore, in repeated sperm samplings of six men over periods of up to three years (Table 4), we observed that intra-subject variation in sperm morphology was considerably smaller than inter-subject variation. We used these differences to determine

**Fig. 1. Schematic representation of the decision process used to classify individual human sperm into shape categories**

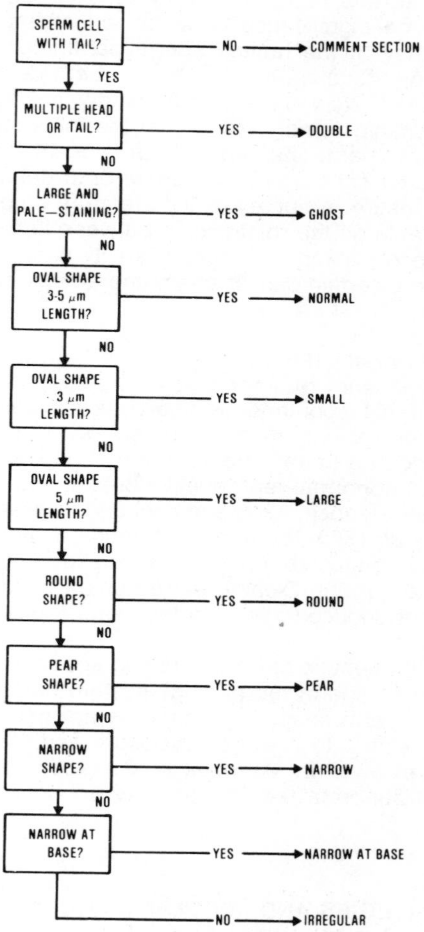


See Wyrobek *et al.* (1982) for details

that sample-size requirements for longitudinal study designs are approximately two to three times smaller than those required for cross-sectional studies. Although the longitudinal study design is statistically advantageous for sperm morphology testing, there is a possible disadvantage in that at least two sperm samples are required from each donor - one sample prior to exposure and one after exposure (although more samples are preferred).

These results demonstrate that (1) reproducible visual criteria for scoring human sperm morphology can be developed, (2) sperm morphology testing is statistically more sensitive to induced change than is sperm concentration, and (3) sperm morphology seems to be particularly sensitive when used in a longitudinal study design.

*Implications for fertility and reproductive outcome*

Although it is generally agreed that major reductions in sperm counts and motility are linked to reduced fertility, the significance of small reductions is unclear. It is uncertain also, whether changes in any of the human sperm parameters are quantitatively related to embryonic failure, heritable genetic abnormalities or birth defects. Human data on these questions are extremely limited. We were unable to identify any agent, for which human data are available, for the relationships among all four of the following: (1) paternal exposure to the chemical; (2) subsequent sperm changes; (3) subsequent fertility changes; and (4) any adverse reproductive outcome such as miscarriages, birth defects and heritable effects in offspring fathered after exposure. A survey of the literature emphasizes the relatively large volume of information available on the relationship between items 1 and 2. Chemically induced sperm changes have been linked to reduced fertility (i.e., linking items 1, 2 and 3) for 1,2-dibromo-3-chloropropane, certain cancer chemotherapeutic agents and several antifertility drugs.

The studies of Furuhjelm *et al.* (1962) suggest a link between poor sperm quality (count and morphology) and the frequency of embryonic failure (i.e., a link between items 2 and 4). In these studies, fathers of 201 spontaneous abortions showed significantly higher sperm abnormalities and lower sperm counts than did fathers of 116 normal pregnancies. No data were given for paternal exposure to any chemical. Although several studies support the link between sperm defects and abnormal reproductive outcome (see Czeizel *et al.*, 1967, for a list of references; Jöel, 1966, Raboch, 1965 and Jöel and Chayen, 1971, for studies on habitual abortions; Furuhjelm *et al*, 1960, for studies of perinatal mortality; and Takala, 1957, for studies of birth defects), some studies found no correlation (e.g., Kneer, 1957; MacLeod & Gold, 1957; Homonnai *et al.*, 1980). Clearly, more human studies are needed to compare exposure of the male parent, induced sperm defects and reproductive outcomes.

Most of the studies on the genetic basis of induced sperm anomalies have been conducted in the mouse. The mouse sperm morphology system has been studied in most detail, and there is evidence that sperm shape abnormalities are genetically controlled, that induced abnormalities can be related to mutagen exposure and that changes in sperm abnormalities can be inherited (reviewed by Wyrobek *et al.*, 1983b). Several studies also support the suggestion that sperm abnormalities in man may be inherited (Nistal *et al.*, 1978; Bisson *et al.*, 1979).

## RATIONALE, STRATEGIES AND PROBLEMS ENCOUNTERED IN THE USE OF HUMAN SPERM ASSAYS IN EXPOSED POPULATIONS

Human sperm studies are considerably more complex than animal sperm studies. Although other approaches for assessing induced testicular dysfunction in man have been investigated (e.g., biopsy and blood gonadotrophin level), sperm assays are the only feasible, direct way to assess chemical effects on the male germ cells. In general, sperm studies in man are warranted if there are (1) sufficient animal data available that suggest testicular effects, or (2) questionnaire data or case reports that suggest a problem with infertility or a pregnancy outcome linked to the male.

The application of human sperm assays may be best suited to a situation in which there is concern over a group that has been exposed to an agent that may affect fertility or reproductive outcome or both. The first level of response might be to survey and compile all human and animal studies of testicular dysfunction associated with the suspected agent. If no animal or human data exist, if there is sufficient concern about the effects on human spermatogenesis and if the suspected agent is available in sufficient quantities, an animal study such as a sperm-morphology assay may be done on several laboratory species to assess the likelihood of antispermatogenic effects. The second level of activity might be a screening questionnaire to determine whether there are changes in human fertility. If there is still sufficient evidence for concern and if the population is sufficiently large, this may be followed by a more detailed questionnaire to determine to what extent the infertility is male- or female-related. It is at this point, either concurrent with or in place of the more detailed questionnaire, that human sperm studies might be considered.

The key steps of a human sperm study are as follows.

(1) *Identifying populations at risk*

Groups of men may be considered at risk if they are exposed to any agent or close analogue of an agent known to be a testicular toxin in any mammal. Also, men exposed to suspect mutagens and carcinogens should be considered at risk for spermatogenic damage. Sometimes groups may be identified more directly by the effects of exposure. For example, the study of men exposed to 1,2-dibromo-3-chloropropane was initiated by workers' complaints of unintentional childlessness (Whorton *et al.*, 1977). Case reports can be useful indicators of potential problem exposures, provided the data are not overinterpreted.

(2) *Study design*

A precise protocol of study is essential. The following items should be considered: confidentiality of medical records, informed consent of subjects, instructions for semen collection procedures, instructions for sample handling and transport and the method of reporting results and interpretations back to the subject. Provision should be made for the follow-up of abnormal test results.

Information on the number of men exposed to the agent, as well as any available estimates of dose, are essential in planning an effective study. Since there is a high variability in semen characteristics among males, a large number of subjects is required to establish differences between control and exposed groups in cross-sectional studies where each individual is sampled only once (Wyrobek *et al.*, 1981). Cross-sectional studies are, therefore, typically large, e.g., approximately 170 men in the study of smoking effects (Viczian, 1969) and 200 in the study of lead effects (Lancranjan *et al.*, 1975).

As stated earlier in this paper, variation of sperm morphology within an individual male is considerably less than variation among individuals (MacLeod, 1965a, 1974; Sherins *et al.*, 1977). Therefore, prospective longitudinal study designs may be more appropriate when fewer men are available for sampling. In these designs, repeated semen samples from the same man before, during and after exposure are compared to assess chemically induced sperm defects. Longitudinal studies have been successfully conducted on men exposed to X-rays and on men receiving drugs for medical reasons (MacLeod, 1965a; Heller *et al.*, 1966; Rowley *et al.*, 1974).

(3) *Gaining access*

Once a population at risk is identified, the investigators need to obtain access to the population. This can be the most time-consuming aspect of the study and may require co-operation from hospital administrators, plant management, labour unions and various governmental agencies. When industrial chemicals are involved, access may become a major issue, since industry is concerned about outside interference in matters that may affect productivity and economics. Other common problems in gaining access are: difficulties in obtaining subject cooperation, lack of an available physician to assist in recruitment for drug-related studies and the frequent high percentage of vasectomies among the test population (as high as 25% in some areas of the USA).

Numerous methods are available for recruiting volunteers (e.g., an announcement at general meetings, the circulation of a form letter, physician-patient interaction, etc.). Regardless of the method of introduction, an effective procedure will include a session between a physician and the prospective donor on matters of information, clarification and motivation. Recruitment of concurrent controls is usually more difficult than recruitment of exposed men.

(4) *Identifying a control group*

For occupational studies, typical control groups are (1) new employees; (2) administrative personnel; (3) workers from another area of the plant; (4) workers in another plant; and (5) historical controls (i.e., data on men who served as controls in previous studies). Usually, each of these groups has some disadvantage due to mismatching in age, socioeconomic factors, numbers of men involved or other chemical exposures, etc. The best approach may be to use two control groups: the best available concurrent control group and a historical group.

For drug studies, unexposed men may be used as concurrent controls. For longitudinal prospective studies, semen samples collected before exposure may be used as controls.

(5) *Assigning men to dosage groups*

In any study, the assessment of dose is vital. The degree of dose-dependent response should be an important criterion in identifying a testicular toxin, and every effort should be made to group men according to dose. Since chemical dosimetry is often inadequate in the workplace, indirect methods of dose grouping need to be considered. For example, Lancranjan *et al.* (1975) grouped men according to the lead concentrations in urine and blood. Men can be grouped by job description, also, which may signify exposure level.

It is considerably easier to assign men to exposure groups in drug studies. With chemotherapy, high doses of chemicals are usually used and recorded in detail. Although such studies may be useful in validating the sperm tests, interpretation is often complicated, because combinations of agents are administered and because the disease itself may affect sperm production. Instances in which only one agent is used, and numerous men are exposed, are of particular interest. A prospective sampling of sperm before, during and after treatment could provide details on dose response, individual susceptibilities to induced damage and recovery.

(6) *Collecting questionnaire data*

(a) *Types of information needed.* Detailed questionnaires are needed to obtain occupational information (dates of employment, types of jobs held, types of agents used, etc.), with special emphasis on the year prior to semen collection. Detailed medical histories are needed, because ill health, recent febrile illness and associated drug consumption, as well as anatomic abnormalities (as determined by the examining physician), may all affect semen quality. Personal data usually include age, smoking and drinking habits, number and birth dates of children, as well as the occurrence of miscarriages, abortions and birth defects.

(b) *Confounding variables*. The following is a brief description of factors other than chemical exposure that are reported to affect human sperm production. They should be considered when the questionnaire is designed and, if a physical examination is done, by the physician.

*Medical factors*. Medical conditions reported to affect sperm production in man include: parotitic orchitis (Bartak *et al.*, 1968; Bartak, 1973), leprosy (Ibrahiem *et al.*, 1979), urogenital tuberculosis (Jimenez-Cruz *et al.*, 1979), renal transplantation (Lingardh *et al.*, 1974), diabetes (Bartak *et al.*, 1975; Bartak, 1979), prostatitis (Yunda *et al.*, 1978), genitourinary infection (Caldamone & Cockett, 1978), varicocele (MacLeod, 1965b, 1969; Czyglik *et al.*, 1973; Greenberg, 1977; Gall *et al.*, 1978; Rodriguez-Rigau *et al.*, 1978), severe allergic response (MacLeod, 1965a), febrile and viral diseases (MacLeod, 1965a) and cancer (Wyrobek *et al.*, 1980; Chapman *et al.*, 1981). Men who have infertile marriages tend to have lower semen quality than fertile men (MacLeod, 1951).

*Age*. Although the effects of age on sperm quality are generally considered insignificant, some suggestion of a small age-related effect is seen in males with an infertile marriage (Atanasov & Tsnakov, 1974), or in men with andrological diseases (Schirren *et al.*, 1975).

*Radiation*. The spermatotoxic and mutagenic effects of radiation on animal testes are well known. Similar patterns of spermatogenic cell killing and subsequent reductions in sperm number have been reported for men receiving testicular radiation (Rowley *et al.*, 1974; Greiner & Meyer, 1977; Ash, 1980).

*Personal habits*. Some relatively common personal habits may affect human sperm production. The effects of tobacco smoking, marijuana use and heavy alcohol consumption are reviewed in this report. Exposure to heat and associated elevations in scrotal temperature should also be considered (MacLeod & Hotchkiss 1941; Procope, 1965; Hendry, 1976). Sas and Szöllösi (1979) showed that professional drivers have poorer sperm quality than controls. There has been a suggestion that major dietary differences (i.e., vegetarian *vs* nonvegetarian) may be related to semen quality (Arora *et al.*, 1961).

(7) *Collecting and analysing semen samples*

In a cross-sectional study, usually one semen sample per individual is analysed. Longitudinal studies may require several samples from each man over a period of several months. Before collection, each donor is given an instruction form emphasizing the importance of obtaining a complete and fresh sample and explaining the methods of collection. Various methods have been used to collect samples, including masturbation, coitus interruptus and collection with a Mylar sheath. Since count varies with different portions of the ejaculate, the entire ejaculate is necessary to evaluate sperm count. The preferred method is masturbation into a clean glass or plastic container.

The accuracy of some of the semen analyses is critically dependent on the collection and care of the sample. It is necessary to control the continence time, if the sperm count is to be accurately evaluated. Morphology, however, does not appear to be affected by continence time (David, 1981). Although frozen samples can be used to assess sperm counts and morphology, it is recommended that fresh samples be used for counts and smears, whenever possible. Morphology can be assessed from air-dried smears which can be shipped and stored for extended periods of time, adding much flexibility to the use of these sperm tests.

(8) *Statistical evaluation and interpretation*

Distribution of sperm data from control and exposed men can be analysed by parametric and nonparametric tests, e.g., the t-test, Mann-Whitney test, Kolmogorov-Smirnov test (as in the 1,2-dibromo-3-chloropropane study of Whorton *et al.*, 1977); or by analysis of the proportion of men with semen pathologies (as in the lead study by Lancranjan *et al.*, 1975). Dose relationships should be assessed whenever possible. The possible effects of age, smoking, illness, medication, drugs and other confounding factors must be analysed statistically (see item 6, above).

## CONCLUSIONS

Human sperm assays have numerous advantages; they can be used to monitor effects directly in exposed men, and large numbers of sperm can be examined easily. Changes in sperm parameters arise, probably, from interference by the test substance with the genetically-controlled differentiation of the sperm cell; therefore, these assays are intrinsically relevant to safety evaluation and for the assessment of potential effects on male fertility and, possibly, on reproductive outcome. The laboratory methods are generally rapid, straightforward and quantitative. Sperm assays have major advantages over other approaches for assessing induced changes in testicular function. Testicular biopsies are impractical: they are traumatic, invasive and may, themselves, affect testicular function. Sole reliance on epidemiological surveys of reproduction function, using questionnaires, requires large cohort sizes and may not be very sensitive. Analyses of gonatrotrophin levels in blood are expensive and generally insensitive to small changes in testicular function. Compared with these methods, sperm assays are noninvasive and inexpensive; they require only small cohort for effective analyses and are sensitive to small changes.

The major disadvantages of sperm assays are: (1) The heritability of the induced damage is not yet clearly understood; (2) limited sperm sampling times and dosage regimens may reduce the sensitivity of the assay (e.g., agents that exert only transient effects may be missed by the use of single sampling times); (3) other factors such as ischaemia, marijuana use, high fever, medication taken, testicular trauma, etc., may produce false positive responses (questionnaire data are of vital importance in identifying these confounding factors); (4) there may be difficulties in obtaining samples, especially in environmental and occupational studies.

## ACKNOWLEDGEMENTS

L. Gordon and G. Watchmaker played major roles in the development of the sperm morphology test used in our laboratory. We thank A. Riggs for formatting and typing this manuscript.

## REFERENCES

Arora, R.B., Saxena, K.N., Choudhury, M.R., Choudhury, R.R. (1961) Sperm studies on Indian men. *Fertil. Steril.*, *12*, 365-366

Ash, P. (1980) The influence of radiation on fertility in man. *Br. J. Radiol.*, *53*, 271-278

Atanasov, A. & Tsankov, T.S. (1974) The influence of the age of men from sterile families on several clinical parameters of the ejaculate. *Akush Ginekol. (Sofiia)*, *13*, 199-203

Bartak, V. (1973) Sperm count, morphology, and motility after unilateral mumps orchitis. *J. Reprod. Fert.*, *32*, 491-494

Bartak, V. (1979) Sperm quality in adult diabetic men. *Int. J. Fertil.*, *24*, 226-232

Bartak, V., Skalova, E. & Nevarilova, A. (1968) Spermiogram changes in adults and youngsters after parotitic orchitis. *Int. J. Fertil.*, *13*, 226-232

Bartak, V., Josifko, M. & Horackova, M. (1975) Juvenile diabetes and human sperm quality. *Int. J. Fertil.*, *20*, 30-32

Bisson, J.P., Leonard, C. & David, G. (1979) Familial character of some morphological abnormalities of spermatozoa. *Arch. Anat. Cytol. Pathol.*, *27*, 230-233

Caldamone, A.A. & Cockett, A.T.K. (1978) Infertility and genitourinary infection. *Urology*, *12*, 304-312

Chapman, R.M., Sutcliffe, S.B. & Malpas, J.S. (1981) Male gonadal dysfunction in Hodgkin's Disease. *J. Am. med. Assoc.*, *245*, 1323-1328

Czeizel, E., Hancsok, M. & Viczian, M. (1967) Examination of the semen of husbands of habitually aborting women. *Orvosi. Hetilap.*, *108*, 1591-1595

Czyglik, F., David, G., Bisson, J.P., Jouannet, P. & Gernigon, C. (1973) Teratospermia in varicocele (Fr.). *Nouv. Presse med.*, *2*, 1127-1130

David, G. (1981) *Factors affecting the variability of semen characteristics*. In: Spira, A. & Jouannet, P., eds, *Human Fertility Factors* (*Colloque INSERM 103*), Paris, INSERM, pp. 57-68

Edmonds, L., Hatch, M., Holmes, L., Kline, J., Letz, G., Levin, B., Miller, R., Shrout, P., Stein, Z., Warburton, D., Weinstock, M., Whorton, M. & Wyrobek, A. (1981) *Guidelines for reproductive studies in exposed populations*. In: Bloom, A.D., ed., *Guidelines for Studies of Human Populations Exposed to Mutagenic and Reproductive Hazards*, New York, March of Dimes Birth Defects Foundation, pp. 37-110

Furuhjelm, J., Jonson, B., Lagergren, C.G. & Lindgren, L. (1960) The quality of the human semen in relation to perinatal mortality. *Acta obstet. gynec. scand.*, *39*, 499-505

Furuhjelm, M., Jonson, B. & Lagergren, C.G. (1962) The quality of human semen in spontaneous abortion. *Int. J. Fertil.*, *7*, 17-21

Gall, H., Schnierstein, J. & Glowania, H.J. (1978) The effect of varicocele on male fertility with particular consideration of progressive motility. *Urologe*, *17*, 317-320

Greenberg, S.H. (1977) Varicocele and male fertility. *Fertil. Steril.*, *28*, 699-706

Greiner, R. & Meyer, A. (1977) Reversible and irreversible azoospermia after irradiation of the malignant tumor of the testicle. *Strahlentherapie*, *153*, 257-262

Heller, C.G., Wootton, P., Rowley, M.J., Lalli, M.F. & Brusca, D.R. (1966) *Action of radiation upon human spermatogenesis*. In: *Proceedings of the Sixth Pan American Congress of Endocrinology* (*Excerpta Medica International Congress Series No. 112*), Amsterdam, Elsevier, pp. 408-410

Hendry, W.F. (1976) Loose pants and cold scrotal douches, effects on spermatogenesis. *Sperm Action Prog. Reprod. Biol.*, *1*, 259-262

Homonnai, Z.T., Paz, G.F., Weiss, J.N. & David, M.P. (1980) Relation between semen quality and fate of pregnancy: retrospective study on 534 pregnancies. *Int. J. Androl.*, *3*, 574-584

Ibrahiem, A.A., Awad, H.A., Metawi, B.A. & Hamada, T.A.Y. (1979) Pathologic changes in testis and epididymis of infertile leprotic males. *Int. J. Lepr.*, *47*, 44-49

Jimenez-Cruz, J.F., de Cabezon, J.S., Soler-Rosello, A. & Sole-Balcells, F. (1979) The spermiogram in urogenital tuberculosis. *Andrologia*, *11*, 67-70

Jöel, C.A. (1966) New etiologic aspects of habitual abortion and infertility, with special reference to the male factor. *Fertil. Steril.*, *17*, 374-380

Jöel, C.A. & Chayen, R. (1971) *Pathological semen as a factor in abortion and infertility*. In: *Fertility Disturbances in Men and Women*, Basel, Karger, pp. 496-507

Kneer, M. (1957) Der habituelle abort. *Dtsch Med. Wochenschr.*, *82*, 1059-1060

Lancranjan, I., Popescu, H.I., Gavanescu, O., Klepsch, I. & Serbanescu, M. (1975) Reproductive ability of workmen occupationally exposed to lead. *Arch. environ. Health*, *30*, 396-401

Lingardh, G., Andersson, L. & Osterman, B. (1974) Fertility in men after renal transplantation. *Acta chir. scand.*, *140*, 494-497

MacLeod, J. (1951) Semen quality in one thousand men of known fertility and in eight hundred cases of infertile marriage. *Fertil. Steril.*, *2*, 115-139

MacLeod, J. (1965a) *Human seminal cytology following the administration of certain antispermatogenic compounds*. In: Austin, C.R. & Perry, J.S., eds, *Symposium on Agents Affecting Fertility*, Boston, MA, Little, Brown & Co., pp. 93-123

MacLeod, J. (1965b) Seminal cytology in the presence of varicocele. *Fertil. Steril.*, *16*, 735-757

MacLeod, J. (1969) Further observations on the role of varicocele in human male infertility. *Fertil. Steril.*, *20*, 545-563

MacLeod, J. (1974) *Male fertility and sterility, effects of environmental factors and of antispermatogenic compounds on the human testis as reflected in seminal cytology*. In: Mancini, R.E. & Martini, L., eds, *Proceedings of the Serono Symposium*, Vol. 5, New York, Academic Press, pp. 123-148

MacLeod, J. & Gold, R.Z. (1957) The male factor in fertility and infertility. IX. Semen quality in relation to accidents of pregnancy. *Fertil. Steril.*, *8*, 36-49

MacLeod, J. & Hotchkiss, R.S. (1941) The effect of hyperpyrexia upon spermatozoa counts in men. *Endocrinology*, *28*, 780-784

Nistal, M., Harruzo, A. & Sanchez-Corral, F. (1978) Toratozoospermia absoluta de presentacion familiar. Espermatozoides microcefalos irregulares sin acrosoma. *Andrologia*, *10*, 234

Procope, B.J. (1965) Effect of repeated increase of body temperature on human sperm cells. *Int. J. Fertil.*, *10*, 333-339

Raboch, J. (1965) Spermiologic findings in repeated spontaneous abortion. *Zentralbl. Gynaekol.*, *87*, 194-197

Rodriguez-Rigau, L.J., Smith, K.D. & Steinberger, E. (1978) Relationship of varicocele to sperm output and fertility of male partners in infertile couples. *J. Urol.*, *120*, 691-694

Rosenberg, M.J., Wyrobek, A.J., Radcliffe, J., Gordon, L., Watchmaker, G., Fox, S.H., Moore, D.H., II & Hornung, R.W. (1984) Sperm as an indicator of reproductive risk among petroleum refinery workers. *Br. J. ind. Med.* (in press)

Rowley, M.J., Leach, D.R., Warner, G.A. & Heller, C.G. (1974) Effect of graded doses of ionizing radiation on the human testis. *Radiat. Res.*, *59*, 665-678

Sas, M. & Szöllösi, J. (1979) Impaired spermiogenesis as a common finding among professional drivers. *Arch. Androl.*, *3*, 57-60

Schirren, C., Laudahn, G., Hartmann, E., Heinze, I. & Richter, E. (1975) The correlation of morphological and biochemical factors in human ejaculate in various andrological diagnoses. I. Relationship between ejaculate volume, number, motility and morphology of the spermatozoa with regard to age. *Andrologia*, *7*, 117-125

Sherins, R.J., Brightwell, D. & Sternthal, P.M. (1977) *Longitudinal analysis of semen of fertile and infertile men.* In: Troen, P. & Nankin, H.R., eds, *The Testis in Normal and Infertile Men*, New York, Raven Press, pp. 473-488

Takala, M.E. (1957) Studies on the seminal fluid of fathers of congenitally malformed children (199 sperm analyses). *Acta obstet. gynecol. scand.*, *36*, 29-41

Viczian, M. (1969) Ergebnisse von spermauntersuchungen bei zigarettenrauchern. *Zschr. Haut-Geschl-Krkh*, *44*, 183-187

Whorton, D., Krauss, R.M., Marshall, S. & Milby, T.H. (1977) Infertility in male pesticide workers. *Lancet*, *ii*, 1259-1261

Wyrobek, A.J., da Cunha, M.F., Gordon, L.A., Watchmaker, G., Gledhill, B., Mayall, B., Gamble, J. & Meistrich, M. (1980) Sperm abnormalities in cancer patients. *Proc. Am. Assoc. Cancer Res.*, *21*, 196

Wyrobek, A.J., Watchmaker, G., Gordon, L., Wong, K., Moore, D., II & Whorton, D. (1981) Sperm shape abnormalities in carbaryl-exposed employees. *Environ. Health Perspect.*, *40*, 255-265

Wyrobek, A.J., Gordon, L.A., Watchmaker, G., Moore, D.H., II (1982) *Human sperm morphology testing: description of a reliable method and its statistical power*. In: Bridges, B.A., Butterworth, B.E. & Weinstein, I.B., eds, *Indicators of Genotoxic Exposure* (*Banbury Report 13*), Cold Spring Harbor, NY, Cold Spring Harbor Laboratory, pp. 527-541

Wyrobek, A.J., Gordon, L.A., Burkhart, J.G., Francis, M.W., Kapp, R.W., Letz, G., Malling, H.V., Topham, J.C. & Whorton, D. (1983a) An evaluation of human sperm as indicators of chemically induced alterations of spermatogenic function: A report of the US Environmental Protection Agency Gene-Tox Program. *Mutat. Res.*, *115*, 73-148

Wyrobek, A.J., Gordon, L.A., Burkhart, J.G., Francis, M.W., Kapp, R.W., Letz, G., Malling, H.V., Topham, J.C. & Whorton, D. (1983b) An evaluation of the mouse sperm morphology test and other sperm tests in nonhuman animals: A report of the US Environmental Protection Agency Gene-Tox program. *Mutat. Res.*, *115*, 1-72

Yunda, I.F., Imshinetskaja, L.P., Karpenko, E.I., Tschernyschow, W.P., Sokolowa, M.N. & Gorpintschenko, I.I. (1978) Prostatis and pathospermia. *Dermatol. Monatsschr.*, *164*, 564-567

# BODY FLUID PROTEINS AND PEPTIDES AS TUMOUR MARKERS IN CLINICAL CANCER RESEARCH AND IN MONITORING EXPOSURE TO CARCINOGENS

**J.O. Järvisalo**

*Laboratory of Biochemistry, Department of Industrial Hygiene and Toxicology, 00370 Helsinki 37, Finland*

**U.H. Stenman**

*Clinical Laboratory, Departments I and II of Obstetrics and Gynecology, Helsinki University Central Hospital, 00290 Helsinki 29, Finland*

## SUMMARY

A brief review is given of the various groups of peptides and proteins that have been applied as tumour markers in human studies. Special attention is given to markers used for detection of bronchogenic lung tumours. In addition, a review is made of the few published studies on the levels of certain tumour markers in working populations that have been exposed to carcinogens.

There is a considerable lack of present knowledge on the applicability of tumour marker assays to monitoring of human exposure to carcinogens. When such studies are planned, it is valuable to combine these analyses with other types of indicators of carcinogen exposure and effects.

## INTRODUCTION

In cancer research, much hope and effort has been given to the development of methods by which simple assays can improve diagnosis, in respect to both the cancer site and its type. Most of the tumour markers studied have been proteins or peptides, which have been either normal (eutopic) to the tissue where the cancer has developed or the products of tumour development (ectopic markers). In addition, bodily reactions towards the cancer invasion, may produce various protein markers, e.g., acid hydrolases, immunoglobulins, immune complexes and acute-phase reactants.

Early reports on the various tumour markers were, typically, rather optimistic regarding specificity for tumour type or tumour. However, later studies have often revealed a lack of specificity in these respects and, consequently, the value of the various tumour markers in cancer diagnosis is still limited. The carcinoembryonic antigen (CEA) is the most commonly used marker, although nonspecific elevation of CEA is fairly common (Anderson *et al.*, 1978; Coombes *et al.*, 1978; Gold *et al.*, 1978; Laurence & Neville, 1978). Presently, the main clinical application of CEA is in follow-up of treated cancer patients (Ladenson & McDonald, 1980; Goldenberg, 1981).

Since the list of reported peptide and protein markers of cancer is currently rather extensive, only relatively few published papers, reviews or monographs are discussed here.

## MAIN TYPES OF PEPTIDES AND PROTEINS USED AS TUMOUR MARKERS

There may be no very logical way to classify the various types of tumour markers, but many authors divide them into two: eutopic (normal products of the tissue at the cancer origin) and ectopic (nontypical products of the original tissue). This classification, however, does not include products like immune complexes or acute-phase reactants (Sell, 1980; Ming Chu, 1982; Sell & Wahren, 1982).

Protein markers may also be classified on the basis of their functional characteristics: (1) Hormones; (2) enzymes; (3) oncofetal antigens (tumour-associated antigens); (4) miscellaneous.

Tumours are responsible for two types of hormone production. In the first, tumours of endocrine-gland origin produce hormones typical of the gland, but in an uncontrolled way: e.g., production of insulin by an insulinoma or calcitonin by medullary carcinoma of the thyroid gland. On the one hand, this hormone production may be very extensive and detection of elevated hormone levels may be essential to the diagnosis of the disease. On the other hand, the incidence of this type of tumour is very low compared to the incidence of cancers of the lung, gastrointestinal tract, breast or uterus.

The other type of tumour-induced hormone production is the so-called ectopic or paraneoplastic type, in which the hormone production is not a typical function of the tissue in which the tumour is formed. The biochemical basis of this hormone production is not clear but, apparently, it reflects the altered gene expression of cancer tissue. The hormones produced may be functionally active but they may also be functionally inactive subunits or precursors. Due to the biological activity of certain of the hormones secreted, the patients may have clinical syndromes caused by increased hormone levels.

### *Enzymes*

The various enzymes that can be applied as tumour markers range from 'tissue-specific' enzymes or isoenzymes (e.g., prostatic acid phosphatase, Regan type of alkaline phosphatase) to various acid hydrolases of lysosomal origin, which probably indicate the reaction of the body to the tumour growth rather than the tumour growth itself.

Even if some enzyme determinations may have value in cancer diagnosis (e.g., Regan isoenzyme of alkaline phosphatase in lung cancer and in gynaecological malignancies, Kottel & Fishman, 1982), in most cases, the enzyme assays do not offer a highly specific diagnostic tool. Much effort has been given to the development of specific and sensitive immunoassays for determination of prostatic acid phosphatase in serum. These methods seem useful for early detection of prostatic cancer (Vihko *et al.*, 1981), but pessimistic views have also been expressed (Gittes, 1983).

*Oncofetal tumour-associated antigens*

Malignant transformation of cells involves the expression of various proteins that are lacking or are minor constituents of normally differentiated cells. Many of these proteins are also expressed during some phases of fetal development and, consequently, have been called oncofetal antigens. More than 20 tumour markers of the oncofetal type have been detected. The best known of these are CEA(s), $\alpha$-fetoprotein, fetal sulfoglycoprotein and pancreatic oncofetal antigen. The number of this type of antigen will probably increase, due to active research in this field.

The usefulness of these markers in primary diagnosis of cancer is currently obscured by their lack of selectivity with respect to location or type of cancer. However, the combined use of several markers may improve the specificity (Seppälä *et al.*, 1982). Certainly, application of tumour markers in the follow-up of cancer patients is now accepted fairly generally.

Recently, radiolabelled antibodies to oncofetal tumour markers have been used for tumour radioimagery (Goldenberg, 1980).

*Other peptide and protein markers*

There are, additionally, miscellaneous groups of proteins and peptides that seem to be associated with clinical tumours. Most of these types of markers in human body fluids may reflect the body's reaction to tumour invasion. In that sense, they may have equivalents in the increase of, for example, acid hydrolases in connection with cancer diseases. Markers in this group include ferritin, $\beta$-$_2$-microglobulin, ceruloplasmin, fibronectin and acute-phase reactants.

*Tumour markers detected in bronchogenic lung cancers*

Lung cancer is one of the main causes of cancer deaths in the Western world, so that there has been a considerable need to develop tumour markers for this type of cancer. The markers have been reviewed recently by Coombes *et al.* (1978), Laurence and Neville (1978) and McIntire (1982).

It seems very important to be able to use tumour markers to monitor populations exposed to carcinogens. Many humans are exposed, at work or elsewhere, to factors or substances that increase the risk of lung cancer - commonly to tobacco smoking, polycyclic aromatic hydrocarbons and certain nickel and chromium compounds (Doll & Peto, 1981).

Some of the markers that have been studied in patients with bronchogenic lung tumours are listed in Table 1. Although the list is fairly comprehensive, very few of these markers have proved to be good predictors of lung tumours, even at the clinical stage of the disease (reviewed in detail by Coombes *et al.*, 1978, and McIntire, 1982). The frequency of elevated levels in these markers ranges from several percent to 20-30 percent. Therefore, diagnosis of lung cancer, in general, or of any specific histological lung-cancer type, cannot be based on the elevation of any of these markers.

**Table 1. Peptide and protein markers in bronchogenic lung cancer[a]**

| | |
|---|---|
| Hormones | *Tropic hormones* |
| | Adrenocorticotropic hormone |
| | Lipotropic hormone |
| | Melanocyte-stimulating hormone |
| | Chorion gonadotropic hormone |
| | Placental lactogen |
| | Thyrotropic hormone |
| | Growth hormone |
| | Prolactin |
| | *Others* |
| | Oxytocin |
| | Antidiuretic hormone |
| | Parathyroid hormone |
| | Glucagon |
| | Insulin |
| | Vasoactive intestinal polypeptide |
| | Calcitonin |
| | Erythropoietin |
| Enzymes | Sialyltransferase |
| | Galactosyltransferase |
| | Regan alkaline phosphatase |
| | Arylsulfatases A and B |
| | Amylase |
| | BB-isoenzyme of creatine kinase |
| | 5'-Nucleotide phosphodiesterase |
| | Isoenzyme V |
| Tumour-associated antigens | Carcinoembryonic antigen |
| | $\alpha$-Fetoprotein |
| | $\zeta$-Fetoprotein |
| | $\beta$-Oncofetal antigen |
| | Lung tumour antigen |
| Others | Immunoglobulins |
| | Immune complexes |
| | Ferritin |
| | $\beta_2$-Microglobulin |
| | Ceruloplasmin |
| | Acute-phase reactants ($\alpha_1$-antitrypsin, C-reactive protein, haptoglobin, orosomucoid) |

[a] Sources: Coombes *et al.* (1978) and McIntire (1982)

At present, there is no published study that has made a really critical multimarker analysis. This would require analysis of several markers, simultaneously in a study group and a sufficiently large reference population. However, considerable analytical and financial resources would be required.

## FACTORS TO BE CONSIDERED WHEN APPLYING TUMOUR MARKER ASSAYS TO DIAGNOSIS AND MONITORING

Application of laboratory methods to clinical studies or to monitoring programmes always involves the use of reference values, to which the results of analysis are to be compared. The recognition of such a need has created a whole field in itself (Gräsbeck & Alström, 1981). In addition, application of the Bayesian theorem to laboratory analyses has deepened, still further, concern about the usability of reference values for clinical studies (Galen & Gambino, 1975). The aim of these approaches has been to create a sound basis for evaluating the capacity of a method to distinguish between health and disease. When any biochemical test is applied for clinical purposes, the reference interval must be defined in a reference population. If sufficient material is available, the central 95% reference interval can be used. Methods based on the mean and the standard deviation are seldom appropriate because, usually, laboratory values are not normally distributed.

The *clinical sensitivity of a test* is used to characterize the incidence of true positive values (outside the reference range) that is obtained when a method is applied to patients known to have a certain disease. Briefly, the term may be described: A good method gives positive results in diseased subjects.

*Specificity, or negativity in health*, is used to characterize the incidence of true negative results when a method is applied to subjects known to be free of the disease. Briefly, the term may be described: A good method gives negative results in healthy subjects.

*Predictive value of a positive result* is defined as the percentage of positive results that are true positives (in the presence of the disease), when the method is applied to a population containing both healthy and diseased subjects. The predictive value depends both on the positivity of the test in disease and its negativity in health and on the prevalence of the disease in the population to which the test method is applied (further described by Galen & Gambino, 1975). The predictive value of a negative result can also be used to characterize a method.

*Efficiency of a method* is then defined as the proportion of the sum of true positives and true negatives to the grand total of the population studied with the method.

Exception for a few markers, like calcitonin in medullary thyroid carcinoma or insulin in insulinoma, most of the known tumour markers can be assumed to have rather low sensitivity, specificity and predictive value when applied to the general population (for calculations on the use of CEA to screen for carcinoma of colon, see Galen & Gambino, 1975). However, their value may considerably improve when they are applied to specific populations in which the probability of the disease is high.

A number of other aspects are important when tumour markers are applied to studies on humans. First of all, the definition of the population under study is important, including factors like gender, age, occupation, medical history and, possibly, environmental sources of exposure that may have causal or modifying effects on human health. A thorough description of these factors would increase understanding of the variability in clinical studies. In monitoring programmes, this type of detailed description is of special importance.

Another important issue, that may considerably weaken the applicability of results from studies on tumour markers, is the poor characterization and/or description of the methods used. The preparation of markers may vary, causing loss of comparability of the results. In the case of immunoassays, the preparation of the antigen, the production of the antibodies and the detection level of the method are all important sources of variability in results between various research groups. Quality control is still a considerable problem in this field, although it appears that the use of monoclonal antibodies may decrease the variability in assays.

There is an important problem to be considered when applying tumour marker analyses to various study programmes. Clearly elevated levels of the markers may occur in individuals who have not been considered ill. Although this may be a result of the nonspecific nature of the test, previously unknown malignant diseases may also be the cause. A method to handle this sort of information must be worked out prior to the analytical phase. Ethical and economical aspects must be taken into account, in a proper and equal way.

## STUDIES PERFORMED ON SPECIFIC GROUPS OF PERSONS EXPOSED TO CARCINOGENS

In his review of lung cancer markers, McIntire (1982) suggests that different applications of tumour marker assays would involve screening for lung cancer in aysmptomatic individuals, usually in populations at high risk owing to factors such as age, family history, smoking and occupational exposure. *Consultation on Health Surveillance of Workers Exposed to Chemicals* (World Health Organization, 1983) listed, under methods that may be applicable to surveillance of workers exposed to carcinogenic and genotoxic agents, the determination of products of tumour development and growth, even with the recognition that present research in the field is still very defective. Furthermore, application of the marker assays seems to find support in experimental studies (e.g., marker concentrations in body fluids may be increased after carcinogen exposure or promoting treatment) (Sell & Becker, 1978).

The only type of exposure to chemical carcinogens for which adequate literature is available, is tobacco smoking. From their experience with CEA determination in smokers, Cullen *et al.* (1976) concluded that the determination may be used to identify people at increased risk. However, only a few published studies have applied tumour marker assays to occupational groups exposed to chemical carcinogens. These studies and some unpublished Finnish studies are reviewed here, briefly.

*Studies on vinyl chloride exposure*. Pagé *et al.* (1976) reported the prevalence of abnormal (above 2.5 μg/l) CEA levels in workers exposed to vinyl chloride monomer. Within this limit, 48% of the study population had abnormal levels of CEA in serum. The authors did not report factors like smoking, medical history or alcohol consumption that may have elevated the serum level of CEA. The reference population results were obtained from the literature cited. In another study, Anderson *et al.* (1978) measured CEA levels in 1115 workers exposed to vinyl chloride monomer in three polymerization plants and 248 workers in a polyvinyl-chloride extrusion plant manufacturing polyvinyl-chloride textile leather. Multiple factors were found that affected the CEA levels in serum; cigarette smoking, past illnesses and alcohol intake had the strongest effects. After the removal of possible confounding factors, the distribution of CEA levels in the polymerization workers differed significantly from that of the extrusion plant group, or from that of an nonexposed control group. Among the various worker subcategories, the maintenance and production work group showed a distribution of CEA levels that was clearly different from that of the controls. The authors also correlated total activity of alkaline phosphatase with CEA levels.

*Studies on styrene-exposed workers*. Anderson *et al.* (1976) reported higher concentrations of CEA in sera of workers exposed to styrene than in those of a reference population standardized for age, sex, smoking and drinking habits, and current and former diseases. In a recent study, the present authors compared levels of CEA, $\beta_2$-microglobulin and ferritin in sera from a group of 31 workers exposed heavily to styrene with those from an industry population without known exposure to toxic chemicals and standardized for age. The styrene-exposed workers had a slightly higher mean level of CEA than the controls (2.8 *versus* 2.0 μg/l), but the difference was not statistically significant. No difference between the two groups was found in the levels of the other two markers (Järvisalo *et al.*, in preparation).

*Asbestosis patients.* In a Finnish follow-up study of asbestosis patients (Suoranta *et al.*, 1982), serum specimens were obtained from 90 patients for assays of CEA, $\beta_2$-microglobulin and ferritin. Applying the cut-off levels, 5 μg/l, 3 mg/l and 400 μg/l, respectively, elevated levels of CEA were observed in 17%, of $\beta_2$-microglobulin in 48% and of ferritin in 22%. Smokers showed a slightly higher level of CEA in serum than nonsmokers, and patients with radiographically detectable progression of the disease showed higher CEA levels than those without such progression. The observed values could not be explained by parenchymal disease of the liver or kidney, which were considered normal (Järvisalo *et al.*, 1983). Although the asbestosis disease, itself, may have had an effect on the results, it is well known that asbestosis patients have a high cumulative risk of malignant diseases. A follow-up study will shed more light on the mechanism of the high prevalence of pathological levels of tumour markers.

In addition, sera from 103 patients of the same cohort have been analysed for the level of tumour-associated trypsin inhibitor (Stenman *et al.*, 1982). However, only three patients showed an elevated level of the inhibitor, and the median value of this study group was similar to that found for a reference population (Järvisalo *et al.*, in preparation).

## CONCLUSIONS

Clearly, the development of tumour marker research will further increase the number of markers that can improve diagnosis of cancer. It is clear, also, that improvement of present techniques will enhance the reliability of the results.

For clinical application, at least two types of studies can be identified: cross-sectional studies to correlate cancer occurrence with the marker levels, and longitudinal studies to evaluate the correlation between progress of the disease and marker levels. Since we already have a large number of well-characterized markers, it would be worthwhile to study how the diagnostic value may be improved by large-scale multimarker studies. These studies would require considerable analytical and economical resources, but if all the relevant background information is gathered also, the information gained may fundamentally improve our chances to use tumour marker assays in cancer diagnosis.

Tumour marker assays could be useful in monitoring humans exposed to chemical carcinogens. However, currently, the background information is limited; therefore, the emphasis should be placed on (multi)marker follow-up studies of people with known high risk, due to their exposure to carcinogens (occupational or otherwise). These assays should be combined, also, with other sorts of indicators of carcinogen exposure and effects. Because of the known nonspecificity of most tumour markers, there will always be a considerable number of false positive findings in the studies. This problem must be faced prior to commencement of the studies, taking ethical and economical aspects into account.

The nonspecificity of many markers, in respect to toxic exposure (alcohol, medicine, smoking) and illnesses, may be useful in detecting those people at higher risk than the general population (Cullen *et al.*, 1976); this nonspecificity may well be related to phenomena such as tumour promotion and cell transformation. As determinants of the risk involved, these factors may, in many cases, be as relevant as the tumour initiation caused by exposure to chemical carcinogens.

## REFERENCES


Anderson, H.A., Lorimer, W., Snyder, J. & Selikoff, I.J. (1976) *Levels of carcinoembryonic antigen in styrene workers*. In: *The Abstract Book of the Medichem 4th International Conference, Haifa, September 7-10*, R37

Anderson, H.A., Snyder, J., Lewinson, T., Woo, C., Lilis, R. & Selikoff, I.J. (1978) Levels of CEA among vinyl chloride and polyvinylchloride exposed workers. *Cancer*, *42*, 1560-1567

Coombes, R.C., Ellison, M.L. & Neville, A.M. (1978) Biochemical markers in bronchogenic carcinoma. *Br. J. Dis. Chest*, *72*, 263-287

Cullen, K.J., Stevens, D.P., Frost, M.A. & Mackay, I.R. (1976) Carcinoembryonic antigen (CEA) and cancer in a longitudinal population study. *Aust. N.Z. J. Med.*, *6*, 279-283

Doll, R. & Peto, R. (1981) The causes of cancer: quantitative estimates of avoidable risks of cancer in the United States today. *J. natl Cancer Inst.*, *66*, 1192-1308

Galen, R.S. & Gambino, S.R. (1975) *Beyond Normality: The Predictive Value and Efficiency of Medical Diagnoses*, New York, John Wiley & Sons

Gittes, R.F. (1983) Serum acid phosphatase and screening for carcinoma of the prostate. *New Engl. J. Med.*, *309*, 852-853

Gold, P., Shuster, J. & Freedman, S.O. (1978) Carcinoembryonic antigen (CEA) in clinical medicine. Historical perspective, pitfalls and projections. *Cancer*, *42*, 1399-1405

Goldenberg, D.M. (1980) An introduction to the radioimmunodetection of cancer. *Cancer Res.*, *40*, 2957-2958

Goldenberg, D.M. (1981) Carcinoembryonic antigen: its role as a marker in the management of cancer. *Br. med. J.*, *282*, 373-375

Gräsbeck, R. & Alström, T., eds (1981) *Reference Values in Laboratory Medicine. The Current State of the Art*, New York, John Wiley & Sons

Järvisalo, J., Juntunen, J., Huuskonen, M.S., Kivistö, H. & Aitio, A. (1984) Tumor markers and neurological signs in asbestosis patients. *Am. J. ind. Med.* (in press)

Kottel, R.H. & Fishman, W.H. (1982) *Developmental alkaline phosphatases as biochemical tumor markers*. In: Ming Chu, T., ed., *Biochemical Markers for Cancer. Clinical and Biochemical Analysis II*, New York, Marcel Dekker, pp. 93-115

Ladenson, J.H. & McDonald, J.M. (1980) Colorectal carcinoma and carcinoembryonic antigen (CEA). *Clin. Chem.*, *26*, 1213-1220

Laurence, D.J.R. & Neville, A.M. (1978) Biochemical tests in diagnosis and monitoring of cancer. *Ann. Rev. clin. Biochem.*, *1*, 185-231

McIntire, K.R. (1982) *Lung cancer markers*. In: Sell, S. & Wahren, B., eds, *Human Cancer Markers*, Clifton, NJ, Humana Press, pp. 359-380

Ming Chu, T., ed. (1982) *Biochemical Markers for Cancer. Clinical and Biochemical Analysis II*, New York, Marcel Dekker

Pagé, M., Theriault, L. & Delforme, F. (1976) Elevated CEA levels in polyvinyl chloride workers. *Biomedicine*, *25*, 279

Sell, S., ed. (1980) *Cancer Markers. Developmental and Diagnostic Significance*, Clifton, NJ, Humana Press

Sell, S. & Becker, F.F. (1978) Alpha-fetoprotein. *J. natl Cancer Inst.*, *60*, 19-26

Sell, S. & Wahren, B., eds (1982) *Human Cancer Markers*, Clifton, NY, Humana Press

Seppälä, M., Rutanen, E.-M., Lindgren, J. & Wahlström, T. (1982) *Multiple markers in the management of cancer patients*. In: Ming Chu, T., ed., *Biochemical Markers for Cancer. Clinical and Biochemical Analysis II*, New York, Marcel Dekker, pp. 321-350

Stenman, U.-H., Huhtala, M.-L., Koistinen, R. & Seppälä, M. (1982) Immunochemical demonstration of an ovarian cancer-associated urinary peptide. *Int. J. Cancer*, *30*, 53-57

Suoranta, H., Huuskonen, M.S., Zitting, A. & Juntunen, J. (1982) Radiographic progression of asbestosis. *Am. J. ind. Med.*, *3*, 67-74

Vihko, P., Lukkarinen, O., Kontturi, M. & Vihko, R. (1981) Effectiveness of radioimmunoassay of human prostate-specific acid phosphatase in the diagnosis and followup of therapy in prostatic cancer. *Cancer Res.*, *41*, 1180-1183

World Health Organization (1983) *Consultation on Health Surveillance of Workers Exposed to Chemicals* (*ICP/WHK 013*), Copenhagen, Regional Office for Europe

# WORKERS EXPOSED TO ETHYLENE OXIDE HAVE INCREASED INCIDENCE OF SISTER CHROMATID EXCHANGE

**F. Sarto, I. Cominato, A.M. Pinton, P.G. Brovedani & C.M. Faccioli**

*Institute of Occupational Health, University of Padova, Padova, Italy*

**V. Bianchi & A.G. Levis**[1]

*Department of Biology, University of Padova, Padova, Italy*

**SUMMARY**

We have determined the frequencies of sister chromatid exchange (SCE) in 41 workers engaged in the sterilization of medical equipment.

Each exposed subject was paired with a control matched for age, smoking habits and sex. The frequencies of SCE were markedly higher in the exposed than the control workers. The exposed subjects were divided into two groups according to time-weighted average concentrations of ethylene oxide (EtO) in the air, that were determined at the work place of each worker during the 5 min following the sterilizer opening, during the entire sterilization cycle or during the 8-h working day. One group (19 subjects) was exposed to a range of 3.7-35.5 ppm (mean, 15.8 ± 9.8 for a sterilization cycle) and showed a mean of 13.0 ± 1.8 SCE compared to 10.2 ± 1.2 in controls (variance, F = 31.2; $p < 0.001$). The second group (22 subjects) was exposed to 0.3 - 2.6 ppm (mean, 1.1 ± 1.0) and had a mean of 11.0 ± 1.6 SCE compared to 9.8 ± 1.4 in controls (F = 4.7; $p < 0.05$). Statistical evaluation of the SCE frequencies correlates SCE induction with the level of EtO exposure and also with smoking and age.

[1] To whom requests for reprints should be addressed

Our data show that EtO can induce human chromosomal damage at very low environmental exposure levels, lower than the present Italian official limits and lower than those found even in the most technologically advanced hospitals.

## INTRODUCTION

Ethylene oxide (EtO) is one of the 25 chemicals produced in the highest volume in the USA (National Institute for Occupational Health, 1981). Only about 0.02% of the annual production is used for sterilization in hospitals. In 1977, the National Institute for Occupational Health estimated that about 75 000 sanitary workers were potentially exposed to EtO. In sterilization facilities, EtO concentration is clearly higher than in chemical manufacturing plants, most of which are in the open air and have closed-cycle processes.

Reports on EtO carcinogenicity in animals and humans have suggested a general lowering of the environmental limits for EtO. Since 1983, the Threshold Limit Value for EtO in the USA has been 10 ppm. It has been proposed that the limit be further decreased to 1 ppm (American Conference of Governmental Industrial Hygienists, 1983). The Italian Threshold Limit Value has been lowered to 3 ppm (Ministero della Sanità, 1983).

Cytogenetic tests on peripheral lymphocytes of workers exposed to hazardous environmental agents make it possible to study the damage caused by environmental exposure. This paper describes a cytogenetic study on lymphocytes of 41 sanitary workers exposed to different levels of EtO that was used to sterilize sanitary equipment. Chromosomal damage was evaluated by comparing the frequency of sister chromatid exchanges (SCE) in the exposed workers and in a control group. Analysis of chromosomal aberrations in these exposed and nonexposed subjects is still in progres.

## MATERIAL AND METHODS

### *Industrial hygiene survey*

In the eight hospitals of the Venetia Region that we examined, EtO sterilization is carried out by varying technologies. Six hospitals used older (first generation) gas sterilizers and two used more recent (second generation) ones. The sterilizing units examined differed in dimensions and in their ventilation systems, the latter being, in most cases, insufficient. Although, in a given unit, the number of sterilizations per day and their duration were reasonably constant, there were appreciable differences among different sterilizing units, e.g., the time required for a sterilization cycle varied from 60 to 180 min.

For each hospital, a standard sterilization cycle was identified, on the basis of the amount and the type of supplies processed; a worker performing standard tasks carried a personal sampler throughout the cycle (from loading to unloading of the sterilizer). All sampling was done during the winter.

EtO was sampled and determined colorimetrically according to Gilli *et al.* (1981), with modifications. The coefficient of variation for the determinations was about 4.8%.

Maximal EtO exposure occurred at the end of the cycle when the sterilizers were opened for unloading. Continuous monitoring of air concentrations was carried out with a photoionization analyser (Hnu, I 101) connected to a Perkin Elmer 56 stripchart recorder. The analyser was calibrated repeatedly with mixtures of 50 and 100 ppm EtO in air, and the probe of the analyser was kept near the operator's mouth. A maximal time of 5 min was required to unload the sterilizer, to place the sterilized equipment in a forced degasser, and then to load the sterilizer again. Concentrations determined with the photoionization analyser overlap those obtained by gas chromatography in the range 5-1000 ppm; values below 5 ppm are unreliable (De Negri *et al.*, 1981). As the EtO exposure was characterized by very high levels that were reached at the opening of the sterilizers, and then quickly declined, three kinds of time-weighted average (TWA) concentrations were calculated:

(1) TWA/5-min conc, i.e., EtO concentration during the 5 min following the sterilizer opening, as the mean of values measured every 15 sec, by the photoionization analyser.

(2) TWA/1 cycle conc, i.e., the time-weighted average concentration of EtO during one sterilization cycle, the duration of which varied among the hospitals. The value obtained by personal sampling thus represented the whole cycle.

(3) TWA/8 h conc, i.e., the time-weighted average concentration of EtO during a standard 8-h working day. For each hospital, the average number of sterilization cycles per day was calculated on the basis of usage in the preceding six months, and the TWA/8-h conc was obtained by assuming that EtO concentration was zero in the intervals between the cycles.

*Subjects examined*

We studied 41 sanitary workers engaged in the sterilizing units of eight hospitals. Haematochemical tests were performed for each subject to assess the blood cell count and liver function. Apart from one subject who had been working for four years with a second-generation sterilizer and showed appreciable leukopenia but normal bone marrow, all subjects were considered healthy. Of the 41 workers, 37 were exposed only to EtO, and the remaining four were potentially exposed also to anaesthetic gases and X-rays in an operating room.

The control group comprised 41 healthy volunteers, sanitary workers who were not exposed professionally to EtO, cytostatic drugs, anaesthetic gases or ionizing radiations, except for four controls working in the operating room, who were matched with the four exposed subjects mentioned above. Because we had demonstrated previously (Sarto *et al.*, 1982) that smoking habits and age significantly affect SCE frequency, controls were chosen by pairing each exposed subject with a suitable control of the same sex, age ($\pm$ 3 yr) and smoking habits ($\pm$ 4 cigarettes per day). Individuals who smoked five or more cigarettes per day were considered smokers. In most cases, the pairing of exposed subjects and controls was done among workers engaged in the same hospital; in the few instances in which this was not possible we chose the controls from our own hospital.

*Sister chromatid exchanges*

Each culture contained 4 ml chromosome Microtest culture medium (Difco) with phytohaemoagglutinin, to which was added 0.2 ml whole blood and 0.1 ml 5-bromodeoxyuridine (Sigma; final concentration, $10^{-4}$M). Lymphocytes were cultured in the dark for 72-74 h, and metaphases were blocked during the last 4 h with 0.25 $\mu$g/ml colchicine (Eurobio). Chromosomal preparations were obtained according to the method of Evans and O'Riordan (1975). Differential staining of sister chromatids was obtained by a modification of the alkaline Giemsa method proposed by Alves and Jonasson (1978): the flame-dried slides were stored for at least 15 d, then stained with Giemsa 2-4% in 0.3 M sodium phosphate, monobasic, pH 10.5, for 6-15 min at 18-22°C. For each subject, SCE were scored in 30 well-spread metaphases on coded slides, taking into account the exchanges in the centromeric region, also.

*Statistical analysis*

The normality of SCE frequency distribution in the different groups of subjects analysed was controlled by the Kolmogorov-Smirnov test (Massey, 1951). The mean values of SCE in the same groups were compared by the F test for variance analysis. Linear and multiple regression tests were performed among the SCE means and the independent variables measured in the single subjects.

## RESULTS

Data obtained from the environmental analyses carried out in the eight hospitals are summarized in Table 1.

The data show that EtO pollution is higher where first-generation sterilizers are employed; in those hospitals, the TWA/8-h conc was very near the 10 ppm Threshold Limit Value suggested by the American Conference of Governmental Industrial Hygienists, but, as indicated by the TWA/5-min conc, the concentration during the workday varied remarkably. With second-generation sterilizers, the TWA/8-h conc was almost 10 times lower than the Italian 3 ppm Threshold Limit Value, yet there were relevant variations (see TWA/5-min conc).

**Table 1. Ethylene oxide concentrations in the eight hospitals studied**

| Sterilizer | | Ethylene oxide (ppm) | | |
|---|---|---|---|---|
| | | TWA[a]/5-min conc | TWA/1-cycle conc | TWA/8-h conc |
| 1st-generation | mean ± SD | 62.5 ± 46 | 15.8 ± 9.8 | 10.7 ± 4.9 |
| | range | 13 - 160 | 3.7 - 35.5 | 3.7 - 20 |
| 2nd-generation | mean ± SD | 8.6 ± 11 | 1.1 ± 1.0 | 0.35 ± 0.12 |
| | range | 5 - 26 | 0.3 - 2.6 | 0.2 - 0.5 |

[a] Time-weighted average

The subjects were grouped according to exposure with first- or second-generation sterilizers, termed the high- and low-exposure groups, respectively. Data for subjects examined and for the mean values of SCE detected in the two exposed groups and in their controls are reported in Table 2; the increase in SCE over that in the controls is highly significant in the high-exposure group and in the total group exposed ($F = 31.2$; $p < 0.001$), but in the low-exposure group the increase is less marked, although still significant ($F = 4.7$; $p < 0.05$).

**Table 2. Data on subject variables and averages of sister chromatid exchanges (SCE) per cell of the subjects exposed to ethylene oxide and their controls**

| Group | Total number | Age in years (mean ± SD) | Length of exposure in years (mean ± SD) | No. of smokers | No. of males | No. of SCE/ metaphase (mean ± SD) |
|---|---|---|---|---|---|---|
| High-exposure group[a] | 19 | 38.8 ± 7.8 | 6.8 ± 3.5 | 12 | 14 | 13.0 ± 1.8[c] |
| Controls | 19 | 38.1 ± 8.1 | -- | 12 | 14 | 10.2 ± 1.2 |
| Low-exposure group[b] | 22 | 34.7 ± 7.8 | 3.0 ± 1.1 | 8 | 14 | 11.0 ± 1.6[d] |
| Controls | 22 | 34.4 ± 8.7 | -- | 8 | 14 | 9.8 ± 1.4 |
| Total exposed | 41 | 36.6 ± 8.0 | 4.9 ± 3.3 | 20 | 28 | 11.9 ± 2.0[c] |
| Total control group | 41 | 36.1 ± 8.5 | -- | 20 | 28 | 10.0 ± 1.3 |

[a] 1st-generation sterilizers

[b] 2nd-generation sterilizers

[c] $p < 0.001$

[d] $p < 0.05$

In order to determine whether a linear relation existed between SCE frequency and EtO exposure in the exposed group, we evaluated the weight of a single variable - exposure, smoking habits and age - on the SCE level. A significant correlation coefficient (r) was found for the regression between SCE and EtO concentrations: For SCE *versus* TWA/8-h conc, r was 0.56 ($p < 0.001$); for SCE *versus* TWA/1-cycle conc, r was 0.50 ($p < 0.001$); for SCE *versus* TWA/5-min, r was 0.44 ($p < 0.01$). In the exposed group, the multiple regression analysis showed a significant correlation between the frequency of SCE, exposure to EtO (TWA/8-h conc), smoking (number of cigarettes per day) and age (years): r was 0.62 ($p < 0.001$).

## DISCUSSION

The mutagenicity of EtO has been widely demonstrated in bacteria, yeasts, *Drosophila*, mice and mammalian cells *in vitro* (IARC, 1982).

The few data that are available on genetic damage induced by EtO in exposed humans (Pero *et al.*, 1982; Högstedt *et al.*, 1983; Yager *et al.*, 1983) indicate concordantly that EtO is active, even at environmental concentrations of around 1 ppm.

The results reported here for the exposed subjects (sanitary workers engaged in medical equipment sterilization) show a significant increase in SCE that is correlated with the environmental concentration of EtO. In particular, exposure evaluated on the basis of the TWA/8-h conc appears to be more strictly correlated with SCE frequency than the acute exposure, based on the TWA/5-min conc.

Our observations suggest that the Italian Threshold Limit Value of 3 ppm for EtO is not low enough to prevent genotoxic effects. In hospitals, EtO exposure occurs within a very short period during the sterilization cycle. It could be reduced by automation of the sterilizing and degassing procedures.

In the absence of sufficient experimental carcinogenicity data and epidemiological observations, the data on cytogenetic effects in exposed humans should be taken into account by the official agencies in establishing Threshold Limit Values. This recommendation has already been made by others (Sorsa *et al.*, 1982).

EtO exposure is a subtle professional risk, as it produces genotoxic effects even at concentrations that give no subjective sign and do not modify the conventional haematological parameters.

## ACKNOWLEDGEMENTS

This work was supported by grants from the National Research Council of Italy (C.N.R., Progetto Finalizzato 'Controllo della Crescita Neoplastica') and the Venetia Region ('Centro Regionale di Alta Specializzazione in Cancerogenesi Ambientale').

## REFERENCES

Alves, P. & Jonasson, J. (1978) New staining method for the detection of sister chromatid exchanges in BrdU-labelled chromosomes. *J. Cell Sci.*, *32*, 185-195

American Conference of Governmental Industrial Hygienists (1983) *Threshold Limit Values for Chemical Substances in the Work Environment Adopted for 1983-84*, Cincinnati, OH

De Negri, E., Alongi, V., Perlangeli, V. & Del Buono, L. (1981) *Determinazione dell'ossido di etilene su materiali sterilizzati: valutazione di uno strumento a fotoionizzazione e confronto con la gas-cromatografia*. In: *Atti 3è Convegno Sulla Sterilizzazione*, Milan, Tecniche Nuove, pp. 19-21

Evans, H.J. & O'Riordan, M.L. (1975) Human peripheral blood lymphocytes for the analysis of chromosome aberrations in mutagen tests. *Mutat. Res.*, *31*, 135-148

Gilli, G., Corrao, G., Scursatone, E. & Defilippi, P. (1981) Primi risultati di una indagine igienistico-ambientale effettuata presso centrali di sterilizzazione ad ossido di etilene. *Med. Lav.*, *3*, 27-31

Högstedt, B., Gullberg, B., Hedner, K., Kolnig, A.M., Mitelman, F., Skerfving, S. & Widegren, B. (1983) Chromosome aberrations and micronuclei in bone marrow cells and peripheral blood lymphocytes in humans exposed to ethylene oxide. *Hereditas*, *98*, 105-113

IARC (1982) *IARC Monographs on the Evaluation of the Carcinogenic Risk of Chemicals to Humans*, Suppl. 4, *Chemicals, Industrial Processes and Industries Associated with Cancer in Humans, Volumes 1-29*, Lyon, International Agency for Research on Cancer, pp. 126-127

Massey, F.J., Jr (1951) The Kolmogorov-Smirnov test for goodness of fit. *J. Am. stat. Assoc.*, *46*, 6-78

Ministero della Sanità - Div.V (1983) *Circolare n. 8. Ossido di etiene: modifica circolare n. 47 dell'1-10-81 e svolgimento dell'indagine epidemiologica*, Rome

National Institute for Occupational Safety and Health (1981) *Ethylene Oxide (EtO)* (*Current Intelligence Bulletin No. 35*), Cincinnati, OH

Pero, R.W., Brungelsson, T., Widergen, B., Högstedt, B. & Welinder, H. (1982) A reduced capacity for unscheduled DNA synthesis in lymphocytes from individuals exposed to propylene oxide and ethylene oxide. *Mutat. Res.*, *104*, 193-200

Sarto, F., Cominato, I., Bianchi, V. & Levis, A.G. (1982) Increased incidence of chromosomal aberrations and sister chromatid exchanges in workers exposed to chromic acid ($CrO_3$) in electroplating factories. *Carcinogenesis*, *3*, 1011-1016

Sorsa, M., Hemminki, K. & Vainio, H. (1982) Biological monitoring of exposure to chemical mutagens in the occupational environment. *Teratog. Carcinog., Mutag.*, *2*, 137-150

Yager, J.W., Hines, C.J. & Spear, R.C. (1983) Exposure to ethylene oxide at work increases sister chromatid exchanges in human pheripheral lymphocytes. *Science*, *219*, 1221-1223

# OVERVIEW

## OVERVIEW

# METHODS OF MONITORING HUMAN EXPOSURE TO CARCINOGENIC AND MUTAGENIC AGENTS

**P.H.M. Lohman, R. Lauwerys & M. Sorsa**

In the first three sessions of this symposium, methods were discussed that are currently available and that can be used to monitor human exposure to carcinogenic and mutagenic agents. During the discussions, difficulty was experienced in differentiating between methods that can be used to detect exposure, those that can be used to detect early adverse effects and those that can be used for quantitative risk estimation. Part of the confusion is due to the fact that many of the methods currently available involve in-vitro test systems, in which the endpoints are adverse biological effects on the organisms or cells used but which may not indicate a similar effect in humans. Extrapolations of such effects in experimental cell systems to adverse health effects in humans *in vivo* are sometimes made too lightly. Another reason for the confusion has been that methods designed to detect *early* adverse effects cannot be used to indicate more detrimental health effects and, therefore should be considered only as methods for monitoring *exposure*.

The definitions presented in the third panel session, 'Biological Monitoring and Health Surveillance', and the continuum between them, may help to resolve difficulties in distinguishing between exposure monitoring and detection of early adverse effects. These definitions were agreed upon at an earlier international seminar on 'Assessment of Toxic Agents at the Workplace', held at Luxembourg in December 1980, and are the following:

'*Biological monitoring* is the measurement and assessment of workplace agents or their metabolites either in tissues, secreta, excreta, expired air or any combination of these to evaluate exposure and health risk compared to an appropriate reference.

'*Health surveillance* is the periodic medico-physiological examination of exposed workers with the objective of protecting health and preventing occupationally related disease. The detection of established disease is outside the scope of this definition.

'The definitions of biological monitoring and health surveillance are separate components of a continuum which can range from the measurement of agents in the body through measurements of metabolites, to signs of early disease.'

Before summarizing the tests that can, in principle, be applied to humans, it was considered worthwhile to review briefly the various aspects of medical surveillance programmes (with special emphasis on occupational health), in an attempt to dispel the confusion. These aspects are represented schematically in Figure 1.

**Fig. 1. Stages in health surveillance programmes**

MEDICAL SURVEILLANCE

| | EXPOSURE MONITORING (primary prevention) | | DETECTION OF EARLY ADVERSE EFFECTS | | | |
|---|---|---|---|---|---|---|
| External exposure → | Internal exposure → | Non-adverse effects | Early adverse effects → | | Late adverse health effects → | Clinical disease |
| air<br>water<br>food<br>etc | - chemicals or metabolites in biological fluids<br>- recent exposure or body burden | reversible or irreversible | - spontaneously reversible if exposure ceases | - non-spontaneously reversible<br>- may be reversible through medical intervention, or<br>- progression to disease may be stopped or slowed after removal from exposure | irreversible progression | cancer |

The first stage comprises exposure monitoring. Methods for monitoring exposure are used in two ways: (1) to ascertain whether potentially carcinogenic or mutagenic agents are present in the environment of the persons under study; and (2), when a potentially carcinogenic or mutagenic agent is known to be present, whether the exposure is below the level (external and/or internal dose) at which a health effect may develop. This kind of monitoring is essentially useful for preventive medicine (primary prevention) and can be done by (i) direct measurement of the concentration or the amount of a chemical present in the environment (air, water, food, etc.); (ii) measurement of the concentration or activity of a chemical or its metabolites in biological media; or (iii) measurement of biological effects (not necessarily adverse to human health), which are related to the internal dose.

The second stage of monitoring that can be carried out is the detection of early adverse effects. Such monitoring programmes involve the use of biological methods that detect effects considered on current scientific grounds to be unwanted. In practice, these methods make it possible to draw a line, often arbitrarily, at levels over which exposure should be considered excessive. Various actions can then be taken on the basis of the criteria indicated in Figure 1.

The third stage of monitoring addresses the problem of quantitative estimation of risk for a specified adverse health effects - in this case, tumour formation, occurrence of heritable defects or adverse effects on reproduction. Such estimates are the ultimate aim of biomonitoring. This aim is extremely difficult to achieve, but must be addressed in order to make

**Table 1. Evaluations of methods available for assessing human exposure to carcinogenic and mutagenic agents**

| Method of biological monitoring | Appropriateness for exposure assessment — qualitative | — recent (< 1 week) internal dose | — long-term body burden | — dose at target site | Appropriateness for health effect assessment — non-adverse (reversible) | — adverse | Interpretation of result — on individual basis | — on group basis | Precision of method — technical reproducibility | — stability of parameter over time | — interlaboratory reproducibility | Sensitivity — for environmental exposures | — for occupational exposures | — for acute exposures | Chemical specificity | Absence of interference of confounding factors | Absence of confounding background levels | Simplicity of analysis | Possibility of easy sample storage | Current applicability — research level | — routine use |
|---|---|---|---|---|---|---|---|---|---|---|---|---|---|---|---|---|---|---|---|---|---|
| 1. Determination of chemical/ metabolite in the body | + | + | ± | − | ± | ± | ± | ± | + | ± | + | + | + | + | + | + | ? | + | ± | + | + |
| 2. Determination of thio-ethers in urine | + | (+) | − | − | − | − | − | + | (+) | − | (−) | ? | ? | ? | − | − | − | − | + | + | (+) |
| 3. Determination of mutagenic activity in excreta | + | (+) | − | − | − | − | − | + | (+) | − | (−) | ± | + | + | − | − | − | − | + | + | ? |
| 4. Detection of blood protein adducts | + | + | +[a] | +[b] | (+) | (+) | + | + | ? | ? | ? | − | + | + | + | ? | ? | − | + | + | (+) |
| 5. Detection of DNA adducts in somatic cells | + | ? | ? | + | ? | ? | (+) | + | ? | ? | ? | (+) | + | + | + | (+) | (+) | − | + | + | ? |
| 6. Detection of protein variants in blood | (+) | ? | − | − | ? | ? | ? | ? | ? | ? | ? | − | + | + | − | − | ? | − | − | + | − |
| 7. Detection of point mutations in blood cells | (+) | ? | ? | ? | ? | ? | − | + | ? | ? | ? | − | + | + | − | − | ? | − | − | + | ? |
| 8. Analysis of chromosomal aberrations in somatic cells | + | (+) | + | − | + | (+) | − | + | (+) | + | (−) | ? | (+) | (+) | − | − | − | − | − | + | (+) |
| 9. Analysis of sister chromatid exchanges in somatic cells | + | (+) | + | − | + | (+) | − | + | (+) | (+) | (−) | ? | (+) | (+) | − | − | − | − | − | + | (+) |
| 10. Analysis of micronucleated cells | + | ? | ? | (+)[c] | ? | ? | − | + | (+) | ? | ? | ? | (+) | (+) | − | − | − | + | − | + | ? |
| 11. Detection of DNA repair in somatic cells | + | ? | ? | ? | ? | ? | − | ? | + | − | − | − | (+) | (+) | − | − | (−) | (−) | − | + | ? |
| 12. Examination of sperm morphology | + | + | − | + | (+) | (+) | + | (+) | + | − | ? | ? | (+) | + | − | − | − | − | − | + | ? |
| 13. Detection of tumour markers | (+) | − | − | − | (+) | (+) | (+) | + | − | − | ? | ? | ? | ? | − | − | − | − | + | + | − |

\+ applicable/true; (+) probably applicable/probably true
− not applicable/not true; (−) not presently applicable/not presently true
± cannot be generalized
? unknown

[a] Plasma proteins
[b] Haemoglobin
[c] In buccal mucosa cells

realistic assessments of hazard and thus avoid the drawing of conclusions by the general public that may be more pessimistic than is actually the case on the basis of the effect measured.

One of the major objectives of this symposium was to evaluate the significance of the available monitoring methods and to assign them to places in the scheme presented in Figure 1. From the various presentations, 13 methods or groups of methods were chosen that can currently be applied or are likely to become available for human monitoring in the near future. A list of criteria was then drawn up by which the performance of each method could be judged. Table 1 shows evaluations of each method made on the basis of those criteria.

This table gives no more than a rough idea of the characteristics of the methods. The value of each method depends heavily on the agent(s) in question and on the particular situation in which it is applied. Epidemiologists, public health officials and physicians responsible for occupational health care should be aware that no 'best' method can be selected. Researchers are reminded that their 'best' method is not necessarily the best for human monitoring.

Our overall view is that the currently available methods can be used successfully as a basis for reducing exposure, even on a personal basis, in cases in which a clearly positive effect is found. Some of the methods are considered on current scientific grounds to give information not only on exposure: for instance, if potent (as yet undefined) effects are found, they may be regarded as health effects. An example of a test that may be used for this purpose is the detection of an increase in chromosomal aberrations. An example of how cytogenetic tests can be used to evaluate effects in an exposed population is given in Figure 2.

**Fig. 2. Cytogenetic monitoring of exposed populations is usually applicable at the qualitative stage**

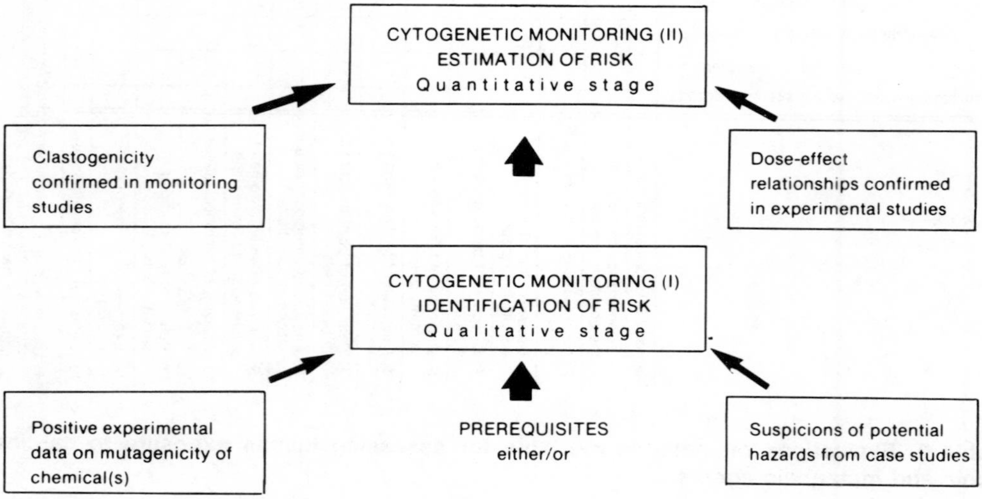


Priorities for carrying out a cytogenetic monitoring study should be set on the basis of previous experimental knowledge or suspicion of a hazard from other sources.

Application of these biological monitoring methods make it possible to determine whether an exposed group 'deviates from normal'. Stringent quality control and appropriate statistical analysis are required, however, before such methods can be used for this purpose. Identification of a 'no-risk' situation after suspected exposure to a known genotoxic chemical can be done with even less confidence, because, in such cases, the method used must be calibrated even more rigorously than when a positive response is obtained. It is essential that any group that is being monitored be informed before the exercise is started that individual fluctuations may be obtained in the results of each test.

The field of quantitative risk estimation is in its infancy; mutagenic and carcinogenic risks in humans cannot be estimated quantitatively other than by thorough calibration. However, newer methods, especially at the molecular level, are being developed that may make possible a beginning of risk estimation, even on an individual basis.

# REPORTS OF PANEL DISCUSSIONS

# PRACTICAL ISSUES IN THE EVALUATION OF MONITORING TECHNIQUES: NEED FOR VALIDATION, QUALITY ASSURANCE AND ESTABLISHMENT OF BASELINE LEVELS

**D. Anderson & M.S. Legator**

The major objective of using short-term testing for human monitoring is to detect exposure to carcinogenic or mutagenic materials of populations at risk, prior to the onset of any adverse health effect. At the present time, results from the procedures cannot be used to predict the health outcome for a given individual; however, some of these procedures, such as cytogenetic evaluation, can indicate that a population has been exposed to a genetically active agent. Other procedures, such as the detection of chemicals in urine, may indicate the presence of active metabolites that can cause gene mutations, although they do not give information about the biological effects of chemicals on exposed populations. Still other procedures, such as detection of increases in protein and/or DNA adducts, signify that DNA has been modified (or may have been modified), but do not supply, either directly or indirectly, any indication of biological effect, except to point to possible target organs.

There are several known examples in which biological monitoring by short-term procedures has led to specific action by government regulatory agencies and industry and to the taking of voluntary measures, e.g., in hospitals to minimize exposure to cytotoxic agents. Biological monitoring played a significant role, also, in the setting of the presently suggested US Occupational Safety and Health Administration standard for exposure to ethylene oxide.

The panel discussed in detail only determination of protein alkylation, urine assays and cytogenetic procedures.

Protein alkylation detection has limited usefulness in biological monitoring because of intra-subject variability, due, for example, to the unexplained presence of alkylated histidine. It was suggested that this might be due to exposure to unidentified agents in the environment, such as ethylene or urethane. At present protein alkylation determination is a somewhat laborious and time-consuming procedure; however, even when background levels of alkylation are present, the development of analytical methods of greater sensitivity and accuracy should allow the detection of low levels of exposure. The suggestion was made that it would be useful to carry out, simultaneously, other biological monitoring procedures, such as cytogenetic studies.

Bacterial mutation assays, such as the *Salmonella typhimurium*/microsomal assay have been adequately validated for in-vitro biological testing of chemicals. Problems associated with the urinary assay include the presence of histidine and tryptophan at levels that could influence the assay. The increased sensitivity of the fluctuation assay may increase the variation due to the possible presence of small amounts of histidine. Another problem is that the presence of toxic materials may reduce the population of viable bacteria. Further, mutagenic agents may be present at levels too low to be detected because the sample is too dilute. Concentration of a urine sample, however, may lead to increased concentrations of histidine and of toxic materials. Additionally, the resin columns that have been used traditionally to concentrate urine, particularly that of smokers, may be inappropriate to detect certain substances, such as cisplatinum, which is polar and, therefore, not retained by the columns, and other concentration procedures should be used.

In gastric fluids, nitrosation of amines can occur by oxidation of nitrogen in ambient air and in the presence of the test bacteria. In faeces, aerobic oxidation of normally anaerobic faecal constituents may lead to spurious results.

In cytogenetic studies, two major types of variation have been documented, one associated with the testing procedure itself and the other with the individuals being tested. The first includes slide-reading discrepancies, culture conditions, e.g., type of medium, temperature, bromodeoxyuridine concentration (for sister chromatid exchange); and time of sampling, which can alter the yields of chromosomal aberrations, and, sometimes, the incidence of sister chromatid exchange (SCE), due, perhaps, to differences in the populations of T and B lymphocytes at different sampling times. The second general source of variability is associated with differences between individual subjects, such as age, sex, use of medications, infections, and also with differences in genotype, in inherent susceptibility to environmental agents, in enzyme inducibility, in immunological competency and in haematological profile. Possibly the largest difference among individuals is in social habits, such as diet, smoking and alcohol consumption.

Given all these sources of variation, it is remarkable that when the results of nine recent, randomly selected studies carried out in different laboratories, using different protocols, were examined for control frequencies, the rate of SCE per cell varied by a factor of only two fold. Similarly, in five studies for various chromosomal aberrations, a variation of only two to four fold was found. This unexpected, relatively close agreement suggests that determinations of both chromosomal aberration and SCE can be used for human monitoring. The data base on chromosomal aberration and SCE in control populations should, however, be extended. A collaborative study between two laboratories, based on a stratified scheme for the selection of approximately 400 employees, is underway currently, to help meet this need. The project will attempt (1) to validate a protocol; (2) to determine chromosomal aberration and SCE frequencies; and (3) to identify and quantify levels of variability in the data. It would also be useful to compare results in these tests after radiation exposure with those after chemical exposure.

In a properly designed study, most confounding factors can be taken into account by using standardized culture conditions and matched controls; and when it is not possible to match controls exactly, the analysis can be adjusted for the relevant factors. Statistical analysis of cytogenetic studies indicates the number of individuals and cells that should be evaluated in order to detect a significant increase above background values; e.g., detection of a 25% increase over controls at the 0.05% or 0.01% level of significance requires evaluation of 600 cells from 63 or 117 individuals, respectively. Similar calculations indicate that fewer subjects and cells are needed for SCE studies; e.g., with a spontaneous level of 6.5 SCE/cell, 30 cells per 22 individuals should be evaluated to obtain a 0.05% level of significance and 30 cells in 40 individuals for the 0.01% level.

Several international projects have been carried out recently to assess the levels of toxic agents in body fluids in people in different countries; these have included determinations of lead and cadmium in blood and in the renal cortex and of DDT, DDE, lindane and polychlorinated biphenyls in breast milk. In order to ensure accuracy, an extensive training programme was set up for workers in the participating laboratories, some of which were located in developing countries, and checks for quality assurance were made throughout the monitoring programme.

International organizations such as the World Health Organization, the United National Environment Programme, the International Programme on Chemical Safety, the Commission of the European Communities and the International Organization for Standardization are addressing the practical issues of establishing valid monitoring techniques, including quality assurance checks, the production of reference material and the establishment of baseline data.

Thus, although biological monitoring by short-term procedures is in its infancy, positive results have been obtained in spite of the many confounding factors, which affect all of the techniques. Such confounding must be anticipated in any study in which human subjects are involved because of the outbred nature of the species and because they are exposed to multiple agents in the environ:; there are also confounding factors characteristic of the methods used, as described above. An understanding of both general and specific confounding factors can lead to optimal experimental design, including proper statistical procedures, which should, in most cases, overcome the difficulties in interpreting these studies.

Some such studies, e.g., cytogenetic studies after exposures to ethylene oxide and vinyl chloride and evaluation of the mutagenicity of alkylating agents in the urine of patients receiving chemotherapy, have prompted decisions to minimize exposure and to provide safer working conditions. Future studies that would evaluate multiple endpoints in the same subject using presently-developing techniques should lead to a significant reduction in exposure of populations at risk to hazardous chemicals.

# HEALTH SIGNIFICANCE OF MONITORED CHEMICAL AND BIOLOGICAL ENDPOINTS

**R. Saracci**

The Panel discussion revolved around four main questions:

1. What does 'health significance' mean?
2. What are the most relevant elements for assessing and establishing health significance?
3. What are the current research needs to establish health significance?
4. What can we say - in the present status of knowledge - about the health significance of measured chemical and biological endpoints?

In answer to the *first question* (What does 'health significance' mean?), it was stressed that the health significance of biological monitoring endpoints lies entirely in the domain of primary prevention. These endpoints are markers (chemical or biological) both of exposure *sensu stricto* and of early effects - namely, effects that remain clinically latent and are wholly reversible upon cessation of the exposure. Monitoring for both classes of marker provides, in conjunction with environmental monitoring and technological assessment (e.g., of a plant's facilities), the necessary scientific information to control, by technological changes, hazardous environmental exposures in order to prevent the occurrence of such late effects as reproductive damage and cancer. It is obvious that monitoring of *any* health effect, including clinically latent but irreversible lesions and overt disease, provides information on the effectiveness of primary prevention by documenting to what degree it has failed. However, only biological monitoring as previously defined is compatible and coherent with a practice of avoiding adverse effects on health and can thus be included among the tools of primary prevention.

With regard to the *second question* (What are the most relevant elements for assessing and establishing health significance?), the variety of elements that could be taken into account spans the whole spectrum of topics of the conference. Attention was drawn particularly, however, to three aspects:

(1) Sensitive methods are currently available to detect adducts of 'initiators' of the carcinogenic process with DNA. Picomole or even femtomole levels can be measured by, e.g., ultra-sensitive enzyme radio-immunoassays and enzyme-linked immunosorbent assays; and validation is possible by exposing cell cultures *in vitro* to known concentrations of carcinogens. In addition, adducts of some carcinogens (benzo[*a*]pyrene, aflatoxin $B_1$, *N*-nitrosoureas) have been detected in exposed human subjects. However, it should be noted that, even with the most sensitive methods, only one adduct per $10^6$ bases can be detected in, say, one-third or one-fourth of an exposed population, and that the possibility of establishing a quantitative relationship between external exposure in man and adduct levels still requires a substantial amount of methodological development.

(2) The situation is less clear for 'cocarcinogens' and 'promoters', as contrasted to 'initiators'. Can one simply exclude indirect effects on DNA of promoters, for example, *via* free-radical generation? Or should any molecule involved in the carcinogenic process be deemed capable of some type of DNA damage? Currently, initiation is regarded as involving irreversible damage to DNA, while promotion would entail reversible effects on gene expression and cell differentiation through mechanisms which - at least as far as the aims of this conference go - are not yet sufficiently clarified to allow the identification of endpoints useful as markers of exposure to promoters.

(3) In the laboratory, a variety of systems (for instance, the occurrence of chromosomal anomalies) are available to separate reversible and irreversible effects, as well as to estimate dose-responses and other quantitative aspects of carcinogenesis and of other biological processes. However, in the real world, additional factors complicate the picture: patterns of exposure that are variable in time; the presence, almost universally, of multiple exposures; the limited persistence in time after exposure (days, weeks) of endpoints like DNA adducts, thus reducing their potential usefulness as measures of exposure.

In response to the *third point* (What are the current research needs to establish health significance?), two broad areas can be outlined in which development is required:

The first is research focused on the individual and on the time periods in the sequence of events leading from external exposure to internal exposure and from early effects to late effects (reproductive damage, cancers). Three areas of research were highlighted:

(1) Investigations using in-vitro systems and animal models, which allow clarification of, for example, the dependence of measured endpoints on factors that determine carcinogen metabolism, and the calibration of applied (external dose to the dose that reaches the target site (cell DNA).

(2) Investigations of interindividual variation (in animals and more relevantly, in man) of response to carcinogenic exposures. Such studies should provide answers to questions such as why only a minority of smokers get lung cancer. One could accept the view that getting lung cancer is inherently a probabilistic phenomenon and that no sub-set of susceptible people exists among smokers, or one could take as a working hypothesis that potential lung cancer cases can be identified among susceptible subjects. That this hypothesis is tenable is suggested by pharmacogenetic studies of carcinogen metabolism, notably in respect to *N*-acetylation and aryl hydrocarbon hydroxylase activity.

(3) Joint utilization of 'exploratory' epidemiological data and of biological monitoring endpoints. For instance, occupational mortality statistics for a number of countries indicate elevated risks for several cancers in certain categories of exposed subject: liver cancer incidence is found to be elevated in statistics from Denmark, the UK and the USA among subjects manipulating - among other chemicals - trichloroethylene; and lung cancer rates are high in silicotics and in foundry workers. Conclusive substantiation of these exploratory findings would require following up cohorts of exposed subjects to determine cancer outcome; this may prove unfeasible in some situations, or require several quinquennia before enough data have accumulated. In these cases, can assessment of early effects of exposure be a reasonable surrogate to support or to refute the exploratory epidemiological data?

The second area in which research is needed focuses on *direct* investigation of the relation between biological monitoring endpoints and late health outcomes, like reproductive effects and cancer. This type of direct research will provide ultimate, firm answers about the health significance of biological monitoring endpoints. Such research can be carried out:

(1) in experimental animals: through lifetime carcinogenesis experiments, such as that currently being conducted in rats by Aitio *et al.*, in which a variety of markers of exposure and early effects are being measured and are to be correlated with cancer incidence once all animals have died.

(2) through determination of cancer incidence in subjects exposed in the past to agents like radiation and benzene and for whom data on early markers, like chromosomal abnormalities, are available; follow-up of cohorts of patients treated with anticancer chemotherapeutic agents affords the opportunity of relating an accurately known administered dose with early endpoints as well as with the incidence of subsequent disease (second primary cancers).

In this type of direct research, reproductive effects may be a more expedient late effect to measure than cancer, since they usually occur much sooner after exposure. For similar considerations of timescale, childhood cancers could be studied in relation to transplacental exposures.

With regard to the *fourth question* (What can we say - in the present status of knowledge - about the health significance of measured chemical and biological endpoints?), the Panel members agreed in expressing a presumption, variable in strength according to the endpoint considered, of the existence of a *qualitative* relationship between endpoints and late health outcomes. Very little can be said from a *quantitative* viewpoint, which remains, however, the real yardstick against which the scientific and practical value of such endpoints must be measured. The establishment of quantitative dose-response relationships in carcinogenesis experiments makes it reasonable to hope that adequate research will determine similar relationships for biological monitoring endpoints. It may well be that, for practical application, not every endpoint requires quantitative validation in respect of every hazardous chemical exposure: rather, such a validation made for certain 'test' chemicals (e.g., some alkylating agents in respect to their haemoglobin adducts), can justify the utilization of the same endpoint, at least qualitatively, for similar chemicals.

In practical terms, biological monitoring must be regarded today chiefly as an element of an integrated occupational health programme. It may, in particular, prove useful - whatever the present limitations - to provide a nonspecific warning of unsuspected and potentially hazardous exposures. The Panel agreed that it would be desirable if a concensus could be reached in the international scientific community concerning the significance of these tests, for groups and for individuals, for the use of medical personnel who inform populations who are being monitored.

# EPIDEMIOLOGICAL RESEARCH AND OCCUPATIONAL HEALTH PRACTICE: PRESENT AND FUTURE ROLES OF GENOTOXIC MONITORING

**T. Norseth**

Biological monitoring methods may have endpoints that can be used in epidemiological research as measures of both exposure and effect. Thus, environmental exposure may be associated with genetic effects, or a genetic effect resulting from exposure (internal exposure) may be associated with disease. The association of an exposure with a genetic effect in an individual may be more rapid in comparison with epidemiological investigation in which endpoints such as cancer or birth defects are the indices. Even if the health significance of the various endpoints of biological monitoring have not been established, the Panel felt that, for population groups, sufficient information is available that such epidemiological studies could be recommended. Epidemiological research and occupational health practice are compatible, provided that the objectives of each are distinguished and that the relevant definitions are agreed upon. Administrative aspects, such as record-keeping and maintenance, confidentiality of data, access to records and record linkage, are important for epidemiological research in general, but are of specific importance to cancer epidemiology and biological monitoring because of the long latency of cancer and because the data in question are of a sensitive nature.

The personal quality of the data constitutes one of the ethical aspects of biological monitoring in general, and of monitoring for epidemiological research and occupational health practice, in particular. It is not self-evident that a worker will put himself forward as a research object; ethical approval and informed consent of subjects must be assured before a monitoring programme is started. Clarification of questions of health significance, risk, other consequences, responsibility and possible compensation for loss of job or adverse health effects, including psychological difficulties, is an important ethical matter that must be taken care of before a monitoring programme is started, either for epidemiological research or to detect an immediate health risk.

While there was general agreement that biological monitoring is an epidemiological research tool and a tool to evaluate suspected exposure to carcinogenic or mutagenic agents when environmental analysis is impracticable (e.g., of complex mixtures), there was some disagreement on the use, at present, of biological monitoring in occupational health practice.

Such monitoring may be used in occupational health practice to clarify a suspicion based on inconclusive epidemiological evidence. The consequences of test results for the individual must be clarified: individual results relate not to exposure, but to the as yet unknown health significance of many of the endpoints of biological monitoring. In order for biological testing to become a useful tool for occupational health care and health surveillance on an individual level, proper epidemiological protocols should be used.

In occupational health programmes, such tests must not be used to indicate whether conditions are acceptable or unacceptable in relation to exposure to carcinogenic or mutagenic agents, and medical advisors should not be pressured to make such statements in negotiations with industrial companies in this respect. Some examples were given in which results from biological monitoring had been used to insist on environmental improvements being made. It was emphasized that a negative result in a biological monitoring test does not make exposure to carcinogenic or mutagenic agents acceptable. Although it was stressed repeatedly, during the discussions, that biological monitoring, in the present state of knowledge can be used on a group basis only, the problems with regard to the use of positive results must be appreciated. In this respect, information to be given to the subjects involved was, again, discussed at some length; it was agreed that workers should be given full information about the theoretical and the known health significance of the results and should be informed, also, about variance of results and about the influence of confounding factors. A plan to follow up subjects in whom positive results have been obtained must be worked out and agreed upon by all parties before a biological monitoring programme is started. Participation in the programme should be determined solely on the basis of informed consent.

Workers in areas where there is high exposure to carcinogenic or mutagenic agents are often rotated, to decrease the burden on each individual. There is limited knowledge regarding the health significance of this practice, but it is not recommended. In such cases, either the exposure should be decreased or the agent(s) in question should be removed.

In most developing countries, strategies for development involve rapid industrialization and modernization of agriculture. An undesirable, but inevitable, consequence of these activities is exposure to chemical pollutants, many of which are known to be carcinogenic or mutagenic. The speed of socioeconomic change, also, renders more people susceptible. Other problems include the export to developing countries of outdated, hazard-prone technology and of toxic substances that are prohibited in other parts of the world. The extensive use of a variety of pesticides in developing countries was discussed; however, difficulties have been encountered in recording the use of and exposure to carcinogenic or mutagenic agents and of effects. Even if, as understood, the emphasis in developing countries is directed more towards rapid industrialization and curative medical efforts than to prevention of long-term health hazards, the World Health Organization and the United Nations Environment Programme were urged to continue support for the latter.

The Panel concluded that more research is needed before all the consequences of the endpoints of biological monitoring are fully understood. In particular, the link between external exposure, internal exposure (as indicated by biological monitoring) and the endpoints of biological monitoring on the one hand, and malignant disease on the other, should be investigated in properly designed epidemiological studies. The Panel concluded that, in the future, biological monitoring will be used in general occupational health practice, in conjunction with environmental monitoring.

# POSTERS

# LIST OF POSTERS PRESENTED

G. AHLBORG
Mutagenic activity in urine of workers exposed to nitrotoluenes

A. AITIO
Urinary excretion of chromium as an indicator of exposure to trivalent chromium sulphate in leather tanning

L. ARINGER
Evaluation of some factors affecting determination of thioethers in urine

E.G. ASTRUP
Case-inquiry study of 5 cases of lung cancer in the Norwegian mineral wool industry. A preliminary report

A. CARERE
Comparative biological monitoring in a rubber plant: parallel detection of chromosomal aberrations, sister chromatid exchanges and DNA repair ability

E. CLONFERO
Urine extracts of workers professionally exposed to polycyclic aromatic hydrocarbons (PAHs), but not smokers, are not mutagenic

E. ELOVAARA
Activation and inactivation reactions of m-xylene and ethylbenzene as indicated by urinary metabolite patterns

G. FICSOR
Human sperm enzymes for the detection of reprotoxic agents

R.C. GARNER
Mutagenicity of gastric juice: the importance of controlling histidine concentration when using *Salmonella* tester strains

R.C. GRAFSTRÖM
Aldehydes found in tobacco smoke cause DNA damage, inhibit DNA repair, and are cytotoxic in cultured normal human bronchial epithelial cells

J. JÄRVISALO
Urinary thioethers in styrene exposed workers of reinforced plastics industry

T. KAUPPINEN
Registration of employees occupationally exposed to carcinogenic substances in Finland

C. LEGRAVEREND
Highly sensitive fluorescent ELISA (HSF-ELISA) for the detection of cisdiamminedichloroplatinum modified DNA

A. LIIMATAINEN
Rapid changes in peat fly ash mutagenicity after release into atmosphere

K. LINNAINMAA
Sister chromatid exchanges in peripheral lymphocytes of workers exposed to phenoxy acid herbicides

N.N. LITVINOV
Modification of blastomogenic and mutagenic effect by environmental chemicals

J. LÄHDETIE
Micronucleated cells in human semen

H. NORPPA
Vinylacetate is a potent inducer of chromosome damage in mammalian cells

H. OHSHIMA
Monitoring of human exposure to aldehydes and nitrate/nitrite by measurement of *N*-nitrosothiazolidine 4-carboxylic acid derivatives in the urine

E. PELTONEN
Methods of monitoring exposure to suspect carcinogenic food additives

W.K. DE RAAT
Adsorption of organic compounds from urine in mutagenicity testing; choice of adsorbent

F. RUSPOLINI
Metabolism some aromatic amine used in a rubber article producing factory. Comparison between working exposure and level of contamination of workings

S. SALOMAA
SCE induction by HPLC fractions from wood stove emission samples

F. STENBÄCK
Two-stage carcinogenesis in hamster liver: classification of preneoplastic changes

K. VAHAKANGAS
Synchronous fluorescence spectrophotometry to assess benzo[*a*]pyrene-DNA in human cells

N.P.E. VERMEULEN
GLC-MS: A selective analytical technique for the estimation of mercapturic acids in biological fluids

M.D. WATERS
Evaluation of genotoxic effects in human populations

M.I. WILLEMS
Mutagens in rat urine and faeces after treatment with genotoxic carcinogens

B.H. WOOLLEN
Dinitrotoluene: Assessment of absorption during explosives manufacture

M. ZICHE & P. DOLARA
Angiogenesis as a marker of neoplastic transformation in humans

# AUTHOR INDEX

# SUBJECT INDEX

## A

**D**

## F



## G

Composition, impression et façonnage
Groupe MCP-Mame
Dépôt légal : Avril 1985